# OPTICAL MICROSCOPY

# OPTICAL MICROSCOPY

## EMERGING METHODS AND APPLICATIONS

Edited by

**Brian Herman**

**John J. Lemasters**

Department of Cell Biology & Anatomy
Laboratories for Cell Biology
University of North Carolina at Chapel Hill
Chapel Hill, North Carolina

**ACADEMIC PRESS, INC.**
Harcourt Brace Jovanovich, Publishers
San Diego  New York  Boston  London  Sydney  Tokyo  Toronto

*Front cover photograph:* Intracellular pH in a cultured cardiac myocyte. (Adapted from Figure 12.7; courtesy of John J. Lemasters *et al.*)

This book is printed on acid-free paper. ∞

Academic Press, Inc.
1250 Sixth Avenue, San Diego, California 92101-4311

*United Kingdom Edition published by*
Academic Press Limited
24–28 Oval Road, London NW1 7DX


Library of Congress Cataloging-in-Publication Data

Optical microscopy : emerging methods and applications / edited by
    Brian Herman and John J. Lemasters.
        p.   cm.
    Includes bibliographical references and index.
    ISBN 0-12-342060-1
    1. Microscopy.   I. Herman, Brian.   II. Lemasters, John J.
QH205.2.O68   1992
5578–dc20                                        92-18139
                                                      CIP


PRINTED IN THE UNITED STATES OF AMERICA
92   93   94   95   96   97   EB   9   8   7   6   5   4   3   2   1

# CONTENTS

Contributors    xi
Preface    xiii

**1**    NEW FLUORESCENT PROBES FOR CELL BIOLOGY
Martin Poenie and Chii-Shiarng Chen    1

   I.    Introduction    2
  II.    An Introduction to Fluorescence    2
 III.    Why Is Fluorescence Useful?    3
  IV.    Fluorescent Probes for Proteins and Enzymes    4
   V.    Fluorescent Lipid and Membrane Probes    12
  VI.    Synthetic Fluorescent Ion Indicators    13

**2**    PHOTOSENSITIVE CAGED COMPOUNDS: DESIGN, PROPERTIES,
AND BIOLOGICAL APPLICATIONS
Joseph P. Y. Kao and Stephen R. Adams    27

   I.    Introduction    28
  II.    Design and Use of Caged Compounds    29
 III.    Preparation and Properties of Caged Compounds    39
  IV.    Biological Applications of Caged Compounds    60
   V.    Concluding Remarks    76

**3**    FLUORESCENCE RATIO IMAGING: ISSUES AND ARTIFACTS
Gary R. Bright    87

   I.    Introduction    87
  II.    Fluorescent Probes    89
 III.    Basic Instrumentation    95
  IV.    Experimental Issues    103
   V.    Conclusion    109

**4  SAMPLING CHARACTERISTICS OF CCD VIDEO CAMERAS**
Richard J. Bookman and Frank T. Horrigan    115

  I.    Introduction    115
 II.    Some Terms and Methods    117
III.    Implementation of CCDs in RS-170 Cameras    122
 IV.    Summary and Prospects    130


**5  MEASUREMENT AND MANIPULATION OF OSCILLATIONS IN
CYTOPLASMIC CALCIUM**
C. S. Chew and M. Ljungström    133

  I.    Introduction    134
 II.    Current Models of Intracellular Calcium Oscillations    135
III.    Characteristics and Limitations of Cell-Permeant Calcium-
        Sensitive Fluorescent Probes    137
 IV.    Equipment Suitable for Detection of Calcium Oscillations    149
  V.    Future Directions    167


**6  SIMULTANEOUS MULTIPLE DETECTION OF FLUORESCENT
MOLECULES: RAPID KINETIC IMAGING OF
CALCIUM AND pH IN LIVING CELLS**
Stephen J. Morris    177

   I.    Introduction: Why Study Simultaneous Kinetics
         by Imaging?    178
  II.    Multiparameter Imaging    181
 III.    "Simultaneous" Intracellular Calcium–pH Measurements    182
  IV.    Basic Properties of Ratio-Type Ion Indicator Dyes    183
   V.    Imaging of Excitation-Type Dyes versus Emission-Type Dyes for
         Rapid Kinetic Fluorescence Video Microscopy    184
  VI.    Advantages of Emission Ratio Dyes    186
 VII.    Improvements in Design for Multi-Image Experiments    186
VIII.    The Need for More than Two Simultaneous Images of the Same
         Cell, and the Dependence of Indo and Fura Calcium $K_d$
         on pH    187
  IX.    Four-Channel Video Microscope for Simultaneous Imaging of
         Two-Ratio Dyes    188
   X.    Choice of Dyes for Sensing Calcium and pH    190
  XI.    Multiple-Wavelength Epifluorescence Excitation    190
 XII.    Simultaneous Collection of Four Video Images of the Same
         Microscopic Field    192
XIII.    Formation of Low-Light Images    194
 XIV.    Imaging Hardware–Software    194

XV. Improving Spatial and Temporal Resolution of the Data   195
XVI. Increasing the Data Collection Speed to 60 Fields
per Second   197
XVII. Calculations   197
XVIII. Some Further Considerations   201
XIX. Applications   202

## 7   REAL-TIME FLUORESCENCE MICROSCOPY IN LIVING CELLS: FLUORESCENCE IMAGING, PHOTOLYSIS OF CAGED COMPOUNDS, AND WHOLE-CELL PATCH CLAMPING
### E. Niggli, R. W. Hadley, M. S. Kirby, and W. J. Lederer   213

I. Introduction   214
II. Real-Time Imaging and Quantitative Fluorescence in Living
Cells: General Considerations   215
III. Patch Clamp in Whole-Cell Mode: Voltage-Clamp
Technology   223
IV. Photolysis: Improved Time Resolution   229
V. Combined Photolysis and Fluorescence   231
VI. Summary   232

## 8   SIMULTANEOUS DIFFERENTIAL INTERFERENCE CONTRAST AND QUANTITATIVE LOW-LIGHT FLUORESCENCE VIDEO IMAGING OF CELL FUNCTION
### J. Kevin Foskett   237

I. Introduction   237
II. Correlation of Fluorescence with Cell Structure   238
III. Summary   259

## 9   FLUORESCENCE MICROSCOPIC IMAGING OF MEMBRANE DOMAINS
### William Rodgers and Michael Glaser   263

I. Introduction   263
II. Fluorescence Imaging of Membrane Domains   265
III. Discussion   279

## 10   MEASUREMENT OF MEMBRANE GLYCOPROTEIN MOVEMENT BY SINGLE-PARTICLE TRACKING
### Michael P. Sheetz and Elliot L. Elson   285

I. Introduction   285
II. Single-Particle Tracking   287
III. Comparison of SPT with FPR   287

IV. Cellular Requirements and Limitations    289
 V. Statistical Analysis    290
VI. Applications    291
VII. Summary    293

**11**  TOTAL INTERNAL REFLECTANCE FLUORESCENCE MICROSCOPY
         Lukas K. Tamm    295

  I. Introduction    296
 II. Theory    297
III. Instrumentation    303
IV. Applications    311
 V. Prospectus    331

**12**  LASER SCANNING CONFOCAL MICROSCOPY OF LIVING CELLS
   John J. Lemasters, Enrique Chacon, George Zahrebelski,
     Jeffrey M. Reece, and Anna-Liisa Nieminen    339

   I. Introduction    339
  II. Optical Principles of Confocal Microscopy    340
 III. Pinhole Size and Confocality    341
 IV. Comparison of Conventional and Confocal Microscope
       Images    342
  V. Practical Considerations for Viewing Living Cells by Confocal
       Microscopy    344
 VI. Imaging Cell Volumes and Surfaces    345
VII. Visualization of Organelles    347
VIII. Electrical Gradients in Living Cells    347
 IX. Ion Imaging    350
  X. Conclusion    352

**13**  INTRAVITAL MICROSCOPY
         Robert S. McCuskey    355

  I. Introduction    355
 II. *In Vivo* Microscopy of Organs    356
III. Conclusion    366

**14**  TIME–RESOLVED FLUORESCENCE LIFETIME IMAGING
         M. vandeVen and E. Gratton    373

  I. Introduction    374
 II. Recent Technological Advances in Spatial and Temporal Imaging:
       The Emerging Field    375

III. Instrumentation Used for Time-Resolved Fluorescence Imaging    378
IV. Approaches to Time-Resolved Imaging    381
V. Theory for Phase and Modulation Sensitive Detection    382
VI. Time-Resolved Imaging with CCD Camera: Subframe Scanning and Gating    384
VII. Multichannel Photon Counting    388
VIII. Imaging Photon Detectors    388
IX. Microchannel Plate Multipliers with Position Sensitive Detectors (PSD)    389
X. Synchroscan Streak Camera    390
XI. Framing Camera    392
XII. Two-Photon Time-Resolved Imaging    393
XIII. Conclusion    393

**15**  AUTOMATED FLUORESCENCE IMAGE CYTOMETRY AS APPLIED TO THE DIAGNOSIS AND UNDERSTANDING OF CERVICAL CANCER
Stephen J. Lockett, Majid Siadat-Pajouh, Ken Jacobson, and Brian Herman    403

I. Summary    404
II. Human Papilloma Viruses and Cervical Cancer    404
III. Methods of Identification of Human Papilloma Viruses    406
IV. Automated Fluorescence Image Cytometry    408
V. Application of Image Cytometry for Studying HPV and Cervical Cancer    411
VI. Future Prospects for Automated Fluorescence Image Cytometry    419

Index    433

# CONTRIBUTORS

Numbers in parentheses indicate the pages on which the authors' contributions begin.

Stephen R. Adams (27), Department of Pharmacology, University of California, San Diego, La Jolla, California 92093

Richard J. Bookman (115), Department of Molecular and Cellular Pharmacology, University of Miami School of Medicine, Miami, Florida 33136

Gary R. Bright (87), Department of Physiology and Biophysics, School of Medicine, Case Western Reserve University, Cleveland, Ohio 44106

Enrique Chacon (339), Department of Cell Biology & Anatomy, University of North Carolina at Chapel Hill, Chapel Hill, North Carolina 27599

Chii-Shiarng Chen (1), Department of Zoology, University of Texas at Austin, Austin, Texas 78712

C. S. Chew (133), Department of Physiology, Morehouse School of Medicine, Atlanta, Georgia 30310

Elliot L. Elson (285), Department of Biochemistry and Molecular Biophysics, Division of Biology and Biomedical Sciences, Washington University Medical School, St. Louis, Missouri 63110

J. Kevin Foskett (237), Division of Cell Biology, Hospital for Sick Children, Toronto, Ontario M5G 1X8, Canada

Michael Glaser (263), Department of Biochemistry, University of Illinois at Urbana-Champaign, Urbana, Illinois 61801

E. Gratton (373), Laboratory for Fluorescence Dynamics, Department of Physics, University of Illinois at Urbana-Champaign, Urbana, Illinois 61801

R. W. Hadley (213), Department of Physiology, University of Maryland School of Medicine, Baltimore, Maryland 21201

Brian Herman (403), Department of Cell Biology & Anatomy, Laboratories for Cell Biology, Lineberger Comprehensive Cancer Center, University of North Carolina at Chapel Hill, Chapel Hill, North Carolina 27599

Frank T. Horrigan (115), Department of Molecular and Cellular Pharmacology, University of Miami School of Medicine, Miami, Florida 33136

Ken Jacobson (403), Department of Cell Biology & Anatomy, Laboratories for Cell Biology, Lineberger Comprehensive Cancer Center, University of North Carolina at Chapel Hill, Chapel Hill, North Carolina 27599

Joseph P. Y. Kao (27), Medical Biotechnology Center and Department of Physiology, School of Medicine, University of Maryland at Baltimore, Baltimore, Maryland 21201

M. S. Kirby (213), Department of Physiology, University of Maryland School of Medicine, Baltimore, Maryland 21201

W. J. Lederer (213), Department of Physiology, Medical Biotechnology Center, University of Maryland School of Medicine, Baltimore, Maryland 21201

John J. Lemasters (339), Department of Cell Biology & Anatomy, University of North Carolina at Chapel Hill, Chapel Hill, North Carolina 27599

M. Ljungström (133), Department of Medical and Physiological Chemistry, Biomedical Center, University of Uppsala, Uppsala, Sweden

Stephen J. Lockett (403), Department of Cell Biology & Anatomy, Laboratories for Cell Biology, University of North Carolina at Chapel Hill, Chapel Hill, North Carolina 27599

Robert S. McCuskey (355), Department of Anatomy, College of Medicine, University of Arizona, Tucson, Arizona 85724

Stephen J. Morris (177), Division of Molecular Biology and Biochemistry, School of Biological Sciences, University of Missouri at Kansas City, Kansas City, Missouri 64110

Anna-Liisa Nieminen (339), Department of Cell Biology & Anatomy, University of North Carolina at Chapel Hill, Chapel Hill, North Carolina 27599

E. Niggli (213), Department of Physiology, University of Bern, Bern, Switzerland; and Department of Physiology, University of Maryland School of Medicine, Baltimore, Maryland 21201

Martin Poenie (1), Department of Zoology, University of Texas at Austin, Austin, Texas 78712

Jeffrey M. Reece (339), Department of Cell Biology & Anatomy, University of North Carolina at Chapel Hill, Chapel Hill, North Carolina 27599

William Rodgers (263), Department of Biochemistry, University of Illinois at Urbana-Champaign, Urbana, Illinois 61801

Michael P. Sheetz (285), Department of Cell Biology, Duke University Medical School, Durham, North Carolina 27710

Majid Siadat-Pajouh (403), Department of Cell Biology & Anatomy, Laboratories for Cell Biology, University of North Carolina at Chapel Hill, Chapel Hill, North Carolina 27599

Lukas K. Tamm (295), Department of Physiology, University of Virginia School of Medicine, Charlottesville, Virginia 22908

M. vandeVen (373), ISS Inc., Champaign, Illinois 61820

George Zahrebelski (339), Department of Cell Biology & Anatomy, University of North Carolina at Chapel Hill, Chapel Hill, North Carolina 27599

# PREFACE

The last decade has seen substantial growth in the use of optical microscopic techniques to explore questions in both the life sciences and the material sciences. This rapid technological growth has spawned many new discoveries of importance to biologists, biochemists, and materials scientists. In this book, we present the most recent and comprehensive descriptions of these new techniques, with special emphasis on the life sciences. This book is intended to be a synopsis of recent developments in optical microscopy, not a how-to approach. We hope the information we have chosen will anticipate the next generation of development of optical microscopic techniques. Cell biologists, biophysicists, neuroscientists, physiologists, pathologists, and materials scientists should all find the information useful.

We organized the book to reflect the variety of technologies used in optical microscopy. Beginning chapters address recent technological advances in monitoring and altering cellular physiology, along with requisite hardware and software concerns. Next come discussions of newer approaches for monitoring multiple cellular activities. Some of these newer approaches use simultaneous or serial measurements with different optical techniques or probes; others combine optical microscopy with nonmicroscopic approaches. The third section deals with means to monitor membrane cytoarchitecture. These chapters range from the submicroscopic detection of lipid domains in membranes to the movement of membrane proteins and the interaction of cell membranes with their substrates. The fourth section describes both the confocal imaging of living cell function and the observation of intact organ physiology using optical techniques. The book ends with a discussion of lifetime imaging and automated clinical image cytometry.

In compiling this book, the hardest decisions were (as always) what to include. We focused on areas that, in our opinion, show greatest potential to yield novel data about cellular physiology. We avoided in-depth presentation of any one area of optical microscopy, as international meetings and other texts provide this information to the scientific community. These chapters, we hope, will trigger scientists' imaginations and lead to new applications and technical advances in optical microscopy.

We thank the authors for their contributions and adherence to our deadlines. We especially appreciate the work of Sonee Young, Deirdre Switzer, and Jan Thompson, who provided excellent secretarial assistance and diligence in coordinating our work with the authors.

Brian Herman
John J. Lemasters

# 1

# NEW FLUORESCENT PROBES FOR CELL BIOLOGY

## Martin Poenie and Chii-Shiarng Chen

Department of Zoology, University of Texas at Austin

I. INTRODUCTION
II. AN INTRODUCTION TO
FLUORESCENCE
III. WHY IS FLUORESCENCE
USEFUL?
IV. FLUORESCENT PROBES FOR
PROTEINS AND ENZYMES
  A. Visualization of Protein Kinase
  C Translocation Using a
  Fluorescent Ligand
  B. Fluorescent Analog
  Cytochemistry
  C. Fluorescent Peptides
  D. Photobleaching and
  Photoactivation
  E. Proteins as Fluorescent
  Indicators
  F. A Fluorescent Indicator for
  cAMP Based on Resonance
  Energy Transfer between
  Catalytic and Regulatory
  Subunits

  G. MeroCaM, a Fluorescent
  Calcium Indicator Based on
  Calmodulin
V. FLUORESCENT LIPID AND
MEMBRANE PROBES
  A. Vesicle Transport
  B. Fluorescent Membrane Potential
  Indicators
VI. SYNTHETIC FLUORESCENT
ION INDICATORS
  A. Fluorescent Calcium Indicators
  B. pH Indicators
  C. Chloride
  D. Magnesium
  E. Sodium
  F. Future Improvements in Ion
  Indicators
REFERENCES

1

## I. INTRODUCTION

Our understanding of cell biology depends a great deal on the tools and techniques we use to get information about the cell. If one could only study cells by grinding them up, we would know little about such things as the electrical activity of cells, ion gradients, or a cell's leading and trailing edges associated with movement. Without powerful molecular techniques we might understand cytoplasm of cells only in terms of its physical characteristics. Ultimately, we seek to understand how molecules inside the cell give rise to observable cell structures and activities. The development of fluorescence techniques has greatly aided this effort. Few other molecular techniques offer the combination of sensitivity, spatial and temporal resolution, convenience, and design versatility available with fluorescent sensors. In this chapter, we seek to highlight a few of the current and emerging uses of fluorescence in cell biology.

## II. AN INTRODUCTION TO FLUORESCENCE

Fluorescence arises from the interaction of light with molecules. The absorption of a photon can temporarily promote an electron from its lowest energy level (ground state) to a higher energy level (an excited state). The difference in energy between ground state and excited state is dependent on the structure of a molecule and corresponds to its excitation spectrum. After a short interval ($\sim 10^{-9}$ s) in the excited state, the electron returns to the ground state with the emission of a photon of light. The average time a molecule spends in the excited state is referred to as the fluorescence lifetime. Numerous interactions can take place between a molecule in the excited state and its environment. The environment can stabilize or destabilize the excited state. Part of the energy of an absorbed photon can be lost as heat or through chemical reactions as the excited molecule collides with molecules in its environment. Loss of energy during the excited state means that the emitted photon will have a lower energy and thus a longer wavelength than the photon originally absorbed. This difference between the wavelength that is most efficient at exciting the molecule and the wavelength at which most photons are emitted is referred to as the Stokes shift. The Stokes shift associated with fluorescence is one reason that fluorescence techniques are so useful and sensitive. Using a combination of bright lamps or lasers and monochromators or filters, one can efficiently excite the sample at one wavelength and detect the emission at a longer wavelength.

The fluorescence microscope is designed to maximize transmission of emitted light uncontaminated by the much more intense excitation light. The ability to see the emitted signal in the absence of the excitation light gives fluorescence a relatively good signal to background. Aided by sensitive video cameras or photomultiplier tubes, one can currently detect as few as 100–1000 fluorescent molecules.

## III. WHY IS FLUORESCENCE USEFUL?

There are several reasons fluorescence has become such an important tool for biologists. One reason is its sensitivity. By using special equipment it has been possible to detect a single phycocyanin molecule by fluorescence (Mathies and Stryer, 1986; Nguyen and Keller, 1987). In more conventional microscope-based systems one can detect between 100–1000 fluorescently labeled molecules. More importantly, the sensitivity of fluorescence permits one to obtain an adequate signal from cells using relatively low concentrations of probe. For example, cells loaded with 10–50 $\mu$M of a fluorescent indicator, such as fura-2, fluoresce brightly enough to saturate a sensitive video camera, such as a silicon intensified target (SIT) camera or an intensified charge-coupled device (CCD). Since there is a possibility that any indicator will be toxic or will significantly perturb the cell, the less indicator needed the better.

Another important aspect of fluorescence is its compatibility with microscope optics. Many fluorophores excite and emit in the visible spectrum or near ultraviolet (UV). In many cases, where fluorescence can be measured one can also image the signal. In addition, because brightly fluorescent molecules contrast strongly with the dark background, one can see the fluorescence from cellular structures too small to be resolved by the light microscope.

A third important feature of fluorescence is its ability to sense fast biological events. Fluorophores have fluorescence lifetimes that are short relative to most biological events so that the response of the fluorophore is usually not an issue. Most issues concerning the speed of a fluorescence indicator do not directly relate to fluorescence itself but rather to the operation of the fluorescent probe as a whole. For example, fluorescent ion indicators shift back and forth between free and bound forms as the free ion concentration changes. However, if the free ion concentration were to change faster than the free and bound forms equilibrate, the probe would not properly sample the event being measured. In this case, the kinetics of ion binding and dissociation are at fault, not the response of the fluorophore.

Finally, fluorescence is sensitive to the molecular structure of the fluorophore and many different kinds of environmental factors, including viscosity, quenching by other molecules, pH, and solvent polarity. A viscous environment that immobilizes a fluorophore can cause a nonfluorescent molecule to fluoresce. This effect is seen when phenophthalein and malachite green are immobilized. Quenching is the opposite effect where collision of the excited fluorophore with other molecules drains off the energy of the excited state. In some cases, quenching results from specific interactions. Quinine, for example, fluorescences in sulfuric acid but not in hydrochloric acid because collisions with chloride ions quench its fluorescence. Differences in solvent polarity can cause dramatic spectral shifts or changes in brightness of a fluorophore. This is seen with some of the dyes used for membrane potential, such as the strylpyridinium dyes, which hardly fluoresce in water but fluoresce strongly in hydrophobic solvents. These different effects can be both useful and troublesome. They are useful because they provide a great deal of versatility in adapting fluorescence for use as biological tracers and sensors. On the down side, these effects can show up whether one wants them or not. For example, the fluorescence spectra of the ratioable calcium indicators, such as fura-2, are distorted by the viscosity of cytoplasm. In order to get an accurate estimate of intracellular free calcium levels, one must take into account the influence of viscosity (Poenie, 1990).

## IV. FLUORESCENT PROBES FOR PROTEINS AND ENZYMES

Fluorescence techniques have historically been important and versatile tools in the study of protein distribution and function in cells. Fluorgenic enzyme substrates provide the sensitivity needed to measure enzymatic process in single living cells while fluorescent antibody techniques are the standard way to visualize and localize cellular proteins. With sensitive video detectors, such as the SIT, ISIT (intensified silicon intensified target), and cooled CCD cameras, one can obtain real-time images of a relatively small number of fluorescently labeled proteins in living cells. In this section on proteins, we seek to illustrate the varied uses of fluorescently labeled proteins and how they can shed light on many different cellular structures and activities.

The most fundamental use of fluorescence with proteins is as a label to single out the protein of interest in the cell. Some proteins are intrinsically fluorescent due to the presence of tryptophan or tightly bound cofactors and this has been exploited as a means of imaging proteins. More often, the goal is to label a particular protein either indirectly using fluorescent ligands, which bind noncovalently to the protein of interest, or directly by making fluorescent derivatives of purified proteins. The fluorescent derivatives of the fungal toxin phalloidin, which selectively stain actin filaments, are an example of the indirect approach. Along similar lines, we provide a preliminary description of a new fluorescent ligand for protein kinase C that should be useful for visualizing the distribution of this enzyme in cells. The alternative, direct labeling of purified proteins has been referred to as fluorescent analog cytochemistry (Taylor *et al.*, 1984). This is the surest way to avoid the problems of nonspecific binding of fluorescent ligands. Furthermore, since the fluorophore is attached covalently, one can make use of photobleaching and photoactivation of fluorescence for purposes of studying the dynamics of protein movements within the cell or cell structure. The only problem here is obtaining a stable and suitably pure preparation of the protein. New applications are on the horizon where fluorescence is used not only to tag proteins but to turn an enzyme into a fluorescent indicator. A change in fluorescence signals binding of enzyme to its substrate or to another protein. This is a notable achievement because it opens the door to developing probes for complex ligands.

## A. Visualization of Protein Kinase C Translocation Using a Fluorescent Ligand

As an example of an indirect-labeling approach, we present some preliminary studies using a new fluorescent probe for protein kinase C. Protein kinase C (PKC) is a family of kinases whose activity requires phospholipid, diacylglycerol, and in some cases calcium (Kikkawa *et al.*, 1989). These enzymes have been implicated in the regulation of many different cellular processes, such as secretion, growth control, and cell motility. One of the fascinating characteristics of PKC is its translocation from cytosol to membranes in the presence of activators such as diacylglycerol or phorbol esters. There is also evidence that, as a later event, a fragment of the enzyme moves to the nucleus where it may affect transcription of some genes. The mechanism of this translocation is not understood. For example, it is not clear whether diacylglycerol binds to PKC before it translocates to the membrane or if diacylglycerol joins with calcium and phosphatidylserine to produce a PKC binding site in the membrane.

To develop a fluorescent PKC probe, we synthesized a series of fluorescent bisindolyl-maleimide PKC inhibitors, which exhibit strong selectivity for PKC over cAMP (cyclic adenosine-3′,5′-monophosphate) dependent kinase (Davis *et al.*, 1990a,b). Our strategy was to make fluorescent molecules that bound to the PKC catalytic domain with high affinity but did not interfere with the properties of the membrane-binding domain. The structure of one the inhibitors tested, fim-1, is shown in Fig. 1.1. This probe resembles a portion of the staurosporine molecule but differs in its molecu-

FIGURE 1.1

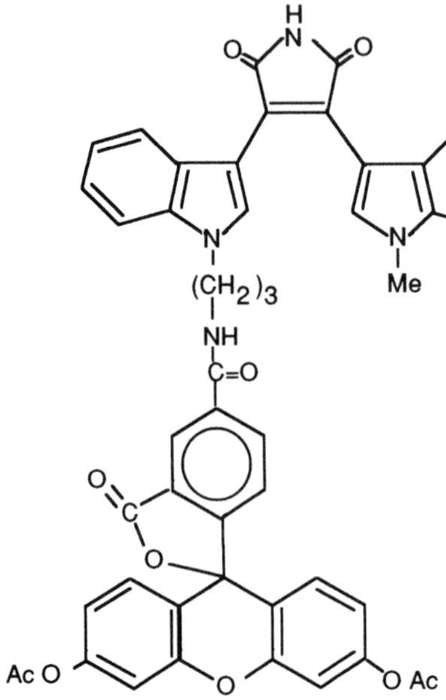

The structure of fim-1, a fluorescent protein kinase C ligand based on bis indolylmaleimide protein kinase C inhibitors.

lar geometry and selectivity for PKC. Preliminary studies indicate that fim-1 does not compete with ATP, diacylglycerol, or phospholipid and inhibits PKC with a 10 nM $IC_{50}$. As an optical probe, fim-1 can be used as a stain for fixed cells and potentially for living cells. As an example of results obtained using fim-1 as a stain, we show cytotoxic T lymphocytes (CTL) bound to target cells (Fig. 1.2) and macrophages before and after treatment with phorbol myristate acetate (Fig. 1.3). The staining of cytotoxic T cells reveals a bright band in the CTL, at target cell contact site. In CTLs alone, treatment with phorbol ester causes staining to accumulate in a spot corresponding to the region of the microtubule organizing center (MTOC). With macrophages, exposure to phorbol ester causes a gradual increase of nuclear or perinuclear fluorescence over 30 min.

One hopes that the high affinity of fim-1 for PKC would provide a high degree of specificity of the probe for PKC or at least for kinases as opposed to nonspecific binding to membranes in the absence of PKC. Even with specific binding one has difficulty knowing that brighter staining automatically reflects enzyme concentration since the bright and dark regions could result from local differences in cell thickness. One can compensate for these

FIGURE 1.2

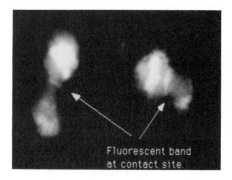

Fluorescence-staining pattern of cytotoxic T lymphocyte target cell conjugates fixed and stained with fim-1. A bright band of fluorescence is observed at the lymphocyte target cell contact site.

FIGURE 1.3

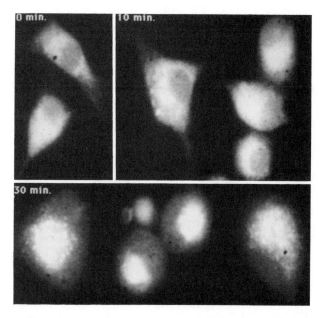

Macrophages fixed and stained with fim-1, before and after treatment with phorbol myristate acetate. Over a time course of 30 min, fluorescence staining becomes concentrated near or in the nucleus.

factors using confocal fluorescence microscopy or a double-staining technique using a second fluorophore, which stains the cytoplasm indiscriminately. Here, the ratio of the fluorescence signals from the two different fluorophores could compensate for differences in cell thickness.

## B. Fluorescent Analog Cytochemistry

A direct and perhaps more satisfying method for visualizing the intracellular distribution of proteins is to covalently label the protein of interest and then microinject it back into the cell. This approach has been referred to as fluorescent analog cytochemistry (Taylor *et al.*, 1984) and has been used to study a variety of cytoskeletal proteins. The technique is conceptually simple but depends on one's ability to obtain pure samples of a protein and preserve functionality throughout the various manipulations. The general approach outlined below was adapted from Taylor *et al.* (1984).

1. Purification of the molecule.
2. Label and purify molecules with a defined ratio of dye to molecule.
3. Compare biochemical, biophysical, and physiological properties of the fluorescent analog with its unlabeled counterpart *in vitro*.
4. Characterize the spectroscopic properties of the analog *in vitro*.
5. Incorporation of the analog into living cells followed by testing the functional capability of the analog *in vivo*.
6. Analysis of the cells containing both the fluorescent analog and a soluble control labeled with a distinct fluorophore.

Fluorescently labeled proteins that meet the criteria outlined above eliminate problems of nonspecific binding and functional perturbations that are inherent with the indirect approach. Indeed, the success of fluorescent analog cytochemistry depends on finding ways to covalently label a protein without destroying its function. This approach has been quite successful with actin, tubulin, and other cytoskeletal proteins. Actin, for example, contains a sulfhydryl group, which reacts with a number of thiol reagents. Actin monomers, fluorescently modified at this sulfhydryl group, still assemble into filaments and interact with cellular proteins. By using fluorescent actin analogs, one can image the dynamic behavior of actin in living cells. Fluorescent tubulin derivatives can be prepared using DTAF, a fluorescein derivative that reacts with amino groups. The same methodology has been applied to $\alpha$-actinin, vinculin, myosin, fibronectin, and calmodulin. With a few exceptions, these structures are

largely invisible when one observes most cells through the light microscope. Fluorescent analogs of cytoskeletal proteins enable one to make
movies of these proteins as they function in living cells, which is a great
step forward for studies of cell motility.

## C. Fluorescent Peptides

Fluorescent analog cytochemistry can, in some cases, be streamlined by the
use of fluorescent peptides. Molecular biology has revealed many short
amino acid sequences in proteins, which serve as specific ligands, substrates,
or processing sites. These peptides often can functionally substitute for a
complete protein. Since peptides are readily synthesized, the tedious process
of protein purification can be avoided. For example, Newmeyer and Forbes
(1988) recently utilized fluorescent peptide-BSA conjugates to study amino
acid sequences needed for protein transport through the nuclear envelope.
Fluorescent derivatives of peptides should provide convenient tools for
studying many important processes in cells and organelle preparations.

## D. Photobleaching and Photoactivation

The ability to visualize particular proteins as they function in living cells
opens the door to many novel experimental approaches, including photobleaching techniques, photoactivation of fluorescence, and energy transfer
studies. Photobleaching permits one to measure the dynamic movements of
protein subunits within a small region of a cell or subcellular structure.
Photoactivation of fluorescence is a complementary technique utilizing
caged fluorophores. With caged fluorophores, a fluorescent molecule is rendered nonfluorescent by a photosensitive protecting group. Exposure to
ultraviolet light removes the protecting group, and fluorescence is then observed. An example of a caged fluorophore is C2CF, a derivative of fluorescein described by Mitchison (1989). Both photobleaching studies and
photoactivation of fluorescence have been used to study dynamic movements of tubulin subunits in fluorescent microtubules. Using photobleaching, Gorbsky *et al.* (1988) showed that kinetochore microtubules exhibit
slower rates of exchange or turnover than most spindle microtubules. By
bleaching a small region within the microtubules that span between kinetochore and the pole, they could observe movement of the bleached spot
with respect to the pole. In their studies, using living cells, the bleached
spot did not move, suggesting that chromosomes move along stationary
microtubules. A different result was obtained by Sawin and Mitchison
(1991) using C2CF-labeled tubulin, which was assembled into spindles in
cell extracts. In this system, they could see a poleward flux of tubulin sub-

units within the kinetochore microtubules. The reason for these differences is not clear but might relate to measurements made *in vivo* versus spindles assembled *in vitro*. Thus, there are differences in both technology and the preparation, which might account for the different results. The use of photobleaching and caged fluorophores in combination with fluorescent analog cytochemistry has much potential. Caged fluorophores may some day be used for fate mapping and tracking the movements of particular cells, organelles, or molecules.

## E. Proteins as Fluorescent Indicators

Development of a fluorescent indicator usually depends on the availability of a specific ligand for the target molecule. For small ions, synthetic ligands could be designed and synthesized with suitable affinity and specificity. For more complex metabolites, this is not feasible at present and one must rely on the specificity inherent in enzyme–substrate binding. There are two strategies whereby enzymes or proteins have successfully been converted to fluorescent indicators and the list will no doubt grow quickly. One strategy uses resonance energy transfer while the other uses a fluorescent probe sensitive to solvent polarity to detect changes in protein conformation.

## F. A Fluorescent Indicator for cAMP Based on Resonance Energy Transfer between Catalytic and Regulatory Subunits

Molecular proximity or binding of two different fluorescent analogs can be studied using fluorescence resonance energy transfer. Resonance energy transfer is the process of transferring energy from one molecule in the excited state (donor) to another molecule (acceptor) without a collision. In order for energy transfer to take place, two molecules must be very close together since energy transfer is inversely proportional to the sixth power of the distance separating the two fluorophores. In addition, energy transfer requires that the donor and acceptor be in resonance. This means that the absorption spectrum of the acceptor must overlap the excitation spectrum of the acceptor. If the acceptor is not fluorescent, it simply quenches the fluorescence of the donor. If the acceptor is fluorescent, then excitation of the donor results in fluorescence emission from the acceptor. The exquisite dependence of energy transfer on proximity makes this a good approach for sensing when two molecules are bound together.

Resonance energy transfer was exploited by Adams *et al.* (1991) in the development of a fluorescence ratio indicator for cAMP. The indicator, called "FlCRhR," is derived from cAMP-dependent kinase, a multisubunit enzyme containing regulatory and catalytic subunits. The kinase regulatory subunits bind tightly to catalytic subunits in the absence of cAMP and dissociate in the presence of cAMP. As an indicator, the regulatory subunit of the kinase provided a high affinity, selective ligand for cAMP while fluorescence energy transfer provided an optical response to the binding of cAMP. Their approach was to label the catalytic subunit of cAMP-dependent kinase with fluorescein and the regulatory subunit with tetramethylrhodamine. When the two subunits were mixed in the absence of cAMP, the subunits associated and resonance energy transfer was observed. In the presence of cAMP, the subunits dissociated and energy transfer was lost. For fluorescence ratio-imaging experiments, they injected fluorescently labeled holoenzyme into cells and excited at 480–495 nm. Emission from fluorescein was used as a measure of free regulatory subunits while rhodamine emission (570–620 nm) reflected complexes of regulatory and catalytic subunits. An increase in cAMP causes subunit dissociation and a loss in energy transfer, which increases the fluorescence from fluorescein (500–570 nm). The ratio of emission intensities at 500–570 nm and 570–620 nm then provide an estimate of cAMP concentration. One surprising result from experiments using FlCRhR is that the indicator did not lose its cAMP response after the first cAMP transient. One might have thought that after the fluorescent subunits dissociated the first time, they would reassociate with endogenous unlabeled subunits destroying the indicator. This does not appear to be as big a problem as it might have been. The success of FlCRhR will no doubt spur other attempts at developing other indicators utilizing resonance energy transfer.

## G. MeroCaM, a Fluorescent Calcium Indicator Based on Calmodulin

A number of fluorescent probes, such as the new MeroCaM, rely on solvent effects associated with changes in protein conformation to cause a change in fluorescence (Hahn *et al.*, 1990). MeroCaM is a fluorescent derivative of calmodulin responding to calcium binding with a change in fluorescence. When calmodulin is bound to calcium, a hydrophobic portion of the protein becomes exposed. When calcium-saturated calmodulin was covalently labeled with the merocyanine dye, Hahn and colleagues obtained a functional calmodulin, which gave a fluorescence response to changes in $[Ca^{2+}]$. Calcium binding caused an increase in the peak excitation efficiency as well as a shift to longer wavelengths. Maximum differences between excitation and emission spectra were obtained at 532 nm and 608 nm. The actual usefulness of MeroCaM for measurements of $[Ca^{2+}]$ inside cells has not been established although such efforts are underway. However, since many proteins exhibit similar conformational changes, this approach could find many other applications.

## V. FLUORESCENT LIPID AND MEMBRANE PROBES

Fluorescence has traditionally been an important tool for studying the biophysical properties of membranes and membrane lipids. There has also been considerable effort aimed at developing fluorescent indicators for sensing the plasma membrane potential. Progress here has been steady but slow. Unfortunately, we do not yet fully understand how the membrane potential indicators sense and respond to the change in membrane potential. In some cases, these are metabolized in the same manner as the corresponding natural lipid. A variety of fluorescent lipids have been synthesized by Pagano and colleagues. Many of these utilized NBD (4-nitrobenzo-2-oxa-1,3-diazole) as the fluorophore. These lipids could be used to stain lipid compartments though often without a great deal of contrast between different compartments from the standpoint of fluorescence microscopy. Recently, however, Pagano et al. (1991) have introduced a BODIPY-ceramide that shifts its emission toward the red as it aggregates in the Golgi. As a result, using filters selectively for either the monomer form (515 nm) or aggregate form (620 nm), one can readily image the Golgi apart from the endoplasmic reticulum or other vesicles.

### A. Vesicle Transport

Much cellular activity involves the movement and fusion of vesicles. Fluorescent ligands have been popular for studying events accompanying endocytosis. Using soluble small fluorescent molecules, fluorescein-labeled dextrans, or labeled cell surface ligands, one can monitor pinocytosis or receptor-mediated endocytosis. This provides a means for studying vesicle movements (Herman and Albertini, 1984), sorting and delivery of vesicle contents (Wang and Goren, 1987; Tabas et al., 1990), and changes in pH (Maxfield, 1982). Recently, new indicators have extended the use of fluorescence to monitor other kinds of vesicle transport. One exciting new discovery is the uptake and release of certain styryl pyridinium dyes (Betz and Bewick, 1991; Betz et al., 1992). These compounds fluoresce in membranes and specifically label the recycling vesicles of both frog and mammalian motor nerve terminals. A second promising development involves the mating of the power of yeast genetics with a simple fluorescence assays. Weisman and Wickner (1988) showed that depriving yeast of certain nutrients caused accumulation of a fluorescent pigment in the yeast vacuole. Alternatively, the yeast vacuole could be visualized by vital fluorescent stains even in the absence of the endogenous pigment. With these tools, Weisman was able to visualize the contents of vesicles transported from the parent vacuole to that of the budding daughter cell. This provided a convenient and selective screen for mutants defective in transport. A third new approach is the use of fluorescence to study fusion of viruses with cells. In a recent study, Lowy *et al.* (1990) visualized single virus–cell fusion events using fluorescence dequenching of a fluorophore trapped in the virus.

## B. Fluorescent Membrane Potential Indicators

Fluorescent probes that respond to membrane potential have made slow but steady progress over the last decade. Many of the probes first used for measuring membrane potential were discovered by the slow and arduous task of screening. Membrane potential probes are currently grouped as either slow dyes (response time of seconds) or fast dyes (response time of milliseconds). The slow indicators accumulate in organelles in response to $\Delta\Psi$. Rhodamine 123, for example, accumulates in the mitochondria based on membrane potential and acts as a selective mitochondrial stain. As they accumulate, some slow indicators form aggregates with altered fluorescence properties. Recently, Reers *et al.* (1991) characterized a slow carbocyanine-based indicator called JC-1, which shows both excitation and emission shifts when it becomes concentrated enough to form aggregates. *In vitro,* partitioning of JC-1 into mitochondria and formation of the red–fluorescing aggregates were linearly dependent on mitochondrial membrane potential. In living cells, JC-1 reveals heterogeneity in mitochondrial membrane potential between different mitochondria in a single cell (Smiley *et al.*, 1991).

The fast membrane potential probes respond to membrane potential changes at the cell surface and do not transfer across the cell membrane. These probes simplify the task of recording and imaging membrane potential of single cells (Gross *et al.*, 1986) and the complex patterns of neuronal activity in the brain. There has been considerable effort directed at finding membrane potential probes suitable for these tasks.

## VI. SYNTHETIC FLUORESCENT ION INDICATORS

Changes in intracellular ion concentrations often serve as a link between receptor–ligand interactions at the plasma membrane and changes in the activity of cellular enzymes. The development of fluorescent ion indicators for all the physiologically important ions in the cell has provided a relatively easy means for measuring the intracellular concentration of these ions. For some of these indicators, ion binding causes a shift in their peak excitation or emission wavelengths rather than a simple change in intensity. This wavelength shift is useful because ion concentration can be determined from the ratio of fluorescence at two different excitation or emission wavelengths, independent of pathlength or absolute dye concentration. As a result, ratioing indicators are uniquely suited for single-cell measurements where the

TABLE 1.1
FLUORESCENT ION INDICATORS

| | |
|---|---|
| Ca$^{2+}$ | fura-2[a], indo-1[a], fluo-3[b] |
| K$^+$ | PBFI[c] |
| Na$^+$ | SBFI[c] |
| Mg$^{2+}$ | Furaptra[d], (mag-fura-2, mag-indo-1) |
| H$^+$ | BCECF[e], SNARF[f], SNAFL[f], dihydroxphthalonitrile[g] |
| Cl$^-$ | SPQ[h] |

[a] Grynkiewicz et al. (1985)
[b] Minta et al. (1989)
[c] Minta and Tsien (1989)
[d] Raju et al. (1989)
[e] Rink et al. (1982)
[f] Whitaker et al. (1991)
[g] Brown and Porter (1977); Kurtz and Balaban (1985)
[h] Krapf et al. (1988)

fluorescence pathlength and local dye concentration can vary widely. When combined with fluorescence video microscopy or confocal microscopy and digital image processing, one can obtain two- or even three-dimensional images of the ion concentration inside the cell. In Table 1.1, we briefly review the characteristics of the different fluorescent ion indicators and how they are used.

## A. Fluorescent Calcium Indicators

The development of fluorescent indicators for intracellular-free calcium ions began with BAPTA and quin-1 (Tsien, 1980). These compounds contained a tetracarboxylate calcium-coordination site similar to the calcium chelator, EGTA. While quin-2 was a breakthrough, it also had many problems. The weak fluorescence of quin-2 made it necessary to load cells with nearly 1 mM indicator in order to obtain adequate signals. This posed serious problems of buffering and potential formaldehyde toxicity when cells were loaded using the acetoxymethyl ester derivative of quin-2. While the spectra of quin-2 exhibit an isoexcitation point at 360 nm, it was never widely used as a ratiometric indicator because the fluorescence at 360 nm is so dim and relatively sensitive to magnesium.

The development of fura-2 and indo-1 followed and offered many improvements over quin-2 (Grynkiewicz et al., 1985). These indicators were much brighter so that cells loaded with 50–100 $\mu$M fura-2 were still brightly fluorescent. Their selectivity for calcium over a variety of other metal ions was also much improved. Finally, when fura-2 and indo-1 bound calcium, their peak excitation wavelength shifted; and for indo-1, calcium binding also shifted the emission peak. The spectral shifts these dyes exhibited were a useful feature because it permitted calcium measurements based on the ratio of fluorescence intensities at two wavelengths. With a single-wavelength indicator, one must take into account both illuminated pathlength and dye concentration in order to relate fluorescence intensity to calcium concentration. These variables cancel out in a fluorescence ratio. This greatly simplified the problem measuring calcium in single cells where pathlength and local dye concentration might vary greatly. In effect, the fluorescence ratios report the relative concentrations of free and bound indicator from each unit volume of cytosol. Fluorescence ratio measurements required either alternating excitation wavelengths or dual-emission detectors. Indo-1 became popular for flow cytometry because it could be excited by an argon laser, and flow cytometers already were set up for monitoring multiple-emission wavelengths. For most other applications, fura-2 tended to predominate, partly because of its slightly longer excitation wavelengths.

The uses of fura-2 spread quickly from fluorescence ratio measurements of cell populations in a cuvete to single-cell microspectrofluorimetric measurements and finally to fluorescence ratio-imaging systems (Tsien et al., 1985; Poenie et al., 1985; Williams et al., 1985; Tsien and Poenie 1986). Prior to the introduction of fura-2, imaging free calcium was limited primarily to large cells injected with the photoprotein aequorin (Gilkey et al., 1978). Fluorescent ratio imaging of fura-2 provided images of intracellular calcium physiology for a diverse collection of cell types and cell activities (Tsien and Poenie, 1986). The extensive application of fura-2 has shed light on many aspects of cell signaling as well as revealed problems with the indicator itself. Data from fura-2 experiments have shown waves of calcium spreading across the fertilized egg similar to results obtained using aequorin (ibid.). Calcium oscillations, first reported in cells injected with aequorin, are commonly seen using fura-2. Calcium gradients associated with directional cell motility have been observed by several laboratories (Poenie et al., 1987; Brundage et al., 1991). Finally, one can monitor events in two different cells simultaneously as they interact (Poenie et al., 1987).

Until recently, fura-2 has been much more popular for imaging applications than indo-1. While in principle one can base calcium measurements on either excitation or emission ratios, several factors mitigate against the use of indo-1 and emission ratioing. These include the shorter excitation wavelengths and the rapid bleaching of indo-1. In addition, ratioing two images at different emission wavelengths presents special problems of image alignment that are not encountered using excitation ratios.

Despite these problems, there are overriding advantages to the use of emission ratioing that have led some researchers to develop imaging systems adapted for ratioing indo-1. One advantage is speed. With excitation ratioing, one needs some source of alternating wavelengths, which usually involves a mechanical chopper. Choppers are typically slow and, at best, one can only obtain a ratio every 66 ms. With indo-1 and two cameras as detectors, a chopper is not needed and ratio images can be acquired at video rates. In addition to speed, two camera systems can acquire images simultaneously. This is especially beneficial for studying moving cells. Any movement of the cell between acquisition of a ratio pair renders a ratio image meaningless.

Since the introduction of fura-2 and indo-1, several new calcium indicators have appeared. These indicators, such as fluo-3, rhod-2, calcium-green, calcium-red, calcium-orange, and calcium-crimson, do not exhibit useful shifts in either peak excitation or emission wavelengths upon binding calcium (Minta *et al.*, 1989; Haugland, 1992). However, all of these newer indicators are excited by visible light. This is an advantage over fura-2 and indo-1, which require UV excitation. The long wavelength indicators are also compatible with the use of caged calcium compounds such as nitro-5 (Tsien and Zucker, 1986). With caged calcium compounds, a bright flash of UV light (~350 nm) converts a molecule with high affinity for calcium into a molecule that binds calcium weakly. When caged calcium is loaded into the cytoplasm of cells, they hold onto calcium tightly. The flash of UV light causes a sudden release of calcium. In order to measure the change in calcium caused by the light flash, one must use an indicator that absorbs poorly at 350 nm or else the indicator itself could be destroyed. Fluo-3 and presumably the other long-wavelength calcium indicators are useful for these types of experiments (Kao *et al.*, 1989 and Chapter 2 of this volume).

## B. pH Indicators

Cytosolic pH is an important factor in controlling transitions from dormant to metabolically active states of the cell. Sea urchin sperm shows an increase in pH in response to factors contained in the jelly surrounding the egg while the egg alkalinizes at fertilization. In fibroblasts, cytoplasmic pH rises in response to serum growth factors. Localized changes in pH are also observed for some subcellular organelles, such as endosomes. Fluorescent pH indicators have been a popular tool for measuring intracellular pH. The most popular indicators have been fluorescein derivatives, such as 6-carboxyfluorescein and BCECF (Rink *et al.*, 1982). The mixed acetate–acetoxymethyl ester derivatives of BCECF can be readily loaded into many types of cells. In addition, BCECF exhibits an isoexcitation point at 439 nm that permits pH measurements based on fluorescent ratios at 490 nm/ 430 nm. This feature was utilized by Paradiso *et al.* (1987) to image pH in gastric glands. Use of video-imaging techniques made it possible to observe the differences in pH regulation and ion transport between adjacent oxyntic and chief cells.

More recently, pH indicators have been introduced based on naphtho-fluorescein and naphthorhodamine (Whitaker *et al.*, 1988). These indicators work in the visible spectrum and shift excitation (SNAFL) or emission spectra (SNARF) in response to changes in pH. The emission shift exhibited by SNARF should make this probe potentially useful with laser-scanning microscopy. In one study, Bassnett *et al.* (1990) used SNARF-1 together with the confocal laser-scanning microscope to measure pH in lens tissue. In this system, they noted that dye bleaching and leakage were problems, but pH measurements were feasible. The problem with bleaching is common using the strong laser illumination needed with confocal laser-scanning microscopy. Dye-loading problems and leakage are common with many of the ion indicators. The long wavelengths of SNARF and SNAFL come at the expense of increased hydrophobicity due to the added aromatic ring. Conjugation of SNARF to dextran prevents dye leakage and compartmentation resulting in a well-behaved indicator. In a recent study, Thiebaut *et al.* (1990) obtained good results by microinjecting SNARF-dextran conjugates into NIH3T3 cells. Here, they showed a correspondence between cytosolic pH and the activity of a multidrug transporter in cells transformed with the multidrug resistance gene. This was not seen with the nontransformed parental NIH3T3 cells.

## C. Chloride

Chloride ions are also important in controlling water transport in kidney tubules and pH in some cells through $Cl^-/(OH^-$ or $HCO_3^-$) exchange. In an effort to develop fluorescent chloride indicators, Illsley and Verkman (1987) have explored quaternized quinoline derivatives. These compounds exhibit excited-state quenching by halide ions and thiocyanate. One of these compounds, called SPQ [6-methoxy-N-(3-sulfopropyl)quinolinium; Illsley and Verkman, 1987], was used to measure chloride ions in living cells. The zwitterionic character of SPQ makes it difficult to load this compound into cells. In the presence of millimolars of external SPQ, cells gradually accumulate the indicator so that measurements are feasible. However, once the extracellular dye is washed away, the dye rapidly leaks out of the cells. Efforts are underway to make new versions of SPQ that contain one or more carboxyl groups derivatized as acetoxymethyl esters. This will not totally alleviate loading problems since the molecule still contains a quaternized nitrogen. Still, SPQ is a promising start and improvements should be possible.

## D. Magnesium

The selectivity of BAPTA for calcium over magnesium depends in part on the difference in size of the two ions. Calcium ions can coordinate with all four carboxyl groups of BAPTA while magnesium can only coordinate with two. Magnesium ions, in effect, bind to only one half of the BAPTA calcium coordination pocket. Synthesis of hemi-BAPTA molecules modeled after fura-2 or indo-1 fluorophore gave fluorescent magnesium indicators with optical properties similar to their calcium-binding counterparts (Raju et al., 1989).

## E. Sodium

Fluorescent indicators for sodium and potassium ions have recently been developed by Minta and Tsien (1989). These indicators selectively coordinate the monovalent cation inside a modified crown ether cage. Preference for the sodium or potassium ion is determined by the size of the crown ether. The fluorescence response depends on the two benzofuran isophthalate moieties, which gives these indicators spectral properties similar to fura-2. Compared to fura-2, however, SBFI exhibits a much smaller range of ratios ($R_{max}/R_{min}$ = 3.0 for SBFI and roughly 30 for fura-2).

In one recent study, Wong and Foskett (1991) used SBFI to measure changes in sodium in rat salivary glands treated with cholinergic agonists. They were able to load these cells with the acetoxymethyl ester derivative but noted that some of the dye was trapped in organelles. However, they found that SBFI in organelles also responded to changes in cytosolic free sodium. They reported resting sodium levels of approximately 10 mM, which rose to 50 mM at the peak of the periodic oscillations induced by carbachol. In another study, Harootunian et al. (1989) were able to image intracellular concentration of free sodium in fibroblasts and lymphocytes. These studies revealed differences in fluorescence ratios between nucleus and cytoplasm, which was attributed to compartmentalization of some SBFI into acidic cytoplasmic vesicles around the nucleus. These differences were abolished when weak bases, such as ammonia, were used to dissipate the pH gradients.

It is too early to judge how useful these new indicators will be. In principle, they should be as easy to use as fura-2. Problems with compartmentalization are common to all the fluorescent ion indicators though with SBFI they appear to be a little worse. Perhaps SBFI can be modified so that it will resist compartmentalization in a manner similar to the modifications made to fura-2 (see below). Poor hydrolysis of the acetoxymethyl esters can also give rise to cells that fluoresce brightly but respond poorly to sodium. This problem is bypassed by microinjection. However, the responses obtained by Wong and Foskett (1991) suggest that some cells do a good job of hydrolyzing the acetoxymethyl ester derivatives and results will simply vary with the type of cell used.

## F. Future Improvements in Ion Indicators

While the fluorescent ion indicators are important tools for studying cell signaling and cell physiology, none of the current generation of indicators are totally satisfactory. The ratioable indicators, such as fura-2 and indo-1, must be excited by UV light while the longer wavelength indicators, such as fluo-3, calcium-green, calcium-red, etc., do not shift wavelengths when they bind calcium. Loading these indicators into cells also presents problems. The most convenient method for loading the fluorescent calcium indicators, as well as SBFI, PBFI, and BCECF, relies on the acetoxymethyl (AM) ester derivatives of these compounds. Fluorescent indicators derivitized as AM esters pass through the cell membrane and are cleaved by cytoplasmic esterases. Unfortunately, the quality of loading can vary the type of tissue used. Unhydrolyzed ester trapped in organelles fluoresces brightly but does not bind calcium. This can seriously dampen calcium–dependent changes in fluorescence ratios. The fully deesterified indicator is also subject to transport into organelles or out of the cell, presumably by a probenecid-sensitive anion transporter. This process, often referred to as compartmentalization, depends on temperature (Poenie *et al.*, 1986). As compartmentalization proceeds, indicator accumulates in perinuclear vesicles, which contain relatively high levels of calcium. These localized regions of apparent high calcium give the deceptive appearance of calcium gradients, making it difficult to discern true gradients of free calcium.

One solution to loading problems and dye compartmentalization is the microinjection of indicators conjugated to dextran. Dextran conjugates of fura-2 and other indicators were recently introduced by Molecular Probes and are preferred where single-cell microinjection is feasible. As a dextran conjugate, fura-2 remains in the cytosol for hours. However, microinjection is not always a suitable method for loading, especially for small cells. As an alternative approach, we have recently synthesized analogs of fura-2 that resist compartmentalization and can be loaded as AM esters. The unique feature of the new indicators is the addition of hindered amine groups, which makes the new indicators zwitterionic. One example of these new indicators is fura-PE1, shown in Fig. 1.4. When loaded as an acetoxymethyl ester, fura-PE1 remains cytosolic far longer than fura-2 (Figs 1.5 and 1.6). The acetoxymethyl ester of fura-PE1 is more water soluble than fura-2-AM, which reduces the amount of residual unhydrolyzed ester present after loading.

Keeping fura-2 in the cytosol is important for some types of experi-

FIGURE 1.4

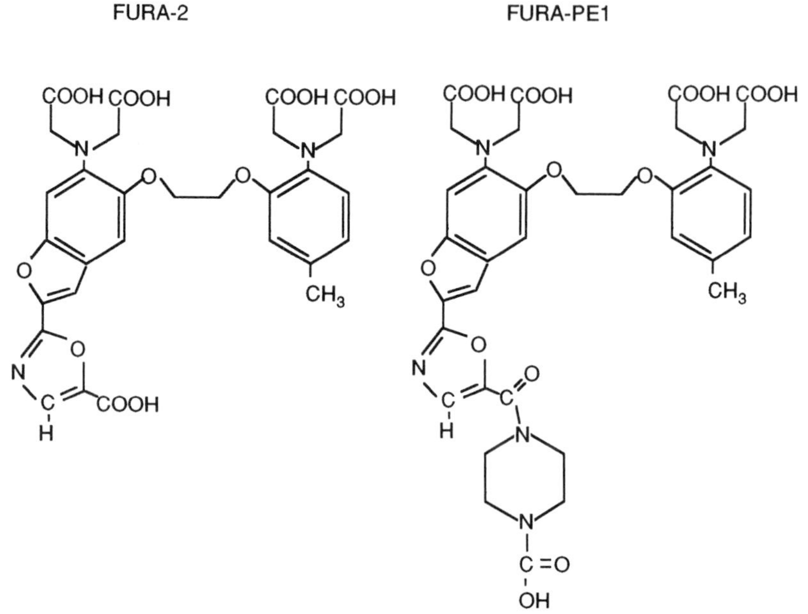

FURA-2                                      FURA-PE1

A comparison of fura-2 and fura-PE1. The addition of a piperazine group linked to fura-2 makes fura-PE1 zwitterionic at physiologic pH.

FIGURE 1.5

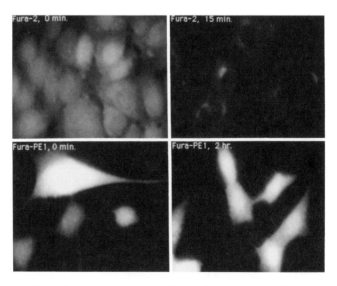

A comparison of retention and compartmentalization characteristics of fura-2 and fura-PE1.

FIGURE 1.6

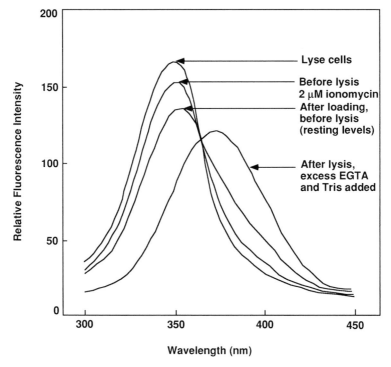

Response of intracellularly trapped fura-PE1, loaded into T lymphocytes as the acetoxymethyl ester, to changes in calcium. The first trace was obtained after 322 T-lymphoma cells were incubated with 5 uM fura-PE1-AM for 30 min at 37°C and then washed. Subsequently, 2 $\mu$M ionomycin was added and a second trace was recorded. A third trace was taken after addition of Lubrol to lyse the cells, releasing the indicator into the extracellular medium. The last trace was recorded after excess EGTA and Tris base was added to chelate free calcium.

ments. In other cases, one would like to target the indicator to specific locations in the cell. For example, ion concentrations at or near the plasma membrane might exhibit extreme fluctuations that are not seen in the cytosol at large. One way to answer this question is to produce analogs of the calcium or other ion indicators that target to specific regions of the cell. Efforts at developing these type of indicators are underway and initial results are promising.

REFERENCES

Adams, S. R., Harootunian, A. T., Buechler, Y. J., Taylor, S. S., and Tsien, R. Y. (1991). Fluorescence ratio imaging of cyclic AMP in single cells. *Nature* 349, 694–697.

Albertini, D. F., and Herman, B. (1984). Cell shape and membrane receptor dynamics. Modulation by the cytoskeleton. *Cell and Muscle Motil.* 5, 235.

Bassnett, S., Reinisch, L., and Beebe, D. C. (1990). Intracellular pH measurement using single excitation-dual emission fluorescence ratios. *Am. J. Physiol.* 258, C171–178.

Betz, W., and Bewick, G. S. (1991). Optical analysis of synaptic vesicle recycling at the frog neuromuscular junction. *Science* 255, 200–202.

Betz, W., Mao, F., and Beswick, G. S. (1992). Activity-dependent fluorescent staining and destaining of living vertebrate motor nerve terminals. *J. Neurosci.* 12, 363–375.

Brown, R. G., and Porter, G. (1977). Effect of pH on the emission and absorption characteristics of 2,3-dicyano-p-hydroquinone. *Chem. Soc. Faraday Trans. I 73*, 1281–1285.

Brundage, R. A., Fogarty, K. E., Tuft, R. A., and Fay, F. S. (1991). Calcium gradients underlying polarization and chemotaxis of eosinophils. *Science* 254, 704–706.

Davis, P. D., Bit, R. A., and Hurst, S. A. (1990a). A convenient synthesis of bis-indolyl- and indoaryl maleic anhydrides. *Tetrahedron Lett.* 31, 2353–2356.

Davis, P. J., Hill, C. H., Keech, E., Lawson, G., Nixon, J. S., Sedgwick, A. D., Wadsworth, J., Westmacott, D., and Wilkinson, S. E. (1990b). Potent selective inhibitors of protein kinase C. *FEBS Lett.* 259, 61–63.

Gilkey, J. C., Jaffe, L. F., Ridgeway, E. B., and Reynolds, G. T. (1978). A free calcium wave traverses the activating egg of the medaka, *Oryzias latipes*. *J. Cell Biol.* 76, 448–466.

Gorbsky, G. J., Sammak, P. J., and Borisy, G. G. (1988). Microtubule dynamics and chromosome motion visualized in living anaphase cells. *J. Cell Biol.* 106, 1185.

Gross, D., Loew, L. M., and Webb, W. W. (1986). Optical imaging of cell membrane potential changes induced by applied electric fields. *Biophys. J.* 50, 339–348.

Grynkiewicz, G., Poenie, M., and Tsien, R. Y. (1985). A new generation of fluorescence $Ca^{2+}$ indicators with greatly improved fluorescence properties. *J. Biol. Chem.* 260, 3440–3450.

Hahn, K. M., Waggoner, A. S., and Taylor, D. L. (1990). A calcium-sensitive fluorescent analog of calmodulin based on a novel calmodulin-binding fluorophore. *J. Biol. Chem.* 265, 20355–20345.

Harootunian, A. T., Kao, J. P., Eckert, B. K., and Tsien, R. Y. (1989). Fluorescence ratio imaging of cytosolic free $Na^+$ in individual fibroblasts and lymphocytes. *J. Biol. Chem.* 264, 19458–19467.

Haugland, R. P. (1992). "Handbook of fluorescent probes and research chemicals." Molecular Probes Inc., Eugene, OR.

Herman, B., and Albertini, D. F. (1984). A time-lapse video image intensification analysis of cytoplasmic organelle movements during endosome translocation. *J. Cell Biol.* 98, 565.

Illsley, N. P., and Verkman, A. S. (1987). Membrane chloride transport measured using a chloride sensitive fluorescent probe. *Biochemistry* 26, 1215–1219.

Kao, J. P. Y., Harootunian, A. T., and Tsien, R. Y. (1989). Photochemically generated calcium pulses and their detection by fluo-3. *J. Biol. Chem.* 264, 8179–8184.

Lowy, R. J., Sarkar, D. P., Chen, Y., and Blumenthal, R. (1990). Observation of single influenza virus-cell fusion and measurement by P.N.A.S 87, 1850–1854.

Kikkawa, U., Kishimoto, A., and Nishizuka, Y. (1989). The protein kinase C family and its implications. *Ann. Rev. Biochem. 58*, 31–44.

Krapf, R., Illsley, N. P., Tseng, H. C., and Verkman, A. S. (1988). Structure–activity relationships of chloride-sensitive fluorescent indicators for biological application. *Analyt. Biochem. 189*, 142.

Kurtz, I., and Balaban, R. S. (1985). Fluorescence emission spectroscopy of 1,4-dihydroxyphthalonitrile. A method for determining intracellular pH in cultured cells. *Biophys. J. 48*, 499.

Mathies, R. A., and Stryer, L. (1986). Single-molecule fluorescence detection: A feasibility study using phycoerythrin. In "Applications of Fluorescence in the Biomedical Sciences." (D. L. Taylor, A. S. Waggoner, R. F. Murphy, F. Lanni, and R. R. Birge, eds.) pp. 129–140, Alan R. Liss, N.Y.

Maxfield, F. R. (1982). Weak bases and ionophores rapidly and reversible raise the pH of endocytotic vesicles in culture mouse fibroblasts. *J. Cell Biol. 95*, 676.

Minta, A., and Tsien, R. Y. (1989). Fluorescent indicators for cytosolic sodium. *J. Biol. Chem. 264*, 19449–19457.

Minta, A., Kao, J. P., and Tsien, R. Y. (1989). Fluorescent indicators for cytosolic calcium based on rhodamine and fluorescein chromophores. *J. Biol. Chem. 264*, 8171–8178.

Mitchison, T. J. (1989). Polewards microtubule flux in the mitotic spindle: Evidence from the photoactivation of fluorescence. *J. Cell. Biol. 109*, 637–652.

Newmeyer, D. D., and Forbes, D. J. (1988). Nuclear import can be separated into distinct steps *in vitro*: Nuclear pore binding and translocation. *Cell 52*, 641–653.

Nguyen, D. C., and Keller, R. A. (1987). Detection of single molecules of phycoerythrin in hydrodynamically focused flows by laser-induced fluorescence. *Anal. Chem. 59*, 2158–2167.

Pagano, R. E., Martin, G. C., Kang, H. C., and Haugland, R. P. (1991). A novel fluorescent ceramide analogue for studying membrane traffic in animals cells: Accumulation at the Golgi apparatus results in altered spectral properties of the sphingolipid precursor. *J. Cell Biol. 113*, 1267–1279.

Paradiso, A. M., Tsien, R. Y., Demarest, J. R., and Machen, T. E. (1987). Na-H and CL-HCO3 exchange in rabbit oxyntic cells using fluorescence microscopy. *Amer. J. Physiol. 253*, C30.

Paradiso, A. M., Tsien, R. Y., and Machen, T. E. (1987). Digital image processing of intracellular pH in gastric oxyntic and chief cells. *Nature 325*, 447–50.

Poenie, M. (1990). Alteration of intracellular fura-2 fluorescence by viscosity: A simple correction. *Cell Calcium 11*, 85–91.

Poenie, M., Alderton, J., Tsien, R. Y., and Steinhardt, R. A. (1985). Changes of free calcium with stages of the cell division cycle. *Nature 315*, 147–149.

Poenie, M., Alderton, J., Steinhardt, R. A., and Tsien, R. Y. (1986). Calcium rises abruptly and briefly throughout the cell at the onset of anaphase. *Science 233*, 886–889.

Poenie, M., Tsien, R. Y., and Schmitt-Verhulst, A. M. (1987). Sequential activation and lethal hit measured by $[Ca^{2+}]i$ in individual cytolytic T cells and targets. *EMBO J. 6*, 2223–2232.

Raju, B., Murphy, E., Levy, L. A., Hall, R. D., and London, R. E. (1989). A fluorescent indicator for measuring cytosolic free $Mg^{2+}$. *Am. J. Physiol. 256,* C540–C548.

Reers, M., Smith, T. W., and Chen, L. B. (1991). J-aggregate formation of a carbocyanine as a quantitative fluorescence indicator of membrane potential. *Biochem. 30,* 4480–4486.

Rink, T. J., Tsien, R. Y., and Pozzan, T. (1982). Cytoplasmic pH and free $Mg^{2+}$ in lymphocytes. *J. Cell Biol. 95,* 189–196.

Salmon, E. D., Leslie, R. J., Saxton, W. M., Karow, M. L., and McIntosh, J. R. (1984). Spindle Microtubule dynamics in sea urchin embryos: Analysis using a fluorescein-labelled tubulin and measurements of fluorescence redistribution after photobleaching. *J. Cell Biol. 99,* 2165.

Sammak, P. J., and Borisy, G. G. (1987). Direct observation of microtubule dynamics in living cells. *Nature 332,* 724–726.

Sawin, K. E., and Mitchison, T. J. (1991). Poleward microtubule flux mitotic spindles assembled *in vitro. J. Cell Biol. 112,* 941–954.

Smiley, S. T., Reers, M., Mottola-Hartshort, C., Lin, M., Chen, A., Smith, T. W., Steele, G. D., Jr., and Chen, L. B. (1991). Intracellular heterogeneity in mitochrondrial membrane potentials revealed by a J-aggregate-forming lipophilic cation JC-1. *Proc. Nat. Acad. Sci. USA 88,* 3671–3675.

Storch, J., Lechene, C., and Leinfield, A. M. (1991). Direct determination pf free fatty acid transport across the adipocyte plasma membrane using quantitative fluorescence microscopy. *J. Biol. Chem. 266,* 11068–11077.

Tabas, I., Lim, S., Xu, X. X., and Maxfield, F. R. (1990). Endocytosed beta-VLDL and LDL are delivered to different intracellular vesicles in mouse peritoneal macrophages. *J. Cell Biol. 111,* 929–940.

Taylor, D. L., and Wang, Y. L. (1980). Fluorescently labelled molecules as probes of structure and function of living cells. *Nature 284,* 405.

Taylor, D. L., Amato, P., and Luby-Phelps, K. (1984). Fluorescent analogue cytochemistry. *Trends Biochem. Sci. 9,* 88.

Thiebaut, F., Currier, S. J., Whitaker, J., Haugland, R. P., Gottesman, M. M., Pastan, I., and Willingham, M. C. (1990). Activity of the multidrug transporter results in alkalinzation of the cytosol: Measurement of cytosolic pH by microinjection of a pH sensitive dye. *J. Histochem. Cytochem* 685–690.

Tsien, R. Y. (1980). New calcium indicators and buffers with high selectivity against magnesium and protons: Design, synthesis, and properties of prototype structures. *Biochem. 19,* 2396–2404.

Tsien, R. Y., and Poenie, M. (1986). Fluorescence ratio imaging: A new window into intracellular ionic signaling. *Trends Biochem. Sci. 11,* 450–455.

Tsien, R. Y., and Zucker, R. S. (1986). Control of cytoplasmic calcium with photolabile tetracarboxylate 2-nitrobenzhydrol chelators. *Biophys. J. 50,* 843–853.

Tsien, R. Y., Rink, T. J., and Poenie, M. (1985). Measurement of cytosolic free calcium in individual small cells using fluorescence microscopy with dual excitation wavelengths. *Cell Calcium 6,* 145–157.

Uster, P. S., and Pagano, R. E. (1986). Resonance energy transfer microscopy: Observations of membrane bound fluorescence probes in model membranes and living cells. *J. Cell Biol. 103,* 1221.

Vandenbunder, B., and Borisy, G. G. (1986). Decoration of microtubules by fluorescently labelled microtubule associated protein 2 (MAP2) does not interfere with their spatial organization and progress through mitosis in living fibroblasts. *Cell Motil. Cytoskel. 6*, 570.

Wadsworth, P., et al. (1983). Microinjection of fluorescent tubulin into dividing sea urchin cells. *J. Cell Biol. 97*, 1249.

Wadsworth, P., and Salmon, E. (1986). Preparation and characterization of fluorescent analogs of tubulin. *Meth. Enzymol. 134*, 519.

Wadsworth, P., and Salmon, E. D. (1986). Analysis of the treadmilling model during metaphase of mitosis using fluorescence redistribution after photobleaching. *J. Cell Biol. 102*, 1032.

Wang, Y., and Goren, M. (1987). Differential and sequential delivery of fluorescent lysosomal probes into phagosomes in mouse peritoneal macrophages. *J. Cell Biol. 104*, 1749.

Weissman, L. S., and Wickner, W. (1988). Intervacuole exchange in the yeast zygote: A new pathway in organelle communication. *Science 241*, 589–91.

Whitaker, J. E., Haugland, R. P., and Prendergast, F. G. (1991). Spectral and photophysical studies of benzo[c]xanthene dyes: Dual emission pH sensors. *Analyt. Biochem. 194*, 330–344.

Williams, D. A., Fogarty, K. E., Tsien, R. Y., and Fay, F. S. (1985). Calcium gradients in single smooth muscle cells revealed by the digital imaging microscope using fura-2. *Nature 318*, 558–561.

Wong, M. M., and Foskett, J. K. (1991). Oscillations of cytosolic sodium during calcium oscillations in exocrine acinar cells. *Science 254*, 1014–1016.

# 2

# PHOTOSENSITIVE CAGED COMPOUNDS

## Design, Properties, and Biological Applications

**Joseph P. Y. Kao**
Medical Biotechnology Center and
Department of Physiology, School of Medicine
University of Maryland at Baltimore

**Stephen R. Adams**
Department of Pharmacology, University of California, San Diego

I. INTRODUCTION

II. DESIGN AND USE OF CAGED COMPOUNDS
A. Requirements of Caged Compounds
B. The 2-Nitrobenzyl Rearrangement
C. Nature of Caging Groups
D. Photolysis Equipment
E. Incorporation of Caged Compounds into Biological Systems
F. Quantitation of Photorelease

III. PREPARATION AND PROPERTIES OF CAGED COMPOUNDS
A. Caged Nucleotides
B. Caged Second Messengers
C. Caged Neurotransmitters
D. Related Photoactivatable Compounds

IV. BIOLOGICAL APPLICATIONS OF CAGED COMPOUNDS
A. Muscle Physiology
B. Regulation of Neurotransmitter Release by $Ca^{2+}$ and Membrane Depolarization
C. $Ca^{2+}$ "Triggers" in Mitosis
D. Mechanisms of Calcium Oscillations
E. Signal Transduction Processes Leading to Closure of Stomatal Pores in Plants
F. Microtubule Dynamics in the Metaphase Mitotic Spindle
G. Actin Microfilament Dynamics in Locomoting Cells
H. Localized Quantal (All-or-None) $Ca^{2+}$ Release Induced by $Ins(1,4,5)P_3$
I. Other Biological Applications

V. CONCLUDING REMARKS
REFERENCES

Optical Microscopy: Emerging Methods and Applications

27

## I. INTRODUCTION

"Caged" compounds, as the term is used in biology, are photosensitive precursors in which the presence of a photolabile masking group abolishes the bioactivity of an effector molecule. Upon photolytic removal of the masking or "caging" group, the fully bioactive molecule is liberated. In most instances, the caging group is covalently attached to the effector molecule [as in caged ATP (adenosine $-5'-$ triphosphate)]. Where the species being caged is a metal ion (e.g., $Ca^{2+}$), or a buffer for a metal ion, the caging group is a photosensitive chelator whose affinity for the metal ion can be decreased or increased, respectively, by photolysis. Because photochemical transformations are generally very rapid, fast changes (jumps) in the concentration of a bioactive molecule can, in principle, be achieved by photolysis of caged compounds.

The use of caged compounds offers several advantages over conventional means of applying effectors to a biological system. First, because photolysis of caged compounds generates effectors *in situ,* much faster and more spatially uniform concentration jumps can be effected than is possible with flow or turbulent mixing techniques, where diffusion of effectors into a thick or structurally complex specimen can cause considerable delay and spatial inhomogeneity in biological response. By using caged compounds, time-resolved analysis of fast biological processes becomes possible. Second, because a caged compound ideally has no biological activity prior to photolysis, it can be introduced into living cells long before when the active effector is actually photoreleased. If one were to apply the native effector directly when needed, by microinjection for example, impalement damage due to microinjection would produce effects that could disturb the phenomenon under study. The ability to load a functionally "silent" photosensitive precursor allows cells to recover from any transient negative effects of the loading procedure. Finally, because a light beam can be shaped and focused through optics, photolysis of caged compounds can be confined to a small volume, even within a single cell. This rather unusual capability allows the properties of subcellular microenvironments to be examined (Parker and Ivorra, 1990b) and also makes it possible to study spatial aspects of dynamic processes underlying cellular motility (Theriot and Mitchison, 1991) and transport (Mitchison, 1989).

Since the introduction of caged cAMP (cyclic adenosine-3',5'-monophosphate) and caged ATP in the late 1970s (Engels and Schlaeger, 1977; Kaplan *et al.,* 1978), a wide variety of caged compounds has been synthesized, and these have found application in a broad range of biological studies. In the sections that follow, we discuss the design, synthesis, properties, as well as the biological applications, of caged compounds.

## II. DESIGN AND USE OF CAGED COMPOUNDS

The successful use of a caged compound to investigate a biological system requires understanding their properties and especially their limitations. This is made more difficult by the commercial availability of an almost bewildering variety of caging groups for each bioactive molecule; for example, four different versions of caged GTPγS [guanosine-5'-O-(3-thiotriphosphate)] are available from one company alone. In this section, the essential properties of an ideal caged compound, and how close chemists and biochemists have come to achieving these, are discussed.

### A. Requirements of Caged Compounds

A caged compound, in general, must at least partially satisfy the following criteria:

1.  Prior to photolysis, the caged compound must be nontoxic, chemically stable, and biologically inert. For example, cyclic AMP caged with the 2-nitrobenzyl group on the phosphate diester was too sensitive to hydrolysis in some cell types so that there was steady low-level release of free cAMP, and cAMP-activated responses were observable even before photorelease (Lester and Nerbonne, 1982). More recently, DMNPE [1-(4',5'-dimethoxy-2'-nitrophenyl) ethyl] caged ATP, which is hydrolytically stable, has been found to inhibit the ATP-sensitive $K^+$ channel in rat heart cells (Nichols *et al.,* 1990) and in mouse pancreatic β-cells (Ämmälä *et al.,* 1991). In rat lung membranes, NPE [1-(2'-nitrophenyl)ethyl] caged ATP was *more* effective than ATP itself in potentiating guanylate cyclase activity stimulated by atrial natriuretic factor (Chang *et al.,* 1990). These examples illustrate that chemical instabilities or prephotolysis bioactivities are potential problems in using caged compounds, and that a bioactive molecule that is caged with respect to one application may not be with respect to another.
2.  The caged precursor must be photoreleasable by light at wavelengths compatible with the biological system being studied. Generally, light at wavelengths below 300 nm can cause photodamage and has limited penetrability through biological material. Other restrictions are imposed by any optics used to irradiate the sample (e.g., if the photolytic light is to be delivered to the specimen through a microscope), as expensive quartz lenses are necessary for using wavelengths shorter than

340 nm. All of the caged compounds presently available are photo-lyzable by near ultraviolet (UV) light in the 300–400-nm range. Opti-mally, a caged compound should absorb the most light (i.e., have its highest extinction coefficient) at the wavelengths used for photolysis. The quantum efficiency (Q) of the photoreaction, defined as the ratio of the number of bioactive molecules released to the number of pho-tons absorbed by the caged precursor, should be as high as possible (maximum quantum efficiency = 1, or 100% release) to minimize the possible deleterious effects of irradiation. In light of the foregoing, the quantity $Q\epsilon$, the product of the quantum efficiency and the extinction coefficient for the caged compound at the desired wavelength of irra-diation, is the parameter that allows the effectiveness of different caged compounds to be compared.

3.  Upon photolysis, the bioactive product should be released at a rate that does not limit the biological system being studied. The initial photo-chemical steps usually occur very rapidly (within nanoseconds) because they arise from short-lived singlet and triplet species, so that the rate-limiting steps are usually the subsequent dark reactions. Release rates vary widely between different caged compounds ·even with the same caging group and are often sensitive to particular experimental condi-tions, such as pH.

4.  Photolysis should not generate toxic side products that perturb or re-act adversely with the biological system. Irradiation of 2-nitrobenzyl caged compounds results in the formation of nitrosobenzaldehydes or nitrosoacetophenones, which can react with biologically important thiol groups. The ill effects of these photolysis products may be de-creased by the inclusion of exogenous protective compounds, such as thiols.

## B. The 2-Nitrobenzyl Rearrangement

### 1. History and Scope

The majority of caged compounds presently in use relies on the photo-chemical rearrangement of 2-nitrobenzyl groups as the means of photo-release. A notable exception is the diazo series of caged calcium buffers (Adams et al., 1989), which will be described in more detail in a later sec-tion. The rearrangement (Fig. 2.1) is an intramolecular redox reaction involving the transfer of an oxygen atom from the 2-nitro group to the adjacent benzylic carbon, resulting in the loss of X–H and the formation of a 2-nitrosobenzaldehyde or a 2-nitrosoacetophenone.

The photosensitivity of 2-nitrobenzyl systems has been known for al-most 100 years. In the 1960s and 1970s, organic chemists applied this to

FIGURE 2.1

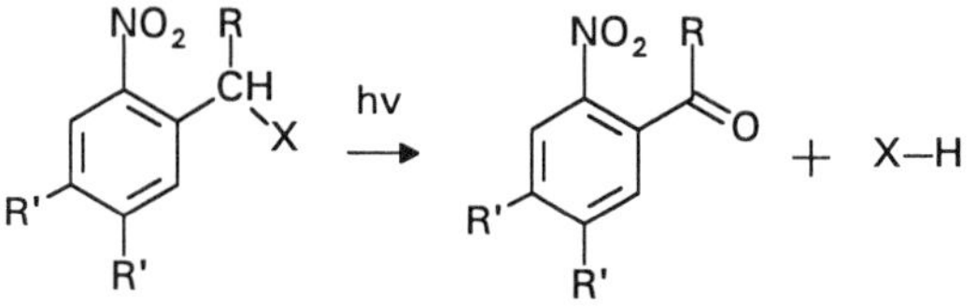

X = OPO$_3$R''; *e.g.*, ATP, InsP$_3$

= OR'', *e.g.*, diacylglycerol

= NHR'', *e.g.*, glycine

= NCOR'', *e.g.*, carbamoylcholine

R = H, CH$_3$ ;   R' = H, OCH$_3$

= OCOR'', *e.g.*, GABA

General applicability of the 2-nitrobenzyl group as a photoremovable caging group.

devise photosensitive protecting groups for peptide synthesis (Pillai, 1980, 1987; Binkley and Flechtner, 1984). The research groups of Kaplan and Engels independently extended their use to the photorelease of ATP (Kaplan *et al.*, 1978) and cyclic nucleotides (Engels and Schlaeger, 1977) in the late 1970s. Since that time, the number of caged compounds has grown steadily to reach the present wide variety, including second messengers, such as Ca$^{2+}$ and inositol trisphosphate, and neurotransmitters, such as glycine and GABA.

The popularity of the 2-nitrobenzyl group stems partly from the insensitivity of the photochemistry to the nature of the leaving group (i.e., the group being caged). Originally applied to phosphate esters, it has now found use (Fig. 2.1) for the photorelease of alcoholic or phenolic compounds (e.g., diacyl glycerol and fluorescein), carboxylic acids, amines (e.g., neurotransmitters, such as glycine), and amides (e.g., carbamoylcholine). Generally, the efficiency of photorelease of these compounds is similar when the same type of caging group is used, whereas the rates of release are much more variable. The reasons for this are still obscure despite considerable efforts to elucidate the mechanism of the photochemical and subsequent "dark" reactions.

2. Mechanism

The general mechanism of the 2-nitrobenzyl rearrangement has been the subject of many studies over the last 30 years (reviewed by De Mayo and Reid, 1961; Morrison, 1969). More recent work has examined the nature of reactive intermediates, such as triplet biradicals and *aci*-nitro or nitronate structures through the use of picosecond flash spectroscopy (Yip *et al.*, 1985; Zhu *et al.*, 1987; Schupp *et al.*, 1987; Schneider *et al.*, 1991; Gravel *et al.*, 1991). These spectroscopic studies do not directly aid the biologist who uses caged compounds but could lead to substantial improvements in the design and properties of new caging groups. The extensive studies by

Trentham and co-workers on the release of ATP from caged ATP [adenosine 5'-triphosphate-P³-1-(2'-nitrophenyl)ethyl ester] have been more biologically relevant (reviewed by McCray and Trentham, 1989).

The first intermediate of the reaction following absorption of a photon of ultraviolet light is the singlet excited state (Fig. 2.2). This short-lived species (lifetime <10 ps; Yip et al., 1985) may revert very rapidly to the ground state through radiationless decay or convert, through intersystem crossing, to the triplet excited state, which may also deactivate thermally to the ground state (Fig. 2.2). In either the singlet or triplet excited states, abstraction of a benzylic hydrogen atom by the adjacent nitro group leads, respectively, to a singlet or triplet biradical (Gravel et al., 1991). Very recently, the triplet biradical (lifetime of a few nanoseconds) formed from the photolysis of a 2-nitrobenzyl ether has been detected by picosecond flash spectroscopy (Yip et al., 1991); the much more reactive singlet biradical has not been observed directly. Both biradicals rearrange to form the aci-nitro, or nitronic acid, species, which is the first long-lived intermediate. As

FIGURE 2.2

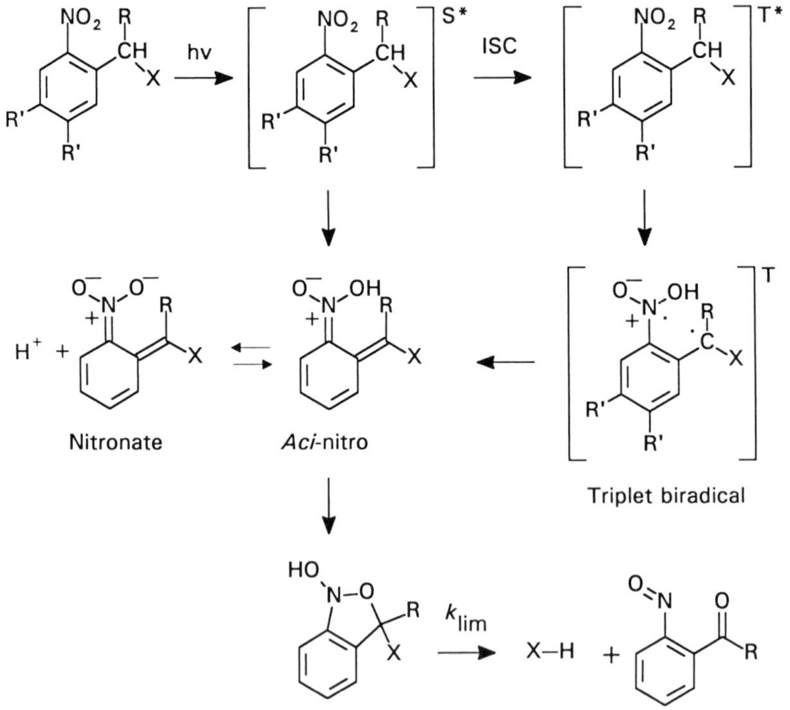

Mechanism for photocleavage of a 2-nitrobenzyl-caged species. S* and T* denote the excited state singlet and triplet species, respectively. ISC is intersystem crossing.

the name suggests, these intermediates are moderately strong acids with $pK_a \approx 3.7$. Once formed, they rapidly dissociate to yield nitronate anions at physiological pH. This process can be detected either by a change in solution conductivity or through the absorbance change of a pH indicator (Zhu *et al.*, 1987). Nitronate anions have a characteristic absorbance with a maximum near 400 nm, and usually decay exponentially with rate constants, which appear to depend on the nature of compound that is caged (Zhu *et al.*, 1987). The products of the decay are the bioactive compound and a nitroso aldehyde or ketone. This final reaction probably proceeds through a cyclic intermediate, which accounts for the intramolecular transfer of one of the oxygen atoms of the nitro group to the benzylic carbon. The cyclic intermediates have yet to be detected, however.

## 3. Kinetics

In the few cases in which it has been measured, the release of the caged compound appears to be concomitant with the decay of the absorbance of the *aci*-nitro or nitronate species. Reactions subsequent to this step do not appear to be rate limiting. The process of product release was most thoroughly investigated with NPE-caged ATP (Walker *et al.*, 1988). Using a fast biochemical assay for ATP and spectroscopic detection of the appearance of the nitroso functional group, Walker and colleagues determined that the photoproducts, ATP and nitrosoacetophenone, were formed simultaneously with the decay of the *aci*-nitro intermediate. These results rule out a proposed mechanism in which ATP is released from the cyclic intermediate *before* the cyclic intermediate collapses to the nitrosoketone (McCray and Trentham, 1989).

## C. Nature of Caging Groups

A number of caging groups have been used in the design of photoreleasable compounds, and their chemical structures are depicted in Fig. 2.3. They are all variations on the basic 2-nitrobenzyl system and contain either an additional methyl group at the benzylic carbon or methoxy groups substituted on the benzene ring; the DMNPE group has both of these modifications. Many biomolecules are commercially available with one or more of these caging groups although there have been few detailed comparisons of their relative advantages.

### 1. 2-Nitrobenzyl (NB) Cage

This was one of the first photosensitive protecting groups to be used successfully in the chemical synthesis of peptides and caged nucleotides (Pillai,

FIGURE 2.3

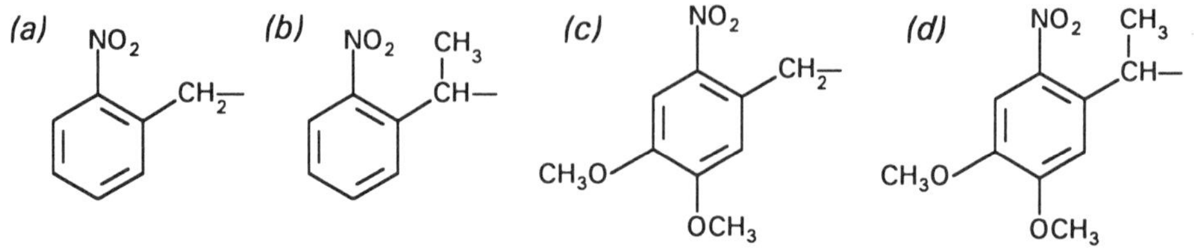

| 2-Nitrobenzyl (NB) | 1-(2'-Nitrophenyl)-ethyl (NPE) | 4,5-Dimethoxy-2-nitrobenzyl (DMNB) | 1-(4',5'-Dimethoxy-2'-nitrophenyl)ethyl (DMNPE) |

Commonly used 2-nitrobenzyl caging groups.

1980; Fig. 2.3a). As previously described, it was used by Kaplan *et al.* (1978) and Engels and Schlaeger (1977) for the first photorelease of ATP and cyclic nucleotides, respectively. Its use, however, was limited in each case: with ATP, the photoproduct 2-nitrosobenzaldehyde appeared to react with the photoreleased nucleotide and thus limited the yield of ATP to only ~25% (Kaplan *et al.*, 1978). With NB-caged cAMP, the compound activated cAMP-sensitive pathways even before photolysis although the compound was deemed moderately stable to hydrolysis in aqueous buffer at physiological pH (Lester and Nerbonne, 1982).

## 2. 1-(2'-Nitrophenyl)ethyl (NPE) Cage

To decrease the reactivity of the nitrosocarbonyl photoproduct, Kaplan *et al.* (1978) introduced the NPE cage in which a methyl group has been added at the benzylic carbon (Fig. 2.3b) and which photolyzes to yield the less reactive 2-nitrosoacetophenone. With the inclusion of protective thiols [e.g., dithiothreitol (DTT) or glutathione], the injurious effects of the photochemically generated 2-nitrosoacetophenone could be further reduced (Kaplan *et al.*, 1978). Another advantage of using thiols is the resulting solubilization of the photochemically generated 2-nitrosoacetophenone, which may otherwise precipitate (Walker *et al.*, 1988).

The NPE group, like the NB group, absorbs maximally in the ultraviolet ($\lambda_{max}$ = 260 nm), with the tail of the spectrum extending into the >300-nm range (extinction coefficient of only ~500 $M^{-1}$ $cm^{-1}$ at 347 nm). Therefore, despite high quantum yields for photorelease (0.6 for NPE-caged ATP), $Q\varepsilon$ is not optimal at the wavelengths normally used for photorelease experiments. This necessitates the use of high-energy lasers and flash lamps if the caged compound is to be significantly photolyzed in flash experiments (e.g., with light pulses of less than 1 ms duration). In some situations a low extinction at the irradiating wavelength is an advantage: when high concentrations (many millimolar) of a caged compound are used, low extinction ensures that the specimen is not optically dense. This means that equal extents of photolysis can be achieved throughout the irradiated volume. Although release of NPE-caged compounds following flash photolysis is usually sufficiently fast ($t_{1/2}$ of a few milliseconds) as not to limit the subsequent response of the biological system, the rate of release does depend strongly on the compound caged and is often pH-sensitive, as in the case of phosphate esters.

### 3. 4,5-Dimethoxy-2-nitrobenzyl (DMNB) Cage

To increase the absorbance of the nitrobenzyl group in the near-UV and visible range, the benzene ring can be substituted with electron-releasing groups, such as methoxy. Di-substitution as with the DMNB-cage (Fig. 2.3c) results in the maximum absorbance band moving to about 360 nm with an extinction coefficient of ~5000 $M^{-1}$ $cm^{-1}$; however, the quantum yield for the photorearrangement is often greatly decreased. As a result, $Q\varepsilon$ for DMNB-caged compounds is often less than, or just comparable to, the $Q\varepsilon$ of NPE-caged compounds at the same wavelength of irradiation. Because of the shift of the absorbance spectrum to long wavelengths, photolysis at near 400 nm is possible with the DMNB group while it is not at all feasible with the NPE cage. A further consequence of the electron-releasing groups on the benzene ring is the increased stabilization of the benzyl carbocation, which may result in greater susceptibility of the DMNB-caged compound to hydrolysis. This is particularly noticeable with caged cyclic nucleotides, which, as phosphate triesters, are inherently several orders of magnitude more prone to hydrolysis than negatively charged phosphate esters like caged ATP. Their sensitivity to other nucleophiles, notably thiols, is also enhanced, which is potentially problematic as thiols are often used in photorelease experiments to minimize the toxicity of the photogenerated nitrosobenzaldehyde.

### 4. 1-(4′,5′-Dimethoxy-2′-nitrophenyl)ethyl (DMNPE) Cage

This caging group (Fig. 2.3d) was recently introduced by Trentham and co-workers in an attempt to combine the greater release rate and long-wavelength UV absorbance of the DMNB group with the higher quantum yield of the NPE group (Wootton and Trentham, 1989). This approach was not completely successful with caged ATP as the DMNPE version released ATP more slowly and with lower quantum efficiency than NPE-caged ATP. Once again, the presence of the two methoxy groups shifts the UV absorbance spectrum of DMNPE-caged compounds to longer wavelengths so they should be photolyzable at wavelengths up to ~400 nm, although this has not been documented. With DMNPE-caged cyclic nucleotides, the further electron-releasing effect of the methyl group on the benzylic carbon results in significant hydrolysis and thiolysis under physiological conditions. As a consequence, DMNPE-caged cyclic nucleotides may not be practical for use in many biological systems. The DMNPE group has recently been used to cage several amino acids (Wilcox *et al.* 1990); these will be discussed in a later section.

## D. Photolysis Equipment

The two types of ultraviolet light source that have been used most frequently in photorelease experiments with caged compounds are lasers and flashlamps. In this section we briefly discuss the relative merits of each method and any new development since 1989, when this subject was thoroughly reviewed by McCray and Trentham.

### 1. Lasers

The type of laser most frequently used is the frequency-doubled ruby laser with pulsed output of up to 300 mJ at 347 nm. Extensive studies on the control of muscle contraction, using caged ATP, caged $Ca^{2+}$, caged $IP_3$, and many other caged compounds, have used this type of source. As a rough guide to the energies needed in photorelease experiments, McCray and Trentham (1989) indicated that 20 mJ of 347-nm light is sufficient to generate 2 mM ATP from 5 mM NPE-caged ATP over a 10-$mm^2$ area with a thickness of 0.1 mm. Other laser sources that could potentially be used for photolysis include xenon halide excimer lasers, Nd-YAG lasers, and tunable frequency-doubled dye lasers. For more specific details see McCray and Trentham (1989).

The advantages of lasers are their monochromaticity and highly collimated output, which is easy to manipulate and focus. Furthermore, their high power coupled with the very short pulse duration makes it possible to obtain good time resolution in photochemical experiments. Disadvantages include their high expense (tens of thousands of dollars), large size, and complexity, so their use requires considerable experience.

### 2. Flash-Lamps and Continuous Lamps

Continuous mercury arc lamps have a line emission spectrum with a convenient emission line at 365 nm. High-pressure xenon lamps produce light output that extends essentially continuously from near ultraviolet to infrared. Suitably filtered to isolate the desired photolytic light in the 300–400-nm range, and with the filtered output focused and directed through a fluorescence microscope, these lamps are quite adequate for photorelease experiments in single cells. Irradiation times ranging from ~10 ms to many tens of seconds can be easily controlled through an electromechanical shutter.

Xenon flash lamps are commercially available, which, when suitably filtered, can deliver 100–200 mJ of light in the 300–400-nm range. Pulse widths ranging from ~100 $\mu$s to ~1 ms are possible. In practice, uniformity of irradiation over the field of illumination can be difficult to achieve when a flash lamp is used. For fast photochemical kinetic studies, at least a flash lamp is required.

3. Two-Photon Photolysis

Two-photon excitation of fluorophores has recently been adapted to laser scanning microscopy by Webb and his colleagues (Denk *et al.*, 1990). This technique has also been applied to the photolysis of caged compounds by scanning the beam of a colliding pulse, mode-locked dye laser capable of 100-femtosecond ($10^{-13}$ s) pulses of 630-nm light, at repetition rates of 80 MHz, through a confocal scanning microscope. In such a system, in order for a caged compound absorbing only in the ultraviolet to undergo photolysis, it has to absorb two photons of red light simultaneously (i.e., two 630-nm photons are equivalent to one 315-nm photon), a nonlinear process that requires very high photon densities and hence beam strengths. Therefore, photolysis only occurs in the most focused part of the beam and thus provides a means of generating highly localized jumps in the concentration of photoreleased bioactive molecules. Such tight spatial control of photorelease is hard to achieve with conventional methods where photolysis occurs not only in the focal plane but also above and below it. Another potential advantage with two-photon photolysis is the use of red light, which causes less cell damage and has greater tissue penetrability than ultraviolet light.

At present, the two-photon technique is still at the developmental stage, but substantial photolysis of caged ATP in illuminated volumes of about $10^4$ $\mu$m$^3$ has been achieved with a laser power of ~3 mW at the focal plane although 600 s of irradiation were required. Much faster rates and smaller volumes of photolysis are probably achievable by tighter focusing of the laser beam. At present, the high cost of the required dye laser and argon ion pump laser, or the recently available alternative tunable titanium–sapphire laser prototypes, and the need for fuller development, limit this new technology.

## E. Incorporation of Caged Compounds into Biological Systems

Once the optimal caged compound has been chosen, it must be introduced into the biological system to be studied. Most of the widely used caged compounds contain multiple negative charges and therefore do not readily cross biological membranes. Notable exceptions include the caged cyclic nucleotides and caged diacyl glycerol, which are uncharged and readily partition into and equilibrate across membranes. Incorporation of charged caged compounds in many cases is trivial. For example, in cell-free preparations or cells that have been mechanically or chemically skinned, as is frequently the case in studies on striated or smooth muscle cells, there is no barrier to passive diffusion of the caged compound into the experimental system. Similarly, if photorelease is being used in conjunction with patch clamping or variations of this technique, including the caged compound in the bath or patch pipet is sufficient to ensure its presence outside or inside the cell.

Incorporating charged caged compounds into intact cells can be a considerable problem and has been approached in a variety of ways. One of the most attractive solutions must be a membrane-permeant derivative of the caged compound in which charged groups have been masked chemically and which readily crosses cell membranes. Once inside the cell, the masking groups are removed by cellular enzymes to liberate the charged caged molecules, which, being now membrane-impermeant, accumulate intracellularly. Such an approach was introduced by Tsien (1981) for loading fluorescent $Ca^{2+}$ indicators into cells. Masking the negatively charged carboxyl groups on the indicator with acetoxymethyl (AM) groups produces the acetoxymethyl ester of the indicator, which is able to cross cell membranes. The indicator molecules become trapped inside cells when intracellular esterases cleave off the AM groups. In this way, millimolar concentrations of $Ca^{2+}$ indicators have been loaded into cells. The same approach has been successfully applied to several caged compounds that also contain carboxyl groups: nitr-5, a caged calcium (Adams *et al.,* 1988); diazo-2, a caged calcium buffer; and diazo-3, a caged proton (Adams *et al.,* 1989). The loading level can be adjusted, to some extent, by varying the concentration of the cell-permeant species in the incubation medium and by controlling the time and temperature of loading. However, optimal loading conditions can vary among different acetoxymethyl esters and different biological preparations and therefore must be empirically determined for each case.

Another approach to the loading problem is the selective permeabilization of the cell membrane to permit entry of low molecular weight species such as most caged compounds, while limiting the loss of intracellular metabolites and enzymes. Kitazawa and colleagues (1989) have used staphylococcal $\alpha$-toxin to permeabilize smooth muscle cells to compounds, such as caged ATP, GTPγS, GDPβS [guanosine-5′-O-(2-thiodiphosphate)], and IP$_3$ [InsP(1,4,5)P$_3$, D-*myo*-inositol-1,4,5-trisphosphate], which have molecular weights less than 1000. Treating cells with $\beta$-escin, a saponin ester, allows the passage of compounds with molecular weights less than 17,000, although loss of some cellular enzymes also occurs (Kobayashi *et al.,* 1989). Use of $\alpha$-toxin or $\beta$-escin has been shown to be superior to previous methods of permeabilization, because these two reagents appear to disrupt cellular physiology to a much lesser extent than permeabilizing agents used in earlier protocols.

Finally, microinjection and iontophoresis have frequently been used to introduce caged compounds into cells. In each case, the amount injected can be estimated quite accurately, but the techniques do require considerable skill and experience, as well as specialized equipment, such as micromanipulators and micropipet pullers.

### F. Quantitation of Photorelease

Following photolysis of a caged compound in some biological system, it is often necessary to confirm that the expected concentration jump did occur and at a sufficient rate. In general, it is very difficult to monitor the release rate and the quantity of photoreleased material in the same experiment where a biological response is also being studied. The difficulty arises from a general lack of suitable indicators that can be used intracellularly for specifically detecting the photoreleased bioactive molecules. Notable exceptions are $Ca^{2+}$ and $H^+$, for which highly specific fluorescent indicators do exist. Thus, it has been demonstrated that photochemical perturbations of intracellular $Ca^{2+}$ concentration can be monitored through the fluorescent $Ca^{2+}$ indicators fluo-3 (Kao *et al.,* 1989) or fura-2 (Delaney and Zucker, 1990). The $Ca^{2+}$ sensitive dye, Arsenazo III, has also been used for monitoring photoreleased $Ca^{2+}$ although this indicator produces only a $Ca^{2+}$-dependent absorbance change and is thus less sensitive than the fluorescent indicators. Furthermore, it is also harder to calibrate (Lando and Zucker, 1989; Delong *et al.,* 1990).

Instead of following the photorelease *in situ,* it is often necessary to infer the extent of the photoreaction in a biological experiment by comparison with samples of the caged compound that have been photolyzed in the same experimental setup but in the absence of the biological specimen. These calibration samples are amenable to subsequent analysis by such methods as high-performance liquid chromatography (HPLC) to determine the extent of photolysis. Other methods used to estimate the photorelease are computational models that have been used with some success with nitr-5 (Lando and Zucker, 1989; Lea and Ashley, 1990).

### III. PREPARATION AND PROPERTIES OF CAGED COMPOUNDS

In this section, we review and discuss the properties of most of the caged compounds currently available. For ease of comparison, pertinent physical properties of caged nucleotides, caged second messengers, caged amino acid neurotransmitters, and caged calciums and calcium buffers are summarized in Tables 2.1–2.4, respectively. When they exist, the commercial suppliers

TABLE 2.1
Caged Nucleotides and Their Properties

| Species caged | Caging group | Q | $k$ (s⁻¹)[c] | $t_{1/2}$ (ms)[c] | Commercial sources[a] | Literature reference |
|---|---|---|---|---|---|---|
| ATP | NB | —[b] | — | — | MP | Kaplan et al. (1978) |
| | NPE | 0.63 | 86 | 8 | C, MP, S, B | Walker et al. (1988) |
| | DMNPE | 0.07 | 18 | 39 | MP | Wootton and Trentham (1989) |
| ATPγS | S-NPE | 0.57 | 35 | 20 | — | Walker et al. (1988) |
| | O-NPE | 0.49 | 105 | 6.6 | — | Walker et al. (1988) |
| ATPβ,γNH | NPE | 0.52 | 250 | 2.8 | — | Walker et al. (1988) |
| ADP | NPE | — | 80[d] | 8.7[d] | — | Homsher and Millar (1990)[d] |
| AMP | NPE | — | 200 | 3.5 | — | Walker et al. (1988) |
| GTP | NPE | — | — | — | C | Schlichting et al. (1989) |
| GTPγS | NB | — | — | — | MP | Dolphin et al. (1988) |
| | NPE | 0.35 | 35[e] | 10[e] | MP | Dolphin et al. (1988)[e] |
| | DMNB | — | — | — | MP | Dolphin et al. (1988) |
| | DMNPE | — | — | — | MP | Dolphin et al. (1988) |
| GMP-PNP | NPE | 0.6 | 230 | 3 | — | Dolphin et al. (1988) |
| ddTTP | NPE | 0.62 | — | — | — | Meldrum et al. (1990) |
| araCTP | NPE | — | — | — | — | Meldrum et al (1990) |
| P$_i$ | NPE | 0.54 | 80,000 | 0.009 | — | Kaplan et al. (1978) |
| | DMNPE | — | 21,000 | 0.033 | — | Wootton and Trentham (1989) |

[a] B, Boehringer; C, Calbiochem; G, Gibco; MP, Molecular Probes; S, Sigma.
[b] —, Not available.
[c] pH 7.0–7.1, except where explicitly indicated.
[d] pH 7.6
[e] pH 7.2.

## TABLE 2.2
### Caged Second Messengers and Their Properties

| Species caged | Caging group | Q | $k$ (s$^{-1}$) | $t_{1/2}$ (ms) | Commercial sources[a] | Literature reference |
|---|---|---|---|---|---|---|
| cAMP | NB | —[b] | 1.5 | 460 | — | Engels and Schlaeger (1977) |
| | NPE | 0.39 | 4.2 | 165 | C, MP | Wootton and Trentham (1989) |
| | DMNB | — | >200 | <3.5 | MP | Nerbonne et al. (1984) |
| | DMNPE | — | — | — | — | Wootton and Trentham (1989) |
| cGMP | NB | — | 3[d] | 230[d] | — | Nargeot et al. (1983) |
| | NPE | — | — | — | C, MP | — |
| Ins (1,4,5)P$_3$ | DMNB | 0.004 | >3000 | <0.23 | C, MP | Nerbonne et al. (1984) |
| sn-1,2-dioctan-oylglycerol | NPE | 0.65 | 225,280[c] | 3.1, 2.5[c] | C | Walker et al. (1989b) |
| | DMNB | 0.09 | — | — | — | Harootunian et al. (1991b) |

[a] C, Calbiochem; MP, Molecular Probes.
[b] —, Not available.
[c] Values for P$^4$- and P$^5$-caged species, respectively.
[d] pH 8; all other values determined at pH 7.0–7.2.

TABLE 2.3

Caged Amino Acids–Neurotransmitters and Their Properties

| Species caged | Caging group | Q | $k$ (s$^{-1}$) | $t_{1/2}$ (ms) | Commercial sources[a] | Literature reference |
|---|---|---|---|---|---|---|
| Carbamoyl-choline | NB | —[b] | 410 | 1.7 | — | Walker et al. (1986) |
| | NPE | 0.25 | 10,000 | 0.07 | C, MP | Walker et al. (1986) |
| | α-carboxy-NB | 0.8 | 17,000 | 0.04 | MP | Milburn et al. (1989) |
| Glycine | O-DMNB | — | — | — | MP | — |
| | O-DMNPE | — | ~1 | ~700 | — | Wilcox et al. (1990) |
| | N-DMNB | — | — | — | — | Wilcox et al. (1990) |
| Aspartate | O-DMNB | — | — | — | MP | — |
| | O-DMNPE | — | — | — | — | Wilcox et al. (1990) |
| Glutamate | O-DMNB | — | — | — | MP | — |
| | O-DMNPE | — | — | — | — | Wilcox et al. (1990) |
| GABA | O-DMNB | — | — | — | MP | — |
| | O-DMNPE | — | — | — | — | Wilcox et al. (1990) |

[a] C, Calbiochem; MP, Molecular Probes.
[b] —, Not available.

## TABLE 2.4
### CAGED CALCIUMS AND CAGED CALCIUM BUFFERS AND THEIR PROPERTIES

| Compound | $K_d^{Ca}$ (μM) Pre-photol. | $K_d^{Ca}$ (μM) Post-photol. | $K_d^{Mg}$ (mM) Pre-photol. | $K_d^{Mg}$ (mM) Post-photol. | Q | $\varepsilon_{max}$ (M⁻¹ cm⁻¹) | $k$ (s⁻¹) | $t_{1/2}$ (ms) | Commercial sources[a] | Literature reference |
|---|---|---|---|---|---|---|---|---|---|---|
| nitr-5 | 0.145 | 6.3 | 8.5 | 8 | 0.035 | 5500 | >3000 | <0.23 | C, G | Adams et al. (1988) |
| nitr-7 | 0.054 | 3 | —[b] | — | 0.042 | 5500 | 550 | 1.26 | C | Adams et al. (1988) |
| nitr-8 | 0.45 | 1370 | — | — | 0.026 | 1100 | — | — | — | S. R. Adams and R. Y. Tsien, unpublished results |
| nitr-9 | — | — | — | — | — | — | — | — | — | S. R. Adams and R. Y. Tsien, unpublished results |
| DM-nitrophen | 0.005 | 3000 | 0.005 | 3 | 0.18 | 4330 | 3000 | 0.23 | C | Kaplan and Ellis-Davis (1988) |
| diazo-2 | 2.2 | 0.073 | 5.5 | 3.4 | 0.030 | 22,200 | 2300 | 0.3 | MP | Adams et al. (1989) |
| diazo-3 | — | — | — | ~20 | 0.048 | 22,800 | 4200 | 0.17 | MP | Adams et al. (1989) |
| diazo-4 | 89 | 0.055 | — | 2.6 | 0.015 | 46,000 | — | — | — | Adams et al. (1989) |

[a] C, Calbiochem; G, Gibco; MP, Molecular Probes.
[b] —, Not available.

are indicated. A primary literature reference, generally one in which preparation of a compound is reported, is also given for each table entry. In Tables 2.1–2.4, extinction coefficients ($\varepsilon$) for the tabulated compounds are not listed, because for any compound, the UV absorption band of interest derives from the caging group whose UV-absorption is generally little affected by the molecule to which it is attached. As a rough guideline, both NB and NPE caging groups have $\varepsilon$'s of ~1000 $M^{-1} s^{-1}$ at ~310 nm and ~500 $M^{-1} s^{-1}$ at ~350–360 nm, while the dimethoxy-substituted caging groups DMNB and DMNPE have $\varepsilon \approx 5000$ $M^{-1} s^{-1}$ at ~350–360 nm.

## A. Caged Nucleotides

### 1. Preparation

ATP was one of the first biomolecules to be caged, and since its introduction in 1978, it has been used extensively in a variety of experiments, including kinetic studies of ATP-driven ion pumps and the regulation of muscle contraction (reviewed by McCray and Trentham, 1989; Somlyo and Somlyo, 1990; Homsher and Millar, 1990). One of the major limitations in early work was the limited availability of NPE-caged ATP, or adenosine-5′-triphosphate-$P^3$-1-(2′-nitrophenyl)ethyl ester, since its synthesis from caged phosphate and ADP (adenosine-5′-diphosphate) is tedious and requires multiple steps and chromatographic separations. In 1988, Walker *et al.* described an alternative strategy for converting any compound containing phosphoric acid groups (including phosphoric anhydrides like ATP and ADP) to its NPE-caged derivative. This method exploits the reaction of a diazo derivative of the caging group with the free acid of the phosphate compound (Fig. 2.4) and is analogous to the well-known esterification of carboxylic acids with diazomethane. This type of reaction was first applied to the synthesis of caged cyclic nucleotides by Engels and Schlaeger (1977) and Nerbonne *et al.* (1984), but its general applicability was not demonstrated until 1988. The diazoalkane derivatives of the caging group are gen-

FIGURE 2.4

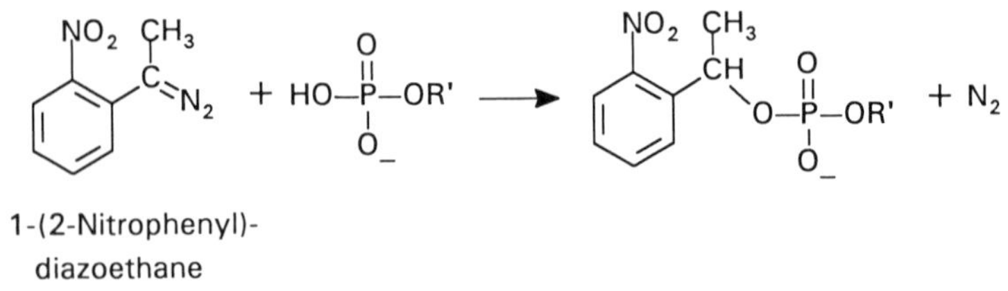

1-(2-Nitrophenyl)-
diazoethane

Synthesis of caged phosphate esters and nucleotides via the nitrophenyl diazoalkane. The 1-(2′-nitrophenyl)diazoethane is shown here as a specific example for illustrative purposes.

erally produced from the readily available carbonyl compound either by reaction with hydrazine and subsequent oxidation with manganese dioxide or by base elimination of a sulfonyl hydrazide. The diazoalkanes are generally heat-unstable, decomposing slowly at room temperature and are therefore often prepared fresh as a solution in an inert solvent (diazoalkanes are potentially explosive as solids). They react with the protonated form of a phosphoric (or carboxylic) acid directly to form the caged compound with only nitrogen gas generated as a by-product. By judicious control of the pH of the reaction, it is possible to cage only one of the acidic OH groups of a phosphate ester or anhydride. Because of the limited aqueous solubility of the diazoalkane derivatives of the caging group, the reactions are generally carried out in a mixed aqueous–organic phase for optimal selectivity or in a polar, aprotic solvent, such as dimethyl sulfoxide if selectivity is not required, as would be the case with cyclic nucleotides in which only one phosphoric OH is available for reaction. Workup usually requires at least one chromatographic step to remove any traces of unreacted starting material. Descriptive methods for preparing NPE-caged ATP, GTPγS, ATPγS, ATPβ,γNH (5′-adenylyl imidodiphosphate) have been published (Walker *et al.*, 1988; 1989a), and the first two compounds are commercially available. The synthesis and reactions of the relevant diazoalkanes for the preparation of NB-, DMNB-, and DMNPE-caged compounds have been described (Nerbonne *et al.*, 1984; Wootton and Trentham, 1989; Wilcox *et al.*, 1990).

When GTPγS is reacted with a nitrophenyl diazoalkane, a mixture of the S-caged and O-caged GTPγS is obtained. Although the S-caged and O-caged GTPγS can be separated by HPLC, some commercial preparations are a mixture. This may be important as the O- and S-caged forms could release GTPγS at different rates, as was found with ATPγS (Walker *et al.*, 1989a). Moreover, when using either the NPE or the DMNPE group, each of the O- and S-caged species is, in turn, a mixture of two diastereomers, because introduction of the 1-(2′-nitrophenyl)ethyl group creates a new chiral center. Indeed, any chiral biomolecule caged with either of these two groups would yield a diastereomeric mixture. Although no detailed studies have been reported, the two diastereomeric components could potentially show different photochemical and biochemical behavior.

## 2. Chemical and Physical Properties

Caged ATP and caged GTPγS have been used most extensively in biological studies and are both commercially available. ATP is obtainable with three different caging groups (NB, NPE, DMNPE), whereas there are four versions of caged GTPγS (NB, NPE, DMNB, DMNPE). As previously described, the nature of the caging group affects the most important properties of the caged compound: its chemical stability, its sensitivity to light, and its rate of release of the bioactive compound. Despite this, there is little hard data on the relative merits of different versions of the same compound caged. Wootton and Trentham (1989), in comparing the properties of ATP and several analogs caged with either the DMNPE or NPE group, found that the NPE-caged ATP was released more than four times faster ($k = 84$ s$^{-1}$ compared to 18 s$^{-1}$ at pH 7.0) and with a nine-fold higher quantum efficiency (0.63 compared to 0.07) when photolyzed at 347 nm. When photolyzing equally absorbing solutions of the two compounds with a 50-ns laser pulse at 347 nm, however, the product yield was only twofold greater for the NPE-caged ATP, suggesting that multiple excitations of DMNPE-caged ATP are occurring during the 50-ns pulse. The DMNPE group might be advantageous if photolysis at wavelengths greater than 370 nm was necessary and if the reduced release rate could be tolerated.

Nitrobenzyl (NB)-caged ATP was reported by Kaplan *et al.* (1978) to release only 20% of the ATP upon complete photolysis, probably because of the reaction of the photoproduct, 2-nitrosobenzaldehyde, with the released ATP. Addition of glutathione or DTT generally protects biological material against any pernicious effects of nitroso compounds and also prevents them from precipitating in aqueous solutions (Kaplan *et al.* 1978; Walker *et al.*, 1988).

Although only the NPE- and DMNPE-caged versions have been examined in detail, all of the above caged ATPs probably release ATP at a rate that is sensitive to pH, as the rate-determining decay of the intermediate *aci*-nitro species is acid catalyzed (Walker *et al.*, 1988).

The synthesis of caged forms of two dideoxynucleotides, dideoxyribosyladenine triphosphate (ddATP) and dideoxyribosylthymine triphosphate (ddTTP) and a nucleotide analog, arabinoslycytosine (araC)-triphosphate, has recently been reported (Meldrum *et al.*, 1990).

## B. Caged Second Messengers

Many second messengers have been the target for synthetic conversion into caged compounds because they are often produced transiently and locally within a cell, are destroyed quickly, and are capable of eliciting a plethora of measurable biological responses—properties that can be exploited experimentally through their controllable release from caged precursors.

### 1. Caged Cyclic Nucleotides

The photolabile NB ester of cyclic AMP (cAMP) was prepared nearly 15 years ago (Engels and Schlaeger, 1977) as a potential membrane-permeant nucleotide analog, which, like the well-known dibutyryl cAMP, would release cAMP upon intracellular hydrolysis. Early attempts to use this as a caged compound revealed several problems, including partial hydrolysis in the absence of light and inefficient photorelease (reviewed by Lester and Nerbonne, 1982). Subsequently, Nerbonne *et al.* (1984) introduced DMNB-caged cAMP and cGMP, which showed two advantages over the NB-caged species. First, the DMNB-caged cyclic nucleotides hydrolyzed much more slowly in the dark, with rate constants, at pH 7, of $0.48 \times 10^{-3}$ $\text{min}^{-1}$ and $2.7 \times 10^{-3}$ $\text{min}^{-1}$ for the axial- and equatorial-caged isomers, respectively (half-lives of 36 h and 100 h, respectively). Second, the rate of release following a flash was much faster, with a rate of at least $200\ \text{s}^{-1}$, which was the limit of time resolution of the instrument used (Nerbonne *et al.*, 1984; the real rate is probably closer to that for DMNB-caged cGMP; see below). DMNB-caged cGMP has also been shown to release cGMP rapidly ($>12,500\ \text{s}^{-1}$; Karpen *et al.*, 1988) but without the side reactions that accompany the photolysis of NB-caged cGMP (Nerbonne *et al.*, 1984).

   More recently, the synthesis of caged cAMP and caged GMP containing the NPE or DMNPE group has been reported (Wootton and Trentham, 1989). NPE-caged cAMP released cAMP with a high quantum efficiency (0.39) but slow kinetics ($k = 4.2\ \text{s}^{-1}$ for an aqueous mixture of axial and equatorial isomers). It was assumed in these kinetics that release rates were the same as the rate of decay of the transient *aci*-nitro absorbance at 406 nm. This assumption was verified in the case of NPE-caged formycin-3′,5′-cyclic-monophosphate [cyclic formycin-3′,5′ monophosphate (cFMP) Wootton and Trentham, 1989]. The intrinsic fluorescence of formycin is quenched by the NPE group in the caged molecule. Upon photocleavage of the caging group, cFMP release could be monitored by the increase in fluorescence. For NPE-caged cFMP, the release rate was $4.9\ \text{s}^{-1}$ and was steeply pH-dependent below pH 7. DMNPE-caged cAMP hydrolyzed too rapidly in the absence of light to allow accurate measurement of its rate of photorelease and is therefore probably unusable as a caged cAMP.


## 2. Caged Calcium

$Ca^{2+}$ is one of the most important and ubiquitous second messengers in eukaryotes. Virtually all of the calcium in a cell is sequestered in compartments or bound to cellular components, with only a fraction of a percent present as free cytosolic $Ca^{2+}$. Increases in cytosolic free $Ca^{2+}$ concentration can result from $Ca^{2+}$ entry across the cell membrane or from release of intracellular stores, and many hormones, growth factors, and neurotransmitters can elicit such responses. Unlike other biologically important molecules that have been caged, $Ca^{2+}$, being a metal ion, cannot be covalently attached to a nitrobenzyl group, which could then be photochemically removed at the desired moment. Instead, free $Ca^{2+}$ concentration can only be controlled by manipulating the affinity with which $Ca^{2+}$ binds to chelators, which are typically polycarboxylate diamines. Two strategies have been used to effect light-induced weakening of the affinity of a chelator for $Ca^{2+}$, so that $Ca^{2+}$ can be released from the chelator following irradiation with UV light. Both approaches rely on the familiar photochemical rearrangement of 2-nitrobenzyl groups described previously; however, the implementation is different in each case.

*a. The Nitr Series*

This series of photolabile chelators was introduced by Tsien and co-workers in the mid-1980s as the first practical caged $Ca^{2+}$ (Tsien and Zucker, 1986; Adams *et al.*, 1988); the most widely used members of the series are nitr-5 and nitr-7. Structurally, they are all based on the $Ca^{2+}$ chelator, BAPTA, a homolog of EGTA in which ethylene bridges connecting oxygen, and nitrogen donor atoms in the EGTA backbone are replaced by aromatic phenylene bridges (Tsien, 1980). Incorporation of aromatic rings into the EGTA backbone weakens the basicity of the nitrogen donor atoms and thus lowers their $pK_a$'s. As a consequence, BAPTA and its derivatives do not suffer from the pH sensitivity and slow kinetics of $Ca^{2+}$ binding of its parent chelator, EGTA. The introduction of aromatic rings into the chelator structure also made it possible to control the affinity of BAPTA for $Ca^{2+}$ by adding electron–donating or electron–withdrawing substituents onto the aromatic rings in order to raise and lower the $Ca^{2+}$ affinity, respectively (Tsien, 1980). The possibility of manipulating the affinity of the chelator for $Ca^{2+}$ through substituents on the aromatic rings provides a mechanism by which a caged $Ca^{2+}$ can be devised: a photolabile substituent that changes photochemically from electron donating to electron withdrawing will result in a decrease in $Ca^{2+}$ affinity and hence release of $Ca^{2+}$ upon photolysis.

All of the members of the nitr series of caged $Ca^{2+}$ have a 2-nitrobenzyl substituent in conjugation with at least one of the chelating nitrogens (Figs. 2.5 and 2.6). On photolysis, the 2-nitrobenzyl group rearranges to the 2-nitrosobenzoyl group, which is strongly electron withdrawing (Fig. 2.5). The affinity of the chelator is thus reduced ~40-fold upon photolysis, which will result in release of free $Ca^{2+}$ and a jump in $[Ca^{2+}]$.

Nitr-5 is the most widely used and available of the nitr chelators. A methylene-dioxy group incorporated into the 2-nitrobenzyl substituent shifts the long-wavelength absorbance peak to 355–360 nm, and thus facilitates photolysis with long-wavelength UV light. A hydroxyl group attached to the benzylic carbon resulted in a fast $Ca^{2+}$ release rate of 3700 s$^{-1}$. Before photolysis, nitr-5 has a $K_d$ for $Ca^{2+}$ of 145 nM, similar to resting cytosolic free $Ca^{2+}$ concentration ($[Ca^{2+}]_i$) in most cells; after UV irradia-

FIGURE 2.5

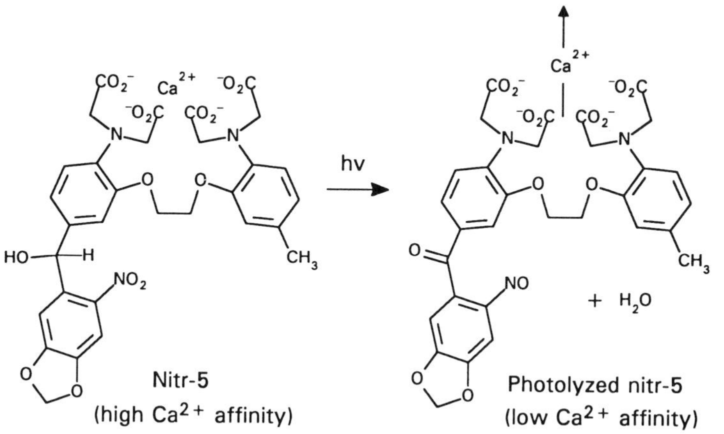

Photolysis of nitr-5, a caged calcium based on the BAPTA structure.

tion, the $Ca^{2+}$ affinity decreases ~40-fold, resulting in a $K_d$ of 6.3 $\mu$M. The quantum efficiency for the photoreaction is modest (0.035).

Other members of the nitr series (Fig. 2.6) include nitr-7, which, by having a *cis*-cyclopentane ring bridging the two chelating ether oxygens, shows higher $Ca^{2+}$ affinity both before and after photolysis ($K_d$'s of 54 nM and 3 $\mu$M, respectively). Therefore, compared to nitr-5, at a typical resting $[Ca^{2+}]_i$ of ~100 nM, more $Ca^{2+}$ is bound and thus photoreleasable from nitr-7. Nitr-8 contains a 2-nitrobenzyl group on each of the two aromatic rings in the chelator. Such a design leads to a much lower postphotolysis

FIGURE 2.6

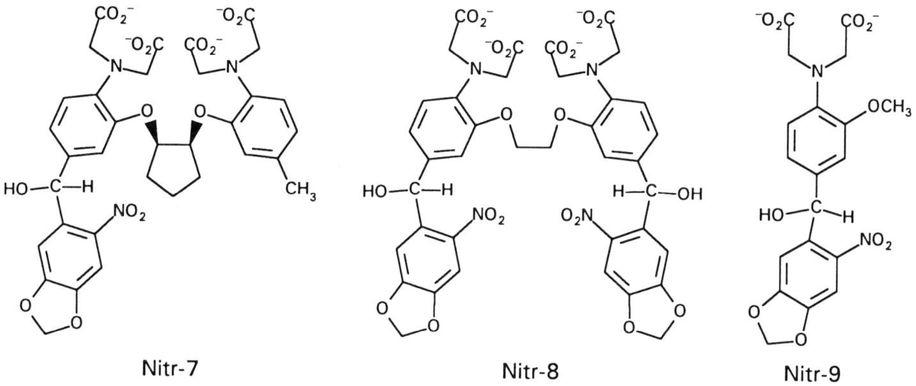

Other members of the nitr series of caged calciums.

$Ca^{2+}$ affinity than is possible with only one 2-nitrobenzyl group although the prephotolysis affinity is also slightly weakened. The greater difference between pre- and postphotolysis affinities means that larger jumps in $[Ca^{2+}]$ can be achieved with nitr-8. However, in order to achieve the largest possible photorelease, *both* nitrobenzyl groups in nitr-8 must be photolyzed; otherwise, if only one nitrobenzyl group is photolyzed, the magnitude of the $Ca^{2+}$ jump is only comparable to that obtainable with nitr-5. Designed as a control for any potential toxic effects of the photoproducts arising from photolysis of the nitr series of chelators, nitr-9 has negligible $Ca^{2+}$ affinity (before or after photolysis) but undergoes the same photochemical changes as nitr-5, -7, or -8.

Nitr chelators can be introduced into biological specimens by microinjection, or through the patch pipet, or by permeabilization. When these techniques are used, the nitr chelators are usually prepared so as to be partially saturated with $Ca^{2+}$ and at such a concentration as to make the chelator the predominant $Ca^{2+}$ buffer in that system. The amount of $Ca^{2+}$ added determines the resting $[Ca^{2+}]_i$ and can be varied as required. The degree of saturation of a nitr chelator by $Ca^{2+}$ can easily be determined in stock solutions by monitoring the characteristic changes in the absorbance spectrum of nitr on binding $Ca^{2+}$ (Adams *et al.*, 1988).

Because the release rates, quantum efficiencies, extinction coefficients, and $Ca^{2+}$ affinities of nitr-5 and -7 before and after photolysis are known, the magnitude of photorelease and the resulting spatial variation of $[Ca^{2+}]$ in a sample can be modeled with some accuracy (Lando and Zucker, 1989; Lea and Ashley, 1990). More directly, the $[Ca^{2+}]_i$ can be measured simultaneously using fluorescent $Ca^{2+}$ indicators, such as fluo-3 (Minta *et al.*, 1989) and fura-2 (Grynkiewicz *et al.*, 1985) although care has to be taken to ensure accurate calibration. Fluo-3 has the advantage of being excitable by visible light that does not photolyze nitr and is much less susceptible to inner-filtering effects arising from the high absorbance of nitr and its photoproduct (Kao *et al.*, 1989). However, fluo-3 is not a ratiometric indicator and is therefore harder to calibrate accurately. Fura-2, on the other hand, can be easily calibrated, and although it is excited in the near UV, $[Ca^{2+}]_i$ can apparently be monitored without significant photolysis of nitr-5 (Delaney and Zucker, 1990). Corrections to the usual calibration procedures for fura-2 may be necessary to account for errors stemming from the high absorbance of nitr-5 and its photoproduct, when present near millimolar concentrations, at the excitation wavelengths of fura-2 (Zucker, 1992).

Nitr chelators can also be loaded into cells as the membrane-permeant acetoxymethyl (AM) esters, which are cleaved by intracellular esterases to liberate the membrane-impermeant chelator (Kao *et al.*, 1989; Harootunian *et al.*, 1991a). The advantages of this method are the ability to load small or fragile cells not amenable to loading by other techniques, as well as the ease with which large numbers of cells can be loaded simultaneously. Potential problems in using AM esters of nitr chelators include difficulty in controlling and quantitating the amount of chelator inside the cell and, in some cases, inability to load sufficient nitr to overcome the intrinsic $Ca^{2+}$ buffering capacity of the cell.

### b. DM-nitrophen

This alternative caged calcium was introduced by Kaplan and Ellis-Davies (1988; Ellis-Davies and Kaplan, 1988) and uses a strategy that is different from the one followed in the nitr series. DM-nitrophen is a derivative of EDTA (ethylenediamine-$N,N,N',N'$-tetraacetic acid) that contains a 2-nitrobenzyl group adjacent to one of the chelating amines (Fig. 2.7). Photorearrangement of the nitrobenzyl group results in physical disruption of the metal coordinating site. The final products of photolysis are iminodiacetic acid and a 2-nitrosoacetophenone derivative, both of which show quite low affinity for $Ca^{2+}$. Thus, almost complete loss of $Ca^{2+}$ affinity is achieved after photolysis in the case of DM-nitrophen as compared to the more modest affinity change observed with the nitr series. The difference is especially dramatic as the initial affinity of DM-nitrophen (about 5 nM) is an order of magnitude higher than that of nitr-5. In DM-nitrophen, the nitrobenzyl group contains two methoxy substituents to shift the long-wavelength absorbance peak to ~350 nm. The quantum efficiency of photolysis is 0.18, and the rate of $Ca^{2+}$ release is greater than 3000 $s^{-1}$.

One major limitation of DM-nitrogen as a caged calcium is its poor selectivity for $Ca^{2+}$ over $Mg^{2+}$ [$K_d$ ($Ca^{2+}$) = 5 nM, $K_d$ ($Mg^{2+}$) = 2.5 $\mu M$], which results from using a hexacoordinate, EDTA-type chelating site as

FIGURE 2.7

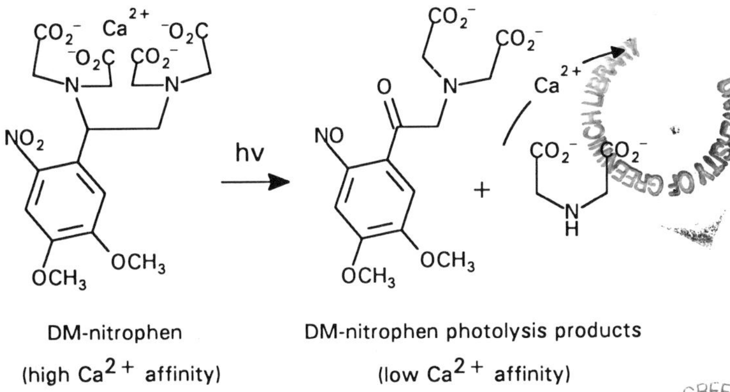

DM-nitrophen
(high $Ca^{2+}$ affinity)

DM-nitrophen photolysis products
(low $Ca^{2+}$ affinity)

Photolysis of DM-nitrophen, a caged calcium/magnesium based on the EDTA structure.

opposed to an octacoordinate EGTA- [ethylene glycol-$O,O'$-bis(2-amino-ethyl)-$N,N,N',N'$-tetraacetic acid] or BAPTA-[1,2-bis(2'-aminophenoxy) ethane-$N,N,N',N'$-tetraacetic acid] type site. Ellis-Davies and Kaplan (1988) also synthesized an EGTA-based DM-nitrophen, but its affinity for $Ca^{2+}$ before photolysis was only $\sim 25$ $\mu M$ and therefore would have been of little use in most biological applications. In a typical resting eukaryote cell with $[Ca^{2+}]_i$ at $\sim 100$ nM and $[Mg^{2+}]_i$ at $\sim 3$ mM, most of the DM-nitrophen will be bound to $Mg^{2+}$ before photolysis. Photolysis will then cause photo-release of free $Mg^{2+}$ (i.e., DM-nitrophen will actually function as a caged magnesium rather than as a caged calcium although this property can be exploited to study the role of $Mg^{2+}$ in modulating cellular processes). In cell-free or permeabilized systems, it is possible to manipulate the experimental buffer to keep DM-nitrophen predominantly in the $Ca^{2+}$-bound form. This can be accomplished either by leaving $Mg^{2+}$ out altogether or by providing sufficient competing chelators (e.g., ATP) for $Mg^{2+}$. In a living cell, such manipulations are much harder to accomplish in a controlled fashion. Other minor disadvantages of DM-nitrophen are that its $Ca^{2+}$ affinity is pH sensitive in the physiological range (similar to EDTA and EGTA but not BAPTA), and that upon photorelease of $Ca^{2+}$, it also takes up a $H^+$, thus raising solution pH. However, considering that cells have high pH-buffering capacity, this is likely to be unimportant. Like EDTA and EGTA, whose nitrogen donor atoms are also protonated, intact DM-nitrophen probably buffers $Ca^{2+}$ slowly at physiological pH (e.g., Smith et al., 1984). This property may be useful, however. Because $Ca^{2+}$ release from photolyzed DM-nitrophen is much faster than $Ca^{2+}$ binding by intact DM-nitrophen, photolysis of a DM-nitrophen sample partially saturated with $Ca^{2+}$ can lead to a transient rise in $[Ca^{2+}]$ of very short duration (of the order of a few milliseconds), due to rapid release of $Ca^{2+}$ from photo-lyzed DM-nitrophen, which then decays more gradually as the photo-released $Ca^{2+}$ is taken up by the remaining unphotolyzed DM-nitrophen on a slower time scale. However, to have free unbound DM-nitrophen at sufficient concentration to rebind the released $Ca^{2+}$, the $[Ca^{2+}]_i$ must be near the apparent $K_d$ of DM-nitrophen, (i.e., $\approx 5$ nM). The nitr series of caged calciums is also capable of generating $Ca^{2+}$ transients, with durations on the order of seconds, in cells where homeostatic mechanisms rapidly return the $[Ca^{2+}]_i$ to resting levels after photorelease (Kao et al., 1989; Harootunian et al., 1991a).

c. Comparison of Nitr and DM-nitrophen

The ideal caged calcium has yet to be realized although some features of nitr and DM-nitrophen come close. Nitr-5 and nitr-7 suffer from the limited $\sim 40$-fold change in $Ca^{2+}$ affinity on photolysis. Nitr-8, with two photolabile nitrobenzyl groups, shows a 1600-fold change in $Ca^{2+}$ affinity on photolysis, but such a change is possible only if both nitrobenzyl groups are photolyzed. Additionally, the relatively high prephotolysis $K_d$ for $Ca^{2+}$ of nitr-5 and nitr-8 means that at typical resting $[Ca^{2+}]_i$ of $\sim100-200$ nM, the chelator is only partially saturated with $Ca^{2+}$ and therefore only that fraction that is $Ca^{2+}$-bound will be able to release $Ca^{2+}$ when photolyzed. Furthermore, unless the $Ca^{2+}$-free chelator is also photolyzed, it can buffer and thus partially counteract the $Ca^{2+}$ photorelease. Nitr-7, with its significantly higher prephotolysis affinity for $Ca^{2+}$ ($K_d = 54$ nM), is one remedy to the above problems, but it still only decreases its affinity $\sim55$-fold on photolysis. All of the nitr chelators can be conveniently loaded into cells as their AM esters.

DM-nitrophen has a high $Ca^{2+}$ affinity before photolysis and gives a much more substantial reduction of affinity on photolysis. Its major limitation is its lack of selectivity for $Ca^{2+}$ over $Mg^{2+}$, which, unless care is taken to control the relative concentrations of DM-nitrophen, $Ca^{2+}$, $Mg^{2+}$, and other $Mg^{2+}$ chelators (e.g., ATP), can cause DM-nitrophen to act instead or simultaneously as a caged $Mg^{2+}$. Because ATP is essentially used exclusively by enzymes as its $Mg^{2+}$ chelate, the presence of a strong $Mg^{2+}$ chelator, such as DM-nitrophen, may also have profound consequences on metabolism. In the physiological range, the affinity of DM-nitrophen is strongly affected by pH. Furthermore, under typical experimental conditions where DM-nitrophen acts as a caged $Ca^{2+}$ or $Mg^{2+}$, photolysis results in products that can take up protons and thus elevate pH. Adequate pH buffering is therefore needed to minimize these effects when using DM-nitrophen. A recent review by Kaplan (1990) also provides a discussion and comparison of nitr-5 and DM-nitrophen.

## 3. Caged Calcium Buffer: Diazo

Caged calcium buffers are calcium chelators that *increase* $Ca^{2+}$ affinity upon photolysis and are therefore opposite in action to nitr or DM-nitrophen (Adams *et al.,* 1989). They may be used to truncate or prevent transient or continuous elevations in $[Ca^{2+}]_i$ by photogeneration of calcium buffer. These compounds have also been referred to as caged "anticalcium."

Two chemical approaches have been used to make caged calcium buffer. The first and most obvious is to cage a calcium chelator, such as EGTA or BAPTA, on one of the chelating carboxyl groups, thereby greatly reducing its $Ca^{2+}$ affinity. Photocleavage of the caging group yields EGTA or BAPTA. The second approach is analogous to that adopted for the nitr series of caged calciums (i.e., manipulating the $Ca^{2+}$ affinity of BAPTA through the electronic effects of substituents on the aromatic rings).

Extensive efforts were described by Adams *et al.* (1989) to mask a carboxyl group of BAPTA with a variety of photolabile protective groups. At best, only very low yields of BAPTA could be photoreleased, possibly because of inhibition of the 2-nitrobenzyl photochemistry by the close proximity of the amine groups of BAPTA. In model compounds, the quantum yields for photorelease of carboxylic acids from DMNB-caged derivatives were considerably lower than the release of other functionalities, such as amines, alcohols, or amides. In comparison, the photorelease of caged BAPTAs was, at best, an order of magnitude lower. Ferenczi *et al.* (1989), in a preliminary communication, described an NPE-caged BAPTA but gave little information regarding its quantum efficiency except that some photolysis could be achieved with high-powered laser pulses.

A more successful alternative to caged BAPTA was substitution of one (or both) of the aromatic rings of BAPTA with a *p*-diazoacetyl group to make the diazo series of calcium chelators (Fig. 2.8). Adding this strongly electron-withdrawing group results in a low-affinity chelator, which, on photolysis, increases $Ca^{2+}$ affinity as the diazo group is converted to an electron-donating carboxymethyl substituent (Fig. 2.8). The photochemistry involved is the well-known Wolff rearrangement of diazoketones to carboxylic acids via hydrolysis of an intermediary ketene. The initial photochemical steps, including loss of nitrogen and rearrangement to the still electron-withdrawing ketene, are known to occur quickly (hundreds of nanoseconds). The increase in $Ca^{2+}$ affinity is concomitant with hydrolysis of the ketene to the carboxymethyl group, with release of one proton.

Diazo-2, being commercially available, is the most widely used member of the series. It contains one diazoacetyl group and changes $Ca^{2+}$ affinity from 2.2 $\mu$M to 73 nM upon photolysis. After photolysis, the high-affinity chelator is generated with a rate constant of 2300 $s^{-1}$. The rate of $Ca^{2+}$ binding to photogenerated chelator is $8.0 \times 10^8 \ M^{-2} \ s^{-1}$ so that calcium

FIGURE 2.8

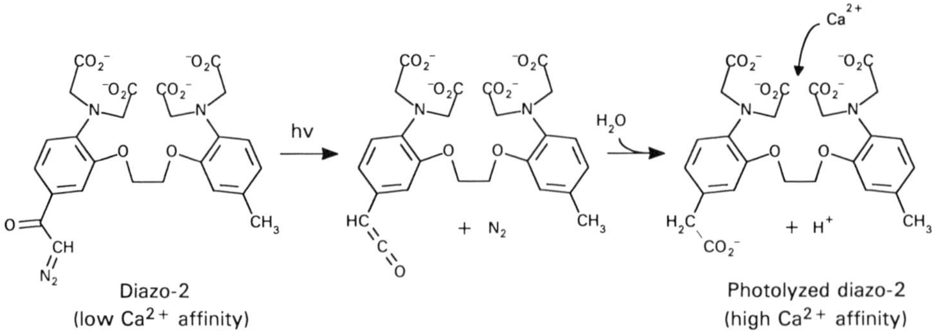

Diazo-2
(low $Ca^{2+}$ affinity)

Photolyzed diazo-2
(high $Ca^{2+}$ affinity)

Photolysis of diazo-2, a caged calcium buffer.

increases with rise times of a few milliseconds can be effectively buffered. With a quantum efficiency of 0.03 and high absorption at 375 nm (extinction coefficient of 22,200 $M^{-1}$ $cm^{-1}$), diazo-2 is considerably more light-sensitive than nitr-5 in the near-UV ($Q\varepsilon \approx 670$ and 190 $M^{-1}$ $cm^{-1}$, respectively).

Diazo-3 is a control for any adverse side effects of the photochemistry of the diazoketone group and the pH drop which may also occur in the absence of sufficient buffering. The main expected side reaction is attack by the photochemically generated ketene intermediate on free amino groups, a process that might lead to protein modification in cells. In control experiments, however, when 100 $\mu$M diazo-3 was photolyzed in the presence of 100 mM lysine (roughly the cellular concentration of lysyl residues), only 6 in $10^5$ amino groups were covalently modified. Therefore, cell damage by the short-lived ketene intermediate is not anticipated.

By incorporating a $p$-diazoacetyl group on each of the benzene rings of BAPTA (to make diazo-4), a change in $Ca^{2+}$ affinity from 89 $\mu$M before photolysis to 55 nM after photolysis can be achieved. As was the case with nitr-8, this large change in affinity on photolysis can be achieved only if *both* diazoacetyl groups are photolyzed. A further advantage arising from the weak prephotolysis affinity of diazo-4 is its lack of buffering even when $[Ca^{2+}]_i$ is elevated. Therefore, before photolysis, the presence of diazo-4 would not interfere with physiological $Ca^{2+}$ movements in living cells. As in the nitr series, diazo chelators can be either microinjected into cells or loaded as the membrane-permeant AM esters. Because of its low molecular weight and low hydrophobicity, compared to nitr-5, diazo AM esters load very efficiently. Millimolar intracellular concentrations of the free caged chelator are probably obtainable.

## 4. Caged Inositol Phosphate

Considerable research during the last decade has focused on the hormone-activated hydrolysis of the membrane lipid, phosphatidylinositol-4,5-bisphosphate ($PIP_2$) to the second messengers, inositol-1,4,5-trisphosphate ($InsP_3$ or $IP_3$) and diacyl glycerol (DG). $InsP_3$ triggers the release of $Ca^{2+}$ from intracellular stores and can also be phosphorylated to yield $Ins(1,3,4,5)P_4$, potentially a second messenger, whose role is still obscure. Caged precursors of $InsP_3$ have proven quite useful in studies into the second messenger functions of $InsP_3$.

The first and the only commercially available caged inositol phosphate, NPE-caged inositol-1,4,5-trisphosphate, was prepared by Walker *et al.* (1987, 1989b) by reacting $Ins(1,4,5)P_3$ with 1-(2′-nitrophenyl)diazoethane and separating the resulting mixture of three mono-NPE esters by HPLC. The $P^1$-NPE-caged isomer proved to be still active at releasing intracellular

$Ca^{2+}$ stores, but the $P^4$- and $P^5$-NPE isomers were devoid of $Ca^{2+}$-mobilizing activity. They were also not substrates or inhibitors of polyphosphoinositol phosphatases, the major $InsP_3$-degrading enzymes in cells, although the $P^5$ isomer was an inhibitor of bovine brain $InsP_3$ 3-kinase, which phosphorylates $InsP_3$ to $InsP_4$. The $P^4$- and $P^5$-NPE caged $InsP_3$ isomers are usually not separated for most biological studies.

Similar to other NPE-caged phosphates, NPE-caged $InsP_3$ shows high quantum efficiency of photorelease (0.65) but low absorbance at wavelengths beyond 320 nm. The $P^4$ and $P^5$ caged $InsP_3$ release $InsP_3$ at rates of 225 and 280 $s^{-1}$, respectively, at pH 7.1. As in the case of NPE-caged ATP, these rates are also pH-sensitive, being 10-fold slower at pH 8.1.

Because caged $InsP_3$ is a pentaanion at physiological pH, it is usually incorporated into cells by microinjection, internal perfusion through a patch pipet, or permeabilization. One recent report, however, claimed that caged $InsP_3$ was sufficiently membrane-permeant to gain entry into cells (Patton *et al.*, 1991).

## 5. Caged Diacyl Glycerol

Harootunian *et al.* (1991b) described the synthesis and biological use of *sn*-1,2-dioctanoylglycerol caged with the DMNB group. It was prepared by the diazoalkane approach used for caging the more acidic OH groups of phosphates. Reaction of the diazoalkane with the weakly acidic hydroxyl group of the diacylglycerol required $BF_3$ catalysis, however. The rate of photorelease of dioctanoyl glycerol from the caged precursor was not determined although the quantum efficiency for removal of the DMNB-caging group was found to be modest ($Q = 0.09$).

## 6. Caged Proton

Although $H^+$ is not usually considered a second messenger in itself, pH changes often result from the effects of other second messengers, such as $Ca^{2+}$. A number of compounds have been described that will release protons upon photolysis at physiological pH. The first, introduced by McCray and Trentham (1985), is the NPE ester of 2-hydroxyphenyl phosphate, which was one of the intermediates in the early synthesis of caged ATP. It photolyzes to produce 2-hydroxyphenyl phosphate (catechol phosphate, $pK_a = 5.3$), which releases a proton.

More recently, 4-formyl-6-methoxy-3-nitrophenoxyacetate has been proposed as a caged proton that, because of the pendant negative acetate moiety, showed limited ability to cross lipid bilayers (Janko and Reichert, 1987). This compound undergoes the 2-nitrobenzaldehyde photorearrangement to form 2-nitrosobenzoic acid, by a mechanism essentially analogous to the 2-nitrobenzyl photoreaction. The photogenerated nitrosobenzoic acid is quite strong ($pK_a \approx 0.75$); therefore, one $H^+$ is released for each molecule photolyzed, at a rate of $\geq 10^6 \, s^{-1}$. The quantum efficiency for the photoconversion is high ($Q = 0.18$, determined at 308 nm) and the extinction coefficient in the UV is 5130 $M^{-1} \, cm^{-1}$ at 345 nm.

Diazo-3, a control compound for the photochemistry of the diazo series of caged calcium buffers, is also a caged proton. Its properties are described in Section III.B.3. Diazo-3 offers several advantages. It has high absorbance in the long-wavelength UV ($\varepsilon = 2.3 \times 10^4 \, M^{-1} \, cm^{-1}$ at 374 nm). Its two negative charges before photolysis prevent it from readily crossing biomembranes. Finally, it can be readily loaded into cells as a membrane-permeant AM ester. However, it has yet to be used as a caged proton despite its commercial availability.

## C. Caged Neurotransmitters

### 1. Caged Carbamoylcholine

The first photogenerated acetylcholine receptor (AchR) ligand, Bis-Q, was developed by Bartels *et al.* in 1971, well before the introduction of the now-conventional caged compounds in 1978. This azobenzene derivative initially appeared to be active as an AchR ligand only in the *trans* form, which can be reversibly photogenerated from the inert *cis*-Bis-Q (reviewed by Lester and Nerbonne, 1982). Later work showed that the *trans* form was inert towards some receptors while the *cis*-isomer caused inactivation or desensitization (Delcour and Hess, 1986).

A different approach was taken by Walker *et al.* (1986) by caging carbamoylcholine with an NB or NPE group. This required direct caging of the carbamoyl nitrogen as an *N*-nitrobenzyl amide, a functionality that had yet to be used with conventional 2-nitrobenzyl photochemistry. NB- and NPE-caged carbamoylcholine both released carbamoylcholine with approximately equal efficiency on photolysis ($Q \approx 0.25$) but showed markedly different rates of release. These *N*-caged compounds were assumed to be similar to *O*-caged compounds in having a release rate that is equal to the rate of decay of the *aci*-nitro photochemical intermediate. This rate had a biphasic dependance on pH, with a minimum at pH 8.5. The half-lives for the *aci*-nitro decay of NB- and NPE-caged carbamoylcholine were 1.7 and 0.067 ms (rate constants of 410 and 10,000 $s^{-1}$), respectively. Unfortunately, neither of these caged compounds was completely inert to the nicotinic acetylcholine receptor from the electric organs of *Electrophorus electricus* and *Torpedo californica* species, as both appeared to cause inhibition and inactivation.

A solution to the problems of NB- and NPE-caged carbamoylcholines was to introduce a carboxyl group into the caging group so the overall charge of the caged compound would now be neutral at physiological pH (Milburn *et al.*, 1989). Placing the carboxyl group on the benzene ring of the 2-nitrobenzyl group inhibited the photorearrangement, but the opposite was found when the carboxyl group was placed at the benzylic carbon. The resulting α-carboxy-2-nitrobenzyl-caged carbamoylcholine released carbamoylcholine with a quantum efficiency of about 0.8 at 347 nm and at a rate of 17,000 s$^{-1}$ at pH 7.0. More importantly, this caged carbamoylcholine did not cause inhibition or inactivation of AchR from the electric organ of *E. electricus* or in $BC_3H1$ smooth muscle cells. Both NPE- and α-carboxy-NB-caged carbamoylcholine are commercially available.

## 2. Caged Phenylephrine

The $\alpha_1$-adrenergic agonist, phenylephrine, has been caged by Walker and Trentham (1988) as an NPE derivative on the phenolic hydroxyl group. It was synthesized by the reaction of 1-(2'-nitrophenyl)diazoethane with phenylephrine in dimethyl sulfoxide, followed by separation of the resulting diasterometers by HPLC. One of these had minimal biological activity as a receptor agonist or antagonist and was used in further studies. Phenylephrine was photoreleased from the caged precursor with quantum efficiency of 0.4 and at a rate of 3 s$^{-1}$ (pH 7, 21°C).

## 3. Caged Amino Acid Neurotransmitters

The synthesis of a number of caged derivatives of the amino acid neurotransmitters, glycine, aspartate, glutamate, and GABA (γ-aminobutyric acid), has been described (Wilcox *et al.*, 1990). The caging group used was either the DMNB group (attached to carboxyl groups on GABA and glutamate) or the DMNPE group (attached to carboxyl groups on glycine, aspartate, glutamate, and GABA, or to the amino group of glycine). The synthetic routes involved alkylation of the carboxylic acid group of the amino acid with the diazoethane or benzyl bromide derivative of the caging group. The N-caged glycine was prepared by N-alkylation of 1-(2'-nitrophenyl)ethylamine by chloroacetate. The DMNPE-caged amino acids, with the caging group attached to the carboxyl group, are all commercially available. However, only the DMNPE-caged glycine has been shown to release glycine on photolysis, and only at a rate of ~1 s$^{-1}$ with unknown quantum efficiency. The suitability for biological use of any of these caged amino acid neurotransmitters has yet to be demonstrated.

### D. Related Photoactivatable Compounds

Photochemistry has been used to control the activity or effect the release of a number of compounds or agents other than the low molecular weight bioactive molecules described in the preceding sections. A detailed description of these applications is beyond the scope of this review. However, two additional classes of photoactivatable compounds have either found application in biology or are potentially useful in biological studies. These will be described below.

## 1. Caged Fluorophores

The technique of fluorescence photoactivation and dissipation (FPD), or photoactivation of fluorescence (PAF), was introduced by Ware and collaborators (Ware *et al.*, 1986; Krafft *et al.*, 1986) as a means of monitoring the dynamic distribution of a probe molecule within living cells. The probe is a fluorophore that has been caged in such a way as to render it nonfluorescent. Upon photolysis, the fluorophore is unmasked to restore its fluorescence, and may then be used as a marker for the protein or other biomolecule to which it is covalently attached. This method is the positive complement to fluorescence recovery after photobleaching, or FRAP, experiments. In FRAP, photobleaching of fluorescently tagged biomolecules creates a negative signal (a dark photobleached zone) against a bright fluorescent background of unbleached molecules. Reappearance of fluorescence in the photobleached zone can be monitored as bleached and intact probe molecules diffuse out of and into the zone, respectively. In FPD, irradiation creates a population of fluorescently tagged biomolecules, which is seen as a bright positive signal against a dark background. The dynamic distribution and dissipation of this initially spatially distinct population of tagged molecules can then be monitored.

A number of fluorophores have been caged with 2-nitrobenzyl caging groups. Cummings and Krafft (1988) described the synthesis and photochemical properties of several amine-containing fluorophores (e.g., amino derivatives of pyrene, coumarin naphthalene, and anthracene) caged with various methoxy substituted 2-nitrobenzyl groups. The fluorescence of the caged fluorophores was partly quenched by the presence of the nitrobenzyl group; on photolysis this quenching was relieved.

An alternative and more successful approach has been used to cage fluorescein. Krafft *et al.* (1988) synthesized protein-reactive fluorescein derivatives that contain a 2-nitrobenzyl group (variously substituted with methoxy and hydroxy) attached by an ether linkage to one of the phenolic OH groups. An auxiliary linker is attached to the remaining phenolic OH group via an ether linkage to change the solubility characteristics of the fluorophore or to impart reactivity towards proteins. The resulting fluorescein diether is nonfluorescent until photocleavage of the nitrobenzyl group permits formation of the moderately fluorescent fluorescein monoether.

A further development of this approach was made by Mitchison (1989), who synthesized the bis(2-nitrobenzyl) ether of carboxy fluorescein, which is rendered protein reactive as the N-hydroxy-2-sulfosuccinimide. A limitation of this molecule is that it requires photolysis of both nitrobenzyl groups to liberate the fluorescein. Removal of only one group produces a fluorescein monoether similar to the photoproduct of Krafft et al. (1988) but which is much less fluorescent than fluorescein. More recently, Theriot and Mitchison (1991) introduced caged resorufin, which contains a single caging group and thus requires only a single photochemical cleaveage to generate the fluorophore. Caged fluorescein and resorufin have been successfully applied in studies on the dynamics of kinetochore microtubules in mitotic cells and of actin microfilaments in locomoting cells, respectively (Mitchison, 1989; Theriot and Mitchison, 1991).

## 2. Caged Enzymes

The photochemical control of enzymic activity has been the subject of considerable study (Turner *et al.*, 1988; and references cited therein). Notable recent developments have included the photoreactivation of inhibited serine proteases in solution or in enzyme crystals. In the latter instance, the reactivation process has been followed by X-ray diffraction (Stoddard *et al.*, 1990a,b). Application of azobenzene photochemistry for reversible control of papain and $\alpha$-chymotrypsin activity has recently been reported (Willner *et al.*, 1991a,b).

## IV. BIOLOGICAL APPLICATIONS OF CAGED COMPOUNDS

Many biological studies have benefited from the temporal and spatial resolution made possible by photorelease techniques. In this section, we discuss some experimental studies from a variety of disciplines to show the wide applicability of photochemical techniques. The examples are meant to be illustrative rather than exhaustive.

### A. Muscle Physiology

Flash photolysis of caged compounds has been used extensively in investigations of muscle physiology. A number of thorough reviews have appeared recently, covering studies in striated/skeletal muscle (Homsher and Millar, 1990; Ashley *et al.*, 1991) and smooth muscle (Somlyo and Somlyo, 1990).

### 1. Modulation of Cardiac $Ca^{2+}$ Currents by Intracellular cAMP and $Ca^{2+}$

Gurney *et al.* (1989) and Charnet *et al.* (1991) used DMNB-caged cAMP and nitr-5 to study the modulation of calcium currents in frog atrial myocytes by cAMP and $Ca^{2+}$, respectively. Frog atrial myocytes show two $Ca^{2+}$ current components. The two components are clearly distinguishable by their kinetics of activation and by their steady-state inactivation properties when $Ba^{2+}$ is the charge carrier: a minor, low-threshold, or low-voltage-activated (LVA) current that peaks around $-40$ mV and shows 50% steady-state inactivation at $-75$ mV, and a dominant, high-threshold, or high-voltage-activated (HVA) current that peaks at around 0 mV and is 50% inactivated at $-34$ mV. The HVA current is dihydropyridine (DHP) sensitive and is modulated by $\beta$-adrenergic stimulation, in contrast to the LVA component. Using the whole-cell, tight-seal voltage clamp technique, the investigators found that stimulation with 1 $\mu$M isoproterenol, a $\beta$-agonist, increased the HVA current $\sim 4.1$-fold, with $Ba^{2+}$ as charge carrier. The onset of current enhancement was observed after a lag of a few seconds following application of isoproterenol; enhancement increased to a maximum over 40–60 s. The lag time was attributable to diffusion and binding of the agonist to receptors, as well as to subsequent processes, such as G-protein activation of adenylate cyclase, which lead to generation of cAMP. Indeed, photoreleasing cAMP intracellularly ($\sim 2.5$ $\mu$M) eliminated the delay that was seen following isoproterenol treatment. Instead, augmentation of the HVA current could be observed immediately following the photoinduced cAMP concentration jump, and increased to a maximum of $\sim 3.1$ times the control amplitude over 40–50 s. Photoreleasing $Ca^{2+}$ from intracellular nitr-5, and thus elevating $[Ca^{2+}]_i$ from $\sim 150$ to $\sim 500$ nM, increased HVA $Ba^{2+}$ current maximally to $\sim 3.6$-fold. Again, although current enhancement could be detected immediately after the jump in $[Ca^{2+}]_i$, maximum enhancement was not achieved until 40–50 s after $Ca^{2+}$ photorelease.

That the effect induced by photoreleased cAMP required 40–50 s to develop fully suggests that augmentation of the HVA current by cAMP occurs through phosphorylation by A-kinase. Potentiation of HVA current induced by photoreleased $Ca^{2+}$ followed a time course essentially parallel to that observed for the cAMP effect. This argues against direct modulation of HVA channels by $Ca^{2+}$. The fact that potentiation by $Ca^{2+}$ required MgATP suggests instead a mechanism based on phosphorylation (Gurney et al., 1989). Interestingly, the effects of $Ca^{2+}$ and cAMP are not additive. Photoreleasing $Ca^{2+}$ in the presence of a saturating concentration (100 $\mu$M) of cAMP in the patch pipet, or after pretreatment of cells with 1 $\mu$M isoproterenol, produced no additional effect on HVA current (Charnet et al., 1991). Conversely, photoreleased $Ca^{2+}$ largely blocked the effect of isoproterenol (Gurney et al. 1989). The results collectively suggest that $Ca^{2+}$ and cAMP can both augment $Ca^{2+}$ channel currents in cardiac myocytes by activating kinase activity. Existing evidence, however, does not permit one to determine whether the $Ca^{2+}$- and cAMP-dependent processes both independently lead to phosphorylation of similar sites on channel proteins, or whether phosphorylation results from positive modulatory interactions of $Ca^{2+}$ with the cAMP pathway.

## 2. Calcium-Induced Calcium Release (CICR) in Cardiac Cells

In the mammalian myocardium, $Ca^{2+}$ released from the sarcoplasmic reticulum (SR) is largely responsible for activating myofibrils. While $Ca^{2+}$ influx during the slow inward current, in the absence of SR $Ca^{2+}$ release, is ineffective in triggering contraction, it has been proposed as a trigger for inducing release of $Ca^{2+}$ from the SR. Although CICR had been convincingly shown to act in skinned cardiac cells (Fabiato 1983), direct demonstration of the process in intact cells appeared only recently. Valdeolmillos *et al.* (1989) loaded nitr-5, via the AM ester, into intact rat ventricular myocytes. They showed that 100-ms UV irradiation of a loaded cell to photorelease $Ca^{2+}$ from intracellularly trapped nitr-5 could elicit a twitch in 76% of cells tested. When ryanodine was used to disrupt $Ca^{2+}$ regulation by the SR, photoreleased $Ca^{2+}$ elicited twitches in only 6% of the cells. When the inward $Ca^{2+}$ current was blocked with $Ni^{2+}$, electrically stimulated twitches were abolished, but photoreleased $Ca^{2+}$ was still very effective in inducing contraction.

That photoreleased $Ca^{2+}$ could trigger contraction in control cells but not in ryanodine-treated cells suggests that (1) the amount of $Ca^{2+}$ released from nitr-5 was, in itself, insufficient to cause contraction, and (2) the SR was the source of the $Ca^{2+}$ that activated contraction—results in agreement with preliminary findings in experiments performed with nitr-5 in combination with caffeine in saponin-permeabilized multicellular preparations from guinea pig ventricle (Barsotti *et al.,* 1988). Additionally, results of the photorelease experiment with $Ni^{2+}$ imply that (1) the inward $Ca^{2+}$ current is normally required for CICR from the SR, and (2) even after $Ni^{2+}$ blockage of $Ca^{2+}$ influx had abolished electrically stimulated twitches, the SR still had sufficient $Ca^{2+}$ content that could be released by CICR to cause contraction.

In a separate study, Näbauer and Morad (1990) also examined aspects of CICR by photoreleasing $Ca^{2+}$ from DM-nitrophen in whole-cell clamped ventricular myocytes from guinea pig and rat. The authors found that $Ca^{2+}$ photoreleased at a holding potential of $-80$ mV could elicit contractions of a magnitude similar to those induced by depolarization. The same photorelease in cells in which the SR $Ca^{2+}$ content had been reduced by treatment with 2 mM caffeine was ineffective in eliciting significant contraction. These findings parallel those of Valdeolmillos *et al.* and admitted similar conclusions. Näbauer and Morad extended their study by observing the effect of $Ca^{2+}$ photoreleased either at the onset of, or at variable intervals (50–100 ms) after, depolarization of the cell. It was found that regardless of the time of release relative to the onset of depolarization, $Ca^{2+}$ photorelease always potentiated and never attenuated contraction. These results, the authors concluded, failed to support the existence of $Ca^{2+}$-induced inactivation of $Ca^{2+}$ release from the SR (Fabiato 1985) in this system. The authors cautioned, however, that rejection of the hypothesis of $Ca^{2+}$-induced inactivation of $Ca^{2+}$ release was not warranted because, in order to use DM-nitrophen, $Mg^{2+}$ had to be eliminated from the intracellular dialyzing solution in the pipet—a condition that may have prevented $Ca^{2+}$-induced inactivation of $Ca^{2+}$ release.

Elaborating earlier results (Barsotti et al., 1988), Kentish et al. (1990) found that $Ca^{2+}$ photoreleased from nitr-5 could induce rapid force development in saponin-skinned rat ventrical trabeculae. Pretreatment with caffeine or ryanodine, however, abolished the response included by $Ca^{2+}$ photorelease—findings in accordance with those previously described. These studies serve to establish the reality of CICR in cardiac muscle.

## 3. Role of Ins(1,4,5)P₃ in Excitation–Contraction Coupling

Ins(1,4,5)P₃ was first proposed as a mediator of excitation–contraction (E–C) coupling in muscle by Vergara et al. (1985) who reported that electrical stimulation of frog skeletal muscle could release $IP_3$, and that exogenous $IP_3$ applied to chemically skinned muscle fibers could cause contraction. Subsequent studies by others using topical application of $IP_3$ to skeletal muscle preparations gave inconsistent results, with some researchers confirming the ability of $IP_3$ to release $Ca^{2+}$ from the SR and cause contraction (e.g., Volpe et al., 1985), while others failed to observe positive effects of $IP_3$ (e.g., Lea et al., 1986). In 1987, Walker and colleagues reported using NPE-caged $IP_3$ to investigate the kinetics of smooth and skeletal muscle activation by $IP_3$. They found that photolytic liberation of $\sim 0.5$ $\mu M$ free $IP_3$ from caged $IP_3$ in situ could cause full contraction of saponin-permeabilized smooth muscle from rabbit main pulmonary artery, with $t_{1/2}$ to peak tension of $\sim 3$ s—comparable or slightly less than $t_{1/2}$ for stimulation by noradrenaline. The kinetics of the rise of tension were consistent with the rate of myosin light-chain phosphorylation by myosin light-chain kinase, the $Ca^{2+}$-dependent process that activates smooth muscle contraction. In contrast, in permeabilized skeletal muscle fiber from frog semitendinosus, photogeneration of $<25$ $\mu M$ $IP_3$ was insufficient to induce contracture. $IP_3$ releases in the $25-80$ $\mu M$ range, being from $\sim 50$- to 100-fold higher concentration than required in smooth muscle, did elicit contractions although $t_{1/2}$ to peak tension ($>10$ s) was some three orders of magnitude longer than that for electrical stimulation. Furthermore, even at the elevated levels of photoreleased $IP_3$, $\sim 20\%$ of the fibers tested did not respond. The differences in response between the smooth and skeletal muscle preparations were not the result of high levels of $IP_3$-5-phosphatase in skeletal muscle, which would be expected to degrade $Ins(1,4,5)P_3$ to $Ins(1,4)P_2$, and thus attenuate the effect of photoreleased $IP_3$. Indeed, in the skeletal muscle preparation, $IP_3$-5-phosphatase activity was actually $\sim$30–40 times lower than in the smooth muscle preparation. The photorelease studies, in conjunction with the phosphatase data, provide strong support for $IP_3$ as second messenger in regulating smooth muscle contraction, while arguing against $IP_3$ as the mediator of E–C coupling in skeletal muscle.

In the same paper in which studies on CICR in rat ventricular preparations were reported, Kentish *et al.* (1990) also reported on experiments designed to test the effect of $IP_3$ in cardiac muscle. In those experiments, photoreleasing 2.5 $\mu$M $IP_3$ induced $Ca^{2+}$ release from permeabilized rat ventricular trabeculae, but the response was slow. A lag of about 1 s was observed before the onset of the force response, and the maximum force developed was considerably less than that produced by $Ca^{2+}$ photoreleased from nitr-5. Increasing the amount of photogenerated $IP_3$ to 20 $\mu$M did give a larger and faster response although there was still a noticeable lag, and the response was still $\sim$3-fold smaller and slower when compared with that elicited by photoreleased $Ca^{2+}$. The $IP_3$ response was also inconsistent: in some samples, photoreleased $IP_3$ gave little or no response even though photoreleased $Ca^{2+}$ continued to be effective in eliciting a force response. Finally, the feebleness of the $IP_3$ response was not due to diminished SR $Ca^{2+}$ content, as subsequent application of caffeine was still able to induce substantial $Ca^{2+}$ release. These results suggest that $IP_3$, when it did release $Ca^{2+}$ from the SR, only released a small fraction of the content of the SR. The authors concluded that the limited extent of release, the slowness of the response in comparison to that elicited by $Ca^{2+}$ photorelease, the existence of a lag time, as well as the inconsistency of the responses, together suggested that $Ins(1,4,5)P_3$ is unlikely to be, by itself, a mediator of E–C coupling in heart muscle.

## B. Regulation of Neurotransmitter Release by $Ca^{2+}$ and Membrane Depolarization

There are two hypotheses that seek to explain how neurons release neurotransmitter. In the calcium hypothesis, presynaptic action potentials activate $Ca^{2+}$ influx through $Ca^{2+}$ channels and thus increase presynaptic $[Ca^{2+}]_i$. Elevated $[Ca^{2+}]$ near transmitter release sites is sufficient to trigger release. In the calcium-voltage hypothesis, depolarization of the presynaptic terminal serves two functions: (1) to permit $Ca^{2+}$ influx and hence elevation of $[Ca^{2+}]_i$, and (2) to induce conformational change of a membrane protein to an active, calcium-sensitive form that, when it binds $Ca^{2+}$, directly triggers transmitter release. A critical test of the two hypotheses requires the ability to manipulate intracellular free $Ca^{2+}$ concentration independently of membrane potential—a task for which caged calciums are well suited.

Attempting to delineate the roles of presynaptic potential and $Ca^{2+}$ influx in transmitter release, Zucker and Haydon (1988) performed a series of experiments on inhibitory synapses formed in culture between neurons from the buccal ganglion of the snail *Helisoma trivolvis*. As an indicator of transmitter release, inhibitory postsynaptic current (i.p.s.c.) was monitored using the whole-cell patch-clamp technique. In calcium-containing medium, depolarization of the presynaptic neuron caused the rate of transmitter release, as evidenced by miniature i.p.s.c.'s, to rise. In medium containing no added $Ca^{2+}$ and extra $Mg^{2+}$ to prevent $Ca^{2+}$ entry, presynaptic depolarization did not affect transmitter release, whereas photorelease of $Ca^{2+}$ sharply increased transmitter release. Furthermore, in the same 0-Ca medium, when presynaptic $[Ca^{2+}]_i$ had already been elevated by photolysis of nitr-5, depolarizations of the presynaptic neuron produced no effect on the frequency of m.i.p.s.c.'s. It was also found that transmitter release occurred in the presence of external $Ca^{2+}$ only when the presynaptic neuron was depolarized to $\geq +15$ mV. In contrast, once presynaptic $[Ca^{2+}]_i$ had been increased from 240 nM to $\sim 10$ $\mu$M, transmitter release was elevated but showed no dependence on presynaptic potential over the range from $-120$ to $+40$ mV. Because $Ca^{2+}$ photorelease could elicit transmitter release and, after photochemical elevation of $[Ca^{2+}]_i$, presynaptic depolarization could evoke additional transmitter release only if external $Ca^{2+}$ was present to permit influx through channels, Zucker and Haydon concluded that action potentials evoke transmitter release *only* by activating $Ca^{2+}$ influx through voltage-dependent $Ca^{2+}$ channels.

Hochner and colleagues (1989) suggested that the results of Zucker and Haydon may not have definitively excluded the calcium-voltage hypothesis because the experiments were carried out on a cultured synapse, where transmitter release is slow. Hochner *et al.* chose to examine the two hypotheses in the neuromuscular junction of the crayfish, *Procambarus clarkii*. To eliminate $Ca^{2+}$ influx, the authors first used an external medium containing no added $Ca^{2+}$ and extra $Mg^{2+}$. While the excitatory axon was being electrically stimulated, $Ca^{2+}$ was photoreleased from nitr-5 in the presynaptic terminals. Two observations were made: first, baseline spontaneous release of transmitter was increased more than 6-fold; and second, an additional increase in the rate of transmitter release was observed 2–4 ms after electrical stimulus—a time course comparable to that of normal transmitter release evoked by depolarization. In these experiments, the effect elicited by photoreleasing $Ca^{2+}$ was transient because intracellular homeostatic mechanisms rapidly lowered $[Ca^{2+}]_i$. To counteract cellular calcium regulation, the metabolic uncoupler carbonyl cyanide $m$-chlorophenylhydrazone (CCCP) was used, in a medium containing 0.2 mM $Ca^{2+}$ to improve the ability of the axon to sustain action potentials, and extra $Mg^{2+}$ as well as 2 mM $Mn^{2+}$ to block $Ca^{2+}$ influx. Under CCCP treatment, spontaneous release increased presumably because $[Ca^{2+}]_i$ had been elevated; but more importantly, presynaptic depolarization could evoke transmitter release, as monitored through synaptic potentials. $Ca^{2+}$ photorelease, in turn, enhanced both spontaneous and evoked release in the presence of CCCP. Hochner and collaborators interpreted these results to mean that once $[Ca^{2+}]_i$ had already been elevated, presynaptic membrane depolarization could evoke transmitter release even in the absence of additional $Ca^{2+}$ influx, a conclusion that favored the calcium-voltage hypothesis.

Mulkey and Zucker (1991) reevaluated the hypotheses in the crayfish neuromuscular junction. Using fura-2 ratio imaging, they discovered that neither of the solution conditions used by Hochner *et al.* to block $Ca^{2+}$ entry was effective in doing so under the stimulation protocols used. Such a finding led Mulkey and Zucker to question the conclusion drawn by Hochner *et al.* in favor of the calcium-voltage hypothesis. It was then found that a cobalt-containing Ringer's solution (no added $Ca^{2+}$, 30 mM $Mg^{2+}$, and 13.5 mM $Co^{2+}$) was truly effective in blocking $Ca^{2+}$ influx. In the cobalt Ringer's, photoreleasing $Ca^{2+}$ from DM-nitrophen caused intense transmitter release; however, electrical stimulation evoked no transmitter release either before or after $Ca^{2+}$ photorelease. Mulkey and Zucker thus concluded that normal evoked secretion of neurotransmitters is not directly effected by presynaptic depolarization but is, rather, triggered exclusively by an increase of $[Ca^{2+}]_i$ at transmitter release sites. Similar results in support of the calcium hypothesis, obtained in analogous experiments with DM-nitrophen on the squid giant synapse, have also been published recently (Delaney and Zucker, 1990).

## C. $Ca^{2+}$ "Triggers" in Mitosis

A vast body of experimental evidence indicates the existence of $Ca^{2+}$-sensitive mechanisms that operate during mitosis (see Wolniak, 1988; Hepler 1989) and has motivated the search for $Ca^{2+}$ triggers of mitotic events. Using the fluorometric $Ca^{2+}$ indicator fura-2, Poenie *et al.* (1985) observed, in sea urchin embryos, transient rises in $[Ca^{2+}]_i$, the occurrences of which correlated temporally with various mitotic events, including nuclear envelope breakdown (NEB), which marks the transition from prophase to prometaphase, and the transition from metaphase into anaphase (anaphase onset), which is marked by separation and poleward movement of chromosome pairs. A subsequent imaging study performed with fura-2 also revealed $Ca^{2+}$ transients temporally correlated with the onset of anaphase in $PtK_1$ epithelioid cells (Poenie et al., 1986). Because temporal correlation does not imply essentiality of $Ca^{2+}$ transients in causing mitotic events, it was necessary to investigate whether active generation or suppression of $Ca^{2+}$ transients could alter the course of mitotic events.

Using nitr-5 and the $Ca^{2+}$ chelator BAPTA in conjunction with fluorescent $Ca^{2+}$ indicators fluo-3 and fura-2, Kao et al. (1990) investigated the role of $Ca^{2+}$ transients in Swiss 3T3 cells at two mitotic stages: NEB and the onset of anaphase. Artificial $Ca^{2+}$ transients generated by photolysis of nitr-5 caused precocious NEB, whereas in metaphase, photoreleased $Ca^{2+}$ pulses had no significant effect. The length of time required for cells to make the transition from metaphase to anaphase was essentially invariant, regardless of whether a $Ca^{2+}$ transient was photogenerated during metaphase. In complementary experiments, loading prophase cells with BAPTA could delay or block NEB in a dose-dependent manner, while concentrations of BAPTA sufficient to suppress $Ca^{2+}$ transients did not hinder the transition into anaphase. These results from active manipulation of $Ca^{2+}$ transients show that $Ca^{2+}$ takes an active role in the process of nuclear envelope breakdown, but $Ca^{2+}$ signals are likely unnecessary for the metaphase–anaphase transition.

## D. Mechanisms of Calcium Oscillations

Monitoring $[Ca^{2+}]_i$ in cells under stimulation by agonists that activate the phosphoinositide signaling pathway often reveals regular repetitive transient elevations in $[Ca^{2+}]_i$. Four major models have been proposed to explain such oscillations in the cytosolic concentration of free $Ca^{2+}$ in nonexcitable cells (for reviews, see Tsien and Tsien, 1990; Berridge, 1990; Rink and Merritt, 1990). In all four models, receptor occupancy by agonists leads to $PIP_2$ hydrolysis by phospholipase C to yield $Ins(1,4,5)P_3$ and diacylglycerol (DG); from there on, the models differ: (1) In the model based on the observations of Payne et al. (1988, 1990) and Parker and Ivorra (1990a), $IP_3$ induces $Ca^{2+}$ release from $IP_3$-sensitive $Ca^{2+}$ stores to generate a spike in $[Ca^{2+}]_i$; the resulting rise in $[Ca^{2+}]_i$ reduces the ability of $IP_3$ to release more $Ca^{2+}$. The sensitivity of the $Ca^{2+}$ release mechanism to $IP_3$ recovers only after $[Ca^{2+}]_i$ had returned to baseline values. In this model, the sensitivity of the release process oscillates while $[IP_3]$ remains steady. (2) Berridge and co-workers (Berridge, 1988; Berridge et al., 1988; Berridge and Galione, 1988) proposed a model in which receptor activation turns on a constant flow of $IP_3$, which serves only to release $Ca^{2+}$ from, and to prevent reuptake into, the $IP_3$-sensitive $Ca^{2+}$ pool. The resulting elevation of $[Ca^{2+}]_i$ initiates repeated cycles of $Ca^{2+}$-induced $Ca^{2+}$ release from, followed by reuptake into, a second $Ca^{2+}$ pool that is *not* susceptible to release by $IP_3$. In such a model, $Ca^{2+}$ directly induces its own release. (3) Drawing from studies of oscillations in hepatocytes, Cobbold and collaborators (Woods *et al.*, 1986, 1987a,b; Berridge *et al.*, 1988) proposed that $Ca^{2+}$ released by $IP_3$ acts synergistically with DG to activate protein kinase C (PKC), which, in turn, downregulates the phosphoinositide pathway and thus inhibits the production of more $IP_3$ and DG. As $Ca^{2+}$ is resequestered and DG is metabolized, PKC activity declines while phosphatases reverse the effects of phosphorylation by PKC to enable a new cycle of activation of the signaling pathway. The crucial element regulating the timing of oscillations in this model is PKC activity. (4) Incorporating the known cubic dependence of $Ca^{2+}$ release on $IP_3$ concentration (Meyer *et al.*, 1988) as well as the stimulation of phospholipase C (PLC) by $Ca^{2+}$ (Taylor and Exton, 1987; Eberhard and Holz, 1988), the mathematically explicit model of Meyer and Stryer (1988) requires that the two messengers $Ca^{2+}$ and $IP_3$ each positively feeds back on the production of the other. Thus, $IP_3$ releases $Ca^{2+}$, which in turn activates PLC to make more $IP_3$, which leads to yet more $Ca^{2+}$ release, and so on. Such positive feedback underlies the sharp upstroke of each $Ca^{2+}$ spike. With explosive release of $Ca^{2+}$, the content of the $IP_3$-sensitive pool rapidly diminishes to a level that can support no further release. Removal of cytosolic $Ca^{2+}$ lowers $[Ca^{2+}]_i$ and allows $[IP_3]$ to decline. Repletion of the $IP_3$-sensitive store then paves the way for the next cycle of explosive $IP_3$ production and $Ca^{2+}$ release. In this model, $Ca^{2+}$ enhances its own release indirectly through its activating effect on $IP_3$ production.

Two criteria are sufficient formally to distinguish among the models: (1) Does the breakdown of $PIP_2$ to $Ins(1,4,5)P_3$ and DG oscillate? and (2) Does an increase in $[Ca^{2+}]_i$ enhance or suppress further $Ca^{2+}$ release? Therefore, the most direct tests of these models would be, in any given system, to monitor $[IP_3]$ and $[DG]$ (i.e., $PIP_2$ breakdown) to see if they oscillate and to check if $Ca^{2+}$ reinforces or inhibits its own release. Unfortunately, measuring $[IP_3]$ and $[DG]$ in single cells where oscillations are occurring is technically unfeasible at present. Indeed, the only relevant parameter that can be measured routinely in single cells is $[Ca^{2+}]_i$. Therefore, an alternative way to test the models would be to monitor the oscillations in $[Ca^{2+}]_i$ while delivering pulses of $IP_3$, DG, or $Ca^{2+}$ to the cell and to observe changes in the phase of the oscillations in response to the pulse perturbations.

In a series of studies, Harootunian *et al.* (1988, 1991a,b) investigated calcium oscillations in fibroblasts elicited by the combined action of an agonist that activates $IP_3$ production and any of several treatments that depolarize the cell. These long-lasting oscillations are characterized by exceptionally stable periodicity and are, therefore, an excellent candidate for study by the pulse perturbation technique. It was found that a photogenerated pulse of $IP_3$ always produced a spike in $[Ca^{2+}]_i$ but never caused oscillations. In cells that have been induced to show oscillations, it was found that a pulse of $IP_3$ always elicited a $Ca^{2+}$ spike regardless of the time of photorelease relative to the endogenous train of spikes. Furthermore, the interval between the exogenously induced spike and the next endogenous spike was typically greater than or equal to one natural oscillation period. These findings suggest that $IP_3$ is insufficient by itself to cause oscillations and that the $Ca^{2+}$ stores are continually sensitive to release by $IP_3$. The extended interval between the exogenous and endogenous $Ca^{2+}$ spikes is most readily understood in terms of the extra time required for the stores to refill after an unscheduled release by an exogenous pulse of $IP_3$.

In a cell in which oscillations were already underway, photolysis of nitr-7 yielded a pulse of $Ca^{2+}$, which was followed by an amplification phase that peaked ~10 s after the photogenerated pulse and which further elevated $[Ca^{2+}]_i$ (Harootunian et al., 1991a). The amplification phase was absent in unstimulated cells and could be abolished in oscillating cells by the presence of microinjected heparin, a blocker of the intracellular $IP_3$ receptor. Amplification of the artificial $Ca^{2+}$ spike indicates that $Ca^{2+}$ reinforces its own release. That the amplification could be abolished by inhibiting the $IP_3$ receptor shows that positive feedback by $Ca^{2+}$ is effected indirectly, through the $IP_3$ pathway.

Photoreleasing a pulse of sn-1,2-dioctanoylglycerol, a diacyl glycerol, increased interspike interval (Harootunian et al., 1991b). This period lengthening effect was most pronounced on the first oscillation cycle after photorelease. Thereafter, the periods gradually shortened until the natural, prephotolysis period was reattained after several cycles. If the interspike interval corresponds to the time required for reversal of PKC action by phosphatases as well as metabolic removal of DG (Model 3), then a pulse of exogenous DG, causing extra activation of PKC, would require extra time before its effects were reversed. An abnormally long period is therefore expected following a DG pulse. Once the effects of DG are reversed and the next endogenous spike is allowed to occur, no memory of the DG pulse should persist, so subsequent oscillation cycles should all have the natural, prepulse period. Because experimental observations contradict this prediction of Model 3, it is unlikely that PKC has more than a modulatory role in these oscillations. Indeed, when PKC was completely removed from the fibroblasts through downregulation with phorbol esters, oscillations could still be initiated and maintained by the usual treatments (Harootunian et al., 1991a).

These studies with photochemical pulse perturbations, supported by results from a variety of pharmacological experiments (Harootunian et al., 1991b), show that $Ca^{2+}$ oscillations induced by depolarization coupled with activation of the $IP_3$ pathway in REF52 fibroblasts are best explained by the model of Meyer and Stryer. Analogous pulse perturbation experiments should allow elucidation of mechanistic features in other oscillatory systems.

## E. Signal Transduction Processes Leading to Closure of Stomatal Pores in Plants

Stomata, pore-bearing structures at the surface of a leaf, regulate $CO_2$ entry into and water loss from the leaf. Control of stomatal pore size is exerted by a pair of guard cells that embrace the pore. Lowering the turgor pressure of guard cells causes shrinkage of the cells and, as consequence, a reduction in stomatal pore size. Abscisic acid (ABA), a plant hormone, has long been known to be capable of inducing closure of stomata. Recent experiments have revealed that, in the lily, *Commelina communis*, application of ABA induces a rise in the $[Ca^{2+}]_i$ of guard cells prior to stomatal closure (McAinsh *et al.*, 1990), a result that suggests that $Ca^{2+}$ is a second messenger mediating this response to ABA.

Using fluo-3 to monitor $[Ca^{2+}]_i$, Gilroy and collaborators (1990) examined the effects of photoreleasing $Ca^{2+}$ and $IP_3$ in guard cells of *C. communis*. By changing the duration of photolysis and thus photolyzing different amounts of intracellular nitr-5, they could raise $[Ca^{2+}]_i$ by variable increments. Elevating $[Ca^{2+}]_i$ to above ~600 nM led to closure of the stomatal pore. This result indicates that $Ca^{2+}$ by itself can trigger closure and that there is a threshold for activation of closure by $Ca^{2+}$. Interestingly, photolysis of nitr-5 did not lead to an extremely rapid rise in $[Ca^{2+}]_i$, as would have been expected from the fact that $Ca^{2+}$ release from photolyzed nitr-5 is complete within a very few milliseconds (Adams *et al.*, 1988). Instead, nitr-5 photolysis in guard cells triggered a rise in $[Ca^{2+}]_i$ that peaked in ~2 *minutes*. Such a slow time course may reflect the existence, at least in guard cells, of $Ca^{2+}$-dependent $Ca^{2+}$ release mechanisms that could amplify a photogenerated $Ca^{2+}$ signal. Photorelease of $IP_3$ caused a rise in $[Ca^{2+}]_i$ followed by stomatal closure. The $IP_3$-induced $Ca^{2+}$ transient peaked in ~10 s and then declined over 5–10 min. The rise time, although much faster than that found in the case of nitr-5, is still at least an order of magnitude slower than that found in mammalian systems (Kao *et al.*, 1989). Intracellular EGTA or BAPTA could block both the $[Ca^{2+}]_i$ rise and stomatal closure induced by $IP_3$ photorelease, while 1 mM $La^{3+}$ in the bathing medium could not. Together, these studies show that $IP_3$ could be a second messenger in the regulation of stomatal closure and that its effect is exerted through $Ca^{2+}$ that is derived from intracellular stores rather than the extracellular medium. Incidentally, the report of Gilroy *et al.* constitutes the first direct evidence showing that $IP_3$ can mobilize $Ca^{2+}$ stores *in vivo* in intact plant cells.

Control of stomatal pore size depends on guard cell turgor, which is, in turn, dependent on the electrolyte content of the guard cells. ABA has been shown to promote stomatal closure by inducing guard cells to lose osmotically active solutes, principally $K^+$ salts (MacRobbie, 1987). That $IP_3$ could induce $Ca^{2+}$ release from plant vacuoles (for review, Einspahr and Thompson, 1990) and that $Ca^{2+}$ appeared to underlie the stomate-closing response to ABA suggest that $IP_3$ might be a second messenger in regulating ion fluxes across the plasma membrane of guard cells. In guard cells of *Vicia faba*, the fava bean plant, Blatt and colleagues (1990) showed that photoreleased $IP_3$ caused reversible membrane depolarizations to $\sim +15$ mV. Underlying this voltage change was suppression of an inward-rectifying $K^+$ current and simultaneous activation of a time-independent inward leak current. In contrast, outward-rectifying $K^+$ current was little affected. Because microinjected EGTA or BAPTA could counteract the effect of $IP_3$ photorelease, the response was very likely mediated through $Ca^{2+}$. These findings are consistent with the turgor-regulating actions of ABA in guard cells: elevation of $[Ca^{2+}]_i$ by $IP_3$ inhibits $K^+$ uptake through the inward-rectifying $K^+$ channels while activating the leak current to depolarize the guard cells and promote $K^+$ efflux through another class of outward-rectifying $K^+$ channels (Blatt 1988, 1990). The net effect is loss of potassium salts from the guard cells. Blatt's studies show that guard cells can use $IP_3$ to modulate $K^+$ channel activity and that $IP_3$ may be a second messenger in cellular transduction of the ABA hormonal signal.

## F. Microtubule Dynamics in the Metaphase Mitotic Spindle

A highly spatially organized assembly of microtubules constitutes the mitotic spindle, which mediates segregation of separated sister chromatids during cell division. A special subset of microtubules comprises fibers linking the kinetochores of chromosomes to the proximal spindle pole. The dynamic properties of these kinetochore microtubules are important because they underlie the mechanism by which chromosomes are aligned and segregated during mitosis. Experiments where biotinylated tubulin was injected into dividing cells have shown that kinetochore microtubules add subunits at the kinetochore throughout metaphase, and thus suggest that there is a poleward flux of tubulin subunits in the kinetochore fiber (Mitchison *et al.*, 1986). In order to detect specific movement of kinetochore microtubules within the spindle, it is necessary to make marks that remain fixed on those microtubules. One approach to achieve this is to inject fluorescently labeled tubulin into mitotic cells, whose spindles then become uniformly labeled. Localized photobleaching with a laser beam creates a dark, nonfluorescent stripe across the spindle microtubules, the movements of which can now be monitored. Such photobleaching experiments revealed little or no polewards flux, however (e.g., Salmon *et al.*, 1984; Saxton *et al.*, 1984; Wadsworth and Salmon 1986; Gorbsky and Borisy 1989).

Trying to resolve the contradiction between photobleaching and biotinylated tubuline injection experiments, Mitchison (1989) conjugated bis-NB-caged carboxyfluorescein (CF) to tubulin monomers via a $\beta$-alanyl linker. After injection into mitotic PtK2 or LLC-PK1 kidney epithelial cells, sufficient time was allowed for the caged CF-labeled tubulin subunits to be incorporated into all microtubules. Once a cell reached metaphase, a 1-$\mu$m-wide fluorescent stripe was imprinted, by flash UV photolysis through a slit, across the metaphase spindle. Thereafter, the microscopic field was uniformly illuminated with low-intensity 490-nm light to allow movements of the fluorescent stripe pattern to be observed. The fluorescent stripe marked kinetochore as well as other microtubules, but because nonkinetochore microtubules exchange tubulin subunits much more rapidly than kinetochore microtubules ($t_{1/2} \leq 1$ min at 37°C), within 30–90 s, fluorescent tubulin from nonkinetochore microtubules has exchanged out of the photoimprinted zone, thus leaving only the more stable kinetochore microtubules bearing the fluorescent mark. Over a period of minutes, the fluorescent zone moved towards the proximal spindle pole, eventually accumulating at, and finally disappearing from, the pole. Video analysis revealed the velocity of movement to be 0.6 $\mu$m/min in PtK2 and ~0.5 $\mu$m/min in LLC-PK1 cells. That fixed points on the kinetochore microtubules moved continuously polewards during metaphase suggests that these microtubules continually polymerize at the kinetochore while continually depolymerizing at the poleward ends. Thus, these experiments demonstrate the existence of a steady polewards microtubule flux in the metaphase spindle.

It is difficult to reconcile the positive result from photoactivation of fluorescence (PAF) with the negative observations from photobleaching experiments, which are, at least in principle, complementary to PAF. There are several potential complications of photobleaching experiments. Whereas Mitchison's PAF technique marks kinetochore microtubules selectively and presents bright signals against a dark background, monitoring nonfluorescent kinetochore microtubules against a background of other bleached spindle fibers, which are rapidly recovering their fluorescence by exchanging in fluorescent tubulin from unbleached regions, could pose a signal-to-noise problem. Gorbsky and Borisy (1989) tried to remedy this weakness of photobleaching experiments by permeabilizing photobleached mitotic cells under conditions that selectively removed soluble tubulin and nonkinetochore microtubules. Even with the improved contrast in this permeabilized system, the level of polewards flux measured was still less than the variability of the experiments. Photobleaching efficient fluorophores, such as fluoresceins, also requires high light intensities, which can cause photodynamic damage (Leslie *et al.*, 1984; Vigers *et al.*, 1988), a problem not encountered in the PAF experiment (Mitchison 1989). Thus, it is possible that PAF methodology favored observation of a phenomenon that, for kinetic and signal-to-noise reasons, could not otherwise be detected.

## G. Actin Microfilament Dynamics in Locomoting Cells

Polymerized actin filaments are required for movement of cells across solid substrates. Although structural information is readily available on actin organization in motile cells, information on the dynamic behavior of actin in the same cells is relatively scarce. Theriot and Mitchison (1991) extended the PAF technique to the study of actin filament dynamics in actively moving keratocytes (epidermal cells) from goldfish (*Carassius*). Locomoting keratocytes extend a broad, flat, uniform lamellipodium at the cell anterior, while subcellular organelles, including the nucleus, are largely confined to the cell body, which is situated at the cell posterior. Keratocytes move by an actin-dependent but microtubule-independent mechanism and are extremely fast, sometimes reaching speeds in excess of 1 $\mu$m s$^{-1}$ (Euteneuer and Schliwa, 1984).

Theriot and Mitchison synthesized an NPE-caged derivative of resorufin (designated CR), which also carried an iodoacetyl group on a piperazine linker, for covalent attachment to sulfhydryl groups on proteins. CR-labeled actin monomers were able to copolymerize normally with unlabeled actin both *in vitro* and in living cells. The lamellipodia on keratocytes are typically 5–15 $\mu$m wide and have relatively uniform density of actin filaments throughout, as judged by fluorescent phalloidin staining. After injecting CR-labeled actin monomers into keratocytes, the authors allowed 20–60 min for the labeled actin to be incorporated into actin filaments. UV irradiation through a slit imprinted a bright fluorescent bar in the lamellipodium by unmasking resorufin fluorophores on actin filaments. The fluorescent zone did not fragment but rather behaved as a coherent unit in both slow- and fast-moving cells. As the keratocyte moved, the fluorescent bar remained fixed relative to the laboratory reference frame, whereas viewed in the moving reference frame of the cell, the bar moved backwards toward the nucleus. Video analysis revealed that for cells moving over a wide range of speeds (0.001–0.13 $\mu$m s$^{-1}$), the rate of movement of the fluorescent bar towards the nucleus (and away from the anterior edge) was always roughly equal to the speed of the cell. This observation shows that the fluorescent marking remained fixed relative to the substrate while the cell moved over it, regardless of the speed of the cell. By monitoring the dissipation of fluorescence in the photoactivated zone, the half-life of the labeled actin in filaments was determined to be 23 s, a value dependent on neither the rate of cell movement nor the position of the fluorescently marked actin filaments within the lamellipodium.

These experiments show (1) that once actin filaments have formed in the lamellipodium, they remain fixed with respect to the substrate, and (2) that all actin filaments in the lamellipodium behave as a coherent whole. From these observations, the authors infer that any cytoskeletal network filling newly extended regions of the lamellipodium must arise from fresh polymerization of actin and not from movement of existing actin filaments relative to each other. Theriot and Mitchison also used their results on the dynamic behavior of actin filaments to differentiate between two alternative models for explaining actin subunit flux through the lamellipodium. In the "treadmilling" model, actin filaments with lengths comparable to the breadth of the lamellipodium orient with their active (barbed) ends toward the cell periphery. Actin polymerization is expected to take place at the active ends at the cell margin, while depolymerization occurs at the less active (pointed) ends at the rear of the lamellipodium. In the "nucleation-release" model, actin filaments do not need to have any particular orientation. The movement of a meshwork of short actin filaments acting as a unit through the lamellipodium gives rise to a flux of actin. New filaments are expected to be nucleated primarily at or near the leading edge and then released. Polymerization and depolymerization can occur throughout the lamellipodium though the polymerization rate is higher at the leading edge. That the half-life of actin filaments is short ($t_{1/2} = 23$ s) and independent of location in the lamellipodium is consistent with nucleation-release but argues against treadmilling. Because actin-filament density was essentially uniform throughout the lamellipodium, actin must be able to polymerize and depolymerize anywhere in the lamellipodium—a conclusion in conflict with the predictions of the treadmilling model but congruent with nucleation-release. Finally, given the rates of actin depolymerization from *in vitro* studies, and taking the 23-s half-life of the filaments into account, the authors calculated the permissible actin-filament length in the lamellipodium to be, on average, 0.5 $\mu$m or less. Once again, such short filament lengths are concordant with the features of the nucleation-release model and incompatible with treadmilling.

## H. Localized Quantal (All-or-None) $Ca^{2+}$ Release Induced by Ins(1,4,5)$P_3$

In permeabilized cells, low doses of $IP_3$ release only a fraction of the $IP_3$-sensitive calcium stores (Muallem *et al.*, 1989; Taylor and Potter, 1990; Meyer and Stryer, 1990). Such observations have given rise to proposals of the existence of a multiplicity of cellular $Ca^{2+}$ stores, each of which may be triggered, with a different threshold, by $IP_3$ to release $Ca^{2+}$ in an all-or-none fashion. In order to test the all-or-none $Ca^{2+}$ release hypothesis, Parker and Ivorra (1990b) combined photorelease technology with confocal detection of fluorescence signals from $Ca^{2+}$-sensitive indicators to investigate spatiotemporal aspects of $Ca^{2+}$ release induced by $IP_3$ in *Xenopus* oocytes.

Parker and Ivorra used *Xenopus laevis* oocytes loaded with caged $IP_3$ and either fluo-3 or rhod-2 as the $Ca^{2+}$ indicator. In the first type of experiment, $IP_3$ photorelease was effected over a large area of the cell (a spot of $150\mu m$ diameter, with an area of $\sim 1.8 \times 10^4\ \mu m^2$), and $Ca^{2+}$-sensitive fluorescence signal from the indicator was collected from the same large spot. The $Ca^{2+}$-activated $Cl^-$ conductance in the cell membrane was simultaneously monitored. Varying doses of $IP_3$ were delivered by changing the duration of the photolytic UV flash. No $Ca^{2+}$ signal could be detected below a threshold of 8 ms of photolysis. Above the threshold, graded photolysis gave graded increases in $Ca^{2+}$ signal as well as membrane $Cl^-$ conductance. The dose-response curves were approximately linear although the threshold for the response in the $Cl^-$ was somewhat higher, as would be expected if an elevation in $[Ca^{2+}]_i$ above resting level was needed to activate the $Cl^-$ conductance. Results from these experiments represent responses averaged over a large volume of cytoplasm.

In the second type of experiment, $IP_3$ photorelease was effected in a small square area ($\sim 50-100\ \mu m^2$), and the $Ca^{2+}$ signal was collected confocally from an even smaller circular area (2 $\mu m$ in diameter, or $\sim 3\ \mu m^2$ of area) centered within the photolysis square. As before, $Cl^-$ conductance was also monitored. Under these measurement conditions, a threshold was again observed for the appearance of the $Ca^{2+}$ signal in response to $IP_3$. However, after threshold was reached, increasing photolysis time brought no corresponding increases in $Ca^{2+}$ signal. The dose-response curve for the $Ca^{2+}$ signal was approximately a step function, with a discontinuity at the threshold dosage. $IP_3$-induced $Ca^{2+}$ signal, when collected from a small volume of cytoplasm, thus appeared to be all-or-none (standard deviation for responses to stimuli of 2–10 times the threshold was only 8% of the mean). In contrast to the $Ca^{2+}$ response, the $Cl^-$ conductance response was graded although the dependence was not linear. The graded increases in the $Cl^-$ response may have been because with increasing photolysis, more $Ca^{2+}$ stores were recruited to release $Ca^{2+}$ (this explanation is possible because photolysis, unlike fluorescence detection, was not performed confocally). To determine the extent of localization of $Ca^{2+}$ release, the authors kept the confocal detection spot fixed but displaced the square photolysis patch from the detection spot by varying distances. With increasing separation, the $Ca^{2+}$ signal decreased in maximum amplitude ($\sim 50\%$ attenuation at 5 $\mu m$). Furthermore, the $Ca^{2+}$ waveform was also broader (took longer to peak). Such results are just what would be expected for diffusional spread of locally released $Ca^{2+}$.

Most significantly, these experiments show that, when a small region of the cytoplasm is examined, $IP_3$ appears to trigger all-or-none release of the local $IP_3$-sensitive $Ca^{2+}$ stores. Over the larger dimensions of the oocyte (~1 mm in diameter), however, graded responses are obtained. This discrepancy could be explained by the existence of different local stores with differing thresholds, so that increasing concentrations of $IP_3$ can recruit more stores for $Ca^{2+}$ release. The precise nature of the local $Ca^{2+}$ release units and the mechanism underlying quantal release remain to be elucidated.

## I. Other Biological Applications

A number of other reports of use of caged compounds in biological research have appeared recently. These include studies on EGF-induced $Ca^{2+}$ signaling in A431 cells (Hughes et al., 1991), the synergism between PLC- and adenylyl cyclase–linked hormones in hepatocytes (Burgess et al., 1991), the biophysics of the SR Ca-ATPase (Lewis and Thomas, 1991) and the contractile apparatus of muscle (Berger et al., 1989; Fajer et al., 1990; Ostap and Thomas, 1991, Dantzig et al., 1991), the electrophysiology of sensory neurones (Dolphin et al., 1988; Scott et al., 1990) and hepatocytes (Ogden et al., 1990; Noel and Capiod, 1991), relaxation (Niggli and Lederer, 1991) and regulation of $Ca^{2+}$ channel phosphorylation (Backx et al., 1991) in cardiac cells, $Ca^{2+}$ mobilization mechanisms in oocytes (Parker and Miledi, 1989; Miledi and Parker, 1989; Peres, 1990; Peres et al., 1991), and X-ray crystallographic characterization of GTP hydrolysis by p21[c-H-ras] (Schlichting et al., 1989).

## V. CONCLUDING REMARKS

The last several years have seen a rapid growth in the synthesis of new caged compounds, which are almost all based on the 2-nitrobenzyl rearrangement. While many caged molecules incorporating the 2-nitrobenzyl type of caging group have worked satisfactorily, others suffer from slow photorelease or low quantum efficiency, or both. Because photochemical events are fast, slow release usually reflects rate limiting steps in the thermal, or "dark," reactions following the photochemical steps. Thus, a direction for new research efforts is the development of caging strategies that rely on photoreactions involving fast or, better yet, no dark reactions after the photochemical steps. Such developments would extend the time resolution of photorelease techniques.

The examples discussed in Section III amply demonstrate the ready applicability of caged compounds for solving problems from diverse areas of biological research. Most of the applications to date have taken advantage of the time resolution made possible by flash photolysis. In contrast, the high spatial resolution potentially attainable with photolysis has only just begun to be realized. Part of the reason for this disparity is probably the somewhat more complex experimental setup required to achieve spatially restricted photolysis or detection. However, as photorelease techniques become more familiar, more and novel experiments in both the spatial and temporal domains are certain to emerge. Thus the prospects for photorelease techniques becoming a routine experimental tool are pleasingly bright.

## REFERENCES

Adams, S. R., Kao, J. P. Y., and Tsien, R. Y. (1988). Biologically useful chelators that release $Ca^{2+}$ upon illumination. *J. Am. Chem. Soc. 110,* 3212–3220.

Adams, S. R., Kao, J. P. Y., and Tsien, R. Y. (1989). Biologically useful chelators that take up $Ca^{2+}$ upon illumination. *J. Am. Chem. Soc. 111,* 7957–7968.

Ämmälä, C., Bokvist, K., Galt, S., and Rorsman, P. (1991). Inhibition of ATP-regulated $K^+$-channels by a photoactivatable ATP-analogue in mouse pancreatic $\beta$-cells. *Biochim. Biophys. Acta 1092,* 347–349.

Ashley, C. C., Mulligan, I. P., and Lea, T. J. (1991). $Ca^{2+}$ and activation mechanisms in skeletal muscle. *Quart. Rev. Biophys. 24,* 1–73.

Backx, P. H., O'Rourke, B., and Marban, E. (1991). Flash photolysis of magnesium-DM-nitrophen in heart cells. A novel approach to probe magnesium- and ATP-dependent regulation of calcium channels. *Am. J. Hypertens. 4,* 416S–421S.

Barsotti, R. J., Kentish, J. C., Lea, T. J., and Mulligan, I. P. (1988). Laser-induced photolysis of nitr-5 triggers calcium release from the sarcoplasmic reticulum of saponin-treated muscles from guinea-pig ventricle. *J. Physiol. (London) 396,* p. 80.

Bartels, E., Wassermann, N. H., and Erlanger, B. F. (1971). Photochromic activators of the acetylcholine receptor. *Proc. Natl. Acad. Sci. 68,* 1820–1823.

Berger, C. L., Svensson, E. C., and Thomas, D. D. (1989). Photolysis of a photolabile precursor of ATP (caged ATP) induces microsecond rotational motions of myosin heads bound to actin. *Proc. Natl. Acad. Sci. USA 86,* 8753–8757.

Berridge, M. J. (1988). Inositol trisphosphate-induced membrane potential oscillations in *Xenopus* oocytes. *J. Physiol. (London),* 589–599.

Berridge, M. J. (1990). Calcium oscillations. *J. Biol. Chem. 265,* 9583–9586.

Berridge, M. J., and Galione, A. (1988). Cytosolic calcium oscillators. *FASEB J. 2,* 3074–3082.

Berridge, M. J., Cobbold, P. H., and Cuthbertson, K. S. R. (1988). Spatial and temporal aspects of cell signalling. *Phil. Trans. R. Soc. London B 320,* 325–343.

Binkely, R. W., and Flechtner, T. W. (1984). Photoremovable protecting groups. In "Synthetic Organic Photochemistry" (W. M. Horspool, ed.), pp. 375–423. Plenum, New York.

Blatt, M. R. (1988). Potassium-dependent, bipolar gating of $K^+$ channels in guard cells. *J. Membrane Biol. 102,* 235–246.

Blatt, M. R. (1990). Potassium channel currents in intact stomatal guard cells: Rapid enhancement by abscisic acid. *Planta 180,* 445–455.

Blatt, M. R., Thiel, G., and Trentham, D. R. (1990). Reversible inactivation of $K^+$ channels of *Vicia* stomatal guard cells following the photolysis of caged inositol 1,4,5-trisphosphate. *Nature (London) 346,* 766–769.

Burgess, G. M., Bird, G. St. J., Obie, J. F., and Putney, J. W., Jr. (1991). The mechanism for synergism between phospholipase C- and adenylylcyclase-linked hormones in liver. *J. Biol. Chem. 266,* 4772–4781.

Chang, C.-H., Jiang, B., and Douglas, J. G. (1990). Caged ATP potentiates guanylate cyclase activity stimulated by atrial natriuretic factor in rat lung membranes. *Eur. J. Pharm. 189,* 111–114.

Charnet, P., Richard, S., Gurney, A. M., Ouadid, H., Tiaho, F., and Nargeot, J. (1991). Modulation of Ca currents in isolated frog atrial cells studied with photosensitive probes. Regulation by cAMP and $Ca^{2+}$: a common pathway? *J. Mol. Cell. Cardiol. 23,* 343–356.

Cummings, R. T., and Krafft, G. A. (1988). Photoactivable fluorophores. 1. Synthesis and photoactivation of o-nitrobenzyl-quenched fluorescent carbamates. *Tetrahedron Lett. 29,* 65–68.

Dantzig, J. A., Hibberd, M. G., Trentham, D. R., and Goldman, Y. E. (1991). Cross-bridge kinetics in the presence of MgADP investigated by photolysis of caged ATP in rabbit psoas muscle fibres. *J. Physiol. (London) 432,* 639–680.

Delaney, K. R., and Zucker, R. S. (1990). Calcium released by photolysis of DM-nitrophen stimulates transmitter release at squid giant synapse. *J. Physiol. (London) 426,* 473–498.

Delcour, A., and Hess, G. P. (1986). Chemical kinetic measurements of the effect of *trans* and *cis* bis[(trimethylammonio)methyl]azobenzene bromide on acetylcholine receptor mediated ion translocation in *Electrophorus electricus* and *Torpedo californica*. *Biochemistry 25,* 1793–1798.

DeLong, L. J., Phillips, C. M., Kaplan, J. H., Scarpa, A., and Blasie, J. K. (1990). A new method for monitoring the kinetics of calcium binding to the sarcoplasmic reticulum $Ca^{2+}$-ATPase employing the flash-photolysis of caged-calcium. *J. Biochem. Biophys. Meth. 21,* 333–339.

De Mayo, P., and Reid, S. T. (1961). Photochemical rearrangements and related transformations. *Quart. Rev. Biophys. 16,* 393–417.

Denk, W., Strickler, J. H., and Webb, W. W. (1990). Two-photon laser scanning fluorescence microscopy. *Science 248,* 73-73-76.

Dolphin, A. C., Wootton, J. F., Scott, R. H., and Trentham, D. R. (1988). Photoactivation of intracellular guanosine triphosphate analogues reduces the amplitude and slows the kinetics of voltage-activated calcium channel currents in sensory neurones. *Pflügers Archiv. 411,* 628–636.

Eberhard, D. A., and Holz, R. W. (1988). Intracellular $Ca^{2+}$ activates phospholipase C. *Trends Neurosci. 11,* 517–520.

Einspahr, K. J., and Thompson, G. A., Jr. (1990). Transmembrane signaling via phosphatidylinositol 4,5-bisphosphate hydrolysis in plants. *Plant Physiol. 93,* 361–366.

Ellis-Davis, G. C. R., and Kaplan, J. H. (1988). A new class of photolabile chelators for the rapid release of divalent cations: Generation of caged Ca and caged Mg. *J. Org. Chem. 53,* 1966–1969.

Engels, J., and Schlaeger, E.-J. (1977). Synthesis, structure, and reactivity of adenosine cyclic 3′,5′-phosphate benzyl triesters. *J. Med. Chem. 20,* 907–911.

Euteneuer, U., and Schliwa, M. (1984). Persistent, directional motility of cells and cytoplasmic fragments in the absence of microtubules. *Nature (London) 310*, 58–61.

Fabiato, A. (1983). Calcium-induced release of calcium from the cardiac sarcoplasmic reticulum. *Am. J. Physiol. 245*, C1–C14.

Fabiato, A. (1985). Time and calcium dependence of activation and inactivation of calcium-induced release of calcium from the sarcoplasmic reticulum of a skinned canine cardiac Purkinje cell. *J. Gen. Physiol. 85*, 247–289.

Fajer, P. G., Fajer, E. A., and Thomas, D. D. (1990). Myosin heads have a broad orientational distribution during isometric muscle contraction: Time-resolved EPR studies using caged ATP. *Proc. Natl. Acad. Sci. USA 87*, 5538–5542.

Ferenczi, M. A., and Goldman, Y. E., and Trentham, D. R. (1989). Relaxation of permeabilized muscle fibres of the rabbit by rapid chelation of $Ca^{2+}$-ions through laser-pulse photolysis of "caged-BAPTA." *J. Physiol. (London) 418*, 155P.

Gilroy, S., Read, N. D., and Trewavas, A. J. (1990). Elevation of cytoplasmic calcium by caged calcium or caged inositol trisphosphate initiates stomatal closure. *Nature (London) 346*, 769–771.

Gorbsky, G. J., and Borisy, G. G. (1989). Microtubules of the kinetochore fiber turn over in metaphase but not in anaphase. *J. Cell Biol. 109*, 653–662.

Gravel, D., Giasson, R., Blanchet, D., Yip, R. W., and Sharma, D. K. (1991). Photochemistry of the *o*-nitrobenzyl system in solution: Effects of O$\cdots$H distance and geometrical constraint on the hydrogen transfer mechanism in the excited state. *Can. J. Chem., 69*, 1193–1200.

Grynkiewicz, G., Poenie, M., and Tsien, R. Y. (1985). A new generation of $Ca^{2+}$ indicators with greatly improved fluorescence properties. *J. Biol. Chem. 260*, 3440–3450.

Gurney, A. M., Charnet, P., Pye, J. M., and Nargeot, J. (1989). Augmentation of cardiac calcium current by flash photolysis of intracellular caged-$Ca^{2+}$ molecules. *Nature (London) 341*, 65–68.

Harootunian, A. T., Kao, J. P. Y., and Tsien, R. Y. (1988). Agonist-induced calcium oscillations in depolarized fibroblasts and their manipulation by photoreleased $Ins(1,4,5)P_3$, $Ca^{++}$, and $Ca^{++}$ buffer. *Cold Spring Harb. Symp. Quant. Biol. 53*, 935–943.

Harootunian, A. T., Kao, J. P. Y., Paranjape, S., and Tsien, R. Y. (1991a). Generation of calcium oscillations in fibroblasts by positive feedback between calcium and $IP_3$. *Science 251*, 75–78.

Harootunian, A. T., Kao, J. P. Y., Paranjape, S., Adams, S. R., Potter, B. V. L., and Tsien, R. Y. (1991b). Cytosolic $Ca^{2+}$ oscillations in REF52 fibroblasts: $Ca^{2+}$-stimulated $IP_3$ production or voltage-dependent $Ca^{2+}$ channels as key positive feedback elements. *Cell Calcium 12*, 153–164.

Hepler, P. K. (1989). Calcium transients during mitosis: Observations in flux. *J. Cell. Biol. 109*, 2567–2573.

Hochner, B., Parnas, H., and Parnas, I. (1989). Membrane depolarization evokes neurotransmitter release in the absence of calcium entry. *Nature (London) 342*, 433–435.

Homsher, E., and Millar, N. C. (1990). Caged compounds and striated muscle contraction. *Annu. Rev. Physiol. 52*, 875–896.

Hughes, A. R., Bird, G. St. J., Obie, J. F., Thastrup, O., and Putney, J. W., Jr. (1991). Role of inositol (1,4,5)trisphosphate in epidermal growth factor-induced $Ca^{2+}$ signaling in A431 cells. *Molec. Pharmacol. 40,* 254–262.

Janko, K., and Reichert, J. (1987). Proton concentration jumps and generation of transmembrane pH-gradients by photolysis of 4-formyl-6-methoxy-3-nitrophenoxyacetic acid. *Biochim. Biophys. Acta 905,* 409–416.

Kao, J. P. Y., Harootunian, A. T., and Tsien, R. Y. (1989). Photochemically generated cytosolic calcium transients and their detection by fluo-3. *J. Biol. Chem. 264,* 8179–8184.

Kao, J. P. Y., Alderton, J. M., Tsien, R. Y., and Steinhardt, R. A. (1990). Active involvement of $Ca^{2+}$ in mitotic progression of Swiss 3T3 fibroblasts. *J. Cell Biol. 111,* 183–196.

Kaplan, J. H. (1990). Photochemical manipulation of divalent cation levels. *Annu. Rev. Physiol. 52,* 897–914.

Kaplan, J. H., and Ellis-Davis, G. C. R. (1988). Photolabile chelators for the rapid photorelease of divalent cations. *Proc. Natl. Acad. Sci. USA 85,* 6571–6575.

Kaplan, J. H., Forbush, B., III, and Hoffman, J. F. (1978). Rapid photolytic release of adenosine 5'-triphosphate from a protected analogue: Utilization by the Na:K pump of human red blood cell ghosts. *Biochemistry 17,* 1929–1935.

Karpen, J. W., Zimmerman, A. L., Stryer, L., and Baylor, D. A. (1988). Gating kinetics of the cyclic-GMP-activated channel of retinal rods: Flash photolysis and voltage jump studies. *Proc. Natl. Acad. Sci. USA 85,* 1287–1291.

Kentish, J. C., Barsotti, R. J., Lea, T. J., Mulligan, I. P., Patel, J. R., and Ferenczi, M. A. (1990). Calcium release from cardiac sarcoplasmic reticulum induced by photorelease of calcium or Ins(1,4,5)$P_3$. *Am. J. Physiol. 258,* H610–H615.

Kitazawa, T., Kobayashi, S., Horiuti, K., Somlyo, A. V., and Somlyo, A. P. (1989). Receptor-coupled, permeabilized smooth muscle. *J. Biol. Chem. 264,* 5339–5342.

Kobayashi, S., Kitazawa, T., Somlyo, A. V., and Somlyo, A. P. (1989). Cytosolic heparin inhibits muscarinic and $\alpha$-adrenergic $Ca^{2+}$ release in smooth muscle. *J. Biol. Chem. 264,* 17997–18004.

Krafft, G., Cummings, R. T., Dizio, J. P., Furukawa, R. H., Brvenik, L. J., Sutton, W. R., and Ware, B. R. (1986). Fluorescence photoactivation and dissipation. *In* "Nucleocytoplasmic Transport" (R. Peters, and M. Trendelenburg, eds.), pp. 35–52. Springer-Verlag, Berlin.

Krafft, G. A., Sutton, W. R., and Cummings, R. T. (1988). Photoactivable fluorophores. 3. Synthesis and photoactivation of fluorogenic difunctionalized fluoresceins. *J. Am. Chem. Soc. 110,* 301–303.

Lando, L., and Zucker, R. S. (1989). "Caged calcium" in *Aplysia* pacemaker neurons. *J. Gen. Physiol. 93,* 1017–1060.

Lea, T. J., and Ashley, C. C. (1990). $Ca^{2+}$ release from the sarcoplasmic reticulum of barnacle myofibrillar bundles initiated by photolysis of caged $Ca^{2+}$. *J. Physiol. (London) 427,* 435–453.

Lea, T. J., Griffiths, P. J., Tregear, R. T., and Ashley, C. C. (1986). An examination of the ability of inositol 1,4,5-trisphosphate to induce Ca release and tension development in skinned muscle fibres of frog and crustacea. *FEBS Lett. 207,* 153–161.

Leslie, R. J., Saxton, W. M., Mitchison, T. J., Neighbors, B., Salmon, E. D., and

McIntosh, J. R. (1984). Assembly properties of fluorescein-labeled tubulin *in vitro* before and after fluorescence bleaching. *J. Cell Biol. 99*, 2146–2156.

Lester, H. A., and Nerbonne, J. M. (1982). Physiological and pharmacological manipulations with light flashes. *Annu. Rev. Biophys. Bioeng. 11*, 151–175.

Lewis, S. M., and Thomas, D. D. (1991). Microsecond rotational dynamics of spin-labeled Ca-ATPase during enzymatic cycling initiated by photolysis of caged ATP. *Biochemistry 30*, 8331–8339.

McAinsh, M. R., Brownlee, C., and Hetherington, A. M. (1990). Abscisic acid-induced elevation of guard cell cytosolic $Ca^{2+}$ precedes stomatal closure. *Nature (London) 343*, 186–188.

McCray, J. A., and Trentham, D. R. (1985). Rapid release of protons by photolysis of a biologically inert precursor, 2-hydroxyphenyl 1-(2-nitrophenyl)ethyl phosphate, a "caged proton." *Biophys. J. 47*, 406a.

McCray, J. A., and Trentham, D. R. (1989). Properties and uses of photoreactive caged compounds. *Annu. Rev. Biophys. Biophys. Chem. 18*, 239–70.

McRobbie, E. A. C. (1987). Ionic relations of guard cells. *In* "Stomatal Function" (E. Zeiger, G. D. Farquhar, and I. R. Cowan, eds.), pp. 125–162. Stanford University Press, Stanford.

Meldrum, R. A., Shall, S., Trentham, D. R., and Wharton, C. W. (1990). Kinetics and mechanism of DNA repair. Preparation, purification and some properties of caged dideoxynucleoside triphosphates. *Biochem. J. 266*, 885–890.

Meyer, R., and Stryer, L. (1988). Molecular model for receptor-stimulated calcium spiking. *Proc. Natl. Acad. Sci. USA 85*, 5051–55.

Meyer, T., and Stryer, L. (1990). Transient calcium release induced by successive increments of inositol 1,4,5-trisphosphate. *Proc. Natl. Acad. Sci. USA 87*, 3841–3845.

Meyer, T., Holowka, D., and Stryer, L. (1988). Highly cooperative opening of calcium channels by inositol 1,4,5-trisphosphate. *Science (Wash. DC) 240*, 653–656.

Milburn, T., Matsubara, N., Billington, A. P., Udgaonkar, J. B., Walker, J. W., Carpenter, B. K., Webb, W. W., Marque, J., Denk, W., McCray, J. A., and Hess, G. P. (1989). Synthesis, photochemistry, and biological activity of a caged photolabile acetylcholine receptor ligand. *Biochemistry 28*, 49–55.

Miledi, R., and Parker, I. (1989). Latencies of membrane currents evoked in *Xenopus* oocytes by receptor activation, inositol trisphosphate and calcium. *J. Physiol. (London) 415*, 189–210.

Minta, A., Kao, J. P. Y., and Tsien, R. Y. (1989) Fluorescent indicators for cytosolic calcium based on rhodamine and fluorescein chromophores. *J. Biol. Chem. 264*, 8171–8178.

Mitchison, T. J. (1989). Polewards microtubule flux in the mitotic spindle: Evidence from the photoactivation of fluorescence. *J. Cell. Biol. 109*, 637–652.

Mitchison, T., Evans, L., Schulze, E., and Kirschner, M. (1986). Sites of microtubule assembly and disassembly in the mitotic spindle. *Cell 45*, 515–527.

Morrison, H. A. (1969). The photochemistry of nitro and nitroso groups. *In* "The Chemistry of the Nitro and Nitroso Groups" (H. Feuer, ed.), pp. 165–213. Interscience, New York.

Muallem, S., Pandol, S., and Beeker, T. G. (1989). Horomone-evoked calcium release from intracellular stores is a quantal process. *J. Biol. Chem. 264*, 205–212.

Mulkey, R. M., and Zucker, R. S. (1991). Action potentials must admit calcium to evoke transmitter release. *Nature (London)* 350, 153–155.

Näbauer, M., and Morad, M. (1990). $Ca^{2+}$-induced $Ca^{2+}$ release as examined by photolysis of caged $Ca^{2+}$ in single ventricular myocytes. *Am. J. Physiol.* 258, C189–C193.

Nargeot, J., Nerbonne, J. M., Engels, J., and Lester, H. A. (1983). Time course of the increase in the myocardial slow inward current after a photochemically generated concentration jump of intracellular cAMP. *Proc. Natl. Acad. Sci. USA* 80, 2395–2399.

Nerbonne, J. M., Richard, S., Nargeot, J., and Lester, H. A. (1984). New photoactivatable cyclic nucleotides produce intracellular jumps in cyclic AMP and cyclic GMP concentrations. *Nature (London)* 310, 74–76.

Nichols, C. G., Niggli, E., and Lederer, W. J. (1990). Modulation of ATP-sensitive potassium channel activity by flash-photolysis of "caged-ATP" in rat heart cells. *Pflügers Arch.* 415, 510–512.

Niggli, E., and Lederer, W. J. (1991). Restoring forces in cardiac myocytes. Insight from relaxations induced by photolysis of caged ATP. *Biophys. J.* 59, 1123–1135.

Noel, J., and Capiod, T. (1991). Photolytic release of cAMP activates $Ca^{2+}$-dependent $K^+$ permeability in guinea-pig liver cells. *Pflügers Archiv* 417, 546–548.

Ogden, D. C., Capiod, T., Walker, J. W., and Trentham, D. R. (1900). Kinetics of the conductance evoked by noradrenaline, inositol trisphosphate or $Ca^{2+}$ in guinea-pig isolated hepatocytes. *J. Physiol. (London)* 422, 585–602.

Ostap, E. M., and Thomas, D. D. (1991). Rotational dynamics of spin-labeled F-actin during activation of myosin S1 ATPase using caged ATP. *Biophys. J.* 59, 1235–1241.

Parker, I., and Ivorra, I. (1990a). Inhibition by $Ca^{2+}$ of inositol trisphosphate-mediated $Ca^{2+}$ liberation: A possible mechanism for oscillatory release of $Ca^{2+}$. *Proc. Natl. Acad. Sci. USA* 87, 260–264.

Parker, I., and Ivorra, I. (1990b). Localized all-or-none calcium liberation by inositol trisphosphate. *Science (Wash. DC)* 250, 977–979.

Parker, I., and Miledi, R. (1989). Nonlinearity and facilitation in phosphoinositide signaling studied by the use of caged inositol trisphosphate in *Xenopus* oocytes. *J. Neurosci.* 9, 4068–4077.

Patton, W. F., Alexander, J. S., Dodge, A. B., Patton, R. J., Hechtman, H. B., and Shepro, D. (1991). Mercury-arc photolysis: A method for examining second messenger regulation of endothelial cell monolayer integrity. *Anal. Biochem.* 196, 31–38.

Payne, R., Walx, B., Levy, S., and Fein, A. (1988). The localization of calcium release by inositol trisphosphate in *Limulus* photoreceptors and its control by negative feedback. *Philos. Trans. R. Soc. London Ser. B 320*, 359–79.

Payne, R., Flores, T. M., and Fein, A. (1990). Feedback inhibition by calcium limits the release of calcium by inositol trisphosphate in *Limulus* ventral photoreceptors. *Neuron 4*, 547–555.

Peres, A. (1990). $InsP_3$- and $Ca^{2+}$-induced $Ca^{2+}$ release in single mouse oocytes. *FEBS Lett.* 275, 213–216.

Peres, A., Bertollini, L., and Racca, C. (1991). Characterization of $Ca^{2+}$ transients induced by intracellular photorelease of InsP$_3$ in mouse ovarian oocytes. *Cell Calcium 12*, 457–465.

Pillai, V. N. R. (1980). Photoremovable protecting groups in organic synthesis. *Synthesis*, 1-26.

Pillai, V. N. R. (1987). Photolytic deprotection and activation of functional groups. *Org. Photochem. 9*, 225–323.

Poenie, M., Alderton, J., Tsien, R. Y., and Steinhardt, R. A. (1985). Changes of free calcium levels with stages of the cell division cycle. *Nature (London) 315*, 147–149.

Poenie, M., Alderton, J., Steinhardt, R., and Tsien, R. Y. (1986). Calcium rises abruptly and briefly throughout the cell at the onset of anaphase. *Science (Wash. DC) 233*, 886–899.

Rink, T. J., and Merritt, J. E. (1990). Calcium signalling. *Curr. Opin. Cell Biol. 2*, 198–205.

Salmon, E. D., Leslie, R. J., Saxton, W. M., Karow, M. L., and McIntosh, J. R. (1984). Spindle microtubule dynamics in sea urchin embryos: Analysis using a fluorescein-labeled tubulin and measurement of fluorescence redistribution after photobleaching. *J. Cell Biol. 99*, 2165–2174.

Saxton, W. M., Stemple, D. L., Leslie, R. J., Salmon, E. D., Zavortink, M., and McIntosh, J. R. (1984). Tubulin dynamics in cultured mammalian cells. *J. Cell Biol. 99*, 2175–2186.

Schlichting, I., Rapp, G., John, J., Wittinghofer, A., Pai, E. F., and Goody, R. S. (1989). Biochemical and crystallographic characterizations of a complex of c-Ha-*ras* p21 and caged GTP with flash photolysis. *Proc. Natl. Acad. Sci. USA 86*, 7687–7690.

Schneider, S., Fink, M., Bug, R., and Schupp, H. (1991). Investigation of the photorearrangement *o*-nitrobenzyl esters by time-resolved resonance Raman spectroscopy. *J. Photochem. Photobiol. Sect. A: Chem. 55*, 329–338.

Schupp, H., Wong, W. K., and Schnabel, W. (1987). Mechanistic studies of the photorearrangement of *o*-nitrobenzyl esters. *J. Photochem. 36*, 85–97.

Scott, R. H., Wootton, J. F., and Dolphin, A. C. (1990). Modulation of neuronal T-type calcium channel currents by photoactivation of intracellular guanosine 5′-*O*(3-thio) triphosphate. *Neurosci. 38*, 285–294.

Smith, P. D., Liesegang, G. W., Berger, R. W., Czerlinksi, G., and Poldolsky, R. J. (1984). A stopped-flow investigation of calcium ion binding by ethylene glycol bis($\beta$-aminoethyl ether)-*N, N*′-tetraacetic acid. *Anal. Biochem. 143*, 188–195.

Somlyo, A. P., and Somlyo, A. V. (1990). Flash photolysis studies of excitation-contraction coupling, regulation, and contraction in smooth muscle. *Annu. Rev. Physiol. 52*, 857–874.

Stoddard, B. L., Bruhnke, J., Porter, N., Ringer, D., and Petsko, G. A. (1990a). Structure and activity of two photoreversible cinnamates bound to chymotrypsin. *Biochemistry 29*, 4871–4879.

Stoddard, B. L., Bruhnke, J., Koenigs, P., Porter, N., Ringe, D., and Petsko, G. A. (1990b). Photolysis and deacylation of inhibited chymotrypsin. *Biochemistry 29*, 8042–8051.

Taylor, C. W., and Potter, B. V. L. (1990). The size of inositol 1,4,5-trisphosphate-sensitive Ca$^{2+}$ stores depends on inositol 1,4,5-trisphosphate concentration. *Biochem. J. 266*, 189–194.

Taylor, S. J., and Exton, J. H. (1987). Guanine-nucleotide and hormone regulation of polyphosphoinositide phospholipase C activity of rat liver plasma membranes. *Biochem. J. 248*, 791–799.

Theriot, J. A., and Mitchison, T. J. (1991). Actin microfilament dynamics in locomoting cells. *Nature (London) 352*, 126–131.

Tsien, R. Y. (1980). New calcium indicators and buffers with high selectivity against magnesium and protons: Design, synthesis, and properties of prototype structures. *Biochemistry 19*, 2396–2404.

Tsien, R. Y. (1981). A non-disruptive technique for loading calcium buffers and indicators into cells. *Nature 290*, 527–528.

Tsien, R. W., and Tsien, R. Y. (1990). Calcium channels, stores, and oscillations. *Annu. Rev. Cell Biol. 6*, 715–60.

Tsien, R. Y., and Zucker, R. S. (1986). Control of cytoplasmic calcium with photolabile tetracarboxylate 2-nitrobenzhydrol chelators. *Biophys. J. 50*, 843–853.

Turner, A. D., Pizzo, S. V., Rozakis, G., and Porter, N. A. (1988). Photoreactivation of irreversibly inhibited serine proteases. *J. Am. Chem. Soc. 110*, 244–250.

Valdeolmillos, M., O'Neill, S. C., Smith, G. L., and Eisner, D. A. (1989). Calcium-induced calcium release activates contraction in intact cardiac cells. *Pflügers Archiv 413*, 676–678.

Vergara, J., Tsien, R. Y., and Delay, M. (1985). Inositol 1,4,5-trisphosphate: A possible chemical link in excitation–contraction coupling in muscle. *Proc. Natl. Acad. Sci. USA 82*, 6352–6356.

Vigers, G. P. A., Coue, M., and McIntosh, J. R. (1988). Fluorescent microtubules break up under illumination. *J. Cell Biol. 107*, 1011–1024.

Volpe, P., Salviatti, G., Di Virgilio, F., and Pozzan, T. (1985). Inositol 1,4,5-trisphosphate induces $Ca^{2+}$ release from SR of skeletal muscle. *Nature (London) 316*, 347–349.

Wadsworth, P., and Salmon, E. D. (1986). Analysis of the treadmilling model during metaphase of mitosis using fluorescence redistribution after photobleaching. *J. Cell Biol. 102*, 1032–1038.

Walker, J. W., and Trentham, D. R. (1988). Caged phenylephrine: Synthesis and photochemical properties. *Biophys. J. 53*, 596a.

Walker, J. W., McCray, J. A., and Hess, G. P. (1986). Photolabile protecting groups for an acetylcholine receptor ligand. Synthesis and photochemistry of a new class of *o*-nitrobenzyl derivatives and their effects on receptor function. *Biochemistry 25*, 1799–1805.

Walker, J. W., Somlyo, A. V., Goldman, Y. E., Somlyo, A. P., and Trentham, D. R. (1987). Kinetics of smooth and skeletal muscle activation by laser pulse photolysis of caged inositol 1,4,5-trisphosphate. *Nature (London) 327*, 249–252.

Walker, J. W., Gordon, P. R., McCray, J. A., and Trentham, D. R. (1988). Photolabile 1-(2-nitrophenyl)ethyl phosphate esters of adenine nucleotide analogues. Synthesis and mechanism of photolysis. *J. Am. Chem. Soc. 110*, 7170–7177.

Walker, J. W., Reid, G. P., and Trentham, D. R. (1989a). Synthesis and properties of caged nucleotides. *Methods Enzymol. 172*, 288–301.

Walker, J. W., Feeney, J., and Trentham, D. R. (1989b). Photolabile precursors of inositol phosphates. Preparation and properties of 1-(2-nitrophenyl)ethyl esters of *myo*-inositol 1,4,5-trisphosphate. *Biochemistry 28*, 3272–3280.

Ware, B. R., Brvenik, L. J., Cummings, R. R., Furukawa, R. H., and Krafft, G. A.

(1986). Fluorescence photoactivation and dissipation (FPD). *In* "Application of Fluorescence in the Biomedical Sciences." (D. L. Taylor, F. Lanni, R. Murphy, and A. Waggoner, eds.), pp. 141–157. Alan R. Liss, Inc., New York.

Wilcox, M., Viola, R. W., Johnson, K. W., Billington, A. P., Carpenter, B. K., McCray, J. A., Guzikowski, A. P., and Hess, G. P. (1990). Synthesis of photolabile "precursors" of amino acid neurotransmitters. *J. Org. Chem.* 55, 1585–1589.

Willner, I., Rubin, S., and Riklin, A. (1991a). Photoregulation of papain activity through anchoring photochromic azo groups to the enzyme backbone. *J. Am. Chem. Soc.* 113, 3321–3325.

Willner, I., Rubin, S., and Zor, T. (1991b). Photoregulation of α-chymotrypsin by its immobilization in a photochromic azobenzene copolymer. *J. Am. Chem. Soc.* 113, 4013–4014.

Wolniak, S. M. (1988). The regulation of mitotic spindle function. *Biochem. Cell Biol.* 66, 490–514.

Woods, N. M. Cuthbertson, K. S. R., and Cobbold, P. H. (1986). Repetitive transient rises in cytoplasmic free calcium in hormone-stimulated hepatocytes. *Nature (London)* 319, 600–602.

Woods, N. M., Cuthbertson, K. S. R., and Cobbold, P. H. (1987a). Agonist-induced oscillations in cytoplasmic free calcium concentration in single rat hepatocytes. *Cell Calcium* 8, 79–100.

Woods, N. M., Cuthbertson, K. S. R., and Cobbold, P. H. (1987b). Phorbol-ester-induced alterations of free calcium ion transients in single rat hepatocytes. *Biochem. J.* 246, 619–623.

Wootton, J. F., and Trentham, D. R. (1989). "Caged" compounds to probe the dynamics of cellular processes: Synthesis and properties of some novel photosensitive P-2-nitrobenzyl esters of nucleotides. *NATO ASI Ser. C* 272, 277–296.

Yip, R. W., Sharma, D. K., Giasson, R., and Gravel, D. (1985). Photochemistry of the o-nitrobenzyl system in solution: Evidence for singlet state intramolecular hydrogen abstraction. *J. Phys. Chem.* 89, 5328–5330.

Zhu, Q. Q., Schnabel, W., and Schupp, H. (1987). Formation and decay of nitronic acid in the photorearrangement of o-nitrobenzyl esters. *J. Photochem.* 39, 317–332.

Zucker, R. S. (1992). Effects of photolabile calcium chelators on fluorescent calcium indicators. *Cell Calcium* 13, 29–40.

Zucker, R. S., and Haydon, P. G. (1988). Membrane potential has no direct role in evoking neurotransmitter release. *Nature (London)* 335, 360–362.

# 3
# FLUORESCENCE RATIO IMAGING: ISSUES AND ARTIFACTS

**Gary R. Bright**
Department of Physiology and Biophysics
School of Medicine
Case Western Reserve University

I. INTRODUCTION
II. FLUORESCENT PROBES
  A. Probes for Measurement of Ions
    and Other Parameters
  B. Probe Characteristics
III. BASIC INSTRUMENTATION
  A. Wavelength Tuning
  B. Alternate Configurations
  C. Format of Data Acquisition

IV. EXPERIMENTAL ISSUES
  A. Sources of noise
  B. Artifacts
V. CONCLUSION
ACKNOWLEDGMENTS
REFERENCES

## I. INTRODUCTION

The cytoplasm of living cells is a highly organized and dynamic biochemical system. The plasma membrane does not just contain a collection of enzymes that carry out their respective functions but maintains a complex, dynamic environment. Cellular regulation of stimulus–response coupling involves a delicate interrelationship of second-messenger systems, including changes in concentrations of ions and cyclic nucleotides, phosphorylation events, lipid metabolism, and gene expression. An understanding of cell regulation requires an understanding of how this delicate balance is maintained and regulated. Because of this intricate organization, it is important to study specific events within the context of the cellular environment.

Optical Microscopy: Emerging Methods and Applications

Fluorescence spectroscopic imaging techniques provide a unique tool for studying cellular biochemistry at the single-cell level. The combination of sensitive fluorescent probes, selective for specific physiological parameters and the ability to map these parameters in 2, 3, and 4 dimensions, has resulted in new insights in cell biology. The most prominent applications have involved fluorescence ratio imaging for quantitation of concentrations of ions and other second messengers (see Bright *et al.,* 1989b; Tsien 1989).

The ability to measure responses in single cells has lead to the discovery of both sustained levels as well as regular oscillations in cytoplasmic calcium (see Jacob *et al.,* 1988; Chew, Chapter 5, this volume). Most of these oscillatory changes have been average changes over the whole cell. Some cells have been large enough to see subcellular oscillations that vary in position within the cells. These waves may be unidirectional, as in muscle cells (Wier *et al.,* 1987; Takamatsu and Wier, 1990a), or complex patterns, such as spirals, as found in Xenopus (Lechleiter *et al.,* 1991). In addition, the responses of cells within a population can be very heterogeneous (see Millard *et al.,* 1988; Bright *et al.,* 1989a; Gylfe *et al.,* 1991).

A particularly exciting direction is the study of cellular interactions. Poenie *et al.* (1987) followed the changes in calcium concentration upon interaction of cytotoxic cells with target cells. Further studies like this will no doubt open up many new concepts about cellular interactions that are important in cell and developmental biology.

This review is an extension and update of our previous discussion of fluorescence ratio imaging (Bright *et al.,* 1989b; see also Tsien and Harootunian, 1990; Moore *et al.,* 1990), which covered fundamental principles and validation of the ratio imaging approach. We consider issues that affect our ability to make reliable spatial and temporal measurements of cellular biochemical parameters using fluorescence ratio imaging. Progress has been made in applying ratio imaging to new areas, such as development of new dyes for measurement of pH, $Mg^{2+}$, $Cl^-$, and cAMP (cyclic adenosine-3′,5′-monophosphate), as well as approaches for measuring membrane potential, membrane fluidity, fluorescence energy transfer, and differential localization of cellular components.

A fluorescence ratio imaging system is an assemblage of chemistry, computer hardware, and software, basically a fluorescence spectrometer with a microscope as a sample chamber. An understanding of the limits of each of the parts and their degree of interrelationship is important in interpreting the results from such measurements. A basic ratio imaging experiment involves (1) loading cells with a fluorescent probe for the parameter of interest, (2) acquiring at least two images at two different wavelengths on the necessary time scale, and (3) analysis of the results.

## II. FLUORESCENT PROBES

The basis of the selectivity and sensitivity of fluorescence ratio imaging is the fluorescent probe. Ratio spectroscopy requires that the probe be differentially sensitive to the parameter of interest in at least two excitation or emission wavelengths. The excitation or emission at one wavelength may be either nonsensitive, much less sensitive, or sensitive in the opposite direction compared to the excitation or emission at the other wavelength. It is the relationship of the changes in excitation or emission at each wavelength with respect to each other that defines the measured response. Because the excitation or emission originates from the same volume, the ratio relationship normalizes for optical pathlength, local probe concentration, and loss of signal due to photobleaching. Whole spectra are not required since measurements at two wavelengths can reflect changes in the shape of the spectrum. The basic premise of the approach is that the ratio represents the relative proportion of two fluorescent species, although in practice this may not always be true. Another basic principle is that the probe is used as a tracer and the introduction of the probe does not effectively alter the endogenous equilibrium of components.

### A. Probes for Measurement of Ions and Other Parameters

There is a growing list of fluorescent probes suitable for fluorescence ratio imaging (Fig. 3.1; Haugland, 1989; Tsien, 1989; Tsien and Waggoner, 1990; see Poenie, Chapter 1, this volume). Both the types of biochemical parameters able to be measured as well as the variety of probes and wavelengths available for making a given measurement are actively expanding. These applications include simple extensions to the measurement of parameters, such as membrane potential (Montana *et al.*, 1989; Gross and Loew, 1989), chloride concentration (Verkman, 1990; Biwersi and Verkman, 1991), and differential localization of intracellular components (Pagliaro and Taylor, 1988; Luby-Phelps and Taylor, 1988) based on new fluorescent probes, as well as the use of different modes of imaging based on polarization, energy transfer, and fluorescence lifetimes. The availability of a wide range of probes with different wavelengths is important for maximizing our ability to correlate multiple parameters at the single-cell level (DeBiasio *et al.*, 1987; Waggoner *et al.*, 1989; Bright, 1993).

Fluorescence polarization is a technique useful for giving information about the microenvironment around a probe. Effectively, fluorescence polarization provides a measure of the rotational movement of a probe. Thus, anything that affects the rotational mobility, such as viscosity changes or binding of a probe to a cellular component, can be measured. The measure-

FIGURE 3.1

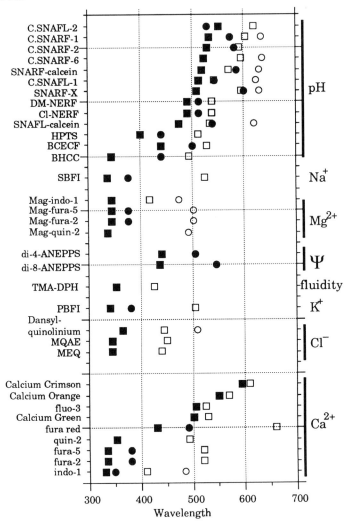

A compilation of a variety of probes for measuring cellular parameters using fluorescence imaging. Wavelengths were compiled from a number of sources (Molecular Probes, Inc; Lambda Probes and Diagnostics; DeBiasio *et al.*, 1987; Gross and Loew 1989; Dix and Verkman, 1990; Whitaker *et al.*, 1991; Biwersi and Verkman, 1991; Bright, unpublished observations). The actual wavelengths used in an experiment may vary from these since much of the data was derived from catalogs of peak wavelengths. The closed and open symbols represent excitation and emission wavelengths, respectively. Where multiple symbols exist, more than one wavelength was reported.

ment of polarization basically involves the excitation of the sample with polarized light and analysis of the emitted light both parallel and perpendicular to the excitation light. The polarization image is the ratio of the resulting images.

Placing a probe within the cytosol allows the viscosity of this compartment to be measured (Dix and Verkman, 1990). Placement of a probe within the membrane provides a means of monitoring changes in membrane fluidity (Fushimi et al., 1990; Florine-Casteel, 1990; see also Axelrod, 1989). One of the drawbacks of this approach is the rather extensive corrections that appear necessary to remove other factors that cause depolarization of the signal (Fushimi et al., 1990).

Fluorescence energy transfer (FET) is another spectroscopic technique that has proved amenable for measurement by ratio imaging. FET provides a means of measuring the distance between donor and acceptor molecules on a molecular scale (see Lakowicz, 1983; Herman, 1989). FET can be useful without measuring actual distances. The relative strength of the signal will be related to the distance. Uster and Pagano (1986) used FET to demonstrate the intermixing and dynamics of lipids in cells. Jovin and Arndt-Jovin (1989) have demonstrated several approaches to measuring FET in studying receptor dynamics and chromosome structure.

Recently, Adams et al. (1991) demonstrated the use of FET to quantitate the levels of cAMP in cells. cAMP-dependent protein kinase is a two-subunit enzyme that dissociates upon binding cAMP. The principle of the assay is to monitor the state of association–dissociation of the enzyme subunits. By placing a donor fluorophore on one subunit and an acceptor fluorophore on the other such that energy transfer is detected in the complex, the proportion of association–dissociation is reflected in the degree of FET. The proportion of dissociated enzyme will be related to the cAMP concentration.

## B. Probe Characteristics

A probe for use in ratio imaging may be as simple as a single free dye, a single dye attached to a carrier molecule, such as a dextran, or more elaborate, such as a carrier molecule or enzyme (see above) labeled with multiple dyes.

Probes composed of single dyes are the easiest to use. Probes composed of multiple dyes impose additional considerations in order to use them properly. For example, if the ratio of fluorescence from two separate dyes is used, differences in photobleaching rates of each dye must be taken into account. If not, this could lead to erroneous ratios.

There are several characteristics of the probe related to loading the probe into the cell, potential toxicity to the cell, compartmentalization, photobleaching dynamics, leakage or active transport from the cell, and environmental effects, which complicate calibration and determine the actual usefulness of a given probe. The appropriate means of delivery into the cell (see McNeil, 1989) will be dictated by the form of the probe. The delivery mechanism may be either chemical, exemplified by the use of acetoxymethyl esters (AM), or physical, exemplified by physical microinjection. A key issue is that the act of delivery does not alter the state of the cell. We have noted both chemical toxicity (Bright, unpublished) and phototoxicity (Bright et al., 1989a) when using AM esters, even very well-behaved ones, such as BCECF-AM. Complete deesterification is also an important problem since this can lead to fluorescence species that show parameter-independent fluorescence, such as the case with fura-2 (Scanlon et al., 1987). Different classes of esterases are likely responsible for the deesterification of different dyes. We have found that serum-starved fibroblasts label more easily with BCECF-AM than nonstarved cells (Bright et al., 1987). Exactly the opposite has been found for labeling of a variety of cells with fura-2-AM (Bright, unpublished observations).

One important goal of the delivery mechanism is to place a probe into the compartment of interest. The cytoplasm is a common target. Unfortunately, the probes do not always remain within this compartment. For example, fura-2-AM can become trapped within mitochondria, nucleus, sarcoplasmic reticulum, secretory granules, and lysosomes (see Moore et al., 1990). BCECF-AM has also been found to label mitochondria under some conditions (Bright et al., 1989a).

The progressive loss of fluorescence from cells means that the signal-to-noise ratio and thus the reliability of the measurement continuously degrades during an experiment. The loss in signal occurs primarily due to photobleaching and leakage or active transport of the probe from the cells (fura-2, DiVirglio et al., 1988; BCECF, Bright et al., 1989a). The ratio usually compensates for these losses (Bright et al., 1989b). Photobleaching can, however, be a more complicated matter if, as has been shown for fura-2 (Becker and Fay, 1987), a new chemical species can be created that is independent of the measured parameter. In this case, a more complicated equation for calculating the ratio may be necessary. The extent of bleaching is approximately proportional to the integral of the exciting intensity (Wells et al., 1990; Tsien and Waggoner 1990).

An approach to avoid leakage is to use dextran-based probes. These probes are too large to be transported through membrane transporters, do not interact with cytosolic components (Luby-Phelps et al., 1986), and are less toxic. The decrease in signal is due solely to photobleaching (Bright et al., 1989).

In many experiments, it is the relative changes in a parameter that is of particular interest rather than the absolute value. Thus, comparison of the raw ratio is adequate. When a good estimate of the actual values of the

FIGURE 3.2

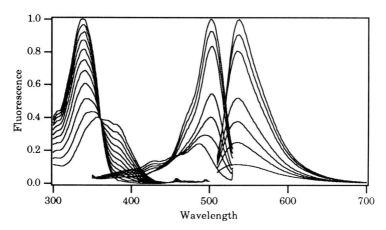

Titration spectra for fura-2 and BCECF. The excitation plus excitation and emission spectra of fura-2 and BCECF, respectively, are presented illustrating the parameter-dependent spectral changes.

parameters is important, then the system must be calibrated. Calibration can be a complicated procedure. The accuracy of these results is directly related to the quality of the calibration. Basically, calibrating involves the titration of the probe over a sufficient range to fully characterize the response of the probe. Hopefully, this can be performed *in situ* (i.e., within the environment where experimental values will be measured). Figure 3.2 illustrates the *in vitro* spectra of fura-2 and BCECF showing the requisite spectral changes in [Ca$^{2+}$] and pH, respectively. Figure 3.3 illustrates the

FIGURE 3.3

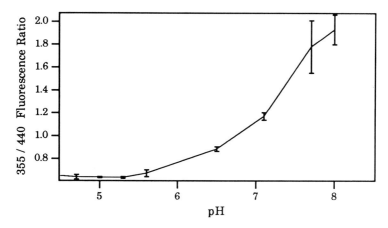

A standard curve for a coumarin-based pH probe illustrating a p$K_a$ ~7.3–7.4.

*in situ* standard curve for a coumarin-based pH probe (Lakshmanan and Bright, unpublished).

One of the major difficulties in calibrating the response is that other environmental factors can significantly affect the probe response. The fluorescence of dyes is affected by temperature, viscosity, competing ions, binding to cellular components, and other environmental factors. All of these have been specifically demonstrated, for example, in the case of fura-2 (see Konishi *et al.*, 1988; Tsien, 1989; Poenie, 1990; Roe *et al.*, 1990; Williams and Fay, 1990; Lattanzio, 1990; Owen, 1991). It may not be easy or feasible to separate some of the parameter-independent from the parameter-specific effects.

With the availability of new probes as well as extensions of the instrumentation (see below), it has become feasible to measure multiple parameters, temporally and spatially, in single-living cells (DeBiasio *et al.*, 1987; Waggoner *et al.*, 1989; Bright, 1993). By measuring other environmental changes that are simultaneously taking place with, for example, fura-2 measurement of $[Ca^{2+}]$, it is feasible to correct the $[Ca^{2+}]$ measurement. The approach has, in fact, been applied to the pH-dependence of fura-2-based $[Ca^{2+}]$ measurements (Lattanzio 1990; Martinez-Zaguilan *et al.*, 1991) in a nonimaging application. Imaging applications are in progress using fura-2 and carboxy SNARF-1 (see Figure 3.4; Bright unpublished; S. J. Morris, personal communication; see Morris, Chapter 6, this volume). As another example, Mag-fura-2, a Mg-sensitive dye, also responds to $[Ca^{2+}]$ changes.

FIGURE 3.4

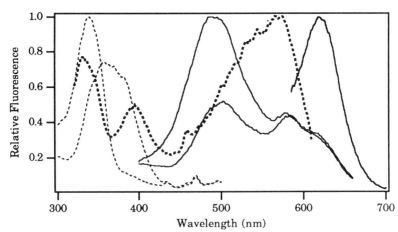

Spectra of a mixture of two probes for measuring $[Ca^{2+}]$ (fura-2) and pH (carboxy SNARF-1). Spectra were recorded at two concentrations of $Ca^{2+}$. Dashed lines indicate excitation spectra and solid lines indicate emission spectra.

Thus ideally one would monitor both $[Mg^{2+}]$ and $[Ca^{2+}]$ independently and correct the changes for $Ca^{2+}$-dependent changes. Data does not yet exist on the pH dependency of Mag-fura-2, but it would seem reasonable to suspect a dependency. How important it is under physiological conditions remains to be determined. The basic requirement to perform these measurements are (1) availability of probes with contrasting wavelengths and (2) ability of the instrumentation to discriminate between each probe (see below).

## III. BASIC INSTRUMENTATION

The fluorescence ratio imaging microscope is an integration of fluorescence microscopy, electronic imaging, digital image processing, and computer visualization. There are several practical discussions of this instrumentation (see Bright *et al.*, 1989b; Tsien and Harootunian 1990; Moore *et al.*, 1990; Bright 1993). The fundamental requirement for a ratio imaging system is the ability to tune the excitation and/or emission wavelengths. This is also one of the most difficult functions to perform simply given the optical train of the microscope (Fig. 3.5). Another very important consideration in the design of a system is the format of the data acquisition. This becomes particularly important when changes in the signal are near the same time scale as the limits of the instrumentation.

### A. Wavelength Tuning

There are at least three components of the optical path that are important in a discussion of tuning wavelengths, the excitation path, dichroic reflector, and emission path. Each of these can be independently tuned for maximal flexibility. The wavelength tuning devices used may vary depending on the type of light source and path. Several criteria are important in selecting a tuning device, including (1) wavelengths available, (2) bandpass, (3) tuning speed, (4) available aperture, and (5) whether the device is suitable for an imaging light path. The two fundamental approaches to wavelength tuning is either to use a series of static devices, such as interference filters, and to mechanically place the appropriate element in place or to use an acousto-optic or electrooptic device.

1. Excitation

The excitation path requires a device that can choose the appropriate wavelength from the light source. A variety of devices exists for tuning wavelengths from the source. The type of device depends on the design of the excitation path and the position within the system that the device is placed.

FIGURE 3.5

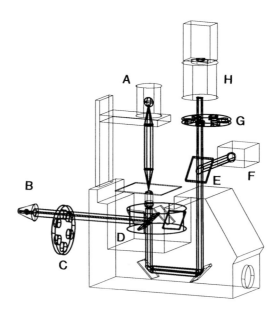

Diagramatic representation of the optical path through a fluorescence ratio imaging micro-scope. (A) Transmitted light source; (B) excitation light source; (C) excitation wavelength selection; (D) dichroic filter for separating excitation and emission wavelengths; (E) dichroic filter for separating fluorescence from transmitted light; (F) video-enhanced contrast camera; (G) emission wavelength selection; and (H) intensified camera. The transmitted light path op-erates simultaneously with the fluorescence and is limited to wavelengths greater than 700 nm (Foskett, 1988; see Foskett, chapter 8, this volume).

The excitation path is the easiest path to tune, and thus, the most common approach to ratio imaging has been to measure an excitation ratio.

The simplest means of selecting a narrow wavelength range from a source is to use an interference filter. These filters transmit a selected wave-length range by the generation of multiple constructive and destructive in-terferences. Interference filters have secondary transmission peaks at wave-lengths that are harmonics of the principal wavelength and must be blocked by additional coatings. High-quality filters are additionally blocked from ultraviolet to ~1.2 $\mu$m. A discussion of interference filters and associated issues can be found in Marcus (1988). The specifications of a filter include center wavelength and bandpass. The wider the bandpass, the greater the amount energy that is transmitted. Interference filters are easily designed for about any wavelength and bandpass. Filters are available in large aper-tures and are well suited for imaging light paths.

Filters used for excitation light paths are exposed to high light energy and heat that can cause cracks. In addition, the interference coatings can be damaged, leading to transmission of wavelengths outside the designed range and creation of a pattern in the excitation light. The object will then be excited with a pattern rather than a uniform field.

For tuning, several discrete filters with appropriate center wavelengths and bandpasses are necessary. The simplest means of automatically selecting the excitation wavelength is to place a filter wheel between the lamp housing condenser and the microscope. A filter wheel is a multiposition wheel with a different interference filter at each position. Most condenser lens systems supplied with microscopes have a sufficient adjustment range to accommodate the introduction of a filter wheel. The goal of the filter wheel is to position the appropriate filter into the light path on command. The wavelength tuning speed is related to the speed of the wheel. Filter wheels are of moderate speeds. The fastest we are aware of is ~30–50 ms for switching between adjacent filters (Sutter Instruments, Novato, CA; Ludl Electronic Products Inc., Hawthorne, NY). A more typical system (Ludl Electronics Products) is on the order of ~100 ms. A characteristic of filter wheels is that different times are required to select a wavelength depending on which position the filter occupies. Thus, when a quick change must be performed, the second filter should be in an adjacent position in the wheel.

Dual monochromators have been popular excitation sources since their introduction by Tsien et al. (1985). The modification added by Tsien et al. was the addition of a chopper mirror that alternately directed light from two lamp-monochromator systems into the microscope. Each monochromator is tuned to a given wavelength and then the object is alternately excited. This type of system provides an easy selection of a wide range of wavelengths. It is capable of switching between the two wavelengths very quickly. These systems are usually designed to switch at video rates (30 frames/s). Interference filters give spatially more uniform illumination than monochromator slits. Monochromators are not suitable for imaging light paths.

Monochromators have some unique advantages. Monochromators are freely and independently adjustable in wavelength and bandpass. It is easy to balance the excitation energy by varying the slit width (bandpass). For filters, which have fixed bandpass, intensity must be adjusted with neutral density filters. Monochromators also permit full spectral scanning, which can be helpful in diagnosing problems with the system and for investigating the environment of the probe within the cell cytoplasm.

A variation on this is to use two monochromators, one lamp and a mirror, either on a solenoid-based transport device (Linderman et al., 1990) or a galvanometer (Ryan et al., 1990), to select into which monochromator to direct the lamp. Newer versions of this approach use a single monochromator and lamp and provide fast wavelength changes based on a faster means for scanning the diffraction grating within the monochromator. Several commercial monochromators implement fast scanning (e.g., SLM Inc.; see also Hartmann and Verkman, 1992). In this device, only one monochromator and lamp is necessary. This approach is more flexible since it is not limited to just two wavelengths.

Another variation is the use of two lamps, two filters, two shutters, and a beam splitter (Marks *et al.*, 1988). The light from both lamps is combined by the beam splitter and projected into the illumination path of the microscope. Appropriate wavelength filters are placed at each lamp. Excitation wavelength alteration is accomplished by opening one shutter and closing the other. The speed of switching is limited only by the speed of the shutters. A 50% beam splitter is wasteful. An alternate approach is to replace the beam splitter with a dichroic reflector for more efficient separation. An extension of this is to cascade several dichroic splitters, separating the excitation light into multiple-wavelength bands with individual shutters.

A second basic mechanism for excitation wavelength tuning is to use an acoustooptic (AO) device. An AO device is effectively a tunable diffraction grating. A standing wave is established within a birefringent crystal using acoustic transducers. By adjusting the frequency and power of the acoustic energy, both wavelength and intensity can be varied on a microsecond time scale without any moving parts.

AO devices have existed for many years for use with lasers. Since they were designed for lasers, the aperture is very small. It is not easy to enlarge the aperture due to the physics involved. Thus, they are limited to being used with lasers and the wavelengths available are limited to those of the multiline laser. These laser-dependent systems are not useful in an imaging light path. These systems work by physically deflecting the diffracted light beam. Tuning occurs by placing a slit at the entrance to the microscope and adjusting the appropriate laser line onto the slit, thus illuminating the specimen. The temperature of this device must be well controlled, otherwise significant drift of the beam can occur. Using a multiline laser possessing the appropriate wavelengths for a given probe, this approach can be used for ratio-imaging experiments (Spring and Smith, 1987; Bright and Taylor, unpublished).

A second type of AO device has recently been described (see Kurtz *et al.*, 1987). This device, the acoustooptic tunable filter (AOTF) provides a large aperture, is suitable for broad-band noncoherent light, fast switching time, and intensity control. A visible light, $TeO_2$, noncolinear AOTF allows tuning without the need for polarizers. Wavelengths can be selected randomly within microseconds. These systems provide a spatially stable first-order beam, thus making it suitable for an imaging light path. They are capable of rapid spectral scans. On the surface they appear to provide an ideal mechanism for wavelength tuning. However, in practice, these devices have an extremely narrow bandpass (~0.5–1 nm in visible range) that varies with frequency. With this narrow of a bandpass, very powerful excitation sources are necessary. Although commercially available, they are currently very expensive. The potentially more useful ones are the small aperture devices for use with fiber optics (Infrared Fiber Systems Inc., Silver Spring, MD). AO devices also provide the ability to act as a multiwavelength device. Application of two frequencies to the device leads to transmission of two wavelengths.

More recently, a new device has become available that combines a birefringent filter and liquid crystals (VariSpec; Cambridge Research and Instrumentation Inc., Cambridge, MA). This device is currently available with a tunable range of 400–740 nm and 15-mm aperture. The bandpass is set at time of manufacture from 2–15 nm. This device may approach many of the promises of the AOTF. To date, this device has not been tested. Unfortunately the need for polarizers necessarily dictates a maximum of 50% transmission of nonpolarized light. Transmission efficiency is ~80% for the remaining 50% of nonpolarized light (Doug Benson, personal communication). Extinction ratios reported in the commercial literature are ~10,000:1. The large aperture makes this device useful for the imaging light path.

## 2. Dichroic Reflector

The dichroic reflector (DR) is an integral part of the reflected fluorescence microscope. It reflects the excitation light onto the specimen and transmits the emitted light to the detector. The DR is typically a long or short pass interference filter designed to operate at 45° in an imaging light path. For most ratio-imaging experiments involving one probe, there is no requirement to tune the DR. However, to extend the analysis to multiple parameters a means of tuning this position may be necessary (see Fig. 3.6 and 3.4). Since this filter must be mounted at 45°, a mechanical mechanism is the only approach available for switching. Of course, manual switching is possible but not very useful. One approach is to mount a filter wheel such that it positions the multiple DRs at 45° (Fig. 3.5).

The speed of wavelength switching is relatively slow since it is based on a filter wheel. The major disadvantage is the movement of an element within the imaging light path. The switching between two DRs can lead to a lateral shift in the images. These translations must be corrected in order to correlate the resulting images. In our device (Eastern Microscope, Raleigh, NC), this translation has been constant, thus the same translation corrections can be used throughout the experiment. An alternative is to include enough adjustments in the design of the filter wheel to align each of the DRs in the wheel. Usually it is more cost effective to make corrections with the image processor after acquisition.

FIGURE 3.6

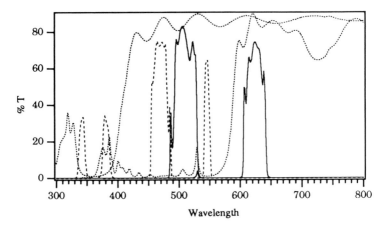

Example of two sets of filters for dual measurement of [$Ca^{2+}$] and pH using fura-2 and carboxy SNAFL-dextran, respectively. Dashed lines represent the excitation filters, dotted lines represent dichroic reflectors, and solid lines represent emission filters. This is an example of a filter set that would be used in combination with excitation, dichroic, and emission filter wheels for dual measurement.

For determining the appropriate registration factors for any combination of spectral images, we use a permanent slide with pan-fluorescent beads. These slides are made by placing Fluoresbrite polychromatic microspheres (#19111, 6 $\mu$m diameter, Polysciences Inc., Warrington, PA) into UV-curable optical cement (Norland Optical Adhesive 60; Edmund Scientific, Barrington, NJ). A drop of this suspension is placed onto a microscope slide. In time, overnight or more, the beads will float to the top of the optical cement since they are less dense than the medium. This process could possibly be accelerated by using a cytocentrifuge. Once the beads rise to the top, a coverslip is very carefully placed over the drop avoiding introduction of bubbles. This may be the point to replace the slide in a cytocentrifuge to force the beads to the region next to the coverslip. At this point, the slide is exposed to UV light. We have used a standard lamp used for visualizing thin-layer chromatographic plates for curing the cement.

An alternative to physically moving the DR is to use a multiple pass DR (Omega Optical, Brattleboro, NH; Chroma Technology Corp.; see Morris 1990). This type of DR reflects multiple wavelengths and transmits multiple wavelengths. Thus, no filter change is necessary if the appropriate reflection and transmission band are available. The advantages are speed, since no changes are needed, and no misalignment of resulting images. The disadvantage is that the combination of wavelengths available is significantly limited compared to the filter wheel system. This necessarily limits the combination of probes that can be used together. The ideal combination is to have a filter wheel with a multiple DR in one of the positions.

An additional approach has been presented by Spring (1990). An appropriately designed DR combined with polarized light can be used to simultaneously record a fluorescence image and a transmitted light, DIC image. The approach avoids the need for insertion of the DIC analyzer and subsequent intensity losses. Such reflectors will likely find utility in certain types of dual fluorescence measurement.

## 3. Emission

When using ratio emission probes, it is necessary to tune the emission wavelength rather than the excitation. All of the methods of tuning the excitation light path that are suitable for imaging, such as a filter wheel and ATOF, are appropriate for the emission light path. Since this path is an imaging path, the alignment and positioning of the filter are much more important than in the excitation path. When optical elements are moved within an imaging light path, translation corrections are typically needed. It may also be necessary to include more serious geometric corrections (Jericevic, *et al.*, 1989) due to chromatic aberrations and lensing effects. In addition, just as multipass DRs are available, multipass transmission filters are also available. Thus, by selective excitation, images of different probes can be recorded without the need to tune this path.

## B. Alternate Configurations

Recently, several alternative configurations beyond those previously mentioned have been described (Tamura *et al.*, 1989; Takamatsu and Wier, 1990b; Morris, 1990; see Morris, Chapter 6, this volume). These configurations were established in order to overcome the inherent speed limitations imposed by the need to sequentially switch excitation or emission wavelengths. The sequential acquisition of wavelength pairs is unsuitable when cellular or subcellular movement occurs on the time scale of the acquisition sequence (see below).

The goal of these configurations is to minimize any active wavelength tuning. This is done by using a ratio emission probe requiring only one excitation wavelength. The emission path of the microscope is split with a DR leading to two independent paths, one for each of the emission wavelengths. Emission filters are inserted for final filtering. Two separate detectors are used, one at each position. This provides simultaneous availability of the pair of images. Morris (personal communication; see Morris, Chapter 6, this volume) has recently extended this approach for measurement of two ratio emission probes involving four imaging paths and four detectors.

The major drawbacks to these approaches involve the complexity of alignment of multiple detectors and the expense of multiple detectors.

One of the considerations in this type of configuration is the way the data are acquired. Most image acquisition hardware provides a means of digitizing image from multiple cameras. However, this is misleading. These devices possess multiplexors that can direct data from any of several imputs to a single analog-to-digital (A/D) converter. Thus, these systems can really only digitize one image at a time. The switching between cameras is fast but, nevertheless, sequential. The acquisition of multiple images still occurs one at a time in sequence. Thus, to truly acquire the data in parallel, multiple digitizers must be triggered simultaneously by the host computer. Takamatsu and Wier (1990a,b) physically linked two complete image processors together. It is also possible to place multiple acquisition systems within one computer.

### C. Format of Data Acquisition

There are two basic types of two-dimensional detectors used in ratio imaging, intensified video-based cameras and cooled, slow-scan charge-coupled device (CCD) cameras. Discussions of the basics of low-light-level detectors and the various characteristics of cameras can be found in several places (Inoue, 1986; Bright and Taylor, 1986; Spring and Smith, 1987; Spring and Lowy, 1989; Bookman, 1990; Tsay *et al.,* 1991).

An important consideration in determining the appropriate detector is the format of the acquisition mechanism. There are two basic formats, sequential and parallel. Sequential is the most common found. What this means is that the image is acquired one pixel at a time in a scanning pattern. This is the principle used in standard video systems and all confocal microscopes. In standard video, the image imposed on the photocathode is read off by scanning an electron beam over the camera target in an overlapping interlaced format. The actual image is imposed on the faceplate in parallel, but the readout format is serial. This readout follows an interlaced pattern (Fig. 3.7). This pattern involves the scanning of two fields. The first field consists of the odd video lines and the second field is composed of the even lines. These interlaced fields make-up a single frame or image. There are several important timing issues inherent in this pattern that are illustrated in Fig. 3.7 (see also Inoue, 1986). This type of mechanism is the basis of most tube-style cameras, such as the SIT and ISIT (see Bright and Taylor, 1986) as well as the video-rate CCD. The improvement in the video-rate CCD is that there is no scanning error due to the solid-state structure of the detector.

An important consideration in this type of detector is that the "live" image is continuously imposed on the faceplate during the readout. This

FIGURE 3.7

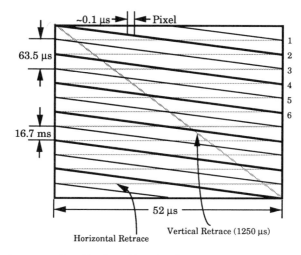

Interlace scanning pattern for video-based imaging detectors.

can lead to motion artifacts if movement in the scene occurs on the same time scale as the readout (see below). The same basic issue with scanning applies to confocal microscopes of all types. The closest similarity to video is the laser-scanning confocal microscope. The laser serially scans across the object one point at time with the signal recorded by a photomultiplier. The main difference is the scan in this case is noninterlaced. The scan rates in confocal microscopes vary depending on the beam deflection mechanism. For a thorough discussion of issues associated with confocal microscopes, see Pawley (1990).

The second basic mechanism is parallel detection as exemplified by the slow-scanned CCD detector. The image is imposed onto the CCD chip by opening a shutter for some period of time. Of course, any movement during the time the shutter is open will lead to blurring in that region. Once the shutter is closed, the data are read out serially. No further change will occur during the readout since the shutter is closed. More detailed discussion of slow-scan CCD detectors can be found in Aikens *et al.*, (1989).

## IV. EXPERIMENTAL ISSUES

The ability to extract quantitative information from the application of ratio imaging requires an assessment of the parameters of the biological question and the assemblage of the appropriate hardware and software appropriate for those parameters. There is no single right configuration of the equipment. There are usually difficult trade-offs involving image acquisition speed, spatial resolution, and image quality. The right compromise depends specifically on the biological question at hand.

The first, and most important component, is the biological question. What kind of data is expected? On what time scale are spectral changes expected to occur, milliseconds, seconds, minutes, or hours? Is there a spatial component to the signal expected? The easiest ratio experiment is to record a temporal series and spatially average the signal over the whole cell and plot a single trace representing the temporal response. This approach provides temporal resolution at the expense of spatial resolution. However, if both temporal and spatial information are desired, several issues must be considered that define the conditions under which the experimental data must be acquired. We consider several of these issues that affect our ability to make an accurate measurement.

The basic question to be addressed is what are some of the factors that affect our ability to quantitate at the single-pixel level. For example, if the cytoplasmic pH of a cell is clamped to a given pH with ionophore, then ideally the ratio of each of the pixels of the whole cytoplasm should be one value. This is never the case. There is always some local variation. These variations ultimately limit our spatial and temporal resolution.

## A. Sources of Noise

Noise limits the amount of information that can ultimately be extracted from an image. There are several sources of noise in an imaging system (Bright and Taylor, 1986). The sources may be electronic, optical, or chemical. Chemical noise refers to the spontaneous fluctuations in concentration of a chemical species that occurs on a microscopic scale. Given a microvolume, the measured intensity varies with the number of fluorescent molecules and thus increases and decreases as these molecules enter and leave that volume. The rate at which molecules enter or leave depends both on the diffusion coefficient (or flow velocity) and on the size of the observation region. If intermolecular interactions occur, the intensity change will also reflect the kinetics of the interaction. This is the basis of fluorescence correlation spectroscopy (see Elson and Qian, 1989).

The smaller the volume measured, the more significant these variations can be. Consider, for example, a pixel with dimensions of 0.3 $\mu$m $\times$ 0.3 $\mu$m $\times$ 3 $\mu$m. Thus, this pixel represents a sampled volume of 2.7 $\times$ $10^{-16}$ liters. For a fura-2-labeled cell, if the fura-2 concentration is on the order of 10 $\mu$M, then there are $\sim 1.6 \times 10^6$ molecules within this volume. Small fluctuations in this concentration of fura-2 will not likely lead to significant intensity changes. However, resting levels of free $[Ca^{2+}]$ are often in the 50–200 nM range. A 100 nM free $[Ca^{2+}]$ means that there are only $\sim 16$ $Ca^{2+}$ ions within this volume. Thus, random fluctuation of only a few

$Ca^{2+}$ ions can lead to a significant change in the number of $Ca^{2+}$ present within the volume. Thus, one might expect an intensity fluctuation to be related to the fluctuation in the number of $Ca^{2+}$ ions present within the volume. Any fluctuation will be a function of the diffusion coefficient for $Ca^{2+}$ and fura-2 and the kinetics of the interaction between $Ca^{2+}$ and fura-2.

The kinetics of association and dissociation of $Ca^{2+}$ with fura-2 have been studied both *in vitro* (Jackson *et al.*, 1987; Kao and Tsien, 1988) and *in vivo* (Baylor and Hollingworth, 1988; Klein *et al.*, 1988). The kinetics of the dye measured *in vitro* predict a response with $t_{1/2}$ of ~3 ms to a sudden jump in [Ca]. Possible interactions with cellular components (Konishi *et al.*, 1988; Klein *et al.*, 1988) may slow the kinetics of both association and dissociation. The viscosity of cytoplasm has been reported to be anywhere from 1.2–1.4 times (Fushimi and Verkman, 1991) to 2–6 times (Luby-Phelps *et al.*, 1986) the viscosity of water. The practical aspects of this are that fluctuations in intensity could occur on a pixel scale based on consideration of (1) concentration of dye and ion, (2) the diffusion coefficient of the species in the cytosol, and (3) the kinetics of the interaction on the time scale of a single video frame.

The real magnitude of these fluctuations cannot be assessed until all of these parameters are measured within the same model system. It seems reasonable, using the above numbers, that at least some of the intensity variation found in a single video image (30-ms sample) may be due to these effects, which we refer to generally as chemical noise. Whether the magnitude of these effects is sufficient to be detected depends on the noise of the detection process. Depending on the focus of the experimenter, these fluctuations may or may not be considered to be noise but, in fact, may be an interesting signal.

The noise of a low-light-level video camera is mostly due to thermionic emission from the photocathode, shot noise, and to subsequent amplifier stages (Csorba, 1985; Bright and Taylor, 1986). A dominant form of noise in imaging systems is thermal noise. This refers to the spontaneous emission of electrons from the photocathode due to the temperature being sufficiently high as to provide enough energy for the electron to overcome the so called work function of the photocathode. This is readily visible in intensified video cameras as a speckle pattern of randomly distributed bright pixels. It is not uncommon to have the thermal noise increase during an experiment due to heat generated by the camera itself, other equipment in the room, as well as by people. Since low-light-level imaging experiments require a closed, dark environment, heat tends to build up in the laboratory due to inadequate ventilation. One of the advantages of the cooled CCD cameras is the built-in temperature regulation. Strict control of the environment of intensified video cameras is important when attempting to make high-precision measurement. Ryan *et al.* (1990) found that the noise in their

[$Ca^{2+}$] measurements was attributable to the shot noise associated with the number of detected photons. This translated into ~7.5% uncertainty in the ratio.

To overcome some of these limits on SNR, the approach has been to sacrifice spatial and temporal resolution for the resultant improvement in signal-detection ability (Bright and Taylor, 1986; Bright *et al.*, 1987; Spring and Lowy, 1989). Averaging can be done temporally by averaging incoming images that will improve the SNR by the square root of the number of frames if the noise is not correlated from image to image as is the case for thermal noise of the detector. Averaging can also be done spatially to reduce the effects of both what we have referred to as chemical noise as well as detector noise. The spatial averaging effectively increases the sampled volume such that the existing fluctuation will have a smaller impact.

The accuracy of a fluorescence measurement improves as the sample area and sampling time increase (Bright *et al.*, 1987; Spring and Lowy, 1989). Spring and Lowy (1989) have illustrated the relationship between the number of frames averaged and the number of pixels spatially averaged within a frame with respect to achieving an acceptably accurate measurement. As expected, the fewer frames averaged, a larger sampling area must be used. Increasing the sampling area will necessarily decrease the spatial resolution. The actual limits will be determined by the intensity of the object. Thus, the appropriate operating conditions must be determined for each type of experiment. Increasing the size of the sample area improved measurement precision without sacrificing speed, but it decreased resolution. A small window required more temporal averaging resulting in high spatial but poor temporal resolution. This data was collected using a structureless object (a solution).

## B. Artifacts

The term "real time" is often used to describe the acquisition rate of an imaging system. The question is what is real time? In standard video, real time typically means 30 frames/s. There are detectors currently available capable of ~1000 frames/s, thus, are these faster than real time? How can you go faster than real time? Real time is a relative term. Its use is very confusing. Typically it means something close to video rates. However, as will be discussed, real time does not necessarily mean the lack of sampling artifacts.

In addition to the basic types of noise, there are several sampling artifacts that serve to decrease the SNR by increasing the variations in a pixel. Living cells are constantly in motion, both globally and on a subcellular level. The organelles are constantly in motion at a minimum due to Brownian motion but also from directed transport. Many aspects of this motion occur on the time scale of the video acquisition and/or the wavelength tuning time. This leads to motion artifacts. Consider a 1–2-$\mu$m fluorescent vesicle within the cytosol that has a viscosity of ~4 cp. In the simplest case, this vesicle undergoes Brownian motion but may also possess some directed motion. If the rate of motion is on the same time scale as the video interlace (see Fig. 3.7), then a local misregistration can occur leading to increased variations where none should exist. Figure 3.8 illustrates that one portion of the vesicle is scanned during the first (odd) field. If we assume, in this example, that the vesicle moves ~25–30% of its diameter during the retrace time, then the second image is shifted to a new position by the time the second (even) field is scanned. This leads to a local misregistration. The magnitude of this effect is related to the rate of movement, magnification, and the format of the data acquisition. Parallel detection, such as with a slow-scan CCD, would not exhibit this artifact, assuming a sufficiently short shutter open time.

A second form of local misregistration can occur on a slightly longer time scale. If the movement of the vesicle is on the order of the time to switch excitation wavelengths, for a ratio excitation probe, then it is possible to have the vesicle image in the numerator image displaced several diameters from the vesicle image in the denominator image. This leads to a total local misregistration. In one, the vesicle pixels are divided by nonvesicle pixels and vice versa in the other area. This leads to two areas that have increased artifactual variations in ratio intensity. This effect is also a function of the rate of movement and magnification but is not a function of the format of the data acquisition. Thus, it cannot be eliminated by using parallel detection. The only way to remove these motion artifacts is to use one

FIGURE 3.8

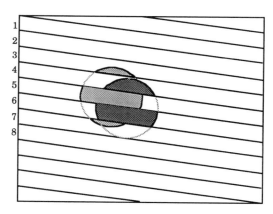

Movement artifact when a vesicle is moving on the same time scale that the video detector is scanning.

of the alternative configurations previously discussed. The configuration would include two simultaneous emission paths coupled to two parallel detection systems.

There are other artifacts that can lead to misinterpretation of the data. With the recent interest in oscillations in cellular parameters, such as $[Ca^{2+}]$, two deserve specific coverage. The first situation involves a conventional filter wheel-based ratio excitation system. Many filter wheel systems provide a retaining ring to hold the filter in place. Use of a set screw to hold a filter is avoided since it can lead to cracking of the filter due to size changes from the heating and cooling of the filter in the excitation light. As the filter wheel oscillates back and forth placing the appropriate filter into the excitation path, the unsecured filter can rotate within the mount. Since the filter is usually exposed to the same pattern of light coming from the excitation source, in time a pattern is often burned into the filter. Thus, there can be some pattern to the excitation light. These are very subtle patterns and usually do not have a dramatic effect on a ratio measurement. This is one of the benefits of the excitation ratio. However, if there is a change in this pattern from image to image due to rotation of the filter, significant systematic variation can occur. For example, Fig. 3.9 illustrates an apparent oscillation in the ratio. In this case, we were expecting to see oscillations in the biological parameter. However, these observed oscillations were totally artifac-

FIGURE 3.9

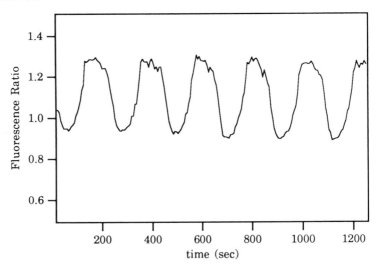

An example of an apparent oscillation in a cellular parameter due to an artifact. This artifact was caused by the excitation filter rotating within its mount while the filter wheel positioned the filter into and out of the excitation path.

tual due to a rotating filter. We normally secure the filters into the holder with pieces of $3 \times 5$ in. index cards. This paper is compressible and a full ring can be used to wedge the filter symmetrically into the mount. A piece of this paper had come loose to cause the effect seen in Fig. 3.9. An alternative approach is to use a small amount of rubber cement. This artifact is an illustration of how a relatively minor variable in the instrumentation could lead to erroneous results. More subtle effects are possible if less dramatic changes occur, such as only a slight shift in the filter.

An alternate means of seeing oscillations when none exist is due to electrical interference. The principle is that any electrical signal that is either in the video frequency range or produces a signal within this range after interfering with another signal will alter the video signal. That is, new signals are mixed in with the video signal. Thus, variations within the video signal may exist that are independent of the object being observed. This has been known to occur when new equipment, such as a large compressor, is added to the same electrical circuits as the laboratory (Bright, unpublished observations) as well as to inadequate grounding of the video camera (Doug Benson, personal communication). Given the age of many research buildings, the stability of the electrical systems is a serious concern. This problem usually manifests itself visually as lines or bands slowly moving down the video image. The size of the band will depend on the nature of the source. If a sampling box is placed on the screen and the values plotted in real time, apparent oscillations occur upon the band traversing the sampling box. Some form of isolation and filtering is desirable for the video detectors. These devices typically have the least sophisticated power regulation systems of the entire ratio imaging system.

## V. CONCLUSION

Fluorescence spectroscopic imaging techniques, specifically fluorescence ratio imaging, are powerful tools for measuring a variety of biochemical parameters at the single-cell level, in essence, single-cell biochemistry. The approach has expanded in recent years, both in terms of applications to new biochemical parameters as well as on extraction of the spatial component of the signals. This access to spatial information has resulted in several new concepts in cell regulation. The ability to extract spatial information requires an understanding of the chemical, optical, electronic, and photophysics factors involved.

ACKNOWLEDGMENTS

I would like to thank Doug Benson and Stephen Morris for discussion of unpublished data. This work has been supported, in part, by grants from the Diabetes

Association of Greater Cleveland (#322-89) and The Council for Tobacco Research, U.S.A. (2683).

REFERENCES

Adams, S. R., Harootunian, A. T., Buechler, Y. J., Taylor, S. S., Tsien, R. Y. (1991). Fluorescence ratio imaging of cyclic AMP in single cells. *Nature 349*, 694–697.

Arndt-Jovin, D. J., Robert-Nicoud, M., Kaufman, S. J., Jovin, T. M. (1985). Fluorescence digital imaging microscopy in cell biology. *Science 230*, 247–256.

Aikens, R. S., Agard, D. A., Sedat, J. W. (1989). Solid state imagers for optical microscopy. *Meth. Cell Biol. 29*, 292–314.

Axelrod, D. (1989). Fluorescence polarization microscopy. *Methods in Cell Biology* 30, 333–352.

Baylor, S. M., and Hollingworth, S. (1988). Fura-2 calcium transients in frog skeletal muscle fibres. *J. Physiol. 403*, 151–192.

Becker, P. L., and Fay, F. S. (1987). Photobleaching of fura-2 and its effect on the determination of calcium concentration. *Am. J. Physiol. 253*, C613-C618.

Biwersi, J., and Verkman, A. S. (1991). Cell permeable fluorescent indicators for chloride. *Biochem. 30*, 7879–7882.

Bookman, R. J. (1990). Temporal response characterization of video cameras. *In* "Optical Methods in Biology. (B. Herman and D. Jacobson, eds.), pp. 235–250, Wiley-Liss, NY.

Bright, G. R. (1993). Multiparameter Imaging of Cellular Function. *In* "Fluorescent Probes for Biological Function of Living Cells—A Practical Guide." (W. T. Mason and G. Relf, eds), Academic Press, NY.

Bright, G. R., and Taylor, D. L. (1986). Imaging at low light level in fluorescence microscopy. *In* "Applications of Fluorescence in the Biomedical Sciences." (Taylor, Waggoner, Murphy, Lanni, and Birge, eds.) pp. 257–288. Alan R. Liss, NY.

Bright, G. R., Fisher, G. W., Rogowska, J., and Taylor, D. L. (1987). Fluorescence ratio imaging microscopy: Temporal and spatial measurements of cytoplasmic pH. *J. Cell Biol. 104*, 1019–1033.

Bright, G. R., Whitaker, J., Haugland, R., and Taylor, D. L. (1989a). Heterogeneity of changes in cytoplasmic pH upon serum stimulation of quiescent fibroblasts. *J. Cell. Physiol.* 140, 410–419.

Bright, G. R., Fisher, G. W., Rogowska, J., and Taylor, D. L. (1989b). Fluorescence ratio imaging microscopy. *Methods in Cell Biology 30*, 157–192.

Csorba, I. P. (1985). "Image Tubes." Howard W. Sams and Co., Inc., Indianapolis, IN.

DeBiasio, R., Bright, G. R., Ernst, L. A., Waggoner, A. S., and Taylor, D. L. (1987). Five parameter fluorescence imaging: Wound healing of living Swiss 3T3 cells. *J. Cell Biol. 105*, 1613–1622.

DiVirglio, F., Steinberg, T. H., Swanson, J. A., and Silverstein, S. C. (1988). Fura-2 secretion and sequestration in macrophages. *J. Immunol.* 140, 915–920.

Dix, J. A., and Verkman, A. S. (1990). Spatially-resolved mapping of fluorescence anisotropy in single cells: Application to cytoplasmic viscosity. *Biophys. J.* 57, 231–240.

Elson, E. L., and Qian, H. (1989). Interpretation of fluorescence correlation spectroscopy and photobleaching recovery in terms of molecular interactions. *Meth. Cell Biol.* 30, 307–332.

Florine-Casteel, K. (1990). Phospholipid order in gel- and fluid-phase cell-size liposomes measured by digitized video fluorescence polarization microscopy. *Biophys. J.* 57, 1199–1215.

Foskett, J. K. (1988). Simultaneous Normarski and fluorescence imaging during video microscopy of cells. *Am. J. Physiol.* 255, C566–C571.

Fushimi, K., and Verkman, A. S. (1991). Low viscosity in the aqueous domain of cell cytoplasm measured by picosecond polarization microfluorometry. *J. Cell Biol.* 112, 719–725.

Fushimi, K., Dix, J. A., and Verkman, A. S. (1990). Cell membrane fluidity in the intact kidney proximal tubule measured by orientation-independent fluorescence anisotropy imaging. *Biophys. J.* 57, 241–254.

Gross, D., and Loew, L. M. (1989). Fluorescent indicators of membrane potential: Microspectrofluorometry and imaging. *Meth. Cell Biol.* 30, 193–219.

Gylfe, E., Grapengiesser, E., and Hellmann, B. (1991). Propagation of cytoplasmic $Ca^{2+}$ oscillations in clusters of pancreatic ß-cells exposed to glucose. *Cell Calcium* 12, 229–240.

Hartmann, T., and Verkman, A. S. (1992). Construction and performance of a rapid scan monochromator for multiwavelength fluorimetry. *Anal. Biochem.* 200, 139–142.

Haugland, R. P. (1989). "Molecular Probes: Handbook of fluorescent probes and research chemicals." Molecular Probes, Inc., Eugene, OR.

Herman, B. (1989). Resonance energy transfer microscopy. *Meth. Cell Biol.* 30, 220–245.

Inoue, S. (1986). "Video Microscopy." Plenum Press, New York.

Jackson, A. P., Timmerman, M. P., Bagshaw, C. R., and Ashley, C. C. (1987). The kinetics of calcium binding to fura-2 and indo-1. *FEBS Lett.* 216, 35–39.

Jacob, R., Merritt, J. E., Hallam, T. J., and Rink, T. J. (1988). Repetitive spikes in cytoplasmic calcium evoked by histamine in human endothelial cells. *Nature* 335, 40–45.

Jericevic, Z., Wiese, B., Bryan, J., and Smith, L. C. (1989). Validation of an imaging system. *Meth. Cell Biol.* 30, 48–84.

Jovin, T. M., and Arndt-Jovin, D. (1989). *In* "Cell Structure and Function by Microspectrofluorometry." (E. Kohen and J. G. Hirschberg, eds.), pp. 99–118. Academic Press, NY.

Kao, J. P., and Tsien, R. Y. (1988). $Ca^{2+}$ binding kinetics of fura-2 and azo-1 from temperature jump relaxation measurements. *Biophys. J.* 53, 635–639.

Klein, M. G., Simon, J., Szucs, G., and Schneider, M. F. (1988). Simultaneous recording of calcium transients in skeletal muscle using high- and low-affinity calcium indicators. *Biophys. J.* 53, 971–988.

Konishi, M., Olson, A., Hollingworth, S. and Baylor, S. M. (1988). Myoplasmic binding of fura-2 investigated by steady-state fluorescence and absorbance measurements. *Biophys. J. 54*, 1089–1104.

Kurtz, I., Dwelle, R., and Katzka, P. (1987). Rapid scanning fluorescence spectroscopy using an acousto-optic tunable filter. *Rev. Sci. Intr. 58*, 1996–2003.

Lakowicz, J. R. (1983). "Principles of Fluorescence Spectroscopy." Plenum Press, New York.

Lattanzio, F. (1990). The effects of pH and temperature on fluorescent calcium indicators as determined with chelex-100 and EDTA buffer systems. *Biochem. Biophys. Res. Commun. 171*, 102–108.

Lechleiter, J., Girard, S., Peralta, E., and Clapham, D. (1991). Spiral calcium wave propagation and annihilation in *Xenopus laevis* Oocytes. *Science 252*, 123–126.

Linderman, J. J., Harris, L. J., Slakey, L. L., and Gross, D. J. (1990). Charge-coupled device imaging of rapid calcium transients in cultured arterial smooth muscle cells. *Cell Calcium 11*, 131–144.

Luby-Phelps, K, and Taylor, D. L. (1988). Subcellular compartmentalization by local differentiation of cytoplasmic structure. *Cell Motil. Cytoskeleton 10*, 28–37.

Luby-Phelps, K., Taylor, D. L., and Lanni, F. (1986). Probing the structure of cytoplasm. *J. Cell Biol. 102*, 2015–2022.

Marcus, D. A. (1988). High-performance optical filters for fluorescence analysis. *Cell Motil. Cytoskeleton 10*, 62–70.

Marks, P. W., Kruskal, B. A., and Maxfield, F. R. (1988). Simultaneous addition of EGF prolongs the increase in cytosolic free calcium seen in response to bradykinin in NRK-49F cells. *J. Cell. Physiol. 136*, 519–525.

Martinez-Zaguilan, R., Martinez, G. M., Lattanzio, F., and Gillies, R. J. (1991). Simultaneous measurement of intracellular pH and $Ca^{2+}$ using the fluorescence of SNARF-1 and fura-2. *Am. J. Physiol. 260*, C297–C307.

McNeil, P. L. (1989). Incorporation of macromolecules into living cells. *Meth. Cell Biol. 29*, 153–174.

Millard, P. J., Gross, D., Webb, W. W., and Fewtrell, C. (1988). Imaging asynchronous changes in intracellular $Ca^{2+}$ in individual stimulated tumor mast cells. *Proc. Natl. Acad. Sci. U.S.A. 85*, 1854–1858.

Montana, V., Farkas, D. L., and Loew, L. M. (1989). Dual-wavelength ratiometric fluorescence measurements of membrane potential. *Biochem. 28*, 4536–4539.

Moore, E. D. W., Becker, P. L. Fogarty, K. E., Williams, D. A., and Fay, F. S. (1990). $Ca^{2+}$ imaging in single living cells: Theoretical and practical issues. *Cell Calcium 11*, 157–179.

Morris, S. J. (1990). Real-time multi-wavelength fluorescence imaging of living cells. *BioTechniques 8*, 296–308.

Owen, C. S. (1991). Spectra of intracellular fura-2. *Cell Calcium 12*, 385–393.

Pagliaro, L., and Taylor, D. L. (1988). Aldolase exists in both the fluid and solid phases of cytoplasm. *J. Cell Biol. 107*, 981–991.

Pawley, JB. (1990). "Handbook of Biological Confocal Microscopy." Plenum Press, NY.

Poenie, M. (1990). Alteration of intracellular fura-2 fluorescence by viscosity: A simple correction. *Cell Calcium 11*, 85–92.

Poenie, M., Tsien, R. Y., and Schmitt-Verhulst, A. M. (1987). Sequential activation and lethal hit measured by $[Ca^{2+}]_i$ in individual cytolytic T cells and targets. *EMBO J. 6*, 2223–2232.

Roe, M. W., Lemasters, J. J., and Herman, B. (1990). Assessment of fura-2 for measurements of cytosolic free calcium. *Cell Calcium 11*, 63–74.

Ryan, T. A., Millard, P. J., and Webb, W. W. (1990). Imaging $[Ca^{2+}]_i$ dynamics during signal transduction. *Cell Calcium 11*, 145–156.

Scanlon, M., Williams, D. W., and Fay, F. S. (1987). A $Ca^{2+}$-insensitive form of fura-2 associated with polymorphonuclear leukocytes. *J. Biol. Chem. 262*, 6308–6312.

Spring, K. R. (1990). Quantitative imaging at low light levels: Differential interference contrast and fluorescence microscopy without significant light loss. *In* Optical Methods in Biology." (B. Herman and D. Jacobson, eds.) p. 513, Wiley-Liss, NY.

Spring, K. R., and Smith, P. D. (1987). Illumination and detection systems for quantitative fluorescence microscopy. *J. Microsc. 147*, 265–278.

Spring, K. R., and Lowy, R. J. (1989). Characteristics of low light level television cameras. *Meth. Cell Biol. 29*, 270–291.

Takamatsu, T., and Wier, W. G. (1990a). Calcium waves in mammalian heart: Quantification and origin, magnitude, waveform, and velocity. *FASEB J. 4*, 1519–1525.

Takamatsu, T., and Wier, W. G. (1990b). High temporal resolution video imaging of intracellular calcium. *Cell Calcium 11*, 111–120.

Tamura, K., Yoshida, S., Fujiwake, H., Watanabe, I., and Sugawara, Y. (1989). Simultaneous measurement of cytosolic free calcium concentration and cell circumference during contraction, both in a single rat cardiomuscular cell, by digital imaging microscopy with indo-1. *Biochem. Biophys. Res. Commun. 162*, 926–932.

Tsay, T-.T., Inman, R., Wray, B., Herman, B., and Jacobson, K. (1991). Characterization of low-light-level cameras for digitized video microscopy. *J. Microsc 160*, 141–159.

Tsien, R. Y. (1989). Fluorescent indicators of ion concentrations. *Meth. Cell Biol. 30*, 127–156.

Tsien, R. Y., and Harootunian, A. T. (1990). Practical design criteria for a dynamic ratio imaging system. *Cell Calcium 11*, 93–110.

Tsien, R. Y., and Waggoner, A. (1990). Fluorophores for confocal microscopy: Photophysics and photochemistry. *In* "Handbook of Biological Confocal Microscopy." (J. B. Pawley, ed.), pp. 169–178. Plenum Press.

Tsien, R. Y., Rink, T. J., and Poenie, M. (1985). Measurement of cytosolic free $Ca^{2+}$ in individual small cells using fluorescence microscopy with dual excitation wavelengths. *Cell Calcium 6*, 145–157.

Uster, P. S., Pagano, R. E. (1986). Resonance energy transfer microscopy: Observations of membrane-bound fluorescent probes in model membrane and in living cells. *J. Cell Biol. 103*, 1221–1234.

Verkman, A. S. (1990). Development and biological application of chloride-sensitive fluorescent indicators. *Am. J. Physiol. 259*, C375–C388.

Waggoner, A. S., DeBiasio, R., Bright, G. R., Ernst, L. A., Conrad, P., Galbraith,

W., and Taylor, D. L. (1989). Multiple Parameter Microscopy. *Methods in Cell Biology 30,* 449–478.

Wells, K. S., Sandison, D. R., Strickler, J., and Webb, W. W. (1990). Quantitative fluorescence imaging with laser scanning confocal microscopy. *In* "Handbook of Biological Confocal Microscopy." (J. B. Pawley, ed.), pp. 27–39. Plenum Press.

Whitaker, J. E., Haugland, R. P., and Prendergast, F. G. (1991). Spectral and photophysical studies of Benzo[c]xanthene dyes: Dual emission pH sensors. *Anal. Biochem. 194,* 330–344.

Wier, W. G., Cannell, M. B., Berlin, J. R., Marban, E., Lederer, W. J. (1987). Cellular and subcellular heterogeneith of $[Ca^{2+}]_i$ in single heart cells revealed by fura-2. *Science 235,* 325–328.

Williams, D. A., and Fay, F. S. (1990). Intracellular calibration of the fluorescent calcium indicator fura-2. *Cell Calcium 11,* 75–84.

# 4
# Sampling Characteristics of CCD Video Cameras

**Richard J. Bookman and Frank T. Horrigan**
Department of Molecular and Cellular Pharmacology
University of Miami School of Medicine

I. INTRODUCTION
II. SOME TERMS AND METHODS
  A. RS-170 Signals
  B. Signal Pathways for Digitized CCD Data
  C. Equipment
  D. Data Acquisition Techniques
III. IMPLEMENTATION OF CCDs IN RS-170 CAMERAS
  A. General Comparison of ILT and FT CCDs

B. Interlacing and Resolution
C. Temporal Sampling Characteristics of an Interline-Transfer CCD
D. Temporal Sampling Characteristics of a Frame-Transfer CCD
IV. SUMMARY AND PROSPECTS
ACKNOWLEDGMENTS
REFERENCES

## I. INTRODUCTION

The rapid advance of many solid-state technologies has made it possible to measure a variety of physiologically important signals in single cells. For many of us in the biomedical research community, the rate of technological improvement often outpaces our ability to understand fully and make optimal use of these new tools. This chapter is written to help those researchers who want to use charge-coupled device (CCD) cameras to detect rapidly changing signals in video microscopy. In particular, we will discuss some of the sampling and signal processing characteristics of these devices that determine and constrain their

Optical Microscopy: Emerging Methods and Applications

use in the measurement of such fast signals, In our laboratory, we are interested in the characterization of the intracellular free $Ca^{2+}$ transient that results from brief membrane depolarization in excitable cells and the exocytotic release of hormones or neurotransmitters (Bookman *et al.*, 1991). In the course of trying to run our imaging detectors at top speed, we have been forced to understand the operation of these CCD cameras in a way that is rarely documented or discussed. The broader questions we will try to address are (1) how can one make the best use of an RS-170 CCD imaging detector? and (2) what are the temporal bandwidth limits of these devices?

In order to answer these questions, it is necessary first to understand the function of two types of CCD chips: interline-transfer (ILT) CCDs and frame-transfer (FT) CCDs. The reader should be aware that the chip itself, although it is the photon receptor, is only the first step in the signal transduction pathway. The organization of the chip, the realization of a camera, the construction of a video signal, and the quantization of that signal by a frame grabber all contribute to or limit the ability to faithfully measure a dynamic scene. The spatial and temporal information represented in a video frame will thus vary with the design of different CCD chips and with various modes of camera operation. These distinctions can become significant and must be understood when attempting to obtain quantitative information at the limits of temporal and spatial resolution provided by such devices.

The focus here is further narrowed in that we will only discuss implementations of these devices in video cameras capable of producing RS-170 video signals. The justification is mainly one of convenience: from acquisition and display through data storage, the equipment for handling RS-170 signals tends to be easier to use and less expensive than that for custom signal formats. In those experiments where signal kinetics are not of interest or where the intrinsic signal bandwidth is very low (0.01–1 Hz), there may be significant advantages in using a special purpose CCD camera. Frame transfer CCD chips have been widely used in cooled cameras (from $-20°C$ to $-100°C$) designed to make full use of the CCD's ability to provide accurate spatial sampling, large signal-to-noise ratio, and a wide dynamic range. As Aikens *et al.* (1989) make clear, such devices are not designed for high-speed imaging. The pixel clock (i.e., time required to digitize each picture element in the final image) places limits on the frame rate. The slower pixel clock rate of these cameras (100–500 kHz versus ~10 MHz in RS-170) places an upper bound on the frame rate. Lasser-Ross *et al.* (1991) have described a strategy for acquiring as many as 250 frames per second with such a cooled CCD by reducing the number of pixels per frame. Lasser-Ross *et al.* (1991) demonstrate the trade-off between pixels per frame and frame rate. In the limit, one is probably better off with a photomultiplier tube.

For the purposes of this discussion, the optical signal could be high or low level corresponding to, for example, Nomarski imaging or fluorescence imaging. In the case of low-light imaging, an intensifier is usually required to provide enough signal gain to overcome various noise sources. The additional signal-distorting characteristics of these intensifiers and other low-light devices will not be considered here. The papers by Spring and Smith (1987), Tsay *et al.* (1990), and Bookman (1990) consider some of these issues. Our particular focus derives from our own laboratory where we try to measure optical signals at rates that are at or near the limit of the RS-170 CCDs. In the course of our investigations, we have started to characterize the detailed sampling behavior of these imaging devices. The description below includes data from measurements of a new CCD camera from Texas Instruments that uses a FT CCD chip with unique capabilities and high resolution (Hynecek, 1989). We believe that this device will be of interest to those investigators who want top speed (i.e., 30–60 Hz) out of an RS-170 device.

There are many excellent sources of information on video cameras and imaging detectors. Inoue (1986) provides a clear description of many aspects of camera operation and video signals. Jain (1989) has a particularly good discussion of problems related to image sampling and quantization although the temporal aspects are somewhat neglected. Aikens *et al.* (1989) explain general features of CCDs and make clear why cooled CCDs are the detectors of choice in low-light applications where the frame rate is not critical.

## II. SOME TERMS AND METHODS

### A. RS-170 Signals

The requirements of broadcast television have determined many of the timing characteristics that are used in RS-170 video cameras today. Although high-definition TV and scientific-imaging devices have proliferated, the large market for devices that do follow a standard often results in cheaper and more convenient RS-170 devices.

RS-170 timing provides a video output signal with 30 frames per second. Each frame is built from two fields, called odd and even, that are transmitted or displayed sequentially. The odd and even fields alternate such that a full frame is displayed during $\frac{1}{30}$ s. Each field is made up of 262.5 horizontal lines. In the final image, the two fields are interdigitated such that the odd and even fields contribute odd and even lines (see Fig. 4.1). Synchronization pulses are added to the image information to provide both horizontal and vertical timing references. Every 63.5 $\mu$s a horizontal synchronization pulse ($H_{sync}$) is generated to start a new line. Every $\frac{1}{60}$ s a vertical synchronization pulse ($V_{sync}$) is generated to indicate that a new field is starting. The $V_{sync}$ pulse occurs during the vertical-blanking interval, which lasts ~1.2 ms.

FIGURE 4.1

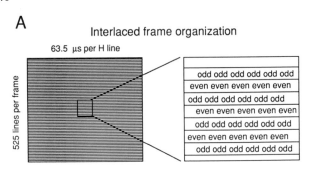

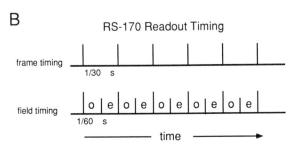

Schematic diagram of RS-170 timing. (A) Spatial relationship of odd and even fields in the video signal. (B) Timing relationship between fields and frames. The labels *o* and *e* indicate odd and even fields, respectively. During each field, 262.5 horizontal lines are transmitted or displayed.

The point to be emphasized is that these timing characteristics apply only to the video signal from the camera. They do *not* imply any particular integration time or sampling period for any particular pixel generated by the image digitizer. Nor do they imply any fixed spatial resolution of the device. The detailed functioning of the detector must be understood in order to know the exact meaning of each pixel. There is nothing inherent in the RS-170 timing definitions, which specify, for example, that all pixels in a frame have sampled light synchronously. In fact, the different devices described below have different temporal-sampling properties.

## B. Signal Pathways for Digitized CCD Data

It is worthwhile to make explicit the signal pathway in a digitizing image acquisition setup with a CCD camera. Figure 4.2 illustrates this path in a general way that is applicable to both ILT and FT devices.

As the photons impinge upon the surface of the CCD, they are spatially sampled by the particular geometry of the CCD chip and its distribution of photoactive sites. (We will not consider further the projection of the light signal onto a two dimensional detector.) Each picture element or pixel of the chip has a size and shape that is presumed uniform over the entire surface of the chip. Photons impinging on the surface generate electron-hole pairs in the initial transduction event. This process can be considered to be instantaneous since an electron is generated in less than a nanosecond. It is this property that is responsible for the lack of lag in CCD cameras. The electronic charges are collected in a potential well in the silicon substrate until the charge is transferred elsewhere or the well overflows. This integration period reflects most (but not all, see below) of the temporal-sampling period for that pixel. It is important to note that the silicon is able to generate electron-hole pairs all the time. The transduction process does not observe the blanking periods of the RS-170 signal. The lack of clock signals for the first two functional blocks in Fig. 4.2 indicates that these processes are operating continuously. Next the charge is shifted towards a part of the chip where both CCD types have some form of on-chip memory. The clocking of this charge-transfer step differs in important ways in ILT and FT CCDs as described below. It is necessary to understand the timing of this event, especially in combining optical and electrophysiological measurements or in ratio imaging where rapid alteration of experimental conditions must be accurately synchronized with image acquisition.

FIGURE 4.2

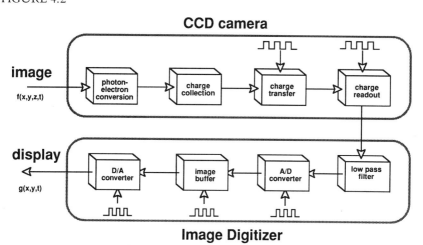

Schematic of the signal path from photon through a CCD camera and acquisition by a digital-imaging system. The clock signals indicate those steps that may have independent timing. The temporal and spatial value of a pixel on the final display can only be understood by tracing the entire signal path from the initial photon-transduction event.

By reading out this on-chip memory, discrete charge packets from the CCD are dumped onto a capacitor and this capacitor's voltage is amplified, generating ~0.5–6.0 $\mu$V/electron. Video amplifiers and clocks take this signal and put together the continuous, analog RS-170 signal with its sync pulses. Note that the video-signal clock only sets the limits for the readout clock. RS-170 timing has no defined number of pixels per line. The CCD itself must provide whatever pixel clocks are necessary to meet the RS-170 standard. The video signal is then sent to an input amplifier on the image digitizer. Typically it is then low-pass filtered continuously. While this filter does serve to prevent aliasing, it may also decrease the system's ability to resolve some of the higher spatial frequencies in a horizontal line (Jain, 1989). The output of the filter is fed to an analog-to-digital (A/D) converter, typically with 256 levels of quantization (8 bits). The A/D converter clock may bear little relation to the video-signal clocks outside of the $V_{sync}$ and $H_{sync}$ restrictions. In particular, an A/D converter will typically be clocked to yield a 512 (horiz.) × 484 (vert.) pixel image. This will be the case whether the CCD imager has more, less, or an equivalent number of pixels. The digitized value is then sent to a framestore that will be clocked to store an interlaced frame appropriately. With yet another potentially independent clock, the information can be readout, converted back to analog form by the D/A converter, and displayed as interlaced RS-170 video. All of these independent clocks make it difficult to know the sampling function that determined a pixel. The important point is that in order to know the time and space value of any one pixel, it is necessary to examine all the clocks and establish when and where photons were sampled.

## C. Equipment

The measurements reported here have been made with two different CCD cameras. The Hamamatsu 2400 (Hamamatsu Photonic Systems, Bridgewater, NJ) is an interline transfer CCD based on the ICX022BL chip manufactured by Sony Corporation (Sony Image Sensing Division, Cypress, CA). This same chip is also found in the Dage 72 camera (Dage-MTI Inc., Michigan City, IN). Tests were also made of a new camera from Texas Instruments (Dallas, TX), the MC-1134P MultiCam, based on TI's TC217 chip. This last device has many novel features that are likely to lead to its use in new cameras from one or more of the instrumentation camera companies.

## D. Data Acquisition Techniques

The RS-170 signals were low pass filtered at 4.5 MHz and digitized at 8 bits by an A/D converter clocked to generate 512 by 484 pixels per image on a Datacube Digimax board. The digitized values were sent to a Datacube

ROIStore, which has triple-ported RAM with independent read and write timing. The images were acquired, analyzed, and displayed by an IC300 DSP/OS Imaging System (Inovision Corp., Research Triangle Park, NC) hosted by a Sun Sparcstation. A particular strength of this system is the *fastq* acquisition program (Inovision Corp.), which allows the user to specify a queue of tasks to be executed every 1/30 s, synchronized with the video clock. Examples of tasks include opening shutters, driving a wavelength selection system, or controlling an optical-disk recorder. For the observation of rapid kinetics, we find it convenient to use *fastq* to obtain images built up by storing only a small rectangular region of interest (ROI) from the incoming frame and have a mosaic resultant image that contains temporal information. If the source ROI is a single column of pixels, then the final image represents time along the x-dimension of the image as shown in Fig. 4.3.

By plotting the intensity of all the pixels in a single row, one has a graphical representation of a one-dimensional signal as a function of time. This more familiar and tractable data set can then be analyzed with more conventional graphing and data analysis tools, such as IGOR (Wavemetrics Inc., Lake Oswego, OR).

The Hg lamp on the microscope was coupled to a rapid-wavelength changing system that is capable of switching wavelengths within 1.0 ms and remaining at that wavelength for an arbitrary dwell time. The system is triggered by the IC300 System within 400 ns of the beginning of the vertical-blanking period. *Fastq* acquisitions using column ROIs were run with a stimulus pattern that turned the excitation system on and off at determined times, synchronous with $V_{sync}$ for a programmed number of frame times.

FIGURE 4.3

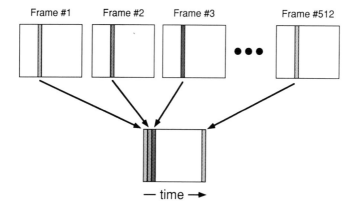

*Fastq* acquisition of pixel columns. The resultant image contains both spatial and temporal information.

## III. IMPLEMENTATION OF CCDs IN RS-170 CAMERAS

### A. General Comparison of ILT and FT CCDs

Figure 4.4 illustrates the overall architectural differences between ILT and FT CCD chips. Both chips contain light-sensing regions and image-storage regions. In an ILT device, there is a masked (i.e., light-insensitive) vertical column next to each light-sensitive column of pixels. In an FT chip, the light-sensitive region is continuous and the masked image-storage region is, for example, in the top half of the chip. When implemented in RS-170 cameras, both FT and ILT devices transfer accumulated pixel charge into their storage registers starting at or near the beginning of a vertical-blanking period.

In the case of the FT chip, this means moving the array of charge, row by row, *through the sensing area* and up into the storage area. The total amount of time required for shifting (during which charges are still accumulating) is on the order of 400 μs. In the ILT device, the charges are shifted to the neighboring *masked* column, a process that only requires a couple of microseconds. The different mechanisms of charge transfer between the two chip types produce differences in the meaning of their pixels. The rela-

FIGURE 4.4

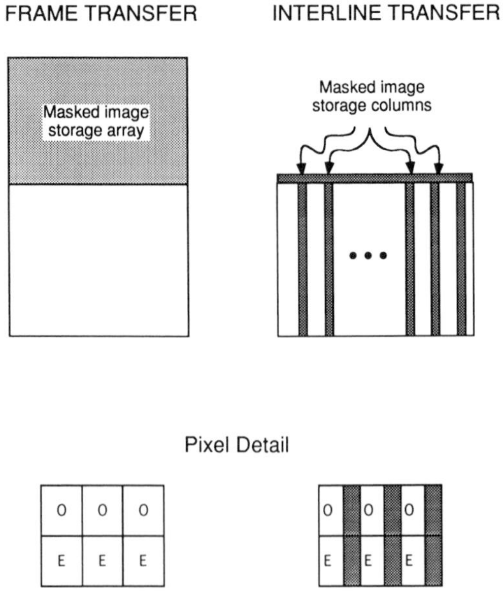

Schematic drawing of CCD architectures for FT and ILT chips. The lower pixel detail shows the difference in the light-capturing area between the two devices.

tive timing is illustrated in Fig. 4.6 and the details are explained in Section III.C. and Section III.D.

The lower part of Fig. 4.4 shows the essential difference in the spatial-sampling characteristics of these two devices. The active (photon-transducing) area of a single pixel in a FT chip is approximately twice as large as in an ILT device. Photons that hit the masked-column area do not contribute electrons to the potential well. This makes the FT device more sensitive than the ILT chip by a factor of ~2, assuming they have equal numbers of pixels. It does not, however, imply that the FT chip has a more equal aspect ratio. Even if the light-collecting area of a pixel in an ILT is long and thin, the pixels may be square. Thus, the size of the sample area is not necessarily the same as the light-collecting area. Square pixels are achieved by having the spacing of the sample grid such that $\Delta x = \Delta y$ (Jain, 1989). Either device type can be manufactured with square pixels, in principle.

Sony Corporation, the world's largest manufacturer of ILT CCDs, has recently addressed the sensitivity loss in ILTs by developing "on-chip-lens" (OCL) technology in which a glass lens element is deposited on each pixel to increase its capture area. This development is unlikely to influence scientific imaging devices since it is part of the effort to lower device costs by using ½ in. instead of ⅔ in. CCDs in consumer video cameras.

## B. Interlacing and Resolution

The RS-170 signal implies that spatial information is output in a serial manner, one horizontal line at a time. As a result of this, the *horizontal* resolution of an image can be compromised by the bandwidth limits inherent in transmitting and sampling this data. Vertical resolution is not affected by the same processes because (1) vertically adjacent pixels are not displayed sequentially and (2) RS-170 timing requires that both camera output and storage devices contain 525 lines. However, the method by which a camera generates these 525 lines can affect both vertical resolution and camera sensitivity. Vertically adjacent pixels displayed by different cameras may represent separate, overlapping, or identical regions of the CCD's image-sensing area.

All RS-170 cameras transmit the two video fields in an interlaced manner such that each field may define a spatially distinct region of the displayed image. For cameras that generate "true interlaced" images, these fields do represent light collected from separate but interdigitated regions of the image-sensing area as illustrated in Fig. 4.5. Therefore, true interlace produces ~500 lines of vertical resolution. Some cameras can be operated in a "noninterlaced" mode, where odd and even fields are acquired sequentially and represent identical spatial information. This mode can only produce a

FIGURE 4.5

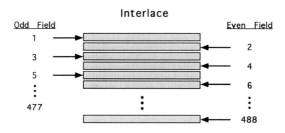

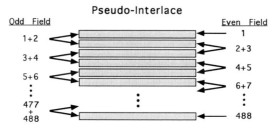

The diagrams illustrate two ways in which a CCD array can be read out in order to construct a 488-line video frame. In interlaced mode, odd and even video fields represent odd and even lines of the sensor array. In pseudointerlace mode, two lines of the sensor array are summed to construct each line of video output. Odd and even fields are acquired sequentially using the illustrated alternate-line summing scheme.

vertical resolution of ~250 lines but effectively doubles the sensitivity of the camera since each pixel can collect light from a sensing area equal to an odd pixel added to an even pixel. Still other cameras use a "pseudointerlace" scheme that retains the high sensitivity of a noninterlaced mode while increasing vertical resolution to approximately 350 lines. Pseudointerlace is implemented by on-chip summing of vertically adjacent photoactive elements. The two fields are acquired sequentially using an alternate line summing procedure as illustrated in Fig. 4.5.

The horizontal resolution of the final image is the result of the total image-acquisition system as previously described. In particular, it is important to note that the clock that controls the A/D converter is responsible for a resampling of the horizontal-line signal. If the CCD camera has, for example, 1000 pixels per horizontal line, then the ability to resolve spatial frequencies is improved. However, if the A/D converter is clocked to produce 512 samples per line, then the benefits of the higher resolution chip are lost even though the additional information is in the camera output. A higher frequency A/D clock on the frame grabber is required in order to retain the high resolution. As long as there are 63.5 $\mu$s per line and 525 lines per frame, RS-170 timing definitions would not be violated. Some 20-MHz digitizing boards are beginning to become available through imaging-board manufacturers.

## C. Temporal Sampling Characteristics of an Interline-Transfer CCD

The Sony ICX022BL ILT chip is capable of operating in a number of different modes. These modes specify the type of interlacing (see above) and whether light is integrated over a frame time or a field time. Some of these capabilities are difficult to characterize since the camera manufacturers have not made it easy to change operating modes. The normal operating mode (interlaced, frame integration) has been used in both the Hamamatsu 2400 CCD camera and the Dage 72 CCD camera. Both cameras operate in a similar fashion, which we will try to describe in detail. The chip contains 768 (horiz.) × 493 (vert.) active-imaging pixels. This means that there are 768 masked vertical-storage buffers in between the sensing columns. There are additional masked pixel columns and rows at the edges of the chip that provide black levels during the blanking periods. When an odd-field $V_{sync}$ signal occurs, the charge contents of the odd pixels are shifted over to the masked vertical-storage buffers. The vertical buffer contains as many vertical elements as the light-sensing column next to it and is designed to operate as a shift register buffer. The shifting operation to transfer charge takes ~10 $\mu$s (see Fig. 4.6), and results in a vertical shift register that has every *other* element filled with charge collected by the pixel to its left. The masked register elements corresponding to pixels from the even field are nominally

FIGURE 4.6

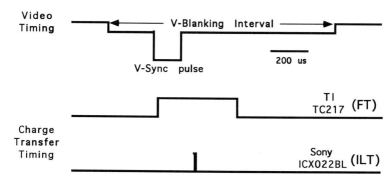

Charge-transfer timing. Timing diagrams indicate the periods during which charge is transfered from the image-sensing region of a CCD to its storage registers. Examples for both frame-transfer (FT) and interline-transfer (ILT) devices are shown in relation to the vertical-blanking interval of the RS-170 video signal.

empty. Given that the phototransduction process goes on continuously, electrons are being generated all the time. The charge-transfer time represents the aperture uncertainty of the temporal-sampling process of this ILT chip since it is not clear during this period whether generated electrons will be in the "last" field or the "next" field. The contents of the potential wells of the even-field pixels in the sensing area are not altered by this odd-field charge transfer in the chip's normal mode of operation. Following the charge-transfer step, the odd-field pixels will continue to collect electrons until they are next emptied $\frac{1}{30}$ s later.

All of the vertical shift register buffers are clocked together such that they each contribute one pixel to a horizontal shift register buffer at the top of the chip in synchrony. Since every other element is empty, there are two vertical register shifts required to fill the horizontal register with odd-field data. The charge content (if there is any) of the shift register locations corresponding to the even pixels is drained to ground. The horizontal shift register buffer feeds the amplifier, which produces the video signal. This process of shifting charge to the vertical, initially advancing all the verticals by one (for the first odd row), shifting the entire horizontal row out through the video amplifier, advancing all the verticals by two, shifting the horizontal, . . . , is repeated until all the odd-field vertical lines are read out. This is accomplished during $\frac{1}{60}$ s (i.e., the time needed to send this odd video field to the display or the frame grabber).

With the arrival of the even $V_{sync}$ pulse, the contents of the even-field pixels are shifted over to the masked vertical shift registers, and the readout process is repeated for these pixels. The critical point to understand is this: every pixel on the ICX022BL chip (operating in frame integration mode) integrates light for 33.33 ms $\pm$ the time during which the charge is being shifted over to the vertical register. However, the even and odd fields are *not* sampling in phase. Rather, they are phase shifted with respect to one another by 180° or $\frac{1}{60}$ s. The sampling-phase relationship is shown in Fig. 4.8.

Figure 4.7A illustrates this behavior of the Sony ICX022BL in the Hamamatsu 2400 camera. A *fastq* acquisition using a column source ROI grabbed the 242nd column of the frame every $\frac{1}{30}$ s. The excitation system on the microscope turned the light on and off during the acquisition. Figure 4.7 shows a detail from two profiles taken from this image, plotting pixel intensity as a function of time. The light was on for two frame times. The odd field "correctly" sampled the signal, while the phase-shifted even field shows two time points with half the intensity and one time point with the true level. The data clearly show that the grey-level values of pixels in the two fields differ during the frame when the light was on. According to the even field, the "on" event lasted for $\frac{3}{30}$ s and had a clear rising and falling phase. The odd field on the other hand shows an event duration of $\frac{2}{30}$ s with no distinguishable rising or falling phases.

FIGURE 4.7

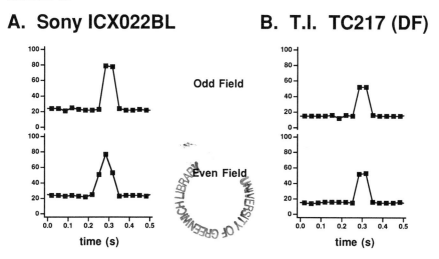

Measured response of ILT and FT CCDs to a two-frame time (²⁄₃₀ s) pulse of light. (A) Profiles of pixel intensity are taken from *fastq* images with the Hamamatsu 2400 CCD camera. Each point represents one frame time. (B) Similar data acquired with the TI MC1134P camera operating in dual-field mode.

This is consistent with the statement that the even and odd fields do not sample the image signal at the same time. It is not possible with this chip to have both fields sampling synchronously. Therefore, operating the camera in its normal frame-integration mode limits its use in 30-Hz imaging applications. If the process under observation is slow or requires many frames to be averaged together, then this sampling distortion may not be a problem.

## D. Temporal Sampling Characteristics of a Frame-Transfer CCD

In frame-transfer devices the active light-sensing elements and masked-off storage registers are segregated to different regions of the chip. The Texas Instruments TC217 frame-transfer chip contains 1134 (horiz.) × 486 (vert.) light-sensitive pixels with additional masked-off dark-reference elements at the periphery of the light-sensitive region. The masked-off storage region can accommodate data from the entire 1154 (horiz.) × 488 (vert.) array representing photoactive and dark reference elements. Additional serial registers and charge detection amplifiers are located next to the storage area and are used to read out the image in an RS-170 compatible form.

The TC217 chip is unique among frame-transfer devices in its ability to acquire and store an entire 488-line video frame. The data are arranged in the storage area such that the separate video fields can be read out independently. Because of these features, the TC217 is the only device that can acquire both video fields simultaneously and display them in an RS-170 compatible form. In addition, the chip can operate in several acquisition "modes" and the TI MC-1134P camera is designed to take advantage of this flexibility.

In the "dual field" or simultaneous field-capture mode the image-sensing region integrates light for $1/30$ s. The entire image (both fields) is then rapidly shifted during the vertical-blanking period into the storage region where it can subsequently be read out in an interlaced manner to provide the two video fields. The fields represent spatially distinct (true-interlaced) regions of the image plane that have integrated light for the same $1/30$ s interval. The advantages of synchronous field acquisition are illustrated by the results of the rapid-illumination test in Fig 4.7B. In contrast to the ILT results, both fields of the FLT image accurately track the amplitude and duration of the signal.

In the "normal-light mode," frame transfers occur during every vertical-blanking period (i.e., every $1/60$ s). After each transfer, one field is read out while the other is discarded. Alternate readout of odd and even fields provides a true-interlaced image. In this mode, each field has integrated light for $1/60$ s, but each field represents a *distinct* time interval and a distinct area of the image plane. Postprocessing could separate the two fields and take advantage of the $1/60$ s time resolution, but the value may be somewhat limited. The fact that sequentially acquired fields represent alternate lines of the image plane may introduce measurement error when the spatial resolution of the image approaches that of the sensor. Since each pixel integrates light for only $1/60$ s, sensitivity is effectively reduced by 50% relative to dual-field mode.

In the "low-light mode," the chip's pixel elements integrate light for $1/60$ s, but during the readout process, pixels in adjacent lines are summed to obtain one field of the video frame. Alternate summing of adjacent lines provides the two fields of the frame and results in a pseudointerlaced image (see Fig. 4.5). As in the normal-light mode, each field output in the low-light mode represents a distinct $1/60$-s time interval. Line summing provides a doubling of light sensitivity over the normal-light mode.

Until recently, frame-transfer CCDs were unable to acquire and store an entire 488-line video frame. Due to this limitation, most frame-transfer cameras actually operate in a manner analogous to the "low-light" mode of the TC217 chip where pseudointerlaced video fields are acquired at $1/60$-s intervals. These devices should perhaps be more accurately termed "field-transfer" cameras as they are unable either to integrate light for $1/30$ s or to produce a true-interlaced image.

One disadvantage of the frame-transfer design is that pixels in the image-sensing region cannot be transferred simultaneously into the storage registers as they are in the ILT design. The TC217 requires about 409 $\mu$s to shift the entire image array into storage and this transfer delay can cause two different problems. First, every pixel will accumulate additional charge during the charge-transfer period as it is moved through the light-sensitive elements of the imaging region. This phenomenon is known as smear and is most significant when a pixel originating from a dark area of the image is shifted through a large, brightly illuminated area. Second, because elements of the image array are not synchronously cleared of charge, each row integrates light during a slightly different period of time even though they all integrate for the same length of time. In other words, each row samples at a unique phase relative to the $V_{\text{sync}}$ clock.

The worst case sampling phase difference is $\sim 4°$ (409 $\mu$s/33.33 ms), comparing the top of the array to the bottom, while the line-to-line sampling phase error is $\sim 0.01°$. The origin of this error can be understood by examining the operation of the chip during charge transfer. Let $P_j(i)$ be the $j$th pixel in a column sampled during the $i$th frame time. Assume that there are $N$ pixels in a column that will be shifted to the storage array, and that $P_1(i)$ and $P_N(i)$ are closest and farthest, respectively, to the storage array. At the beginning of the $i$th charge transfer, $P_1(i)$ is moved into storage during the first step of the transfer. $P_2(i)$ will be shifted ahead by one position. Thus, $P_2(i)$ will integrate light in position 1 for a shift-clock period ($\sim 1$ $\mu$s) before being shifted into the storage array. In general, $P_K(i)$ has been moved up to position $(K - 1)$ in which it will integrate light. Clearly, the pixels farthest from storage will integrate all the way up the column. Meanwhile, location $N$ on the chip is converting photons to electrons all the time. After $P_N(i)$ has been moved to position $(N - 1)$, location $N$ is still collecting electrons. With the next charge shift clock, these electrons will be moved to location $(N - 1)$, and with the next clock to $(N - 2)$, and so on up the column. At the next to last step, when $P_N(i)$ is in position 1, position 2 contains charges collected through the column that were "dragged" up the column by the shift clocking. When $P_N(i)$ is finally clocked into the storage array, the charges that were in position 2 are shifted to position 1, and these charges will be attributed to $P_1(i + 1)$ (i.e., the next frame). Thus $P_1(i)$ integrates light during the $i$th frame time as well as during the $(i - 1)$th charge-transfer time, $P_N(i)$ integrates during the $i$th frame time and the $i$th transfer time. The $k$th line in between integrates during the $i$th frame time as well as during $(k - 1)$ shift clocks of the $(i - 1)$th charge transfer and $(N - k)$ shift clocks of the $i$th transfer. In this fashion, each row of pixels integrates during a unique period.

Since the charge-transfer period is only about 1% of the frame integration period, this sampling error will not be significant under constant illumination conditions. However, under certain experimental conditions, their effects must be taken into account. For example, in fluorescence ratio-imaging applications, unequal image intensities at different excitation wavelengths could produce errors much greater than 1%. In a flash photolysis experiment, a flash delivered during a charge-transfer period will appear in two frames. Such problems can be avoided if the camera or light source is shuttered during the charge-transfer period. In any case, the sampling-phase error of the TC217, operating in dual-field mode, is insignificant compared to the 180° error of the ILT chip.

## IV. SUMMARY AND PROSPECTS

We believe that it is important to understand the sampling and transfer characteristics of the transducers used in video microscopy. This is particularly true in imaging applications when one tries to squeeze the most out of RS-170 devices. The temporal-sampling characteristics are summarized in Figure 4.8. Based on our measurements and the descriptions above, it is clear that the Sony ICX022BL CCD can not accurately perform 30-Hz im-

FIGURE 4.8

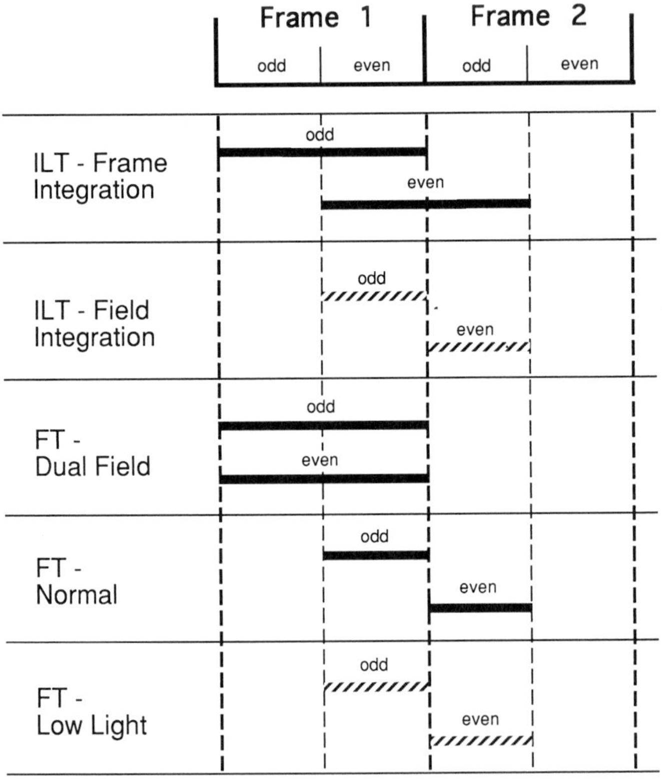

Schematic drawing of the temporal-sampling properties of the ILT and FT CCD cameras. The bar indicates the time period during which light is integrated for display in the indicated field of frame 2. The black bars indicate an acquisition mode with full vertical resolution. The striped bars indicate reduced vertical resolution from pixel summing.

aging when used in its normal mode. At best, the overlapping temporal-sampling reduces the bandwidth to 20 Hz and makes image analysis more difficult. Cameras using the TC217 FT chip, such as the TI MC1134P, are the only RS-170 devices we are aware of that are capable of (nearly) simultaneously sampling both fields of an image. For increased bandwidth, both types of devices can be operated for 60-Hz imaging. Such use within the RS-170 constraints necessarily reduces the vertical resolution and may require greater illumination energies to achieve acceptable signal-to-noise ratios. The ability of both types of devices to sum two pixels help to offset this problem. If the experimental situation permits, time-locked stimuli can permit many 60-Hz images to be signal averaged. It is this strategy that optimizes imaging rapid transients with RS-170 devices.

## ACKNOWLEDGMENTS

Thanks to our colleague Nenad Amodaj for helpful discussions. R. Wick and M. Oshiro at Hamamatsu Photonics, P. Thomas at Dage-MTI, Rich Owoc at Sony, and D. Harmon and K. Shaver at Texas Instruments have all generously shared information with us. Doug Benson and Steve Case at Inovision Corporation have helped to provide us with software tools to make these kinds of measurements both possible and pleasurable. This work is supported by grants from the National Science Foundation, the National Institutes of Health, and The Florida Affiliate of the American Heart Association.

## REFERENCES

Aikens, R. S., Agard, D. A., and Sedat, J. W. (1989). Solid-state imagers for microscopy *Meth. Cell Biol. 29,* 291–313.

Bookman, R. J. (1990). Temporal response characterization of video cameras. *In* "Optical Microscopy for Biology." (B. Herman and K. Jacobson, eds.), pp. 235–250. Wiley-Liss, New York.

Bookman, R. J., Lim, N. F., Schweizer, F., and Nowycky, M. (1991). Single cell assays of excitation–secretion coupling. *Ann. N.Y. Acad. Sci. 635,* 352–364.

Hynecek, J. (1989). A new high-resolution 11-mm-diagonal image sensor for still-picture photography. *IEEE Trans. Elec. Dev. 36*(11), 2466–2474.

Inoue, S. (1986). "Video Microscopy." Plenum Press, New York.

Jain, A. K. (1989). "Fundamentals of Digital Image Processing." Prentice-Hall, Englewood Cliffs, NJ.

Lasser-Ross, N., Miyakawa, H., Lev-Ram, V., Young, S. R., and Ross, W. N. (1991). High time resolution fluorescence imaging with a CCD camera. *J. Neurosci. Meth. 36,* 253–261.

Spring, K. R., and Smith, P. D. (1987). Illumination and detection systems for quantitative fluorescence microscopy. *J. Microsc. 147,* 265–278.

Tsay, T.-T., Inman, R., Wray, B., Herman, B., and Jacobson, K. (1990). Characterization of low-light-level cameras for digitized video microscopy. *J. Microsc. 160,* (2), 141–159.

# 5
# MEASUREMENT AND MANIPULATION OF OSCILLATIONS IN CYTOPLASMIC CALCIUM

**C. S. Chew**
Department of Physiology, Morehouse School of Medicine
**M. Ljungström**
Department of Medical and Physiological Chemistry, Biomedical Center, University of Uppsala

I. INTRODUCTION
II. CURRENT MODELS OF INTRACELLULAR CALCIUM OSCILLATIONS
III. CHARACTERISTICS AND LIMITATIONS OF CELL-PERMEANT CALCIUM-SENSITIVE FLUORESCENT PROBES
   A. Potential Perturbations in Cellular-Response Patterns Induced by Fluorescent Probes
   B. Calibration of the Fluorescent Signal
   C. Temporal–Spatial Characteristics of Calcium Fluxes

IV. EQUIPMENT SUITABLE FOR DETECTION OF CALCIUM OSCILLATIONS
   A. Microscopes and Accessories
   B. Filter-Changer Configurations
   C. Video Camera Considerations
   D. Computer-Operating Systems
   E. Image Processors
   F. Data-Storage Strategies
   G. Software
V. FUTURE DIRECTIONS
ACKNOWLEDGMENTS
REFERENCES

# I. INTRODUCTION

Oscillatory behavior in excitable cells is an established phenomenon; however, the concept that oscillations in intracellular ions, such as calcium, might occur in electrically nonexcitable cells has emerged only recently. Berridge and Rapp (1979) originally suggested that oscillations in intracellular calcium concentrations occurred in *Calliphora* salivary glands based on observations that agonist-induced oscillations in cellular-membrane potentials exhibited calcium dependency. Direct measurement of intracellular calcium with calcium-sensitive microelectrodes lent further support to the hypothesis that, at least in this cell type, intracellular-free calcium levels oscillated upon stimulation with an appropriate secretory agonist (Berridge, 1980). Agonist-induced oscillations in intracellular calcium were later detected in aequorin-loaded rat hepatocytes by Cobbold and colleagues (Woods *et al.*, 1986, 1987).

Following the development of fluorescent, cell-permeant calcium probes, particularly the "second generation" indicators, fura-2 and indo-1 (Grynkiewicz *et al.*, 1985), it became possible to study changes in response patterns in many cell types, including those too small for classic microelectrode–microinjection studies. Initial findings with these probes were limited mainly to studies of cell populations suspended in cuvettes. In such experimental protocols, there is temporal but not spatial information with signal detection limited to the average population response. Typically, cell population studies have defined agonist-stimulated calcium-response patterns as biphasic with a sharp initial rise in intracellular-free calcium concentrations ($[Ca^{2+}]_i$) followed by a lower, sustained elevation. With increased application of microscope-based digitized video–image analysis techniques, it has become apparent that population responses do not accurately reflect calcium-signaling patterns in individual cells and publications depicting a variety of calcium-oscillation patterns in nonexcitable cell types have increased exponentially within the past five years.

The increasing numbers of cell types in which $[Ca^{2+}]_i$ has been found to oscillate and to spread as a transient wave from a localized source suggest that oscillations in intracellular calcium concentrations as well as localized changes in $[Ca^{2+}]_i$ may play an important role in signal-transduction mechanisms. The goal of the future will be to determine whether or not this is indeed the case and, if so, to define the cellular activities controlled by oscillatory mechanisms and to localize the intracellular compartments and biochemical components controlling these events. In this chapter, we will briefly consider current models that seek to explain how intracellular calcium oscillations occur and then focus on video-based approaches to measurement of calcium oscillations in single, nonexcitable cells. Relevant properties of fluorescent calcium probes, the equipment required for digitized video imaging of calcium oscillations and present limitations, including potential sources of artifacts, will be discussed in the context of video-based applications. Problems associated with correlation of calcium oscillations with physiological-response patterns will also be considered.

## II. CURRENT MODELS OF INTRACELLULAR CALCIUM OSCILLATIONS

Following the discovery of oscillatory activity in nonexcitable cells, models seeking to explain this phenomenon have proliferated. These models are in a constant state of evolution as new experimental information is acquired. Several recent reviews have emphasized the correlation between agonist-induced increases in cellular inositol 1,4,5 trisphosphate (InsP$_3$) concentrations and calcium oscillations (Berridge and Galione, 1988; Cuthbertson, 1989; Rink and Jacob, 1989; Berridge, 1990; Petersen *et al.*, 1991; Jacob, 1990a). In general, current models fall into two main categories. One links oscillations in [Ca$^{2+}$]$_i$ to agonist-induced oscillations in InsP$_3$ levels (receptor-controlled). The second category does not require oscillations in InsP$_3$. Models that do not invoke oscillations in InsP$_3$ levels have also been defined as second-messenger control models (Berridge, 1990).

Oscillations in InsP$_3$ were originally proposed as a calcium control mechanism by Woods and colleagues (1987) and Meyer and Stryer (1988). In an expansion of this model, Cuthbertson and Chay (1991) have attempted to account for differences in shape and duration of calcium transients observed with different agonists using mathematical analyses. The biochemical basis of the most recent model includes receptor activation of guanosine 5'-triphosphate (GTP) binding proteins with positive feedforward influences that lead to sudden activation of phospholipase C, the enzyme that breaks down phosphatidylinositol(4,5)bisphosphate to form InsP$_3$ and diacylglycerol (DAG). Once calcium is released from intracellular stores by InsP$_3$, negative feedback mechanisms come into play to reduce [Ca$^{2+}$]$_i$. These negative mechanisms may include phosphorylation of receptors by protein kinase C that is activated by DAG and possibly calcium as well as phospholipase-C-induced activation of GTPase(s), which decrease activated G protein concentrations. Since inositol phosphate intermediates cannot be measured in single cells, there is presently no direct experimental evidence showing that InsP$_3$ levels oscillate. Thus, hypotheses based on this assumption remain speculative.

Second-messenger control models have been reinforced recently by data showing that oscillations in [Ca$^{2+}$]$_i$ can be induced by steady infusion of InsP$_3$ (Wakui *et al.*, 1989) and by data showing that calcium oscillations can be generated independent of InsP$_3$ elevation (Rooney *et al.*, 1991). Current forms of this model usually invoke two or more pools of intracellular calcium. In a popular two-pool model, calcium oscillations are thought to occur because calcium released from an InsP$_3$-sensitive pool(s) evokes further release of calcium from an InsP$_3$-insensitive pool. This secondary calcium release is referred to as calcium-induced calcium release (CICR) and was first proposed as a mechanism in muscle (discussed in reviews by Putney, 1990; Jacob, 1990a). The interval between calcium transients is explained in the two-pool model as the time it takes to refill the different calcium pools (Goldbeter *et al.*, 1990). The two-pool model has recently been expanded to explain how wavelike propagation of $[Ca^{2+}]_i$ might occur within single cells (Dupont *et al.*, 1991). There is some evidence that calcium waves originate from a discrete subcellular location. Since calcium waves appear to be propagated across cells in a nondecremental fashion, involvement of multiple calcium pools rather than a simple diffusional process is the favored interpretation (Rooney *et al.*, 1990; Ryan *et al.*, 1990).

Experimental data has shown repeatedly that although calcium oscillations can occur in the absence of extracellular calcium, sustained oscillations require the presence of extracellular calcium. Influx of calcium into cells is a mysterious event in nonexcitable cells. Recent evidence suggests that calcium influx may be regulated indirectly by an interaction between InsP$_3$-sensitive pools and an undefined calcium-influx pathway. For example, experiments with thapsigargin (a nonphorbol ester tumor promoter that releases calcium from InsP$_3$-sensitive pools by inhibition of an inwardly directed Ca–ATPase or calcium pump) have shown that thapsigargin-induced depletion of agonist-sensitive calcium pools is sufficient to cause calcium influx into the cell, that is, calcium influx can be initiated in the absence of an increase in InsP$_3$ concentrations (Jackson *et al.*, 1988; Thastrup *et al.*, 1990). Extensions of the original thapsigargin experiments have led Putney to modify his original capacitive calcium entry model in which calcium influx depended on calcium movement through an InsP$_3$-sensitive pool within endoplasmic reticulum (ER) located immediately adjacent to the plasma membrane (Putney, 1986). In the revised model, calcium entry still depends on depletion of calcium from InsP$_3$-sensitive pools; however, calcium is not required to move through this pool before entry into the cytosol (Putney, 1990). In a somewhat different approach, Irvine (1990) has suggested that a phosphorylation product of InsP$_3$, inositol 1,3,4,5 tetrakisphosphate (InsP$_4$), may act in concert with InsP$_3$ and calcium to generate oscillations. InsP$_4$ is proposed as a modulator of a plasma membrane calcium channel that is associated with an InsP$_3$-sensitive calcium pool in the ER. Elevated ctyosolic $[Ca^{2+}]_i$ may also serve to modulate the conformation of an InsP$_3$ receptor associated with the plasma membrane allowing calcium influx to occur. Irvine further speculates than an $InsP_4$-sensitive calcium channel might also modulate movement of calcium from $InsP_3$-insensitive to $InsP_3$-sensitive calcium pools. Like the Berridge–Goldbeter–Dupont model, this conformational-coupling model incorporates aspects of earlier calcium entry–release mechanisms proposed in muscle. The relationship between these and earlier muscle models is reinforced by recent data showing sequence homology between the skeletal muscle ryanodine receptor and the $InsP_3$ receptor (Takeshima et al., 1989; Mignery et al., 1989; Furiuchi et al., 1989).

To date, emphasis has been placed on the role of $InsP_3$-sensitive calcium pool(s) in the generation of calcium transients. There are several reports, however, of agonist-induced increases in $[Ca^{2+}]_i$ in the absence of a detectable rise in $InsP_3$ and, in some cases, oscillations in $[Ca^{2+}]_i$ have also been detected (Sistare et al., 1985; Chew and Brown, 1986; Yada et al., 1986; Eckert et al., 1989; McCann et al., 1989; Staddon and Hansford, 1989; Reber et al., 1990; Kelley et al., 1990; Hellman et al., 1990; Ljungström and Chew, 1991). Therefore, current models are probably too simplistic to explain calcium-oscillation mechanisms under different physiological conditions. It is of interest that agonist-sensitive, $InsP_3$-insensitive calcium pools may, under some conditions, partially overlap with or be included within a larger $InsP_3$-sensitive pool (Chew and Brown, 1986; Kelley et al., 1990). Thus, these different intracellular calcium pools may be modulated by multiple input mechanisms. Figure 5.1 depicts a composite model that incorporates aspects of previously proposed receptor and second-messenger calcium-oscillation models discussed above. The model also suggests possible sites of action of second messengers other than $InsP_3$.

## III. CHARACTERISTICS AND LIMITATIONS OF CELL-PERMEANT CALCIUM-SENSITIVE FLUORESCENT PROBES

There are several different calcium-sensitive fluorescent probes available for the characterization of calcium-signaling patterns in single cells. We will focus mainly on the two that are presently the most commonly used in video-imaging experiments, fura-2 and indo-1, considering the advantages and disadvantages of each in video-imaging experiments aimed at characterizing calcium oscillations. These dyes are widely used because they can be loaded into living cells without having to resort to microinjection and their spectral-response patterns are such that ratioing of either excitation or emission wavelength maxima are possible. The ability to ratio changes in fluorescence at different wavelengths is an important attribute because ratioing greatly reduces problems associated with photobleaching, variations in pathlength, illumination intensity, and signal inhomogeneity and offers the

FIGURE 5.1

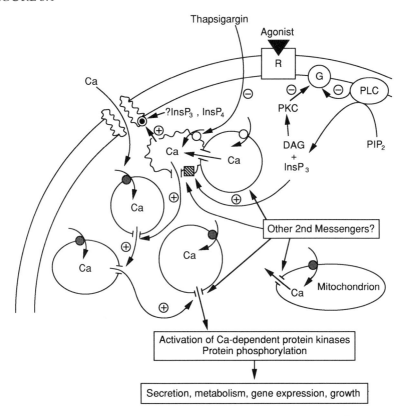

Composite model of potential regulatory mechanisms controlling calcium oscillations and calcium wave spread in nonexcitable cells. This model incorporates enzymatic and ionic control pathways proposed in current receptor control models and further suggests the possibility that second messengers in addition to, or other than, InsP₃ can initiate calcium oscillations. Aspects of current two-pool, second-messenger models are also included. Although not shown in the model, the Ca-ATPase inhibitor, thapsigargin, may also release calcium from InsP₃-insensitive pools. A single agonist-receptor G-protein complex is depicted in the figure; however, more than one G-protein may be activated by the same or different agonists (cf. review by Petersen *et al.*, 1991). See text for details. Symbols are defined as follows: R, receptor; G, GTP-binding protein containing $\alpha,\beta,\gamma$ subunits; PLC, phospholipase C; PIP₂, phosphatidylinositol 1,2 bisphosphate; InsP₃, inositol 1,4,5 trisphosphate; InsP₄, inositol 1,3,4,5 tekrakisphosphate; DAG, 1,2 diacylglycerol.

potential for a more sophisticated approach to quantification and detailed characterization of calcium-response patterns (Tsien *et al.*, 1985; Bright *et al.*, 1989; Tsien, 1989). A disadvantage of fura-2 and indo-1 is that, unlike the "third generation" dye fura-red and the nonratiometric dyes, such as fluo-3 (Kao *et al.*, 1989), Calcium Green, Calcium Orange, and Calcium

Crimson (Molecular Probes, Eugene, OR), both must be excited in the near ultraviolet (UV) (see below). With the exception of fura-red, an advantage of the newer dyes is improved quantum yield as compared to the first generation nonratiometric fluorescent probe, quin-2 (Tsien and Pozzan, 1989). Reduced quantum yield results, in practice, in reduced signal-to-noise ratios (SNR) in measurements of fluorescent intensity with a requirement for increased dye loading into cells in order to obtain a sufficient signal. Since all of these compounds are EGTA [Ethylene glycol-$O,O'$-bis ($\beta$-aminoethyl ether)-$N,N,N',N'$-tetraacetic acid] analogs, they serve as calcium buffers. Increasing intracellular dye concentrations will, therefore, increase calcium buffering and may abolish calcium transients or reduce their periodicity (see Section III.A).

In the absence of better alternatives, fura-2 is presently the dye of choice for most video-imaging protocols. There are, however, certain advantages indo-1 has over fura-2 that should be considered in experiments designed to measure very rapid calcium oscillations. In solution, fura-2 has respective excitation maxima at 340 and 380 nm in the calcium-bound and calcium-free forms with a single-emission maxima of 510 nm. In contrast to fura-2, indo-1 exhibits a shift in the emission rather than the excitation spectrum upon binding to calcium. When excited at 355 nm, the emission maxima for indo-1 are 405 nm in the calcium-bound and 490 nm in calcium-free conditions (Grynkiewicz et al., 1985). The practicality of this different response pattern is that if a beam splitter and dual-emission detectors are employed, more rapid changes in intracellular-free calcium can be recorded (see Morris, Chapter 6, this volume). Calcium oscillations that may be too rapid to be ratioed using fura-2 may, therefore, be detected with indo-1 because "real-time" measurements are possible. In addition, artifacts associated with cell movement are significantly reduced with indo-1. The dual-camera approach has been used, for example, by Takamatsu and Wier (1990), to acquire ratioed images of indo-1–loaded cardiac cells at rates of 16.7 ms/ratio pair.

Although at first glance indo-1 appears to be a more attractive probe as compared to fura-2, technical difficulties and increased costs associated with the use of dual-emission detectors have resulted in more limited usage of this probe. When indo-1 is used in conjunction with video cameras, the images must first be corrected for differences in light path introduced by dichroic mirrors, geometric distortions, and spatial misregistration (resulting from utilization of two different low-light-level cameras) before data can be analyzed accurately (cf. Takamatsu and Wier, 1990). These corrections are complex and computationally intensive. Detailed descriptions of adaptive algorithms can be found in several publications (Castleman, 1987; Inoué, 1986; Steinberg et al., 1987; Jericevic et al., 1989). The difficulties of the dual-camera approach have been further highlighted by Wier and colleagues who recently reported an alternative, fura-2–based method using a single video camera instead of their earlier dual-camera applications (Wier and Blatter, 1991). When excited at 335–365 nm, indo-1 also photobleaches more rapidly than fura-2 (Tsien, 1989; Moore et al., 1990). Wahl and colleagues (1990) have reported that when the 360 nm emission peak of the mercury lamp is blocked and indo-1 excited with a 340-nm narrow band-pass filter ($\lambda\frac{1}{2}$ = 10 nm), bleaching is substantially reduced. It is not clear, however, how much of the emission signal is sacrificed by shifting the excitation wavelength to a lower level. Cellular autofluorescence is also more problematic with indo-1 because of spectral overlap between autofluorescence and indo-1 peak emission wavelengths (Aubin, 1979; Sick and Rosenthal, 1989). Moreover, at least in some cell types, indo-1 appears to bind more avidly to nonsaturable intracellular sites (Blatter and Wier, 1990; Weir and Blatter, 1991). Much like fura-2, indo-1 appears to undergo a spectral shift within the cell when in the calcium-free but not the bound form (Tsien et al., 1985; Popov et al., 1988; Roe et al., 1990; Owen, 1991). In cultured fibroblasts and astrocytes, however, indo-1 leaks more slowly from cells as compared to fura-2 and exhibits less obvious compartmentalization (Steinberg et al., 1987).

The cost of using indo-1 may become comparable to that for fura-2 if a recently described technique, referred to as double-view video microscopy or "W" microscopy (Kinosita et al., 1991), proves to perform as predicted. This dual-image technique allows simultaneous observation of separate fluorescent images that are transformed so that they appear side by side in a single image that can be acquired with a single camera. The application is relatively simple and inexpensive involving the combined addition of a rectangular slit to define the edges of a pair of images with replacement of dichroic mirrors by a pair of polarizing beam splitters placed in the parallel beam between the objective and the video camera. Before the images can be ratioed, however, it is necessary to apply software-based corrections for registration, geometric distortion, and shading as previously discussed. Once corrections are made, image ratio values can be calculated on a pixel-by-pixel basis. This interesting technique is promising but will require additional testing to demonstrate reliability and reproducibility. An important difference between this technique and previously described simultaneous dual-image acquisition techniques (Takamatsu and Wier, 1990; Foskett, 1988; Spring, 1990) is that the W microscopy requires a single camera rather than dual cameras or photomultiplier tube (PMT) detectors.

## A. Potential Perturbations in Cellular-Response Patterns Induced by Fluorescent Probes

A serious and often neglected consideration in calcium-oscillation experiments is the possibility of adverse effects of dye buffering of $[Ca^{2+}]_i$ not only on the pattern of calcium oscillations but also on cell function. Fura-2 has been shown to suppress calcium transients in skeletal muscle fibers when used in high concentrations (Baylor and Hollingsworth, 1988). Rooney et al. (1989) have also shown that increasing fura-2 buffering in hepatocytes leads to reduced frequency of oscillations. Indo-1 exhibits calcium buffering similar to that observed with fura-2. In fibroblasts loaded with 3 $\mu$M indo-1, for example, agonist-induced increases in $[Ca^{2+}]_i$ rose to 800 nM within 17 s. At 5 $\mu$M indo-1, the rate to peak was 25 s, and at 20 $\mu$M, it took 37 s to reach a peak value, which was lower and broader than that obtained at 3 $\mu$M indo-1 (Wahl et al., 1990).

The potential pharmacological effects of artificially altering intracellular calcium response patterns on epithelial and exocrine secretion and other physiological responses have not yet been systematically tested in single cells presumably because, in contrast to skeletal muscle fibers and myocytes in which contractions can be continuously monitored, it is difficult to assess simultaneously secretory activity and fluctuations in $[Ca^{2+}]_i$ responses. Recent morphological and electrophysiological approaches to this problem have been described for rat parotid acinar cells and rabbit gastric parietal cells (Foskett et al., 1989; Foskett and Melvin, 1989; Ljungström and Chew, 1991). One such approach that utilizes dynamic changes in cell morphology as an index of secretory responsiveness is depicted in Fig. 5.2. In exocytotic cells, use of the reverse hemolytic plaque assays to determine whether or not exocytosis of cellular products has occurred is less satisfactory because dynamic changes in secretion cannot be detected. A recently described technique in which exocytosis is monitored by decreased fluorescence of quinacrine (a basic fluorescent molecule that accumulates in acidic secretory compartments) is somewhat more promising (Chiavaroli et al., 1991). This technique must be used cautiously, however, because quinacrine accumulation is not specific for secretory granules, and the quinacrine fluorescence emission spectrum overlaps with that of fura-2. In addition, quinacrine is a nonratiometric dye, therefore, decreased fluorescence due to dye exocytosis is complicated by photobleaching contributions. Another less satisfactory approach is to assess dye effects on responses of cell populations. Although such measurements provide no information on response heterogeneity, they are important in that nonspecific effects can be detected and, if present, hopefully corrected by changing dye-loading conditions before performing single-cell analyses.

## B. Calibration of the Fluorescent Signal

Initially, relatively simple calibration procedures appeared to provide a reasonable estimate of $[Ca^{2+}]_i$ (Grynkiewicz et al., 1985). With an explosion of literature on the subject, however, it has become clear that quantitation of $[Ca^{2+}]_i$ using available fluorescent probes is a complex and difficult task in

FIGURE 5.2

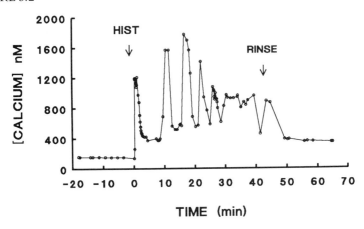

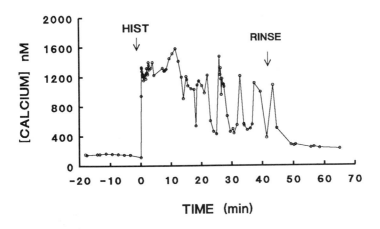

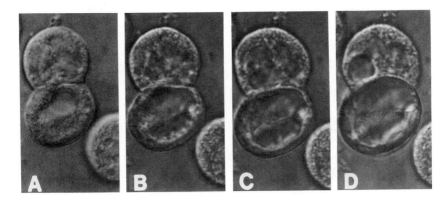

many different cell types (Tsien and Harootunian, 1990; Moore *et al.*, 1990; Williams and Fay, 1990; Roe *et al.*, 1990). In addition to potential effects of ionic strength, temperature, pH, and viscosity, calibration problems may be associated with dye compartmentalization, dye extrusion, spectral shifts in excitation–emission spectra within the intracellular environment, enhancement or suppression of peak maxima by the chemical environment or light scattering–absorption, incomplete clevage of the cell-permeable acetoxymethyl ester (AM) forms, and/or binding to intracellular proteins. Indeed, Blatter and Wier (1990) have reported that approximately one-third of the fura-2 and indo-1 that is loaded into cardiac cells by exposure of the cells to the AM forms of these indicators is not diffusable in the myoplasm. In this same cell type, Spurgeon and colleagues (1990) have suggested that approximately 50% of indo-1 is compartmentalized in mitochondria. Mitochondrial compartmentalization has been reported in endothelial cells, cardiac myocytes, and type-1 astrocytes (Lukacs and Kapus, 1987; Davis *et al.*, 1987; Steinberg *et al.*, 1987; Milani *et al.*, 1990). Compartmentalization in mitochondria-rich gastric parietal cells may also be at least partially associated with mitochondrial accumulation (Chew and Scanlon, 1990). In an earlier publication, Baylor and Hollingsworth (1988) suggested that 60–65% of intracellular fura-2 may be bound to myoplasmic proteins in skeletal muscle. Hydrolysis of fura-2-AM and indo-1-AM with retention of the free acid forms, fura-2 and indo-1, has also been demonstrated in isolated mitochondria from liver (Gunter *et al.*, 1988).

There is other evidence that fura-2 accumulates in secretory granules of mast cells (Almers and Neher, 1985). A similar phenomenon is observed in gastric mucus and peptic cells (unpublished observations). Colocalization studies of vesicles in macrophages and fibroblasts suggested that fura-2 is trapped in the same vesicles that fluoresce red when exposed to the weak base, acridine orange (Malgaroli *et al.*, 1987). Since acridine orange undergoes a spectral shift from green to red upon accumulation or "stacking" in

---

Experiment showing histamine-stimulated morphological transformations and calcium oscillations measured in fura-2–loaded gastric parietal cells. Cells were maintained in primary culture for 24 h then loaded with fura-2 by incubating with fura-2-AM (2 $\mu$M, 30 min, 37°). Fluorescent 340/380-nm images and DIC images were acquired respectively with Dage ISIT 66 and Dage Newvicon cameras. [$Ca^{2+}$]$_i$ was estimated from 340/380-nm ratio values using external calcium standards. A–D in photograph identify images acquired immediately before the addition of 10 $\mu$M histamine (A) and ~5, 15, and 30 min after addition. Note enlargement of secretory vacuoles over time. These changes have been shown to be the result of increased acid secretion that is trapped in intracellular vacuoles (canalicular spaces) in cultured cells. Acid is not trapped this way *in vivo* because canalicular openings allow acid to move from the cell into the gastric lumen. Details of this cellular model and experimental protocols may be found in Ljungström and Chew (1991). Data redrawn from this publication with permission.

acidic compartments, such data suggest that fura-2 also accumulates in acidic lysosomes in these cell types. In cells that have relatively few mito-chondria and secretory granules or in flat cells in which compartments can be excluded using video-imaging techniques, compartmentalization may pose less of a problem. Williams *et al.* (1985), for example, found a good correlation between *in vitro* and *in vitro* fura-2 calibration curves with toad gastric smooth muscle. Cultured corneal epithelial cells may also exhibit near-ideal behavior with respect to dye response (M. Scanlon, personal communication).

In other cell types, compartmentalization problems may be superseded by problems associated with incomplete cleavage of the dye. This phenome-non occurs, for example, in human polymorphonuclear leukocytes, neuro-blastoma cells, and endothelial cells (Highsmith *et al.*, 1986; Scanlon *et al.*, 1987; Oakes *et al.*, 1988). In cells that have the capacity to remove fura-2 from the cytosol, dye extrusion is often rapid and temperature dependent, suggesting involvement of ion-transport processes. In N2A neuroblastoma cells, dye extrusion levels of up to 40% of the total have been reported to occur within 10 min (Di Virgilio *et al.*, 1988). The extrusion is thought to be associated with an organic-anion transport mechanism because probene-cid and sulfinpyrazone, agents that are inhibitors of this transport mecha-nism, inhibit dye extrusion. A method for the use of these inhibitors has been described (De Virgilio *et al.*, 1989). Since transport inhibitors may exert toxic effects on cells, their use may not be appropriate in all situations. This is particularly true in studies directed at defining relationships between calcium oscillations and physiological responses.

If uncorrected, cell-viscosity effects can also lead to erroneous measure-ment of $[Ca^{2+}]_i$. With fura-2, for example, this may be due to viscosity-related enhancement of the 380-nm signal with respect to the 340 nm (Poenie *et al.*, 1986; Tsien and Harootunian, 1990). Roe *et al.* (1990) have suggested that this effect is related to a reduction in the $K_d$ of the dye based on measurements made at levels of viscosity thought to be similar to those in the cell cytosol. Wavelength-dependent light absorption–scattering and solvent polarity have also been suggested to be involved in unequal shifts of fura-2 excitation spectra and indo-1 emission spectra (Roe *et al.*, 1990; Owen, 1991; Owen *et al.*, 1991). Ionic-strength changes may also affect the $K_d$ of fura-2 for calcium (Grynkiewicz *et al.*, 1985; Williams *et al.*, 1985). If not identified and corrected, these nonspecific effects can lead to errors in calibration; however, such miscalibrations will not interfere with detection and comparison of calcium-signaling patterns in single cells within the same sampling population.

Several recent publications have outlined methods for correction of spe-cific fura-2 calibration problems. It is not yet clear, however, whether or not any single correction procedure can be universally applied. Indeed, at this juncture a universal solution to the problem is highly unlikely, particularly since it is probable that more than one calibration problem exists in any given cell type. Ideally *in situ* calibration protocols (Williams *et al.,* 1985) have the potential for overcoming many of these uncertainties. In reality, a careful perusal of the literature indicates that a number of cell types appear to be highly resistant to calcium ionophores even under conditions such as those described by Williams and Fay (1990) in which cellular-calcium regulatory mechanisms (Ca–ATPases, Na–Ca exchangers) are overwhelmed by combined addition of a calcium ionophore, such as 4-Bromo-A23187, a sodium ionophore, such as monensin, and the Na–K ATPase inhibitor, ouabain. In general, therefore, the problem of calibration should be approached with caution and an informed degree of skepticism. In experiments in which the questions relate simply to calcium-signaling patterns within single cells in a population, the emphasis on calibration is less important than in experiments in which an absolute concentration is required.

## 1. Calcium Standards

Whether or not one is fortunate enough to be working with a cell type in which a reasonably straightforward calibration of fura-2 ratios with $[Ca^{2+}]_i$ can be performed, calcium standards are quite useful for characterization and calibration of video-imaging equipment. The preparation of these standards requires a reasonable knowledge of standard chemical titration procedures and a sensitive, accurate pH meter. Titrations must be carefully performed and can be quite time consuming. These preparatory problems may be largely circumvented if one is willing and able to purchase calcium standards. For example, Molecular Probes (Eugene, OR) now offers EGTA-calcium buffers prepared as described by Tsien and Pozzan (1989). This simplified method involves preparation of a concentrated $K_2$ EGTA solution in which EGTA and calcium concentrations are equal using EGTA and $CaCO_3$ and titrating to pH 7–8 with a base, such as potassium hydroxide (KOH). The $K_2$ EGTA stock is used in conjunction with EGTA to prepare calcium standards based on the formula: $[CaEGTA] = [Ca^{2+}][EGTA_{free}]/K_d(Ca^{2+})$ where $K_d(Ca^{2+})$ is determined using the dissociation constants of EGTA for calcium at appropriate pH and temperature.

It is important to be aware that with EGTA-buffered standards, slight variations in pH will significantly alter free-calcium concentrations because protons compete with calcium for EGTA-binding sites. Marks and Maxfield (1991) have recently proposed an alternate approach in which BAPTA [1,2-bis(*o*-aminophenoxy)ethane-*N,N,N',N'*-tetraacetic acid] is used in place of EGTA. BAPTA has a calcium affinity similar to EGTA, does not interfere with fura-2 or indo-1 fluorescence because it excites and emits in the ultraviolet range, and offers the advantage of being much less sensitive to small changes in pH because it exists mainly in an unprotonated form at physiological pH (Tsien, 1980). A disadvantage of BAPTA, in addition to increased cost, is that it is approximately twice as sensitive as EGTA to changes in ionic strength (Marks and Maxfield, 1991). It should be noted that neither EGTA- nor BAPTA-calcium standards appear to be stable for more than a few weeks when stored at 4°C. The problem may be related to bacterial growth, which can alter pH. If so, freezing the solutions may prolong their shelf life. To date, there is no published quantitative data on stability of frozen solutions; therefore, storage conditions should be carefully tested.

## C. Temporal–Spatial Characteristics of Calcium Fluxes

An important consideration in choosing a fluorescent calcium-sensitive probe to define oscillations in intracellular-free calcium is the calcium-binding kinetics of the probe. Unfortunately, there is very little specific information regarding the kinetic performance of fluorescent calcium indicators within the intracellular environment at physiological temperatures. *In vitro* studies of fura-2 and indo-1 using stopped-flow spectroflorimetry (Jackson *et al.*, 1987) and of fura-2 using temperature-jump relaxation measurements (Kao and Tsien, 1988) indicate that the calcium-binding kinetics of these indicators in solution are such that calcium transients reaching a peak of $\sim 1$ $\mu M$ that rise and fall at rates slower than $\sim 5$ ms can be detected at 20°C. The response kinetics are improved $\sim 2-3 \times$ at higher temperature but are reduced at lower calcium concentrations (Kao and Tsien, 1988). It appears, however, that the behavior of these indicators in solution probably does not accurately reflect their behavior in the intracellular environment. In skeletal muscle fibers, for example, fura-2 appears to respond more slowly, with a decrease in the association and dissociation rate constants for fura-2 binding to calcium of approximately 4–8-fold at 10–16° (Baylor and Hollingsworth, 1988; Klein *et al.*, 1988). Thus, within the intracellular environment, the fura-2 response during the falling phase of the calcium transient may be increased to 20 ms or more depending on factors, such as temperature and prevailing free calcium concentrations. We know of no published data on dye kinetics in other cell types. If similar dye-response patterns are found to occur, temporal resolution may be problematic when resolution of very rapid calcium transients is attempted. With most nonexcitable cells, however, calcium oscillations detected thus far generally occur at intervals ranging from approximately 20 s to 10 min (Berridge *et al.*,

1988; Berridge and Galione, 1988; Jacob, 1990a,b). Under these conditions, the response kinetics of fura-2 and indo-1 should be adequate.

## 1. Calcium Waves and Potential Limitations in Spatial Detection of Calcium Fluxes

Recent data suggest that the rising phase of calcium transients is actually a wave that arises at a specific location and spreads rapidly across the cell (Rooney *et al.*, 1990; Jacob, 1990a). In order to resolve transients as they spread across cells and to localize calcium increases within cell compartments, high-resolution, low-lag video cameras are required. Based on current information, the speed of some calcium transients may exceed not only fluorescent probe response kinetics but also the spatial and temporal detection limits of standard video-imaging equipment (see also Section III.C). The speeds at which calcium waves have been estimated to spread across cells ranges from a low of ~0.5 $\mu$m/s detected in aequorin-loaded Xenopus oocytes (Miller *et al.*, 1991) to ~100 $\mu$m/s in fura-2- and/or fluo-3-loaded myocytes (Takamatsu and Wier, 1990; Williams, 1990). Estimates of the rate of spread of calcium waves across single epithelial cells range between 15–50 $\mu$m/s (Rooney *et al.*, 1990; Kasai and Augustine, 1990; Sanderson *et al.*, 1990; Jacob, 1990a). A recent report suggests, however, that calcium waves may move across hepatocyte couplets maintained in short-term culture at rates between 100–200 $\mu$m/s (Nathanson and Bergstahler, 1992).

The ability to resolve calcium waves spatially and temporally as they move across a cell is dictated by the sampling theorem, which states that at least two samples per cycle at the highest frequency response must be acquired to avoid aliasing (Inoué, 1986; Castleman, 1987; Young, 1989). If aliasing is severe, high-frequency components will appear to have low-frequency responses, an example of which can be seen in old movies where wheels on covered wagons appear to be turning in reverse. Given the current limitations on low-light-level video cameras, rapid acquisition of high-resolution images is difficult with small cells (10–20-$\mu$m diameter), particularly if they are rounded and contain fura-2-loaded granules that do not respond to changes in $[Ca^{2+}]_i$ with the same kinetics as the bulk cytosol. For example, if a calcium wave moves completely across a 10-$\mu$m cell at a rate of 20 $\mu$m/s, the wave will cross a cell in 500 ms. To even partially resolve this wave in the ratiometric mode, a minimum of one $\lambda$ pair/250 ms is required. Thus, with fura-2, an image must be acquired at least every 125 ms. If the wave moves at a ratio of 100 $\mu$m/s, an image must be acquired at least every 25 ms, a value that may be within the range of the kinetic limitations of the dye. If calcium is not only released from intracellular storage sites but also enters the cell from the extracellular medium, the spread of calcium waves through the cell will be even more difficult to characterize than a simple wave that arises on one side of the cell and moves to the opposite side. The ability to resolve such rapid changes in $[Ca^{2+}]_i$ will depend on the SNR and image-retention (lag) characteristics of the low-light-level video camera as well as mechanical constraints of the excitation filter changer, the numerical aperture of the objective and image storage time. Some of these limitations can be overcome by using a cooled CCD camera and high-excitation intensities. This approach will produce a sufficiently high SNR to allow for a substantial decrease in total exposure time, hence image acquisition is more rapid (See Section IV.C. below).

Laser-based confocal microscopy offers the potential advantage of allowing for more precise localization of the origins of calcium transients because under optimal conditions a single thin section of the cell can be studied at video rates with rejection of out-of-focus fluorescence (cf. Niggli and Lederer, 1990; Nathanson and Bergstahler, 1992; see Lemasters *et al.*, Chapter 12, this volume). There are, however, several problems with this methodology in its present state of development. For example, the limited availability, high cost, and instability of lasers exciting at wavelengths in the near UV presently force the use of nonratiometric dyes, such as fluo-3, in most situations. BioRad (Cambridge, MA) has recently begun to package a high-powered argon laser that provides useable excitation lines at 351/363 and 488 nm and a helium–cadmium laser that can excite at 325 and 442 nm. The argon laser, therefore, can be used in indo-1 experiments in conjunction with dual PMT detectors and the helium–cadmium and argon lasers can be used in combination to excite fura-red. Since achromatic deflection is required for emission wavelength detection, image registration should not be a problem when dual detectors are used. However, unlike the standard low-powered argon lasers, both of these lasers are quite expensive and short lived. The utility of fura-red in calcium-oscillation experiments has not yet been demonstrated and is questionable because of the relatively low-quantum yield of this probe. Even with fluo-3, which has a substantially higher quantum yield as compared to indo-1 or fura-2, the few confocal studies of calcium waves that have been performed with this probe have utilized a fully opened aperture to increase signal intensity. This aperture setting substantially reduces z-axis resolution.

For flat, thin cells a few microns thick, confocal microscopy presently offers no clear advantage over conventional video-images analysis systems. Even with thicker cells, the utility of confocal microscopy is offset by the ever-present potential for biological photodamage when living cells are exposed to high-intensity lasers for extended periods of time. Such exposures are unavoidable in long-term experiments. It is well known that prolonged exposure of cells to intense light leads to increased elevation of fluoroprobes to the "forbidden" triplet state, a condition that appears to be primarily responsible for dye photobleaching, singlet oxygen production, and cellular photodamage (see also Section IV.C.). Tsien has proposed an interesting potential solution that involves attachment of triplet-state quenchers to fluorophores (Tsien and Waggoner, 1990). Unfortunately, this proposal is not yet a reality. Because of the drawbacks of current confocal microscopy techniques, image-deconvolution methods, such as those described by Agard and colleagues (1989) and Fay and colleagues (1989), in conjunction with faster, more powerful computers and improved video cameras may provide a better solution in many dynamic experimental situations.

## IV. EQUIPMENT SUITABLE FOR DETECTION OF CALCIUM OSCILLATIONS

There are many different factors that should be considered in the design of an appropriate video-based image-analysis system. The basic components include the microscope, an image processor and computer equipped with appropriate software and mass storage capacity, an air table to avoid vibration artifacts, computer-controlled electronic shutters to control light input, a filter changer for alternating excitation wavelengths, and heating and perfusion systems for the microscope stage. (One possible configuration is shown in Fig. 5.3). Although most of these components can now be purchased in a single system, it is useful to consider the qualities that provide for optimal performance. These factors will vary depending on the experimental protocols one plans to employ. For detection of rapid calcium transients, the system must obviously be optimized for speed. Current video camera limitations unfortunately require the sacrifice of spatial resolution in favor of increased temporal resolution. A recent lively and detailed discussion by Tsien and Harootunian (1990) provides an excellent description of desired characteristics and limitations of video-based image-analysis systems.

### A. Microscopes and Accessories

For most cellular studies, an inverted fluorescent microscope is preferred because this configuration allows for direct manipulation of cells. The disadvantage of inverted versus upright microscopes is that more light-absorbing optical components are present in the light path. The authors have direct experience with two different Zeiss inverted microscopes, the older IM35 and the Axiovert 35, Although the IM35 is more stable, the Axiovert has fewer components in the light path. Both microscopes are relatively easy to modify to fit changing experimental requirements; however, the standard nose piece on Axiovert does not accept either the older Zeiss objectives or objectives from other manufacturers because it is designed to support the

FIGURE 5.3

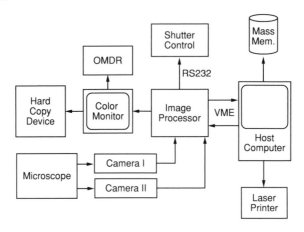

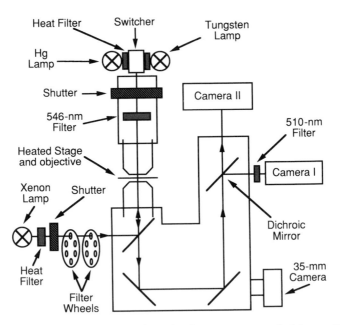

Diagram of one possible equipment configuration for measurement of calcium oscillations in single cells loaded with the fluorescent calcium indicator, fura-2. The top portion of the figure summarizes input–output pathways of the image processor and host computer. Hard-copy device linked to the color monitor may be either a black and white or a color printer. The optical memory disk recorder (OMDR) is an analog device. The lower portion of the figure shows details of the microscope configuration. This particular microscope was modified as described by Foskett (1988) and Ljungström and Chew (1991) to allow use of two cameras at the same time. Camera I in the figure is used for low-light-level detection of fura-2 fluores-

Zeiss infinity corrected optics design. Therefore, if use of other objectives is planned, a modified nose piece must be used. Since fura-2 excites at wavelengths in the near UV, it is important when use of this probe is planned to make certain that lenses in the lamp collector and the telan lens in the nose piece of the older Zeiss microscopes are quartz.

With respect to microscope objectives, in our experience and the experience of others (cf. Moore *et al.,* 1990), Zeiss objectives that are not quartz are inferior to the Nikon UV series in allowing passage of light at 340 nm. The problem is most serious with the older Zeiss 63×, 1.25 NA, Neofluar oil-immersion objective; however, the newer "fura-selected" Zeiss 100×, 1.3 NA and the 40×, 1.3 NA Neofluar oil-immersion objectives have also been found to perform less well in side-by-side comparisons with Nikon 100×, 1.3 NA, glycerol immersion and the 40×, 1.3 NA UV Fluor oil-immersion objectives (unpublished observations). In contrast, the Zeiss infinity-corrected 40×, 1.3 NA acrostigmat performs quite well. If there is a substantial decrease in light passage at 340 nm, one will be forced either to increase dye loading into cells or to increase light intensity to acquire sufficient fluorescent signal to be detected at this wavelength. As previously discussed, increased dye loading will buffer and, therefore, suppress calcium transients and expand peak widths, an undesirable result. Even with reasonably good light passage at 340 nm, it is usually necessary to balance the 340- and 380-nm signals with neutral density filters. In the absence of a reasonable balance, the ability to collect images at the two wavelengths will be seriously compromised either because of the limited dynamic range of the camera in the case of tube and intensified CCD cameras or because of significantly increased exposure times in the case of the wider dynamic range cooled CCDs. Also, when the 340–380-nm signal intensities are well balanced, the SNR at the two wavelengths will be similar. Good balance between the two wavelengths will minimize the error in the calculated ratio values (Moore *et al.,* 1990). (See also Section IV.C.)

The microscope should be equipped with a xenon light source. Xenon lamps have an even-light output in both the UV and visible ranges in contrast to mercury arc lamps, which have sharp peaks at several wavelengths,

---

cence. Camera II is used for acquisition of high-resolution DIC or phase images. One or two filter wheels may be used for changing excitation wavelengths and balancing light throughput with neutral density filters. The tungsten lamp on the top of the microscope is used to provide light for set-up operations and routine use of the microscope. The mercury (Hg) lamp is used for high-resolution DIC analyses. See text for details. For very rapid sampling times, two xenon lamps can be placed on the back of the microscope. In this case, a single excitation filter is placed in front of each lamp and filter wheels are omitted. Unless standard filter holders are modified, excitation filters, neutral density filters, and heat protection filters must be placed between the lamp and electronic, computer-controlled shutters. The shutters open and close in ~5 ms.

including 365 nm, the wavelength that closely coincides with the isosbestic ($Ca^{2+}$-insensitive) point of fura-2 (Grynkiewicz *et al.*, 1985). It is useful, however, to have a mercury arc lamp on the microscope in addition to a xenon to increase the flexibility of the system. For example, a mercury lamp may be needed for differential interference contrast (DIC) microscopy when high-magnification objectives are used in conjunction with high-resolution cameras that have relatively low-light sensitivity, such as the newvicon series, which are used (Ljungström and Chew, 1991). Mercury lamps provide a more stable light source than tungsten–halogen lamps. These lamps are also needed to detect weak fluorescent signals from fluoroprobes that are optimally excited at wavelengths at or near the peak emissions of this lamp [as rhodamine and fluorescein isothiocyanate (FITC)].

Although ozone-free lamps are more expensive, they are well worth the extra cost because other lamps generate unacceptably high levels of ozone even in a relatively well-ventilated room. Unless one plans to use a liquid-light guide or some other device to provide diffuse light, a 75-W xenon lamp produces the appropriate light intensity. Lamps of higher wattage are not generally recommended because they generate more heat and can potentially damage microscope filters and shutters. Even the 75-W xenon is problematic in this regard. In our present system configuration, for example, a UG5 filter (Omega Optical, Brattleboro, VT) is placed between the lamp output and a Unibiltz shutter (Vincent Associates, Rochester, NY), which remains closed except during sampling (Fig. 5.3). Although the UG5 filter is metal coated for added heat protection, cumulative UV-light exposure forces its replacement every few months.

With regard to lamp functionality, xenon lamps are viable for approximately 400 h. New lamps should be left on for several hours when ignited for the first time to stabilize the arc. If arcs are not stabilized, they will jump around on the electrode leading to increased noise and unpredictable experimental conditions. As xenon lamps age, deposits form inside the bulb. These deposits can reduce lamp output and increase lamp temperature. If the temperature reaches unacceptable limits, lamp pressure, which normally is ~10× higher than atmospheric, will increase to levels sufficient to explode the lamp. Lamp explosion is not only dangerous but can also cause serious damage to the lamp house mirror and quartz collector. Since arc lamps can generate radio frequency noise and emit high-voltage transients when ignited, direct experience has shown that digital electronic components can be damaged if they are not properly grounded. To protect cameras, filter-wheel controllers and other electronic equipment, lamps should be ignited before this equipment is turned on. Power supplies should also be placed as far away from this equipment as possible. Furthermore, since unstable power supplies can have significant effects on the fluorescent signal detected by the video camera, only very stable power supplies should be used. If the power supply is sufficiently unstable, artifactual oscillations in the ratioed fluorescent images may occur. A similar phenomenon can occur if the lamp output fluctuates or if the temperature on the microscope stage fluctuates to such a degree that expansion and contraction of the coverslip containing the cells occurs.

In order to avoid vibration artifacts, an air table is essential. Reasonably priced tables can be purchased from, for example, TMC in Peabody, Mass. (Micro-g series) or Kinetic Systems in Roslindale, Mass. Ideally the temperature of cells on the microscope stage should be controlled by heating the microscope objective, which is a major heat sink as well as that portion of the microscope stage occupied by the cells. Accurate temperature control is particularly important in calcium–oscillation experiments because changing temperatures can affect oscillatory patterns. For example, Hajjar and Bonventre (1991) have shown that stepwise increases in temperature lead to incremental increases in oscillation frequency. In an earlier study, Gray (1988) found that sinusoidal calcium oscillations in parotid acinar cells occurred at room temperature but not at 33°. This reverse effect was attributed to a temperature-related suppression of enzyme-regulated feedback mechanisms associated with regulation of calcium homeostasis.

In general, we have found commercially available stage warmers to be either too expensive or too unwieldy to use in experiments. After undergoing many permutations, our present configuration is composed of a temperature-controlled light-weight aluminum carrier, which was fabricated to hold a 35-mm glass-bottomed dish with a plexiglass insert containing input and output orifices. There is a narrow rectangular slit in the insert to allow for rapid media changes as well as constant perfusion with a reasonably laminar flow. The objective is maintained at approximately 37° with a temperature-controlled aluminum collar. Unfortunately, repeated heating and cooling will ultimately damage objectives and we know of no way to avoid this problem. Perfusion is accomplished either by gravity flow or with a pump. Perfusion media is maintained slightly above 37° in a water bath or water-jacketed, hand-blown glass chamber. Unless one plans to set up a controlled $CO_2$ environment, bicarbonate must be omitted from the perfusion media. Otherwise media pH will rapidly become basic providing a distinctly nonphysiological environment for the cells. Although the effect of bicarbonate on pH should be obvious, it is often overlooked by those who routinely use commercially prepared media.

## B. Filter-Changer Configurations

There are a number of possible configurations for controlling excitation wavelengths, fluorescent-light intensity, and duration (cf. Tsien et al., 1985; Kruskal et al., 1986; Kurtz et al., 1987; Ljungström and Chew, 1991; Linderman et al., 1990; Moore et al., 1990; Jacob et al., 1990b). For excitation of fura-2, two different configurations have been widely used. The first employs a filter wheel that can alternate several excitation filters. The better wheels (as, for example, the LEP series, LEP Ltd., Hawthorne, N.Y.) have sufficient space in their carriers to accommodate insertion of neutral density filters, which are used to decrease excitation light intensities and, as previously discussed, are also needed to balance light intensities at different excitation wavelengths. Dual-filter wheels are useful because they allow independent variation of neutral density and excitation filters; however, these configurations are more expensive and operation of two wheels may decrease sampling speed. Since most single-filter wheels can now alternate filters at speeds in the 50–200-ms range, these wheels are sufficiently fast for most video applications.

The second type of commonly used configuration employs dual-excitation lamps either with excitation and neutral density filters or with grating monochromators placed in front of each lamp (Kruskal et al., 1986; Tsien et al., 1985). To control the exposure of cells to light, computer-controlled electronic shutters are placed between the filter wheel and lamp or between the excitation filters in front of the lamp and the sample if filters in the dual-lamp configuration are placed in carriers in front of the lamp. Although the dual-lamp configuration is mechanically simpler and allows for substantially faster switching (~5 ms) between excitation wavelengths, this configuration has major disadvantages. For one, it is difficult to align the two lamps; therefore, shading problems can be severe unless elaborate light diffusion–guide techniques are employed. Another problem is that the most convenient placement of excitation filters is next to the lamp where they are not protected by shutters. Constant exposure of excitation filters, even the newer metal-coated filters, leads to burnt and cracked filters on a regular basis. After working with both configurations, we have settled on the filter wheel as the best alternative at the present time. The speed at which excitation filters can be changed is adequate for the detection of moderately rapid transients, and there is considerably more flexibility in the choice of excitation filters during an experiment. Since a single-excitation wavelength is required for indo-1, acquisition of ratioed images is not limited by excitation procedures. Indeed, it is possible to acquire images from dual cameras as rapidly as signals can be converted from analog to digital and recorded (i.e., at video rates of 30 frames/s with a real-time image processor). Other configurations, such as the rapidly alternating "cube" described by Jacob et al. (1990b), offered increased speed at the expense of decreased flexibility because neutral density filters cannot be used along with excitation filters. The advantages and disadvantages of dual monochromators have been discussed in detail by Tsien and Harootunian (1990). The acoustooptical tunable filter (AOTF) has great potential for producing rapid wavelength changes ($\mu$s resolution, Kurtz *et al.*, 1987); however, there is presently no commercially available AOTF that works well in the UV. Moreover, these filters are not efficient at blocking light that falls outside the wavelength of interest. A liquid crystal tunable laser under development at Cambridge Research Instruments may provide a better solution in the future.

## C. Video Camera Considerations

As previously discussed, exposure of cells to intense light prolonged exposure to low-light levels can cause photodamage to cells and photobleaching of fluorescent probes (see also, Piccilo and Kaplan, 1984; Benson *et al.*, 1985; Plant *et al.*, 1985; Foskett, 1985). With photobleaching there is a resultant reduction in SNR and, in the case of fura-2, generation of calcium-insensitive forms of the probe (Becker and Fay, 1987). In particular, UV-light-induced damage is a major concern in experiments in which cell calcium levels are monitored for extended periods of time. For example, Bals *et al.* (1990) have shown that calcium transients can be monitored continuously in single myocytes for only 9 s using the full-light intensity from a xenon lamp. Upon longer exposure, photobleaching was apparent and the beating frequency was reduced with apparent decreases in both resting and peak $[Ca^{2+}]_i$ measured with fura-2. Such data emphasize the importance of optimizing and balancing light throughput capabilities of the system, particularly when longer sampling times, such as those required for accurate characterization of oscillatory patterns, are anticipated.

There are currently three major categories of commercially available low-light-level video cameras that are of sufficient sensitivity to be utilized in video-image analysis experiments. These include the technologically older image-tube-type cameras (the silicon intensified target cameras, SITs, and intensified silicon target cameras ISITs), the newer intensified charge-coupled devices (CCDs), and the thermoelectrically cooled CCDs. (Detailed discussions of these cameras can be found in the following representative publications: Wampler, 1985; Inoué, 1986; Bright and Taylor, 1986; Connor, 1986; Wick, 1987; Hiraoka *et al.*, 1987; Spring and Lowy, 1989; Aikens *et al.*, 1989; Aikens, 1990; Bookman, 1990; Tsay *et al.*, 1990). General performance characteristics of each camera type have been published and these characteristics are also described to some degree in the manufacturer's literature. As discussed by Spring and Lowy (1989) and Tsien and Harootunian (1990), camera-sensitivity ratings are difficult to compare because performance criteria are not uniform. This is particularly problematic when camera sensitivities are compared at wavelengths that are unrelated to dye-emission wavelength characteristics, and the degree of image retention (lag) after the light input is blocked is defined at high-light intensities rather than low intensities where lag is more pronounced. Since different cameras have different spectral sensitivities and performance varies with individual cameras, specifications provided at longer wavelengths may have little relationship to responses at lower wavelengths. The best approach is to consult the manufacturer's literature then perform side-by-side comparisons of cameras of interest under conditions that closely mimic those to be used during experiments. Based on such comparisons, we and others have formed definite, sometimes conflicting, opinions about camera performance. A point of general agreement is that there is presently no perfect camera available for low-light-level biological experiments.

## 1. SITs, ISITs, and Intensified CCDs

Both the SITs and the ISITs suffer from problems with image retention (lag), shading, nonlinear responses to light, and geometric distortion. In terms of light sensitivity, the ISIT is $\sim 10 \times$ more sensitive than the SIT and $\sim 5 \times$ more sensitive than intensified CCDs (Inoué, 1986; Spring and Lowy, 1989; Tsay et al., 1990). The trade-off is that the ISIT is an electrically noisier camera than the SIT or the intensified CCDs. With all three cameras, however, some frame averaging is needed to reduce noise to acceptable statistical levels. When compared to the SIT, the ISIT is frequently criticized for its nonlinear response to changing light intensities. This is certainly true if internal protection circuitry of the ISIT is not disabled. With such disabling, we have found in direct comparisons at low-light levels, no major difference in image quality or linearity of response between our Dage-MTI model 66 ISIT equipped with a research grade tube (I) and disabled internal gamma and automatic gain control and a Dage-MTI SIT. Although nonlinearity of response and limited intrascene dynamic range are problems with the both the ISIT and the SIT, carefully controlled light levels and camera gain can produce reasonably linear, reproducible responses. In Fig. 5.4, for example, side-by-side comparisons of the ISIT with a Dage Genllsys intensifier coupled to a Dage Model 72 CCD demonstrate that similar 340–380-nm ratio values can be obtained using calcium standards ranging between 0 and 2000 nM calcium concentrations. [Weir and colleagues (1987) have reported a similar correlation between an ISIT and a PMT]. In Fig. 5.4, the limited intrascene dynamic range of both cameras did not allow measurement of calcium concentrations above 2000 nm at the fixed gain settings used in the experiment. A serious problem with some of the intensified CCDs is the automatic shutdown function that is activated in response to light intensities approaching the upper limits of the camera's intrascene dynamic range. Since this response may occur without warning in some makes of cameras, quantitation may be severely compromised.

In experimental situations with living cells, it is often impossible to avoid situations in which nonlinear camera responses occur. For example,

FIGURE 5.4

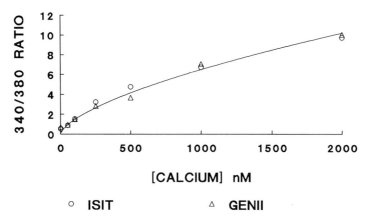

○ **ISIT**         △ **GENII**

Comparison of calcium standard curves obtained with Dage ISIT 66 and Dage GenIsys intensified CCD cameras. Both cameras were mounted on a Zeiss IM35 microscope at the same time using a configuration similar to that in Fig. 5.3 except the 510-nm emission filter was returned to the standard position in the Zeiss carrier in the body of the microscope and 100% of the light was alternately directed to one or the other camera. Images of fura-2 calcium standards prepared as previously described (Ljungström and Chew, 1991) were acquired at 340- and 380-nm excitation wavelengths (Omega Optical, 10-nm bandpass) using Inovision IC300 software on a Sun 3/280 computer. Ratio values computed following background subtractions were plotted versus calcium concentrations of the standards. Note the similarity in the values obtained with the two cameras. Due to the limited intrascene dynamic range of both cameras, ratio values at higher calcium concentrations could not be obtained at the fixed gain settings used in the experiment.

when a relatively flat cell or a cell with fine, thin processes is situated next to a rounded cell that fluoresces more brightly because of the difference in depth-dye concentration, the difference in fluorescent signal from these cells is such that the fluorescent signals from one or the other cell will fall in the nonlinear range of the camera. Since the intrascene dynamic range of the intensified CCDs is approximately the same as that of the ISIT at fixed gain, this camera is also unable to compensate for differences in cell brightness. If both the bright and dim cells are analyzed, they will appear to have very different apparent calcium concentrations because one of the two will fall outside the useable intensity range of the camera. The problem may be exacerbated with cameras that have automatic gain suppression functions. With the SITs, ISITs, and intensified CCDs, the best approach in such situations is to be aware of their limitations at the outset and either use only data that falls in the linear intensity range of the detector or qualify the interpretation of the data to emphasize its nonquantitative nature. Software-controlled thresholding can be used to reject signals with intensities falling in the nonlinear range. It is also possible for nonlinear intensity (within

FIGURE 5.5

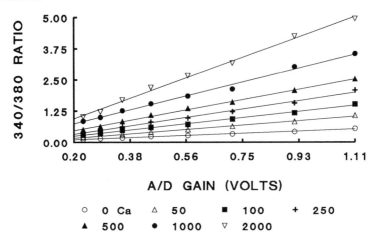

A/D GAIN (VOLTS)

○ 0 Ca    △ 50    ■ 100    + 250
▲ 500    ● 1000    ▽ 2000

Effect of changing gain on a programmable digitizer on 340/380-nm ratio values obtained from fura-2 calcium standards. Data was acquired with a Dage GenIsys intensified CCD as described in Fig. 5.4 except gain settings on the Datacube Digimax digitizer used in this experiment were sequentially increased to cover the entire voltage range of the digitizer for the 340-nm excitation wavelength. The 380-nm wavelength was fixed at 0.45 V. Note that the response of the digitizer was linear over the entire voltage range. Increasing the voltage for full-scale input to the digitizer decreases sensitivity but improves the dynamic range of ratio values obtained at increasing calcium concentrations.

limits) to correct responses by mapping the camera response curve to a linear curve and applying the mapping function to individual images (Bright *et al.*, 1989); however, this approach is quite time consuming and software intensive and, therefore, not practical on a day-to-day basis.

Under some conditions it may be possible to adjust the voltage of the A/D converter to improve the dynamic range of the signal. Data in Fig. 5.5 show that the response of our Datacube Digimax A/D converter is linear between 0.23 and 1.1 V. Similar experiments can be performed with other A/D converters to define their linearity. The data in Fig. 5.5 also emphasize a point that is generally not mentioned in the literature (i.e., the range of ratio values obtained at different calcium concentrations is dependent not only on camera gain adjustments but also on the gain of the A/D converter). Clearly it is better to change the gain on the digitizer rather than on the SITs and ISITs because changing the gain on the digitizer unlike changing the camera gain can be readily normalized to correspond to different settings used in other experiments. An advantage of the intensified CCDs is improved linearity with different camera gain settings.

For detection of relatively slow calcium transients, lag problems associated with the SITs and ISITs (~25–250 ms, depending on light intensity) can be avoided because images do not need to be acquired at rates faster than those where lag is apparent. When long-term experiments are a priority and sampling rates are seconds to minutes apart, the ISIT may be the camera of choice over the SIT or the intensified CCDs because the extreme light sensitivity of this camera allows one to work at lower light levels and thereby reduce light damage and avoid photobleaching problems (see Fig. 5.6 for example).

Important advantages of the microchannel plate intensified CCDs over the tube-based cameras include reduced lag ($\mu$s), reduced shading and geometric distortion, and improved linearity of response. As previously discussed, these cameras offer no significant improvement over the ISIT with respect to static dynamic range. Thus, the ability of these cameras to respond linearly and reproducibility to gain changes is often of little value in a dynamic experimental setting. We have also found the quality of transmitted light images acquired with these cameras to be disappointingly similar to those acquired with our ISIT. Despite these limitations, in experiments in which rapid calcium oscillations are anticipated, the intensified CCD is the obvious choice over the tube-based cameras. If one is willing to sacrifice image resolution, sampling rates of $\sim$16 ms can be achieved using indo-1 (Takamatsu and Wier, 1990). Reduced shading problems with these cameras may also allow sampling of a larger field of cells with improved sampling statistics. It should be noted, however, that there are likely to be

FIGURE 5.6

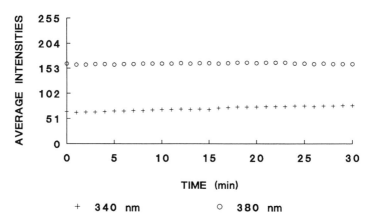

Lack of photobleaching in fura-2–loaded cells during prolonged exposure to very low light levels. Gastric parietal cells in primary culture were loaded with fura-2 as described in Fig. 5.2, rinsed and placed on the stage of a Zeiss IM35 equipped with a Nikon 100×, 1.3 NA UV Fluor objective and Dage ISIT 66 camera. Cells were exposed to constant light for 30 min (380 nm, 97% attenuation with neutral density filters). Ratioed images were acquired using IC300 software at 1-min intervals during the exposure period.

other sources of shading, including lamp illumination inhomogeneity and differences in optical paths introduced, for example, by the use of different dichroic filters, such as those required for indo-1 experiments. Another factor to be considered is that CCDs in general display considerable sensitivity at wavelengths above 600 nm. If this light is not excluded with appropriate cut-off filters, the noise may be so high that appropriate fluorescent signals are obscured (Tsay *et al.*, 1990 and unpublished observations).

## 2. Cooled CCDs

At present there are relatively few laboratories using water or liquid nitrogen cooled CCDs to image calcium transients. The main reasons for this low usage are that the cooled CCDs are more than twice as expensive as the intensified CCDs, are difficult to interface into standard image-analysis systems because they do not utilize the RS170 video standard, and are difficult to focus on the sample because of their slow readout times. In addition, unless the CCDs on these cameras are masked physically or in software, the rate of image acquisition is unacceptably slow. These cameras do, however, offer a number of important improvements over the other two types, including a greatly expanded intrascene dynamic range (up to 50,000:1 versus ~40:1 at fixed gain settings for SITs, ISITs, and intensified CCDs), negligible geometric distortion, and noise (Inoué, 1986; Aikens *et al.*, 1989; Aikens, 1990). There are currently two distributors who offer various cooled CCD configurations for biological use, one in the U.S. (Photometrics, Ltd., Tucson, AZ) and one in Great Britain (AstroMed).

For video-imaging applications there are two relevant architectures, the full-frame and the frame-transfer type CCDs (Aikens *et al.*, 1989; Aikens, 1990; Connor, 1986; Hiraoka *et al.*, 1987). Full-frame CCDs have a single parallel register that collects photons that are then shifted sequentially into a serial register. These charges are then transported to the output amplifier. The process of moving charge into the serial register and out to the amplifier is referred to as "readout." The reason the full-frame CCDs are slower than the other camera types is that readout takes 1–5 s or more to complete depending on the pixel resolution of the CCD. In general, these configurations offer high-quality images but are too slow for detecting calcium oscillations. The frame-transfer CCDs utilize an opaque mask to cover the parallel array next to the serial register (storage array). An equally sized parallel array (image array) is used to acquire the image, which is then rapidly (~1 ms) shifted to the masked storage area from which it is read out via the serial array. This configuration allows for acquisition of two images in rapid succession, but readout times are the same; therefore, temporal resolution is limited to two images or one ratio pair/5 s for a 512 × 512K CCD. Since only half of the storage area is used to acquire the image, spatial resolution is lost. The CCD readout may be programmed so that limited pixel arrays are acquired. Speed of acquisition can also be increased by combining pixels (binning); however, this procedure results in loss of spatial resolution (Lindeman *et al.*, 1990). Since the light sensitivity and pixel sizes (smaller pixels mean improved resolution) of different brands of CCDs can vary substantially, these variables should be carefully considered prior to purchase.

Despite the current limitations of the cooled CCDs, these cameras offer the best potential for acquiring useable, quantitative biological information. The utility of the cooled CCDs is becoming more apparent as laboratories define more creative methods to circumvent their limitations. For example, Fay and colleagues have recently built an ultrafast three-dimensional imaging microscope that utilizes a modified cooled CCD to capture fluorescent images on a ms time scale. Several images in different focal planes can be captured in rapid succession before the slow readout process is activated. Captured images are deconvolved off line to create a final three-dimensional image. With this technique, mitochondrial membrane potentials have been estimated in single, living neuroblastoma cells (Loew *et al.*, 1991). In order to employ this technique with fura-2, a powerful UV excitation source is required to provide sufficient light for the short exposure times employed. At present, this is an expensive proposition because such lasers cost ~$15,000–$20,000 and are difficult to maintain in good working order.

The development of calcium-sensitive fluorescent probes with high-quantum yields that excite at longer wavelengths emitted by the less expensive, more stable lasers will be important to determine whether or not short bursts of high-intensity light as compared to low-intensity, longer exposure cause more or less biological perturbation to cells. This will depend not only on the effects of high-intensity light but also on total exposure times employed with the different techniques (i.e., total dose = intensity × time). Bonhoeffer and Staiger (1988) have found exposure of brain slices to 100 × 200 ms of light with 30-s "rest" intervals to be less harmful than a sustained 20-s exposure. They speculated that light-induced membrane damage caused by generation of free radicals can be repaired during rest intervals. In the absence of such intervals, membrane repair may not occur. More definitive experiments are required to determine how many rapid bursts can be administered to a cell before a rest period is required and the amount of rest time needed for repair.

## D. Computer-Operating Systems

An ideal video-image analysis system is based on a real-time, multitasking operating system capable of simultaneously controlling filter wheels, shutters, video cameras, image processors, image storage, and changing experimental protocols (Tsien and Harootunian, 1990). Unfortunately, real-time operating systems generally have a limited to nonexistent file structure, poor third-party software support, and usually do not have standard graphics interfaces. For these reasons, commercially available software for ratioing applications is typically written to run under either the DOS or the UNIX operating systems, neither of which provides true real-time capabilities. The advantage of using either DOS or UNIX is the availability of a substantial amount of supporting graphics software and broad-based hardware support. However, DOS computers do not perform well in a multitasking environment. These computers are also not good choices for handling and processing of large amounts of image data generated in long-term experiments. The performance of DOS computers can be improved by adding dedicated image processors with large, specialized capabilities, including fixed array processors and boards to synchronize cameras and filter changers (Theler *et al.*, 1991; O'Sullivan *et al.*, 1989); but graphics and image-handling capabilities remain limited. Although UNIX is not a true "real-time" operating system, UNIX does perform well in a multitasking environment, and UNIX programs can be written in such a way that essential operations, such as timed image acquisition and synchronized filter changes, are accomplished within fixed time limits. For example, UNIX can readily manage real-time operations by addressing real-time targets, such as high-ended image processors that are linked to the CPU through a standard VME bus. The recently developed SPARC architecture implemented mainly on Sun workstations has improved CPU performance speed to such a degree that timing problems may no longer be an issue. The cost of these computers has also decreased dramatically. Indeed, it is now possible to purchase a suitably powerful workstation for about the same price as that required to purchase a reasonable DOS machine 3–4 years ago. Major advantages of UNIX computers include multitasking capabilities, the ability to manage large amounts of data, and the ability to display data in high resolution.

## E. Image Processors

There are a bewildering array of image processors on the market. A number of these are characterized in a book edited by Inoué (1986); however, new and better processors have become available since this book was written. In order to provide sufficient image-acquisition speed, image processors must operate as close to real time as possible. Image processors are actually a composite of several different components that serve different functions. During image acquisition, images from analog video cameras are first converted into digital (numeric) information by a video digitizer (A/D converter). The most commonly used digitizers have 8–12 bits of resolution.

The better digitizers can be programmed to accept a full range of image resolutions. (For example, the Datacube MAX-SCAN we currently utilize in our Inovision IC-300 system (RTP, N.C.) will accept images at resolutions ranging from 1 × 1 to 4096 × 4096 pixels). Once images are digitized, they are manipulated by an array processor with digitized image memory, one or more arithmetic logical units (ALUs), and several look-up tables (LUTS). ALUs perform a variety of real-time mathematical operations, such as averaging, addition, subtraction in a pipeline fashion, one pixel at a time. Pipeline processing is typically 16 bits wide. LUTs convert grey-level values of individual pixels into different grey values or a color as specified by a particular table. They serve to transform both input operands and output from the ALUs. Several large LUTs should be available for maximum speed of performance. Our Datacube MAX-MUX contains 64K × 16-bit RAM for LUT applications. The LUTs convert 16-bit input into 16-bit output. This allows computation of the ratios of two 8-bit images at frame rates. Our Datacube MAX-SP also performs frame averaging and image subtractions at frame rates. We have found this configuration to be quite adequate for ratioing applications. Other useful features include framestore modules that have user-programmable video resolutions and the ability to accommodate video signals from non–RS-170 sources. The ability to program video resolutions is invaluable in experiments in which it is desirable to save only a portion of an image in order to reduce image-storage problems. Also, with a good framestore module, smaller images can be processed at speeds faster than video rates (For example, a Datacube ROI-STORE can process a 128 × 128 sized ROI in ~2 ms).

Digitized images can be stored directly on one or more high-capacity magnetic hard disks on digital optical disks or on tape. Banks of hard disks can be used to improve the speed of acquisition (Ryan et al., 1990). Processed images may also be reconverted to analog format for storage on analog recorders (as a Sony U-matic or a Panasonic optical memory disc recorder—see below).

## F. Data-Storage Strategies

Whichever type of computer and operating system one chooses, the importance of acquiring as much memory as possible for image storage cannot be overemphasized. The problem is obvious if one considers that one 512 × 512 × 8-bit image requires ~¼ Mbyte of hard-disk space. In a single experiment designed to characterize calcium-oscillation patterns, hundreds to thousands of images may be acquired. Although the size of each image can be reduced with appropriate software, the storage problems can be monumental. In terms of on-line performance, a large amount of random access memory (RAM) and direct access to magnetic hard-disk storage is presently the best available solution. With sufficient RAM, a number of images can be acquired in real time. With hard disks with nanosecond access times, image data and histories for separate images can be written and accessed almost instantaneously. In addition, image quality is not degraded. Unfortunately, this media is also the most expensive and has a finite capacity. Hard-disk capacity can be increased by 40% or more using image-compression routines. For UNIX users, the "compress" command is recommended rather than the "compact" command because data can be lost with the algorithms used by the compact command. When compress is used, data is restored intact by the "uncompress" command. In the future, it may be possible to compress data as it is collected. These techniques are now under development. Since the hard disk is such a convenient and fast data-storage option, it is expected to continue to occupy a prominent position as an important and necessary device.

After the hard disk, the next best choice is an on-line optical memory disk recorder (OMDR), a device that is significantly slower than hard disks. Both analog and digital recorders are available. Optical disk recorders may also be purchased in the write once, read many (WORM) configuration, as erasable versions or with both options. Currently, reasonably priced analog WORMs, such as the Panasonic TQ-2028F (Secaucus, NJ), are faster and more reliable than similarly priced digital recorders. For WORMs in general, the average access time is 80 ms with a data-transfer rate of 1500 kbyte/ s and the average disk capacity is 3200 Mbyte. The popular TQ-2028F is faster and has a higher disk capacity than average. This recorder can write images to disk at frame rates and stores 16,000 images/disk. The approximate cost of a disk is ~$125. In addition to relatively high data-storage costs, other disadvantages of optical recorders include the loss of digital image-history information that cannot be included with images as on hard disks, a minor loss of resolution when images are transferred back to the computer for processing, and utilization of the NTSC standard that limits the use of these recorders mainly to North America and Japan (Inoué, 1986). As discussed by Tsien and Harootunian (1990), the loss of resolution can be reduced by storing image backgrounds on the recorder then using these backgrounds during data processing to adjust for differences in the recorder offset voltages. Image histories must be stored in separate files on the hard disk or some other digital storage device.

In contrast to the analog OMDRs, erasable optical drives have an average access time of ~200 ms, a data transfer rate of 929 kbyte/s, and an average disk capacity of 8900 Mbyte. Hidden in these averages is a wide range of capabilities. There are presently over 50 different companies who offer erasable drives fabricated by a number of manufacturers (see Sun-Expert, Vol. 2, #7, July 1991 for a recent survey). Disk-storage capacities for these recorders range from a low of 0.65 Gbyte to a high of 1000 Gbyte!

Write speeds vary from 30 kbyte/s to 8900 kbyte/s. Both single-disk and jukebox-type configurations are available. Although most erasable optical drives presently available are slower to write than the analog recorders, are not standardized among manufacturers, and have a reputation for losing data, it is expected that these drives will become more reliable and cost effective in the future. Because they are digital, these drives do not have the disadvantages of the analog recording devices. With improved speed, digital drives may ultimately replace the analog devices as a more desirable image acquisition and storage device.

Unless one is operating on a very limited budget, data cartridge tape media is best reserved for archival storage because the process of writing data to tape is slow and, unlike hard disks and OMDRs, data must be accessed sequentially. In real-life situations this means that when several experiments are stored on the tape, it may take 10–15 min to step through the tape to find and then read out the experiment you wish to analyze. The shelf life of tapes is also thought to be less than that for optical disks, which is thought to be 10 yr or more depending on storage conditions. Several different types of tape media and drives are available. The most common are ¼-in. 80–150-Mbyte tapes, the drives for which are included as an option on most UNIX computers. These tapes are often used for distribution of system software upgrades; however, it is likely that ¼-in. tape drives will be replaced by CD ROM drives in the next few years. The rate of data transmission for data cartridge tapes is slow and the cost is relatively high at ~10¢/Mbyte. Data cartridge tapes with gigabyte storage capabilities are also available from companies, such as 3M. With digital helical scan 8-mm external tape drives (DAT), the data-transfer rate is also slow (~500 kbyte/s).

The average seek time for the 8-mm tapes is 85 s with an average storage capacity of 5 Gbyte. The advantage of these tapes as compared to magnetic data cartridge tapes is their small size, high capacity, and low cost (~2¢/Mbyte). With VHS-based (analog) drives, data-transfer time is rapid (~2–10 Mbyte/s) and tape capacity can be as high as 36 Gbyte. The disadvantage of these drives is that video images must be converted to VHS format for storage then reconverted frame by frame for digitized image analysis. At least two European groups are presently utilizing these recorders in low-cost, high-speed imaging systems (Theler *et al.*, 1991; O'Sullivan *et al.*, 1989).

If sufficient funds are available, a reasonably well-balanced approach to data-storage problems is to purchase as much hard-disk space (1-Gbyte minimum) and video RAM as possible and then supplement this storage medium with a fast analog optical disk recorder and an inexpensive 8-mm DAT. If funds permit, the DAT can be supplemented with or replaced by a more expensive but a faster and more convenient digital erasable optical disk recorder. For image acquisition rates at ms intervals, images can be stored in RAM. The number of images that can be collected in RAM will depend on both the pixel size of the images and the total amount of RAM available. One or more hard disks can be used to store images directly in experiments with acquisition rates of seconds to minutes, particularly when the software allows definition of the image size, which can be reduced when necessary to save memory space. The analog optical disk recorder can be used to collect images in real time if necessary and/or as a second-line image storage device. The DAT is used for long-term archival storage only. The erasable optical disk recorder can serve as both a second-line image storage device and as archival storage for valuable data. In many experiments, once images of individual cells are processed and numeric ratio values calculated, only the numeric information and reference transmitted light or DIC images need to be saved. Numeric data can be transferred to a spreadsheet or graphics plotting program with postscript and slide-maker support for generation of high-quality hard copies of data. If data are transferred from a UNIX to a DOS computer, we have found a good combination to be Lotus Symphony and Slidewrite Plus. Symphony imports ASCii files in spreadsheet format. Slidewrite reads Symphony and Lotus 1-2-3 files and provides postscript printer support as well as support for high-resolution Matrix slidemakers. If the computers are networked, other options are possible depending on the type of network used.

## G. Software

Ideally software should allow users to manipulate and modify every aspect in the system that is under computer control. The software should also be relatively easy to use to avoid time-consuming user learning curves; however, "black box" turnkey systems should be avoided not only because their inflexibility limits their usefulness in complex experiments, but also because such systems rapidly become obsolete. Both command and menu-driven interfaces should be provided to optimize flexibility. The ability to manipulate variables is particularly important in initial set-up experiments when camera settings are optimized, shuttering functions are defined to fit camera response times and to minimize light damage, and signal dynamic ranges are defined in terms of camera performance. The software should also be as flexible as possible to fit changing experimental protocols as new fluorescent probes become available and multiparameter protocols are defined. With commercial software, it is essential that the software code be accessible either directly or through a knowledgeable vendor who is willing to make modifications when the need arises.

A detailed discussion of software needed for acquiring and analyzing rapidly acquired video images is beyond the scope of this chapter. Therefore, we will only list a few of the features we have found to be particularly useful in dynamic experimental situations and in analysis of images once they are acquired. These are as follows:

1. Definition and storage of regions of interest (ROIs) of any size
2. Interactive LUT modification
3. Optional control of optical disk recorders and several video cameras at once as well as interactive modification of A/D gains and offsets for different excitation wavelengths
4. Choice of sampling configurations so that fluorescent images acquired at different excitation wavelengths can be acquired singly and/or any order along with reference images
5. Interactive modification of frame averaging and delay times between samples
6. Viewing and plotting image ratio values on the computer monitor in real time
7. Viewing of raw image information, as backgrounds, numerator, and denominator values in ratioing experiments
8. Viewing of live video images on one monitor and ratioed images on a separate monitor or as a split image on the same monitor
9. The ability to perform unattended time-lapse experiments coupled with the ability to generate and display animated "movies" of fluorescent and other kinds of images
10. Annotation of image header information
11. Application of calibration curves and different thresholding values to data and to display images in pseudocolor with calibration bars
12. Acquisition of images in real time using video frame buffers and CPU RAM
13. Simultaneous calculation of ratio values on multiple cells or areas within a single cell
14. Performance of image contrast and enhancement functions on reference images
15. Generation of image montages with varying numbers of images and application of text labels to images using different fonts and font sizes
16. Generation of three-dimensional plots of ratioed images

## V. FUTURE DIRECTIONS

In this chapter, we have discussed the possible origins of calcium oscillations in nonexcitable cells and considered approaches and limitations of video-based methods that are currently being used to measure oscillations in single cells. With improved fluorescent probes and video-imaging equipment, it should be possible to define more accurately the mechanisms involved in generation of calcium oscillations. In order to address these questions appropriately, serious consideration must be given to the physiological status of the cells under investigation. Since much of the data collected thus far has been acquired from cultured cells, it is important to determine whether or not responses of these cells accurately reflect responses *in vivo*. One potentially useful approach is to compare responses of acutely isolated groups of cells in which polarity is maintained with cells in primary culture maintained in defined media for different lengths of time. Different attachment matrices should be tested as these matrices may also affect cell function. In addition, improved indices of responsiveness for nonexcitable cells are needed. In the future, it will not be sufficient to measure calcium-response patterns in the absence of functional correlations. Better statistical analyses of the relationship between calcium-signaling patterns and cellular activities thought to be regulated by calcium-signaling mechanisms will also be required if the physiologic functions of calcium oscillations are to be defined.

ACKNOWLEDGMENTS

We wish to express our thanks to Dr. Douglas Benson and colleagues at Inovision who have provided us invaluable support and assistance with image-analysis applications for several years. We also wish to thank Dr. Fred Fay for his willingness to offer support and advice and for allowing one of us (C.C.) to visit his lab to view work in progress on new CCD imaging techniques. Finally, we warmly thank Frank Chew Jr. for providing graphics illustrations.

REFERENCES

Agard, D. A., Hiraoka, Y., Shaw, P., and Sedat, J. W. (1989). Fluorescence microscopy in three dimensions. *Meth. Cell Biol. 30*, 353–398.
Aikens, R. S. (1990). CCD cameras for video microscopy. Optical Microscopy for Biology (B. Herman and K. Jacobson, eds.), pp. 207–218, Wiley-Liss, NY.
Aikens, R. S., Agard, D. A., and Sedat, J. W. (1989). Solid-state imagers for microscopy. *Meth. Cell Biol. 29*, 291–313.
Almers, W., and Nehers, E. (1985). The Ca signal from fura-2 loaded mast cells depends strongly on the method of dye-loading. *FEBS Lett. 192*, 13–18.
Arndt-Jovin, D. J., Rober-Nicoud, M., Kaufman, S. J., and Jovin, T. M. (1985). Fluorescence digital imaging microscopy in cell biology. *Science 230*, 247–256.
Aubin, J. E. (1979). Autofluorescence of viable cultured mammalian cells. *J. Histochem. Cytochem. 27*, 36–43.
Bals, S., Bechem, M., Paffhausen, W., and Pott, L. (1990). Spontaneous and experimentally evoked $[Ca^{2+}]_i$-transients in cardiac myocytes measured by means of a fast fura-2 technique. *Cell Calcium 11*, 385–396.
Baylor, S. M., and Hollingsworth, S. (1988). Fura-2 calcium transients in frog skeletal muscle fibres. *J. Physiol. (LOnd.) 403*, 151–192.
Becker, P. L., and Fay, F. S. (1987). Photobleaching of fura-2 and its effect on the determination of calcium concentrations. *Am. J. Physiol. 253*, C613–C618.

Benson, D. M., Bryan, J., Plant, A. L., Gotto, A. M. Jr., and Smith, L. C. (1985). Digital imagining fluorescence microscopy: Spatial heterogeneity of photobleaching rate constants in individual cells. *J. Cell Biol.* 100, 1309–1323.

Berridge, M. J. (1980). Preliminary measurements of intracellular calcium in an insect salivary gland using a calcium sensitive electrode. *Cell Calcium 1*, 217–227.

Berridge, M. J. (1990). Calcium oscillations. *J. Biol. Chem. 265*, 9583–9586.

Berridge, M. J., and Galione, A. (1988). Cytosolic calcium oscillators. *FASEB J. 2*, 3074–3082.

Berridge, M. J., and Rapp, P. E. (1979). A comparative survey of the function, mechanism and control of cellular oscillators. *J. Exp. Biol. 81*, 217–280.

Berridge, M. J., Cobbold, P. H., and Cuthberson, K. S. R. (1988). Spacial and temporal aspects of cell signalling. *Philos. Trans. R. Soc. Lond. [Biol] 320*, 325–343.

Blatter, L. A., and Wier, W. G. (1990). Intracellular diffusion, binding, and compartmentalization of the fluorescent calcium indicators indo-1 and fura-2. *Biophys. J. 58*, 1491–1499.

Bonhoeffer, T., and Staiger, V. (1988). Optical recording with single cell resolution from monolayered slice cultures of rat hippocampus. *Neurosci. Lett. 92*, 259–264.

Bookman, R. J. (1990). Temporal response characterization of video cameras. *In* "Optical Microscopy for Biology." (B. Herman and K. Jacobson, eds.), pp. 235–250, Wiley-Liss, N.Y.

Bright, G. R., and Taylor, D. L. (1986). Imaging at low light level in fluorescence microscopy. *In* "Applications of Fluorescence in the Biomedical Sciences." (D. L. Taylor, A. Waggoner, R. Murphy, F. Lanni, and R. Birge, eds.), pp. 257–288, Alan R. Liss, N.Y.

Bright, G. R., Fisher, G. W., Rogowska, J., and Taylor, D. L. (1989). Fluorescence ratio imaging microscopy. *Meth. Cell Biol. 30*, 157–192.

Castleman, K. R. (1987). Spatial and photometric resolution and calibration requirements for cell image analysis instruments. *Appl. Opt. 26*, 3338–3342.

Chew, C. S., and Brown, M. R. (1986). Release of intracellular $Ca^{2+}$ and elevation of inositol triphosphate by secretagogues in parietal and chief cells isolated from rabbit gastric mucosa. *Biochim. Biophys. Acta 888*, 116–125.

Chew, C. S., and Scanlon. (1990). Can calcium in single cells be quantitated with fura 2? *FASEB J. 4*, 5458.

Chiavaroli, C., Vacher, P., Vacesey, A., Mons, N., Letari, O., Pralong, W., Lagnaux, Y., Whelan, R., and Schlegel, W. (1991). Simultaneous monitoring of cytosolic free calcium and exocytosis at the single cell level. *J. Neuroendocrinol. 3*, 253–260.

Conner, J. A. (1986). Digital imaging of free calcium changes and of spatial gradients in growing processes in single, mammalian central nervous system cells. *Proc. Natl. Acad. Sci. U.S. 83*, 6179–6183.

Cutherberson, K. S. R. (1989). Intracellular calcium oscillators. *In* "Cell to Cell Signalling: From Experiments to Theoretical Models." (A. Goldbeter, ed.), pp. 435–447, London Academic Press.

Cutherberson, K. S. R., and Chay, T. R. (1991). Modelling receptor-controlled intracellular calcium oscillators. *Cell Calcium 12*, 97–109.

Davis, M. H., Altschuld, R. A., Yung, D. W., and Brierly, G. P. (1987). Estimation of intramitochondrial pCa and pH by fura-2 and 2,7-bicarboxyethyl-5(6)-

carboxyfluorescein (BCEFC) fluorescence. *Biochem. Biophys. Res. Commun.* *149*, 40–45.

Di Virgilio, F., Steinberg, T. H., Swanson, J. A., and Silverstein, S. C. (1988). Fura-2 secretion and sequestration in macrophages: A blocker of organic anion transport reveals that these processes occur via a membrane transport system for organic anions. *J. Immunol. 140*, 915–920.

Di Virgilio, F., Steinberg, T. H., and Silverstein, S. C. (1989). Organic-Anion transport inhibitors to facilitate measurement of cytosolic free $Ca^{2+}$ with fura-2. *Meth. Cell Biol. 31*, 453–462.

Dupont, G., Berridge, M. J., and Goldbeter, A. (1991). Signal-induced $Ca^{2+}$ oscillations: Properties of a model based on $Ca^{2+}$-induced $Ca^{2+}$ release. *Cell Calcium 12*, 73–85.

Eckert, R. W., Scherubl, H., Petzelt, C., Raue, F., and Ziegler, R. (1989). Rhythmic oscillations of cytosolic free calcium in rat C-cells. *Mol. Cell. Endocrinol. 64*, 267–270.

Fay, F. S., Carrington, W., and Fogarty, K. E. (1989). Three-dimensional molecular distribution in single cells analyzed using the digital imagining microscope. *J. Microsc. 153*, 133–149.

Foskett, J. K. (1985). NBD-taurine fluorescence as a probe of anion exchange in gallbladder epithelium. *Am. J. Physiol. 249*, 1252–1255.

Foskett, J. K. (1988). Simultaneous Nomarski and fluorescence imagining during video microscopy of cells. *Am. J. Physiol. 255*, C566–C571.

Foskett, J. K., and Melvin, J. E. (1989). Activation of salivary secretion: Coupling of cell volume and $[Ca^{2+}]_i$ in single cells. *Science 244*, 1582–1585.

Foskett, J. K., Gunter-Smith, P., Melvin, J. E., and Turner, R. J. (1989). Physiological localization of an agonist-sensitive pool of $Ca^{2+}$ in parotid acinar cells. *Proc. Natl. Acad. Sci. U.S.A. 86*, 167–171.

Furiuchi, T., Yoshikawa, S., Migawaki, A., Wada, K., Maeda, N., and Mikoshiba, K. (1989). Primary structure and functional expression of the inositol 1,4,5-trisphosphate binding protein P400. *Nature 342*, 32–38.

Goldbeter, A., Dupont, G., and Berridge, M. J. (1990). Minimal model for signal-induced $Ca^{2+}$ oscillations and for their frequency encoding through protein phosphorylation. *Proc. Natl. Acad. Sci. U.S.A. 87*, 1461–1465.

Gray, P. T. A. (1988). Oscillations in free cytosolic calcium evoked by cholinergic and catecholaminergic agonists in rat parotid acinar cells. *J. Physiol 406*, 35–53.

Grynkiewicz, G., Poenie, M., and Tsien, R. (1985). A new generation of $Ca^{2+}$ indicators with greatly improved fluorescence properties. *J. Biol. Chem. 260*, 3440–3450.

Gunter, T. E., Restrepo, D., and Gunter, K. K. (1988). Conversion of esterified fura-2 and indo-1 to $Ca^{2+}$-sensitive forms by mitochrondria. *Am. J. Physiol. 255*, C304–C310.

Hajjar, R. J., and Bonventre, J. V. (1991). Oscillations in intracellular calcium induced by vasopressin in individual fura-2-loaded mesanglial cells. *J. Biol. Chem 266*, 21589–21594.

Harootunian, A. T., Kao, J. P. Y., and Tsien, R. Y. (1988). Agonist-induced calcium oscillations in depolarized fibroblasts and their manipulation by photoreleased Ins(1,4,5)P$_3$, $Ca^{2+}$, and $Ca^{2+}$ buffer. *Cold Spring Harbor Symp. Quant. Biol. 53*, 935–943.

Hellman, B., Gylfe, E., Grapengiesser, E., Panten, U., Schwanstecher, C., and Heipel, C. (1990). Glucose induces temperature-dependent oscillations of cytoplasmic $Ca^{2+}$ in single pancreatic $\beta$-cells related to their electrical activity. *Cell Calcium 11*, 413–418.

Highsmith, S., Bloebaum, P., and Snowdowne, K. W. (1986). Sarcoplasmic reticulum interacts with the $Ca^{2+}$ indicator precursor fura-2/AM. *Biochem. Biophys. Res. Commun. 138*, 1153–1162.

Hirauka, Y., Sedat, J. W., and Agard, D. A. (1987). The use of a charge-coupled device for quantitative optical microscopy of biological structures. *Science 238*, 36–41.

Inoué, S. (1986). "Video Microscopy." Plenum Publishing Corp., New York.

Irvine, R. (1990). "Quantal" $Ca^{2+}$ release and the control of $Ca^{2+}$ entry by inositol phosphates—a possible mechanism. *FEBS Lett. 263*, 5–9.

Jackson, A. P., Timmerman, M. P., Bagshan, C. R., and Ashley, C. C. (1987). The kinetics of calcium binding to fura-2 and indo-1. *FEBS Lett. 216*, 35–39.

Jackson, T. R., Patterson, S. R., Thastrup, O., and Hanley, M. R. (1988). A novel tumor promoter, thapsigargin, transiently increases cytoplasmic free $Ca^{2+}$ without generation of inositol phosphates in NG115-401L neuronal cells. *Biochem. J. 253*, 81–86.

Jacob, R. (1990a). Calcium oscillations in electrically non-excitable cells. *Biochim. Biophys. Acta 1052*, 427–438.

Jacob, R. (1990b). Imaging cytoplasmic free calcium in histamine stimulated endothelial cells and in fMet-Leu-Phe stimulated neutrophils. *Cell Calcium 11*, 241–249.

Jeričevič, Ž., Wiese, B., Bryan, J., and Smith, L. C. (1989). Validation of an imaging system: Steps to evaluate and validate a microscope imaging system for quantitative studies. *Methods Cell Biol. 30*, 47–83.

Kao, J. P. Y., and Tsien, R. Y. (1988). $Ca^{2+}$ binding kinetics of fura-2 and azo-1 from temperature-jump relaxation measurements. *Biophys. J. 53*, 635–639.

Kao, J. P. Y., Harootunian, A. T., and Tsien, R. Y. (1989). Photochemically generated cytosolic calcium pulses and their detection by fluo-3. *J. Biol. Chem. 264*, 8179–8184.

Kasai, H., and Augustine, G. J. (1990). Cytosolic $Ca^{2+}$ gradients triggering unidirectional fluid secretion from exocrine pancreas. *Nature 348*, 735–738.

Kelley, L. I., Blackmore, P. F., Graber, S. E., and Stewart, S. J. (1990). Agents that raise cAMP in human T lymphocytes release an intracellular pool of calcium in the absence of inositol phosphate production. *J. Biol. Chem. 265*, 17657–17664.

Kinosita, K. Jr., Itoh, H., Ishiwata, S., Hirano, K., Nishizaka, T., and Hayakawa, T. (1991). Dual-view microscopy with a single camera: Real-time imaging of molecular orientations and calcium. *J. Cell Biol. 115*, 67–73.

Klein, M. G., Simon, B. J., Szucs, G., and Schneider, M. F. (1988). Simultaneous recording of calcium transients in skeletal muscle using high- and low-affinity calcium indicators. *Biophys. J. 53*, 971–988.

Kruskal, B. A., Shak, S., and Maxfield, F. R. (1986). Spreading of human neutrophils is immediately preceded by a large increase in cytoplasmic free calcium. *Proc. Natl. Acad. Sci. U.S.A. 83*, 2919–2923.

Kurtz, I., Dwelle, R., and Katzka, P. (1987). Rapid scanning fluorescence spectroscopy using an acousto-optical tunable filter. *Rev. Sci. Instrum. 58*, 1996–2003.

Linderman, J. J., Harris, L. J., Slakey, L. L., and Gross, D. J. (1990). Charge-coupled device imagining of rapid calcium transients in cultured arterial smooth muscle cells. *Cell Calcium 11*, 131–144.

Ljungström, M., and Chew, C. S. (1991). Calcium oscillations and morphological transformations in single cultured gastric parietal cells. *Am. J. Physiol. 260*, C67–C78.

Loew, L. M., Carrington, W. A., Fay, F. S., Tuft, R. A., and Wei, M.-d. (1991). Quantitative determination of membrane potential in individual mitochondria within a neurite. *J. Cell Biol. 115*, 300a.

Lukacs, G. L., and Kapus, A. (1987). Measurement of the matrix free $Ca^{2+}$ concentration in heart mitochondria by intrapped fura-2 and quin2. *Biochem. J. 248*, 609–613.

Malgaroli, A., Milani, D., Meldolesi, J., and Pozzan, T. (1987). Fura-2 measurements of cytosolic free $Ca^{2+}$ in monolayers and suspensions of various types of animal cells. *J. Cell Biol. 105*, 2145–2155.

Marks, P. W., and Maxfield, F. R. (1991). Preparation of solutions with free calcium concentration in the nanomolar range using 1, 2-bis(o-aminophenoxy)ethane-N,N,N,N-tetraacetic acid. *Anal. Biochem. 193*, 61–71.

Meyer, T., and Stryer, L. (1988). Molecular model for receptor-stimulated calcium spiking. *Proc. Natl. Acad. Sci. U.S.A. 85*, 5051–5055.

Mignery, G. A., Südhof, T. C., Takei, K., and De Camilli, P. (1989). Putative receptors for inositol 1,4,5 trisphosphate similar to ryanodine receptor. *Nature 342*, 192–195.

Milani, D., Malgaroli, A., Guidolin, D., Fasolato, C., Skaper, S. D., Meldolesi, J., and Pozzan, T. (1990). $Ca^{2+}$ channels and intracellular $Ca^{2+}$ stores in neuronal and neuroendocrine cells. *Cell Calcium 11*, 191–199.

Miller, A. L., McLaughlin, J. A., and Jaffe, L. F. (1991). Imaging free calcium in *Xenopus* eggs during polar pattern formation and cytokinesis. *J. Cell Biol. 115*, 280a.

Moore, E., Becker, P. L., Fogarty, K. E., and Fay, F. S. (1990). $Ca^{2+}$ imagining in single living cells: Theoretical and practical issues. *Cell Calcium 11*, 157–179.

Nathanson, M. H., and Burgstahler, A. D. (1992). Coordination of hormone-induced calcium signals in isolated rat hepatocyte couplets: Demonstration with confocal microscopy. *Mol. Biol. Cell 3*, 113–121.

Niggli, E., and Lederer, W. J. (1990). Real-time confocal microscopy and calcium measurements in heart muscle cells: Towards the development of a fluorescence microscope with high temporal and spatial resolution. *Cell Calcium 11*, 121–130.

Oakes, S. G., Martin, W. J. II, Lisek, C. A., and Powis, G. (1988). Incomplete hydrolysis of the calcium indicator Fura-2 pentaacetoxymethyl ester (Fura-2/AM) by cells. *Anal. Biochem. 169*, 159–166.

O'Sullivan, A. J., Cheek, T. R., Moceton, R. B., Berridge, M. J., and Burgoyne, R. D. (1989). Localization and heterogeneity of agonist-induced changes in cytosolic calcium concentration in single bovine adrenal chromaffin cells from video imaging of fura-2. *EMBO J. 8*, 401–411.

Owen, C. S. (1991). Spectra of intracellular fura-2. *Cell Calcium 12*, 385–393.

Owen, C. S., Sykes, N., Shuler, R., and Ost, D. (1991). Non-calcium environmental sensitivity of intracellular Indo-1. *Anal. Biochem. 192*, 142–148.

Petersen, O. H., Gallacher, D. V., Wauki, M., Yale, D. I., Petersen, C. C. H., and Toescu, E. C. (1991). Receptor-activated cytoplasmic $Ca^{2+}$ oscillations in pancreatic acinar cells: Generation and spreading of $Ca^{2+}$ signals. *Cell Calcium 12,* 135–144.

Picciolo, G. L., and Kaplan, D. S. (1984). Reduction of fading of fluorescent reaction product for microphotometric quantitation. *Adv. Appl. Microbiol. 30,* 197–234.

Plant, A. L., Benson, D. M., and Smith, L. C. (1985). Cellular uptake and intracellular localization of benzo(a)pyrine by digitized fluorescence imagining microscopy. *J. Cell Biol. 100,* 1295–1308.

Poenie, M., Alderton, J., Steinhardt, R., and Tsien, R. Y. (1986). Calcium rises abruptly and briefly throughout the cell at the onset of anaphase. *Science 233,* 886–889.

Popov, E. G., Gavrilov, Y.Iu., Pozin, E.Ya., and Gabrosov, Z. A. (1988). Multiwavelength method for measuring concentration of free cytosolic calcium using the fluorescent probe indo-1. *Arch. Biochem. Biophys. 261,* 91–96.

Putney, J. W. Jr. (1986). A model for receptor-regulated calcium entry. *Cell Calcium 7,* 1–12.

Putney, J. W. Jr. (1990). Capacitive calcium entry revisited. *Cell Calcium 11,* 611–624.

Reber, B. F. X., Somogyi, R., and Stuki, J. W. (1990). Hormone-induced intracellular calcium oscillations and mitochondrial energy supply in single hepatocytes. *Biochim. Biophys. Acta 1018,* 190–193.

Rink, T. J., and Jacob, R. (1989). Calcium oscillations in nonexcitable cells. *TINS 12,* 43–46.

Roe, M. W., Lemasters, J. J., and Herman, B. (1990). Assessment of fura-2 for measurements of cytosolic free calcium. *Cell Calcium 11,* 63–73.

Rooney, T. A., Sass, E. J., and Thomas, A. P. (1989). Characterization of cytosolic calcium oscillations induced by phenylephrine and vasopressin in single fura-2 loaded hepatocytes. *J. Biol. Chem. 264,* 17131–17141.

Rooney, T. A., Sass, E. J., and Thomas, A. P. (1990). Agonist-induced cytosolic calcium oscillation originate from a specific locus in single hepatocytes. *J. Biol. Chem. 265,* 10792–10796.

Rooney, T. A., Renard, D. C., Sass, E. J., and Thomas, A. P. (1991). Oscillatory cytosolic calcium waves independent of stimulated inositol 1,4,5 trisphosphate formation in hepatocytes. *J. Biol. Chem. 266,* 12272–12282.

Ryan, T. A., Millard, P. J., and Webb, W. W. (1990). Imaging $[Ca^{2+}]_i$ dynamics during signal transduction. *Cell Calcium 11,* 145–155.

Sanderson, M. J., Charles, A. C., and Dirksen, E. R. (1990). Mechanical stimulation and intercellular communication increases intracellular $Ca^{2+}$ in epithelial cells. *Cell Regul. 1,* 585–596.

Scanlon, M., Williams, D. A., and Fay, F. S. (1987). A $Ca^{2+}$-insensitive form of fura-2 associated with polymorphonuclear leukocytes. *J. Biol. Chem. 262,* 6308–6312.

Sick, T. J., and Rosenthal, M. (1989). Indo-1 measurements of intracellular free calcium in the hippocampal slice: Complications of labile NADH fluorescence. *J. Neurosci. Methods 28,* 125–132.

Sistare, F. D., Picking, R. A., and Haynes, A. C. Jr. (1985). Sensitivity of the response of cytosolic calcium in quin-2 loaded rat hepatocytes to glucagon, adenine nucleosides, and adenine nucleotides. *J. Biol. Chem. 220,* 12744–12747.

Spring, K. R. (1990). Quantitative imaging at low light levels: Differential interference contrast and fluorescence microscopy without significant light loss. In "Optical Microscopy for Biology." (B. Herman and K. Jacobson, ed.), pp. 513–522, Wiley-Liss, New York.

Spring, K. R., and Lowy, R. J. (1989). Characteristics of low light level television cameras. In "Methods in Cell Biology." (Y. L. Wang and D. L. Taylor, eds.), pp. 269–289, Academic Press, Orlando.

Spurgeon, H. A., Stern, M. D., Baartz, G., Raffaeli, S., Hansford, R. G., Talo, A., Lakatta, E. G., and Copogrossi, M. C. (1990). Simultaneous measurement of $Ca^{2+}$, contraction and potential in cardiac myocytes. *Am. J. Physiol. 258,* H574–H586.

Staddon, J. M., and Hansford, R. G. (1989). Evidence indicating that the glucagon-induced increases in cytosolic free $Ca^{2+}$ concentration in hepatocytes is mediated by an increase in cyclic AMP concentration. *Eur. J. Biochem. 179,* 47–52.

Steinberg, S. F., Bilezikian, J. P., and Al Awquati, Q. (1987). Fura-2 fluorescence is localised to mitochondria in endothelial cells. *Am. J. Physiol. 253,* C744–C747.

Takamatsu, T., and Wier, W. G. (1990). High temporal resolution video imaging of intracellular calcium. *Cell Calcium 11,* 111–120.

Takashima, H., Nishimura, S., Matsumoto, T., Ishida, H., Kanagawa, K., Minamino, N., Ueda, M., Hanaoka, M., Hirose, T., and Numa, S. (1989). Primary structure and expression from complementary DNA of skeletal muscle ryanodine receptor. *Nature 339,* 439–445.

Thastrup, O., Cullen, P. J., Drobak, B. K., Hanley, M. R., and Dawson, A. P. (1990). Thapsigargin, a tumor promoter, discharges intracellular $Ca^{2+}$ stores by specific inhibition of the endoplasmic reticulum $Ca^{2+}$-ATPase. *Proc. Natl. Acad. Sci. U.S.A. 87,* 2466–2470.

Theler, J.-M., Wollheim, C. B., and Schlegal, W. (1991). Rapid on-line image processing as a tool in the evaluation of kinetics and morphological aspects of receptor-induced activation. *J. Receptor Res. 11,* 627–639.

Tsay, T-T., Inman, R., Wray, B., Herman, B., and Jacobson, K. (1990). Characterization of low-light-level cameras for digitized video microscopy. *J. Microscopy 160,* 141–159.

Tsien, R. Y. (1980). New calcium indicators and buffers with high selectivity against magnesium and protons: Design, synthesis and properties of prototype structures. *Biochem. J. 19,* 2396–2404.

Tsien, R. Y. (1989). Fluorescent indicators of ion concentrations. *Meth. Cell. Biol. 30,* 127–156.

Tsien, R. Y., and Harootunian, A. T. (1990). Practical design criteria for a dynamic ratio imaging system. *Cell Calcium 11,* 93–109.

Tsien, R., and Pozzan, T. (1989). Measurement of cytosolic free $Ca^{2+}$ with Quin2. *Meth. Enzymol. 172,* 230–262.

Tsien, R. Y., and Waggoner, A. (1990). Fluorophores for confocal microscopy: Photolysis and photochemistry. In "Handbook of Biological Confocal Microscopy." (J. B. Pawley, ed.), pp. 169–178, Plenum Press, N.Y.

Tsien, R. Y., Rink, T. J., and Poenie, M. (1985). Measurement of cytosolic free

$Ca^{2+}$ in individual small cells using fluorescence microscopy with dual excitation wavelengths. *Cell Calcium 6,* 145–158.

Wahl, M., Lucherini, M. J., and Gruenstein, E. (1990). Intracellular $Ca^{2+}$ measurement with indo-1 in substrate-attached cells: Advantages and special considerations. *Cell Calcium 11,* 487–500.

Wakui, M., Potter, B. V., and Petersen, O. H. (1989). Pulsatile intracellular calcium release does not depend on fluctuations in inositol trisphosphate concentration. *Nature 339,* 317–320.

Wampler, J. E. (1985). Low-light level video systems. "Bioluminescence and Chemiluminescence: Instrumentation and Applications, Vol. II." (K. Van Dyke, ed.), pp. 123–145, CRC Press, Boca Raton, FL.

Wick, R. A. (1987). Quantum-limited imaging using microchannel plate technology. *App. Opt. 26,* 3210–3218.

Wier, W. G., and Blatter, L. A. (1991). $Ca^{2+}$-oscillations and $Ca^{2+}$-waves in mammalian cardiac and vascular smooth muscle cells. *Cell Calcium 12,* 241–254.

Wier, W. G., Cannell, M. B., Berlin, J. R., Marban, E., and Lederer, W. J. (1987). Cellular and subcellular homogeneity of $[Ca^{2+}]_i$ in single heart cells revealed by fura-2. *Science 235,* 325–328.

Williams, D. A. (1990). Quantitative intracellular calcium imaging with laser-scanning confocal microscopy. *Cell Calcium 11,* 589–597.

Williams, D. A., and Fay, F. S. (1990). Intracellular calibration of the fluorescent calcium indicator fura-2. *Cell Calcium 11,* 75–83.

Williams, D. A., Fogarty, K. E., Tsien, R. Y., and Fay, F. S. (1985). Calcium transients in single smooth muscle cells revealed by the digital imaging microscope using fura-2. *Nature (Lond.) 318,* 558–561.

Woods, N. M., Cuthbertson, K. S. R., and Cobbold, P. H. (1986). Repetitive transient rises in cytoplasmic free calcium in hormone-stimulated hepatocytes. *Nature 319,* 600–602.

Woods, N. M., Cutherbertson, K. S. R., and Cobbold, P. H. (1987). Agonist-induced oscillations in cytoplasmic free calcium concentration in single rat hepatocytes. *Cell Calcium 8,* 79–100.

Yada, T., Oiki, S., Ueda, S., and Okada, Y. (1986). Synchronous oscillation of the cytoplasmic $Ca^{2+}$ concentration and membrane potential in cultured epithelial cells (Intestine 407). *Biochim. Biophys. Acta 887,* 105–112.

Young, I. T. (1989). Image fidelity: Characterizing the imaging transfer function. *In* "Fluorescence microscopy of living cells in culture." *Meth. Cell Biol. 30,* 1–45.

# 6
# SIMULTANEOUS MULTIPLE DETECTION OF FLUORESCENT MOLECULES

Rapid Kinetic Imaging
of Calcium and pH
in Living Cells

**Stephen J. Morris**
Division of Molecular Biology and Biochemistry
School of Biological Sciences
University of Missouri at Kansas City

I. INTRODUCTION: WHY STUDY SIMULTANEOUS KINETICS BY IMAGING?

II. MULTIPARAMETER IMAGING

III. "SIMULTANEOUS" INTRACELLULAR CALCIUM–pH MEASUREMENTS

IV. BASIC PROPERTIES OF RATIO-TYPE ION INDICATOR DYES
  A. Advantages of Fluorescence Ratio Imaging for Quantification

V. IMAGING OF EXCITATION-TYPE DYES VERSUS EMISSION-TYPE DYES FOR RAPID KINETIC FLUORESCENCE VIDEO MICROSCOPY
  A. Simultaneous versus Sequential Measurements

VI. ADVANTAGES OF EMISSION RATIO DYES

VII. IMPROVEMENTS IN DESIGN FOR MULTI-IMAGE EXPERIMENTS

VIII. THE NEED FOR MORE THAN TWO SIMULTANEOUS IMAGES OF THE SAME CELL, AND THE DEPENDENCE OF INDO AND FURA CALCIUM $K_d$ ON pH

IX. FOUR-CHANNEL VIDEO MICROSCOPE FOR SIMULTANEOUS IMAGING OF TWO-RATIO DYES

X. CHOICE OF DYES FOR SENSING CALCIUM AND pH

XI. MULTIPLE-WAVELENGTH EPIFLUORESCENCE EXCITATION

Optical Microscopy: Emerging Methods and Applications

XII. SIMULTANEOUS
COLLECTION OF FOUR
VIDEO IMAGES OF THE
SAME MICROSCOPIC
FIELD
XIII. FORMATION OF LOW-
LIGHT IMAGES
XIV. IMAGING
HARDWARE–SOFTWARE
XV. IMPROVING SPATIAL AND
TEMPORAL RESOLUTION
OF THE DATA
XVI. INCREASING THE DATA
COLLECTION SPEED TO 60
FIELDS PER SECOND
XVII. CALCULATIONS
A. Spillover
B. Correction of Geometric
Distortion to Improve
Registration

C. Ratio Analysis
D. Ca–pH Calculations
XVIII. SOME FURTHER
CONSIDERATIONS
A. Is the Equipment Fast
Enough for the Experiment?
B. Are the Dyes Fast Enough?
C. Available Light
XIX. APPLICATIONS
A. Simultaneous Intracellular
$Ca^{2+}$ and pH Measurements
Show Differential Kinetics
B. Depolarization of Pituitary
Intermediate Lobe
Melanotropes Raises $[Ca^{2+}]_i$
and Lowers $pH_i$
C. MDCK Cell $Na^+–H^+$
Antiporter
ACKNOWLEDGMENTS
REFERENCES

# I. INTRODUCTION: WHY STUDY SIMULTANEOUS
KINETICS BY IMAGING?

Quantitative examination of the fluorescence from living cells in a spectrofluorimeter provides interesting data but has drawbacks: (1) the results represent the average from a large number of cells, and (2) a change in the signal (e.g., change in fluorescence intensity, polarization, resonance energy transfer) is required. Quantitation by low-light fluorescence digital video microscopy provides methods to overcome these problems. Of special interest is the use of ultralow-excitation light levels and image intensifiers to overcome problems associated with photodynamic damage and photobleaching (reviewed in Bright and Taylor, 1986; Morris, 1990; Takamatsu and Wier, 1990a; Waggoner et al., 1989). A large number of fluorescent probes, which do not show a signal change in a fluorimeter, can be examined by microscopy, where spatial redistribution can be used as a variable (Bright, 1992). For those dyes that do show spectroscopic changes, the spatial distribution of the change can also be studied

(Herman, 1989; Lemasters *et al.*, 1990; Sarkar *et al.*, 1989; Waggoner *et al.*, 1989).

One can often resolve questions concerning the interplay of temporal and spatial cellular activities from concurrent observations, which could not be resolved observing each activity singly. Examples are reviewed by Foskett (Chapter 8, this volume) for combined phase contrast and fluorescence video microscopy and Niggli *et al.* (Chapter 7, this volume) for video microscopy combined with patch-clamp measurements. The multiparameter approach permits resolution of several fluorophores placed in the same cell system. These could be indicator dyes for pH or ions, or tracers attached to specific components. Recent comprehensive reviews by Waggoner *et al.* (1989) and Bright (1992) discuss probes, equipment, and experimental techniques.

Since activities may be changing nonsynchronously, the changes may be obscured when large numbers of cells are studied simultaneously. Imaging becomes the method of choice. At least two kinetic microscopes capable of imaging ratio dyes at video rates (as fast as 60 images/s) have been described (Morris, 1990; Takamatsu and Wier, 1990a). Takamatsu and Wier (1990b) have imaged the advancing front of calcium waves in cardiac myocytes and measured the propagation rate at about 100 $\mu$m/s at 22°C. Extension of the ratio approach to more than one variable once again raises the complexity of the equipment involved although not by a great deal. We have improved our real-time microscope to simultaneously acquire up to four images at video field rates (Morris *et al.*, 1991). This allows the capture of intensified images of two different dual-emission wavelength "ratio" dyes or four single wavelength dyes. We have used the four-channel video microscope to demonstrate simultaneous intracellular calcium and pH measurements.

This article will deal with use of ratio-type ion indicator dyes for studying the correlation of rapid kinetics of intracellular ion concentrations. It is intended to be a guide for implementing a multi-image kinetic microscope system for simultaneous measurement of intracellular $Ca^{2+}$ and pH. For kinetic studies, the problems of photodynamic damage and photobleaching on one hand and the need for good spatial and temporal resolution on the other press the resolution of the instrumentation. Simultaneous rather than sequential resolution of multiple probes at multiple wavelengths presents a third set of problems. Very rapid data acquisition from multiple dyes presents a fourth.

First, we will discuss why the properties of the indicators require concurrent measurements, followed by several of the problems to be faced in collecting images for kinetic analysis at rapid rates. We demonstrate the utility of this approach with preliminary results from three experiments that show differential $Ca^{2+}-H^+$ kinetics: Ionomycin-induced changes in pH and $Ca^{2+}$ in kidney epithelial cells, $pH_i$ responses of neurosecretory cells to depolarization-induced influx of extracellular $Ca^{2+}$, and increases in $[Ca^{2+}]_i$ from intracellular stores in response to acid loading in Madine-Darby canine kidney (MDCK) cells.

Although the discussion will focus on simultaneous calcium–pH kinetics, the methodology can be applied to any set of probes that pass the basic requirements for truly simultaneous imaging. This in turn can be combined with phase contrast or differential interference contrast (DIC) imaging as well as patch-clamp measurements. At present, the investment in equipment is on the order of $100,000. Thus, the capability for doing this type of spectroscopy is equivalent to the cost fifteen years ago of a first-rate multiwavelength spectrofluorimetric stopped flow system for living-cell measurements. We have consistently aimed for inexpensive solutions whenever possible, and will discuss alternate strategies that can be applied.

Experimental problems to be solved with such a method are intriguing. A highly biased list includes the following:

1. Exocytosis is often triggered by calcium entry and a rise in intracellular calcium (Knight *et al.*, 1989; Thomas *et al.*, 1990). Stimulation of exocytosis by cell-surface receptor activity often leads to a rise in intracellular pH, while pH modulation can affect secretory activity (Kuijpers *et al.*, 1989; Rosario *et al.*, 1991). Is the Δ pH a causal step in exocytotic activity?

2. Mitogenic response to growth factors includes an increase in cell pH (Bright *et al.*, 1989; Maly *et al.*, 1990; Owen *et al.*, 1989; Vicentini and Villereal, 1986). Mitogens also increase intracellular calcium levels (Diliberto *et al.*, 1991; Ladoux *et al.*, 1989; Maly *et al.*, 1990; Roe *et al.*, 1989) and activate protein kinase C (Owen *et al.*, 1989; Vicentini and Villereal, 1986; Moolenaar, 1986). How closely coupled are the events in this cascade and by what mechanisms?

3. Contraction of vascular smooth-muscle cells is regulated both by intracellular calcium and pH (Smallwood *et al.*, 1983; Blackmore *et al.*, 1984). Intracellular modulators, such as protein kinase C activators and $InsP_3$, also regulate internal pH (Berk *et al.*, 1991; Caramelo *et al.*, 1989; Schulz *et al.*, 1989). Again how closely are these events coupled?

4. It has been postulated that many calcium transport systems function as calcium–hydrogen exchangers (Khananshvili, 1990; MacLeod, 1991; Rink and Sage, 1990; Tsukamoto *et al.*, 1990). If this is the case, then the direct coupling should be revealed by studies of the reciprocal relationship of $Ca^{2+}$ and $H^+$ ion kinetics.

5. Besides the growing number of intriguing questions linking intra-

cellular calcium and pH, the sensitivity of indo and fura dyes to pH (Lattanzio, 1990; Martinez-Zaguilan *et al.*, 1991; Roe *et al.*, 1990) makes it imperative under some circumstances to examine these two variables at the same time in order to correct for possible artifacts.

## II. MULTIPARAMETER IMAGING

The multiparameter fluorescence experimental approach takes advantage of a number of different fluorescent compounds whose excitation and emission spectra are sensitive to specific environmental factors and which preferentially accumulate into specific subcellular compartments of living cells. Fortunately, the number of parameter-specific probes keeps expanding, indicating that future research can achieve an even greater capability to make these discriminations. Categories of vital probes have been reviewed in Bright (1992), Haugland (1992), and Waggoner *et al.* (1989). One is not limited to fluorescence observations; a combination of fluorescence and phase contrast or DIC is easy to implement. This allows for imaging of the dynamics of cytoskeleton, intracellular organelles, or other morphological features. Parameters, such as cytosolic-free calcium, sodium and pH, plasma and mitochondrial membrane potential, membrane fluidity, cell surface morphology, cell volume, and cell viability, can be measured. Applications include effects of a sudden, local change in intracellular ion concentration or pH, dynamic relationships between cytoskeletal elements and intracellular organelles, cell locomotion, and intracellular trafficking and compartmentation.

Four or more variables have been studied on a quantitative basis in the same living cells by judiciously selecting parameter-specific fluorophores with nonoverlapping excitation and emission spectra and exchanging the epifluorescence filter sets. Spectral overlaps can often be processed out by careful standardization and subtraction. If one adds the ability to make ongoing recordings, kinetic studies may be done. Cells have been followed for several minutes to hours (Foskett, 1988; Herman *et al.*, 1990; Lemasters *et al.*, 1990; Morris, 1990; Morris *et al.*, 1991; Waggoner *et al.*, 1989). Using this approach, DeBiasio *et al.* (1987) imaged five parameters in 3T3 fibroblasts: nuclei (Hoechst 33342), mitochondria (diIC$_i$-[5]) endosomes (lissamine rhodamine B dextran), actin (fluorescein-labeled), and total cell volume (Cy7-dextran). The first four could be followed for several hours in response to wounding. Lemasters *et al.* (1990) studied the effects of toxic and hypoxic injury on single cells. By following intracellular calcium (fura-2), intracellular pH (BCECF), mitochondrial membrane potential (rhodamine 123), cell viability (propidium iodide), and cell blebbing (phase contrast), they concluded that intracellular calcium levels showed no correlation to blebbing and eventual death, while cells that acidified showed prolonged survival and better recovery. Although time resolution in these studies is relatively slow, they offer tremendous insight into approaches and possibilities.

## III. "SIMULTANEOUS" INTRACELLULAR CALCIUM–pH MEASUREMENTS

The classic example of multiwavelength imaging is imaging of "ratio" indicators for calcium and pH (reviewed by Bright Chapter 3, this volume; see Haugland, 1992 for a review of available probes and extensive bibliography). This subject has become a growth industry with more than 1000 publications concerning the calcium dyes fura-2 and indo-1 since their announcement by Grynkiewicz et al. (1985). The ratio approach can be applied to many different types of dyes. Ratio imaging of pH (Bright et al., 1987) has also become very popular. Many pH dyes are available (Haugland, 1992; Valet et al., 1981). Although rapid kinetic studies are still rare, there is a serious interest in resolving the rapid kinetics of phenomena, such as calcium waves and calcium oscillations (reviewed in Tsien and Tsien, 1990).

There are now a large number of studies imaging single ions; however, studies involving two or more ions are, up to this point, relatively rare and rapid kinetic measurements on two or more ions are rarer still. A number of researchers have loaded cells with either a pH or a calcium dye, recorded the time course of the changes, then attempted to recombine the separate measurements (e.g., Dickens et al., 1989; Malgaroli et al., 1990; Malgaroli and Maeldolesi, 1991; Törnquist and Tashjian, 1992). These studies suffer from the basic criticism that the measurements, in coming from different populations, may not be comparable, especially if the relative rates of change or start times are to be analyzed.

Several groups wishing to study calcium and pH in the same cells have hit upon the expedient of loading the cells with indicator dyes for both ions, and collecting the needed information by sequentially varying excitation and emission at the four wavelength pairs. Probes are usually introduced as the acetoxymethyl (AM) esters. The most popular choice has been fura-2 and BCECF [2′,7′-bis-(2-carboxyethyl)5-(and 6)-carboxyfluorescein] (Aakerlund et al., 1990; Miyata et al., 1989; Simpson and Rink, 1987) although other combinations have been used (Khananshvili, 1990). Simpson and Rink (1987) used fura–BCECF to fluorimetrically monitor suspended platelets. They alternated excitation at 340 and 490 nm and monitored emission at 520 nm. They demonstrated that the small increase in $pH_i$, induced by thrombin, lagged behind the large increase in $[Ca^{2+}]_i$. The ratio approach was not used in this relatively slow experiment; dye leakage was calibrated separately. Miyata et al. (1989) imaged calcium and pH in myocytes using fura–BCECF. Emission was collected with a broad bandpass filter at 510–540 nm and the excitation varied sequentially among 340 nm, 380 nm, 440 nm, and 492 nm. The images were then processed to yield the ratios. No kinetic measurements were attempted.

Martinez-Zaguilan et al. (1991) used fura-2–SNARF-1 and automated measurements in an SLM 8000C spectrofluorimeter. Their measurement cycle included the isosbestic points for the dyes to separately monitor dye loading, leakage, and bleaching. They first measured excitation at 340, 360, and 380 nm with emission at 510 nm, followed by excitation at 534 nm with changing emission at 584,600 and 644 nm. Each cycle of six wavelength pairs required 266 ms. Their most interesting result was the demonstrated need to correct the intracellular calcium values for the pH (see below).

## IV. BASIC PROPERTIES OF RATIO-TYPE ION INDICATOR DYES

### A. Advantages of Fluorescence Ratio Imaging for Quantification

Measurement of ion concentrations in living cells using indicator dyes was originally hampered by a number of problems. Besides the need to find dyes that have high specificity for a given ion and a $K_d$ in the physiological range, quantitation is made difficult by the propensity of the dyes to change concentrations due to leakage, photobleaching, or movement into other compartments in the cell. Many of these problems have been solved by the introduction of "ratio"-type indicator dyes (reviewed in Bright, 1992; Haugland, 1992; Taylor et al., 1988; Tsien, 1989; and Poenie and Chen, Chapter 1, this volume). These compounds not only change quantum yield upon binding an ion, but they shift the peak position of their excitation or emission spectrum (or both). As shown in Fig. 6.1 for the emission dyes indo-1 and SNARF-1, acquisition of information at the two emission wavelengths allows quantitation by means of the ratio of the two intensities, which can be related directly to the ion concentration. The ratio is independent of dye concentration; changes in dye levels will not effect the calculated ion concentration (so long as the dye itself does not change the ion level).

Imaging such dyes requires calculation of the ratio at each pixel after capture of two images. The ratio approach can be applied to any dye with the requisite properties. Such indicators exist for calcium, magnesium, sodium, potassium, and hydrogen ions as well as membrane potential (Haugland, 1992; Tsien, 1989; Waggoner et al., 1989).

Like most fluorescent indicators, the ion-specific ratio probes have negative properties, most of which we will not have the time or space to discuss. Fluorescent dyes are excellent at sensing their immediate environment, thus the dyes usually tell you more than you want to know. For

FIGURE 6.1

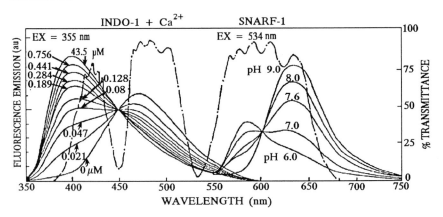

Indo-1 and SNARF-1 emission spectra (cf. Haugland, 1992). Emission peaks are well separated, allowing simultaneous imaging. The transmittance spectrum of the three-wavelength dichroic mirror used in the four-image microscope (Fig. 6.4) is also shown (---- - ----).

example, most ion-sensing dyes are sensitive not only for the target ion but for similar ions in the periodic table, the general ionicity, pH, and polarity of its surroundings, whether it is hindered or free to move, and other even more subtle circumstances, such as presence of other dyes. Each dye or combination of dyes and cells requires careful scrutiny to avoid assaying artifacts (Brecker and Fay, 1987; Roe et al., 1990). A common problem with ion-sensing dyes introduced into cells as acetate or acetoxymethyl esters is that they partition onto other spaces besides the cytoplasm. This may be solved with a new set of fura-type dyes synthesized by Poenie and collaborators (see Chap. 1) that show far less tendency to move from one compartment to another. Other problems with dyes are discussed in other chapters (see Chew and Ljungström, Chapter 5, this volume, Poenie and Chen, Chapter 1, this volume). We will be concerned with pH sensitivity as well as the basic assumptions involved in the ratio method.

## V. IMAGING OF EXCITATION-TYPE DYES VERSUS EMISSION-TYPE DYES FOR RAPID KINETIC FLUORESCENCE VIDEO MICROSCOPY

### A. Simultaneous versus Sequential Measurements

The ratio-type indicators present the archetypal case of multiparameter imaging; although only a single variable is forthcoming, two images are required. For excitation ratio-type dyes (e.g., indo-1, BCECF), the excitation wavelength must be varied alternately between the two wavelengths. This places a number of constraints on the experimental apparatus. Designs for ratio instruments typically use a revolving filter wheel placed in the epifluorescence excitation path, forming images alternately at one or the other wavelength on a single camera (Fig. 6.2), although systems have been constructed using monochrometers and a chopper, or monochrometers, shutters and a bifurcated fiber-optic bundle. If the changes to be measured are very slow, the filters can be changed by hand; otherwise, motor drives are also employed. The time resolution of such composite images is on the order of 0.5 s although commercial wheels with switching times of 0.1 s or less are available (reviewed in Bright, 1992; Tsien and Harootunian, 1990). Besides introducing vibration, rapidly changing wheels or choppers require precise stepping motor drives and sophisticated software to keep track of the wavelength of light being presented to the sample.

Kinetic experiments require continuous formation of microscopic images from the two different wavelengths. Slewing filter wheels or monochrometers are relatively slow but perfectly acceptable of the processes to be studied are slower. However, standard RS170 video equipment can produce and store 30 images per second. With a few tricks this can be extend to 60 images/s. Processes with half times in the 50–500-ms range could be resolved. If rapid sampling is required, then the excitation dyes present intrinsic problems. The fastest time resolution for a filter wheel or chopper system forming ratios from a single dye would be half the frame rate or

FIGURE 6.2

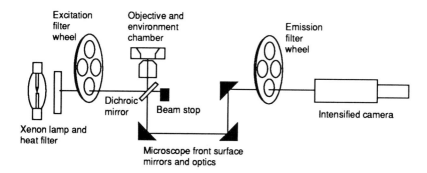

Microscope layout for imaging using filter wheels. For an excitation-type dye, such as fura-2 or BCECF, the excitation wheel would be rotated and emission at a single wavelength imaged. Emission dyes would use a fixed excitation and alternating emission filters. Multiple dyes could be imaged by placing several dichroic mirrors in a revolving wheel (cf. Bright, 1992 and Chapter 3, this volume).

15 frames/s. For two dyes (four images) this drops to 7.5 frames/s. To accomplish this, the filters would have to switch wavelength during the short dead time between video frames. The wheel must be carefully timed with stepping motors and its position known exactly. Wavelength overlap cannot be tolerated; therefore, the frame transmitted during the change may be lost to the data set.

The rotating wheel produces vibration. This can be removed by coupling the output from the wheel to the microscope with a fiber-optic bundle or liquid-light guide. Vibration-free, video-rate switching for excitation-type dyes can be achieved by selecting wavelengths with monochrometers or bandpass filters and using Uniblitz-type mechanical or optoacoustic coupler-type shutters. However the data rate for the ratio is still limited to half the sampling rate.

Finally, the ratio method is based on the basic assumption that the calculations are performed on data collected at the two wavelengths at the same time. If the processes being studied are slow compared to the sampling rate, collecting the data alternately presents no difficulty. However, for rapidly changing systems, the sampling time differences between the images will not give a true instantaneous value for the ratio. While it is possible to correct the data for time lag, this is tedious at best and requires at least a first approximation of the kinetics involved. This may be the information sought.

## VI. ADVANTAGES OF EMISSION RATIO DYES

Dyes, such as indo-1 or SNARF-1, use the same excitation wavelength but have distinct emission peaks; the information to form the ratio is always available (Fig. 6.1). These images can also be resolved by alternating filters in the emission path (Fig. 6.2) although one is then back to the same set of problems as for an excitation wheel.

## VII. IMPROVEMENTS IN DESIGN
## FOR MULTI-IMAGE EXPERIMENTS

We have previously reported on a microscope design that solves these problems (Morris, 1990) (Fig. 6.3). It can acquire fluorescent images of living cells at two excitation wavelengths, generated either by a single excitation wavelength or by dual excitation at two different wavelengths. The emitted light is separated into two images by the dichroic mirror and imaged by two cameras simultaneously. It offers the following improvements over filter-wheel or monochromator–shutter arrangements: (1) There are no moving parts in the optical train. This removes vibration and synchroniza-

FIGURE 6.3

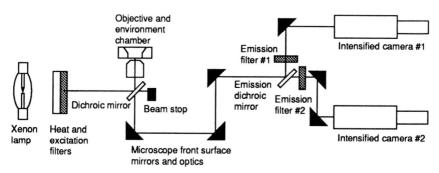

Microscope design for formation of dual images. Standard epifluorescence is used for a dye like indo-1. Excitation is via a standard epifluorescence "cube" except that the emission filter has been removed. The two emission wavelengths are divided by a second dichroic mirror and two barrier filters. Images are formed by two intensified CCD cameras. Adjustments in the image splitter bring the two images into register and focus. For experiments where two single-emission dyes are to be excited simultaneously (Morris, 1990; Morris *et al.*, 1991; Morris *et al.*, 1992), a dual-wavelength excitation filter, and a dual bandpass dichroic mirror (Marcus, 1988) are used in the excitation filter block. See Morris (1990) for further details.

tion problems. (2) By taking two images simultaneously from the same specimen, data collection rates equal camera frame rates. (3) By collecting simultaneous images, the ratios come from simultaneous, not sequential, samples. Thus, there is no temporal sampling error. Sampling the spatial distribution of rapid changes in fluorescence is limited only by camera frame rates, fluorescence emission levels and light losses in the optical train, properties of the dyes themselves, and the sensitivity of the imaging devices. The machine can capture data for 30 ratio images/s using standard RS170 video equipment. The rate will stay the same when multiple dyes are imaged. No special software is needed to control or track the filter wheel or shutters or mark the video frames for the type of excited light.

Similar designs for single excitation–dual emission have recently been reported by Takamatsu and Wier (1990a, 1990b). Noran Instruments (Middletown, WI) has recently announced a commercially available confocal instrument with the same capabilities.

## VIII. THE NEED FOR MORE THAN TWO SIMULTANEOUS IMAGES OF THE SAME CELL, AND THE DEPENDENCE OF INDO AND FURA CALCIUM $K_d$ ON pH

Imaging the activity of multiple probes in the same location at the same time provides the unique opportunity of observing the relative changes in

FIGURE 6.4

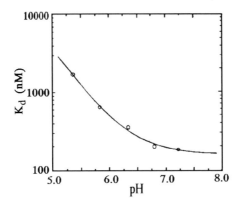

The dependence of the Indo–$Ca^{2+}$ $K_d$ on pH. Three-parameter fit of Eq. (2) to the data of Lattanzio (1990) as described by Martinez–Zaguilan *et al.* (1991). See text for further details.

two or more variables. This becomes necessary for calcium measurements using indo and fura if the pH of the cell is low or changing rapidly. Like the parent compounds on which they are based, these calcium chelators have $K_d$'s for calcium, which are pH dependent (Lattanzio, 1990; Lattanzio and Bartschat, 1991; Martinez–Zaguilan *et al.*, 1991; Roe *et al.*, 1990). Figure 6.4 shows data from Lattanzio (1990) detailing the large change in indo-1 $K_d$ between pH 5.5 and 8. This problem is mentioned in the original publication (Grynkeiwicz *et al.*, 1985), which also noted that this shift was negligible above pH 7.0. This has led to the general strategy of choosing a convenient, published value for $K_d$ and presuming that the $pH_i$ remains stable throughout the experiment. Recent publications point out the danger in such an approach (Martinez–Zaguilan *et al.*, 1991; Roe *et al.*, 1990). Many cells either have resting pH's below 7, or show a shift in pH in the course of the experiment, or both. Thus, if small changes in $Ca^{2+}$ are to be documented, or if the $pH_i$ changes during the experiment, then the calculated calcium level should reflect the prevailing pH. For kinetic experiments this correction should be made using a pH value obtained at the same time as the calcium data. If this is not done, then as for the case of staggering the ratio readings, the data can be skewed by the pH kinetics, which does not need to be following the calcium changes.

## IX. FOUR-CHANNEL VIDEO MICROSCOPE FOR SIMULTANEOUS IMAGING OF TWO-RATIO DYES

The foregoing discussion presents the need for absolutely simultaneous monitoring of both $[Ca^{2+}]_i$ and $pH_i$. To accomplish this, we have built the microscope shown in Fig. 6.5. It will collect two, three, or four simultaneous low-light images. A short discussion of the instrument has been published (Morris *et al.*, 1991). We will describe this instrument and show how the pH measurements can be used to correct the calcium values. All measurements are made simultaneously, so that there is no bias or lag in the correction. With appropriate choices of dyes and the ability to excite simultaneously at multiple wavelengths, it becomes possible to image up to four probes at the same time. These can be a combination of fluorescence and phase or DIC. For example, we have studied cell–cell fusion by simultaneously following redistribution of a membrane and a cytoplasmic probe

FIGURE 6.5

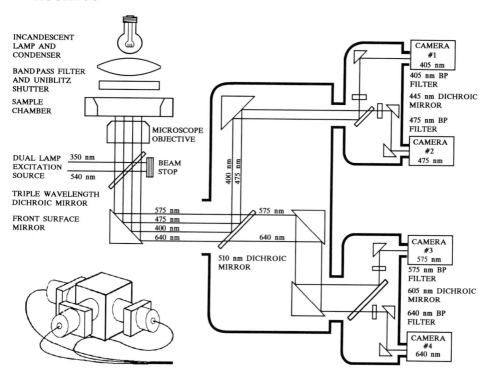

A four-image microscope configured for simultaneous excitation–emission of the two "ratio" dyes indo-1 and SNARF-1 for Ca²⁺ and pH. It can be constructed on either a standard upright or inverted microscope. The design is similar in concept to the two-image machine of Fig. 6.3. It has no moving parts. The dual- and triple-wavelength excitation systems are discussed in the text. Emitted fluorescence is first divided by a 510-nm dichroic mirror. The fluorescence from indo is further split by a second dichroic mirror at 445 nm and imaged by two intensified cameras at 405 and 475 nm. The emission from SNARF-1 is split by a third dichroic mirror at 605 nm and imaged by two intensified cameras at 575 nm and 640 nm.

along with phase contrast images to register shape changes (Morris, 1990; Morris et al., 1991).

## X. CHOICE OF DYES FOR SENSING CALCIUM AND pH

Indo-1 is the dye of choice for emission ratioing. There are a number of ratio-type dyes for pH (reviewed in Haugland, 1992). SNARF-1, an emission ratio-type fluorescein analog, has near-ideal fluorescence spectrum for use with indo (Fig. 6.1), as well as a p$K_a$ of 7.30. Its pH-sensing properties are relatively insensitive to other ions, including calcium, and to indo or fura, making it a good choice to pair with indo. Like fura-2-AM, it is easily loaded into cells as the AM (acetoxymethyl ester) ester along with indo-1-AM (Martinez-Zigulian et al., 1991; Morris et al., 1991).

To create the information to image both of these dyes, we require simultaneous excitation at 350 and 540 nm and images of emission at 405, 475, 575, and 640 nm. This is schematized in Figs. 6.5 and 6.6. Either an upright or inverted microscope can be used; however, examination of cultured cells is most easily made in an environment chamber with a glass bottom, of the type used for patch–clamp studies, fitted to an inverted microscope (Tsien and Harootunian, 1990). The design is similar in concept to the two–image machine (Fig. 6.3). It has no moving parts. Cells containing both indo-1 and SNARF-1 are illuminated at two excitation wavelengths (350 nm and 540 nm). Emitted fluorescence is first divided by a 510-nm dichroic mirror. The fluorescence from indo is further split by a second dichroic mirror at 445 nm and imaged on two intensified cameras at 405 and 475 nm. The emission from SNARF-1 is split by a third dichroic mirror at 605 nm and imaged by two intensified cameras at 575 nm and 640 nm.

## XI. MULTIPLE-WAVELENGTH EPIFLUORESCENCE EXCITATION

A wavelength selectible dual-epifluorescence excitation system was built by placing bandpass filters in front of two Nikon 75-W xenon lamps and collecting the light with a Dolan-Jenner 5-mm-diameter fused silica bifurcated fiber-optic bundle, then mixing the output with a 50-mm by 5-mm-diameter Oriel liquid-light guide. For indo-1 and SNARF-1, we use Chroma Technology 350-nm and 540-nm 10-nm-width bandpass filters. Multiwavelength epifluorescence illumination is made possible using newly developed "multichroic" mirrors (Marcus, 1988). For this experiment, a Chroma Technology multiple-bandpass dichroic mirror is used. The spectrum is shown in Fig. 6.1. This is positioned in the usual place for the epifluorescence dichroic mirror.

This illumination system is relatively inexpensive and very flexible. Ex-

FIGURE 6.6

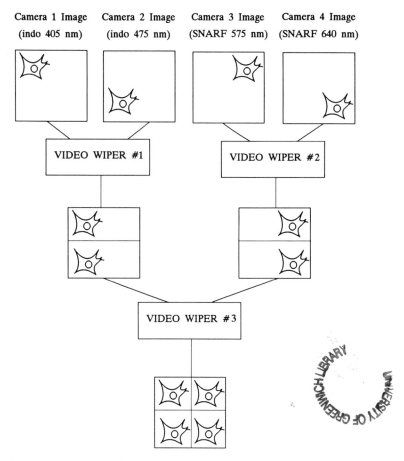

Scheme for recording four simultaneous images of the same object on a single standard video frame. See text for details.

citation wavelengths can be quickly changed. To improve light-intensity fluctuation problems and increase wavelength selectivity, we plan to build a three-wavelength illuminator (Fig. 6.5, lower left) using an Oriel three-port lamp housing, three condensers and filters and a Dolan-Jenner trifurcated fiber-optic light guide. Channels can be selected with Uniblitz shutters under computer control, which could be programmed to select wavelengths alternately for excitation dyes like fura-2. The design is compact and the flexible light guides allow for mounting either on or off the optical table. Compared to using mirrors and beam splitters, alignment problems are minimized.

The three dichroic beam splitters that divide the emitted light are housed in three ganged Nikon multi-image modules. These commercial pieces greatly ease assembly of the microscope. Each module contains a turret, which will hold four Nikon dichroic mirror–emission filter combinations. An $x,y$ translator and $z$-axis focus aid registration of the images. Transmission losses for the module itself are about 5% per unit. The losses due to the dichroics and filters must be added to this.

Four Chroma Technology bandpass filters ensure that no stray light that might have passed through the dichroic splitters enters the cameras. These have been designed to maximize emitted photon throughput without compromising the precision of the ratios. They are multicavity filters to maximize transmission within the chosen band and exclude light on either side. These are housed in the dichroic mirror cubes in the image splitters. So far, we have found it unnecessary to add separate filter holders just in front of the intensifiers. However, changing the high wavelength filter in each cube requires removing it from the housing, which upsets the registration. Also the design of the Nikon cubes limits the thickness of the high wavelength filter to less than 3 mm. The separate filter mounts would greatly increase flexibility.

## XII. SIMULTANEOUS COLLECTION OF FOUR VIDEO IMAGES OF THE SAME MICROSCOPIC FIELD

We have used two approaches to combine four "quarter images" onto one RS170 video frame. Both schemes reduce spatial resolution but allow four images to be stored simultaneously on one standard RS170 video frame. This greatly reduces the cost of the recording equipment involved. The first scheme uses a hard-wired module (#MV 40D, For-A Corp., Newton, MA). An object is placed in the center of the microscopic field and imaged by the four cameras. The four outputs are connected to the MV-40D. This "grabs" four separate inputs in digital image buffers, then for each input, drops every other vertical and horizontal pixel. Each reduced image is then remapped onto a 256 × 240 pixel quarter-image segment of a fifth buffer. This image is converted to RS170 analog and presented to the Matrox MVP-AT real-time array-processing board as a standard video signal. Image distortion is minimal. However, both vertical and horizontal resolution are compromised. The images are digitized at 7-bit resolution, which further degrades the data quality.

We prefer the second approach as being more flexible and less expensive (Fig. 6.6): An object is placed in the center of the microscope field. The four cameras are positioned off axis so that the 256 × 240 pixel region of interest is in the upper left corner of camera 1, the lower left corner of camera 2, etc. The outputs from cameras 1 and 2 are fed into video wiper I (Primebridge PVW-1), and cameras 3 and 4 into video wiper II. These are combined in video wiper III as shown. Some patience is required to initially set all images to the same size and orientation; once set, there is no drift. To our surprise, jitter is minimal. There is a 1–2-pixel "flicker" between images at the edges, making these areas unanalyzable, but experimental information is generally at the center of the images. This approach has the advantage of being able to create any size and placement of the images. For example, the input was divided horizontally into thirds for the membrane-fusion experiments (data not shown) (Morris, 1990; Morris *et al.*, 1992). Until a programmable, digitizing version of the ganged video wipers becomes available, we foresee no difficulty in continuing to use them. Figure 6.7 shows four images of the same cell captured by this method.

Using the ganged video wipers requires that the images be offset from the optical axis of the microscope. We have constructed a tower that holds

FIGURE 6.7

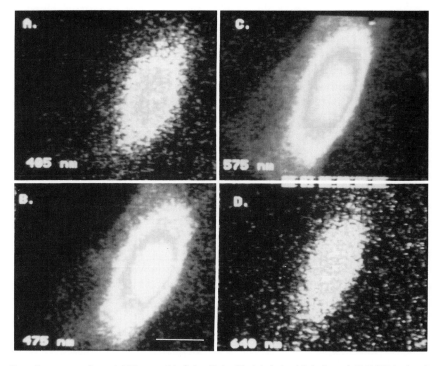

Four-image raw data. A kidney epithelial cell double labeled with indo and SNARF is simultaneously illuminated at 350 nm and 540 nm and imaged simultaneously at (A) 405 nm, (B) 475 nm, (C) 575 nm, and (D) 640 nm, using the ganged video wipers described in Fig. 6.6 and the text. Image represents a false-colored, 16-frame average of the raw data, which has been corrected for background, shading error, and the spillover of the 475-nm image into the 575-nm image. To calculate the raw ratio data, the software will divide image (A) by (B) and (C) by (D). The resulting ratio maps will be corrected by Eq. (1–3) to produce the maps of Fig. 6.8. (Bar = 10 $\mu$m.) (*See Plate 1 for a color version of this figure.*)

all four intensified cameras. Each camera has its own set of $x, y$ and $x$ translators. Magnification differences are compensated with the coupling lenses between the intensifier and the camera.

Both approaches reduce the available area per image. Ideally, one would record all four images, then synchronize playback to analyze each wavelength pair as a full 512 × 480 pixel frame. While this is possible with larger, faster computers and multiple video cassette recorders (VCRs) or optical memory disk recorders (OMDRs), this equipment is presently beyond our budget capability. The advent of high-definition television should reduce the cost of large format VCRs and other special-format equipment.

## XIII. FORMATION OF LOW-LIGHT IMAGES

After passage through the appropriate dichroics and barrier filters, the fluorescence emission is focused onto one of four VideoScope KS-1381 image-intensifer plates coupled to Dage-MTI 72× CCD (charge-coupled device) cameras, set at maximum gain and gamma of 1.0. Automatic compensation of gain and pedestal is shut off. This produces stable, high-contrast images using ultralow-light levels with little persistence or bloom. All cameras are synchronized to the horizontal and vertical outputs of the crystal clock of the Matrox MVP-AT array processor fed to two Sigma PDA-100A pulse-distribution amplifiers.

## XIV. IMAGING HARDWARE–SOFTWARE

The microscope is designed to be modular and independent of the capture and analysis hardware and software; thus, components can be updated as equipment improves or drops in price. We presently use an IBM-compatible 386H/33 PC containing a Matrox MVP-AT imaging board running commercial software (Belvour Consulting MicroMeasure 4000A). The output from the ganged video wipers is presented to the Matrox board. For real-time capture, the combined, background corrected video images are simultaneously displayed on an RGB monitor (in monochrome), and recorded by a 400-line resolution Sony VO 9600 ¾-in. U-matic VCR with time-code generator–reader, or when the expense is warranted, directly onto a 450-line resolution Panasonic TQ 3031F OMDR. VCR data can be passed to the OMDR for further analysis, which will eliminate further degradation.

Other useful software features include

1. *Real-time ratio display and calculation.* The software can calculate and display noncorrected ratio images at 15/s for either alternate full-frame data or multi-image single-frame data. This is very useful for qualita-

tive, on-line assessment of the quality of a prep before investing huge amounts of experimental time (Tsien and Harotoonian, 1990). This could be recorded for analysis if shading error-corrected data is not required (e.g., for qualitative quick screening of drug actions). The experimenter can define up to 15 regions of interest (ROI) of any size and shape (cf. Fig. 6.9). Both the raw data and the ratio of for each ROI can be displayed on the 386 VGA monitor and/or dumped to an ascii file. This is very handy for previewing experiments or for development work. The software will integrate the grey levels for these regions and store the results into an ascii file at up to 15 ROI samples/frame in real time. Background and shading error corrections can be applied off line.

2.  *Full-frame storage.* In the configuration described above, every video frame contains four ¼ frame images to be used for off-line analysis. For excitation ratio dyes or analyses at slow data rates, the software will alternate between two camera inputs, or four inputs, either taking successive frames or averaging 2–999 frames before changing to the next channel. Thus, for relatively slow experiments, one can collect two half-screen images for one dye followed by the two for the other dye, or alternate full-screen images of all four wavelengths. The software will control several Uniblitz shutters, either for excitation ratio dyes or for intermittent observations to save the samples during long experiments. Data can be stored on the VCR or the OMDR (especially handy for averaged data) or directly to the fixed disk (useful for experiments with relatively few images). The array processor and software can perform real-time background subtraction if desired.

## XV. IMPROVING SPATIAL AND TEMPORAL RESOLUTION OF THE DATA

Low-excitation light levels are necessary to reduce photobleaching and photodynamic damage to the cells, especially if experiments last more than a few seconds. This leads to low rates of emitted light and thus the requirement for intensified cameras. The rapid measurement rate we wish to achieve exacerbates this problem. We routinely record for several minutes. Even with perfect optical registration, very low fluxes of fluorescent photons will create problems, which can only be solved by averaging and smoothing. Takamatsu and Wier (1990a,b) have published video-rate images of the spread of $Ca^{2+}$-waves in single myocytes, from which they could calculate the propagation rates. They used a high light level for a very short time period. Our best experiments so far require the averaging of 2–16 frames to reduce noise (Morris, 1990; Morris *et al.*, 1991). These experiments were deliberately performed at very low excitation levels to forestall photobleaching and photodynamic damage over the long recording time; thus, noise could be reduced with increased excitation levels.

A primary goal of this research is to improve image quality to the point where meaningful measurements can be made on individual frames or fields. Thus, we continually strive to increase light throughput for a given excitation level. We have chosen the VideoScope KS-1381 image intensifiers as being the most sensitive and having the best spatial resolution at low-light levels of all such real-time devices presently on the market. Exchanging the coupling lenses for direct fiber-optic coupling can increase intensifier efficiency from the present 13–15% to >60% (TF Lynch, VideoScope International Inc., personal communication). The Nikon Diaphot, chosen for its extremely simple optical design, was ordered with 100/100 split to the eye pieces or camera port. The CCD cameras, while not quite as sensitive as newvicons, have a higher interscene dynamic range and better signal-to-noise ratio (SNR), (Spring and Lowy, 1989, Tsay *et al.,* 1990). We have built a thermostatting sleeve for oil-immersion objectives, allowing contact with the cell chamber without becoming a heat sink. This permits use of larger numerical apertures objectives, which increases the light-gathering power, and use of high-power large numerical apaerature objectives for single-cell observations (Inoue, 1986; Morris, 1990; Tsien and Harootunian, 1990). Mounting the components on an antivibration table has significantly improved image quality.

The infrared filters in the Nikon image splitters and cameras have been removed to increase sensitivity. This resulted in a more than 3-fold gain in intensity at 640 nm. Infrared filters are placed in front of the excitation filters and in the transillumination light path to remove heat and reduce background.

If time and money permits, microscope objectives from various manufacturers can be evaluated for fluorescent-light throughput and point spread (Gibson and Lanni, 1990). Careful adjustment of the conditions for loading the cells can maximize dye concentrations versus toxicity to increase emission (Roe *et al.,* 1990). Excitation levels can be balanced against photobleaching and photodynamic damage. Depending on availability, working time for probes can be increased. Compounds like Bodipy have an order of magnitude longer working time before photobleaching than does fluorescein (Haugland, 1992). The addition of free-radical scavengers like n-propyl gallate (NPG) to the cell buffers will reduce photodynamic damage and vastly increase dye working time (Giloh and Sedat, 1982). This worked well for the study of hemagglutinin-promoted cell–cell fusion (Morris, 1990; Sarkar *et al.,* 1989; Morris *et al.,* 1992) but interfered with calcium-flux measurements. All such procedures must be carefully checked for photodynamic damage and cell toxicity (reviewed in Bright and Taylor, 1986; Bright *et al.,* 1989; Tsien, 1989). Computer-controlled shutters can be periodically closed to reduce light exposure during long experiments (Tsien and Harootunian, 1990).

Low-light video equipment is improving rapidly. New designs for CCD cameras with larger formats and increased sensitivity will become available at low prices, as will video boards and faster computers to handle their output. Direct digital storage of data will increase resolution and reduce noise.


## XVI. INCREASING THE DATA COLLECTION SPEED TO 60 FIELDS PER SECOND

Standard video consists of two interlaced fields per frame. It has always been possible to examine these alternate fields with appropriate playback equipment or software (Inoue, 1986). However, the alternate images are then offset by one line, producing a slight jitter. Unlike scanning-type cameras, modern frame-rate CCD cameras integrate the image for about 15 ms then transfer the image to a second buffer in about 1 ms, where one field is coded into an RS170 wave form while the primary buffer acquires the next image. On the next cycle the alternate field is coded and transferred. Dage/MTI now offers a version of their model 72 CCD camera (Model 72×) that averages adjacent line pairs from the transfer buffer, then outputs this "field." This is repeated for every field. Thus, identical fields are transferred at 60/ s. Vertical resolution drops from 550 to 330 lines. One can use a standard RS170 VCR or OMDR to store the repetitive fields as interlaced frames, then apply software or hardware with a field-selection option for analysis.


## XVII. CALCULATIONS

### A. Spillover

Accurate ratio calculations require that several corrections be applied to the images. Besides the basic need for appropriate correction for background and shading error (Inoue, 1986; Tsien and Harootunian, 1990), simultaneous excitation of both fluorophores creates some special problems. The first is that there is "spillover" of emitted light from indo into the 575-nm image of the SNARF (Fig. 6.1). This can be handled in a straightforward manner similar to scintillation counting spillover. The spillover fraction is calculated by first determining the intensity of fluorescence at 475 nm and 575 nm for indo standards at various calcium concentrations, then performing a two-parameter straight-line fit to this data for the equation

$$I_{575} = C_1 I_{475} - C_2$$

The spillover for any pixel (or ROI) can then be calculated as

$$I_{\text{spillover},575} = C_1 I_{475} - C_2$$

The experimental image is first corrected sequentially for background, shading error, and geometric distortion. Then the spillover is removed for each pixel or ROI by subtracting $I_{\text{spillover},575}$ from the total fluorescence at 575 nm.

Second, like its parent compound fluorescein, SNARF-1 is excited by the UV light used to excite indo-1. We have chosen the UV excitation wavelength of 350 nm both for efficiency of indo excitation and for the fact that this is an isosbestic point for SNARF excitation (Haugland, 1992). This adds a constant amount of fluorescence to both SNARF emission images. Although contributions from 350-nm excitation are small (10–15% of the excitation at 450 nm), they will skew the pH calculations. Thus, one needs to calibrate the pH dye in the microscope using both excitation wavelengths.

## B. Correction of Geometric Distortion to Improve Registration

Pixel-by-pixel overlap of image pairs (registration) is crucial both for accurate ratio calculations from dual-emission dyes and for comparisons of features in multiple dye experiments. CCD cameras were chosen to reduce distortion of the images by analog-scanning devices. The Nikon splitter also removes the need to reinvert the mirror image produced when a simple dichroic mirror is used (Foskett, 1988) and improves light throughput over multiple dichroic mirrors (Takamatsu and Wier, 1990a,b). Small differences in magnification of each image are adjusted using the lenses coupling the image intensifiers to the cameras and the $z$ translators.

However, the dichroic mirrors in the image splitters are optical elements. In additional to changing the $x,y$ position of the image, any distortions in the glass will introduce distortions in one image relative to the other. Antireflection coating and careful choice of mirror surfaces are required. For all multiwavelength experiments, there is the ever present danger of chromatic aberrations, even with well-corrected objectives. This will cause misregistration of the images, especially for areas off the optical axis. To reduce this possibility, we confine our analysis to a small area at the center of the microscope field and have built $x,y,z$ translators for all cameras to aid alignment. The software contains several real-time measurement features, which allow very precise adjustment of size and shape of the input images.

Nonetheless, variable amounts of geometric distortion persist that make true, pixel-for-pixel ratio analysis impossible (Jericevic et al., 1989;

Takamatsu and Wier, 1990a,b). This problem can be solved by off-line correction for geometric distortion using the procedure of Jericevic *et al.* (1989). We have made preliminary tests of commercial software (Inovision "dps/os"), running on a Sun ELC workstation, to perform this correction as well as that for spillover and others discussed below. The corrected image is passed back to the MVP-AT for storage at the rate of 5–10/min. This hardware–software combination provides an inexpensive solution to a number of processing problems, such as recalculation of calcium maps for prevailing pH.

## C. Ratio Analysis

Ratio calculations and display of the corrected four image data are performed off line by dividing the upper two images by the lower two, producing two $240 \times 256$ pixel maps for calcium and pH. These are converted to concentrations using Eqs. 1–3. The resulting quarter-frame maps are then combined using the video editor for comparison. The ratio maps of Fig. 6.8 were thus created from data similar to Fig. 6.7.

## D. Ca–pH Calculations

Depending on the required precision of the data, $[Ca^{2+}]_i$ can be corrected for $pH_i$ on a region-by-region or pixel-by-pixel basis as follows: All images are corrected for background, shading error, spillover, and geometric distortion. For each region or pixel of the calcium image to be corrected, the corresponding region in the pH image is defined. Fortunately, SNARF-1 is not sensitive to either the calcium concentration or the presence of indo (Martinez-Zaguilan, 1991; and unpublished observations) and no correction is required. Therefore, the pH ratio for each region is calculated and converted to a pH value. This can be done either by interpolation from a standard curve of SNARF at several pH values or from the relationship (Martinez-Zaguilan, 1991):

$$pH = pK_a + \log(S_{f2}/S_{b2}) + \log[(R - R_{min})/(R_{max} - R)] \tag{1}$$

where $R$, $R_{max}$, and $R_{min}$ are the experimental, maximum and minimum pH ratios, $(S_{f2}/S_{b2})$ is the ratio of fluorescence values for free and bound forms of the dye at the $R$ denominator wavelength, and $pK_a = 7.30$.

The indo $K_d$ corresponding to this pH ($K_{d,corr}$) is then calculated from the relationship (Martinez-Zaguilan, 1991):

$$K_{d,corr} = [K_{d,max} + 10^{(pH - pK_a)}K_{d,min}]/[10^{(pH - pK_a)} + 1] \tag{2}$$

FIGURE 6.8

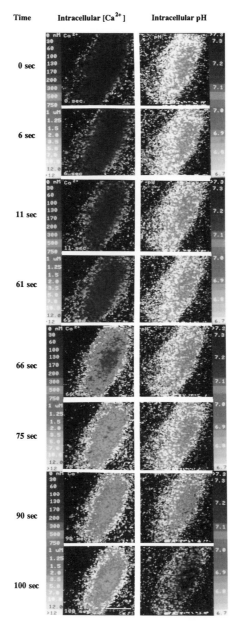

Kidney epithelial cell intracellular calcium and pH response to ionomycin. See text for details. (Bar = 10 μm.) (*See Plate 2 for a color version of this figure.*)

$K_{d,max}$, $K_{d,min}$, and $pK_a$ were calculated by least-squares fit of Eq. 2 to the data of Lattanzio (1990) for indo/EGTA at 37°C as $8.00 \times 10^4$ nM, 116 nM, and 3.99. Finally the pH-corrected $[Ca^{2+}]_i$ is calculated using the relationship (Grynkiewicz *et al.*, 1985):

$$[Ca^{2+}] = K_{d,corr} (S_{f2}/S_{b2}) [(R - R_{min})/(R_{max} - R)] \tag{3}$$

Since these calculations are relatively time consuming for repetitive kinetic data, when analyzing whole-cell $Ca^{2+}$–pH, we define regions of interest, extract integrated gray-level values for these ROIs from uncorrected images, and store to an ASCII file at video rates, using the MicroMeasure software. The ROI values are then corrected off line for total background and shading error before graphing.

## XVIII. SOME FURTHER CONSIDERATIONS

### A. Is the Equipment Fast Enough for the Experiment?

The validity of the entire method depends on all imaging equipment being faster than the video frame or field rates being used for sampling. For the equipment chosen, the persistence of the intensifier and camera must be significantly less than the video capture rate. The newer KS-1381 intensifiers use a low-persistence phosphor ($1/e$ less than 300 $\mu$s) and the transfer time of the CCD is less than 1 ms. Problems can arise during rapid rise from very low-light levels due to the autoprotect circuit in the intensifier power supply. This can be bypassed if there is no danger of overloading the intensifier.

A more insidious problem was pointed out by Bookman (1990): For many analog-scanning cameras, the integration time for each 16.67-ms field is 33.33 ms, the time that has passed since the field was last read. Also, since it takes the flying spot 33.33 ms to return to the same spot, portions of the image at the beginning of the sweep are sampled sooner than portions at the end. Both situations distort the timing of the data. This can be avoided by using CCD-type cameras.

### B. Are the Dyes Fast Enough?

The reporter probes must have kinetics that are faster than the measurements to be made with them. Stopped-flow and temperature-jump measurements give relaxation times at 37°C for fura and indo, which should just begin to distort kinetics with half-times less than 10 ms (Jackson, 1987; Kao and Tsien, 1988; Lattanzio, 1990; Lattanzio and Bartschat, 1991). Thus measurements at faster than field rates will require special care. The kinetics of SNARF have yet to be investigated. Since the pH-dependent change in fluorescein fluorescence is commonly used as a measure of dead time of stopped-flow mixing, we predict no problems in using SNARF, which is a fluorescein derivative.

## C. Available Light

We have discussed some of the trade-offs between high-excitation levels, photobleaching, and photodynamic damage. Low-excitation levels reduce emission levels to the point where even the best image amplifiers will fail to detect photons at all pixels. This can be partially offset by averaging adjacent pixels and temporally averaging two or more frames. If frame averaging cannot be avoided, then running averages can partially mitigate the reduced resolution. "Median" image filters produce less distortion than "averaging"-type kernels for reducing spatial noise.

## XIX. APPLICATIONS

### A. Simultaneous Intracellular $Ca^{2+}$ and pH Measurements Show Differential Kinetics

Kidney glomerulus epithelial cells were grown in culture on $10 \times 10$ mm squares of #0 coverslip glass and labeled with the AM forms of indo-1 and SNARF-1 using modifications of published procedures (Miyata, 1989; Morris et al., 1991). A coverslip of double-labeled cells was placed into low-calcium (0.5 mM EGTA) balanced salt solution at pH 7.34 and 22°C and examined with a thermostatted Nikon Fluor DL $100\times$, 1.3 numerical aperture (NA) phase objective. This first experiment used epi-illumination to excite indo fluorescence and transillumination for SNARF. (The transilluminator had to be refocused after each change in bath volume and was superseded by a dual-wavelength illuminator previously described). A cell exhibiting good fluorescence for both dyes was chosen for examination. The real-time ratio feature of the software was used to check the approximate resting $[Ca^{2+}]$. Raw data for all four wavelengths was then recorded as described (Fig. 6.7). The cells were treated with ionomycin followed by increased intracellular calcium. Fluorescence of all standards was recorded at the end of the experiments at the same camera and intensifier gains as for the experiment and the data analyzed off line.

Standard curves for the relationship of the ratio of SNARF emission at 575 and 640 nm for SNARF-1 were generated by interpolation of ratio measurements at pH 6.6, 6.8, 7.0, 7.2, and 7.4. Since SNARF fluorescence is not affected by either indo or pH (Martinez-Zaguilan, 1991), it is not necessary to correct for their presence. However, simultaneous recording of indo and SNARF results in the inclusion of some indo 475-nm peak fluorescence in the SNARF 575-nm images. This is corrected as previously described. After correcting the images, the pH map is calculated and used to correct the calcium concentration.

The $[Ca^{2+}]_i$ and $pH_i$ kinetics are presented in Fig. 6.8 as false color maps. The cell is placed in 0 $Ca^{2+}$ buffer. At 0 time, the cell has very low internal calcium and a pH around 7.0. Immediately after the addition of 1 $\mu M$ ionomycin (6 s), the calcium increases to the 100–200 nM range, where it remains for some seconds until the addition of 2 mM external $Ca^{2+}$ to the bath (61 s). At this point, internal $Ca^{2+}$ increases rapidly (66 s). The pH that has remained stable suddenly begins to collapse toward the bath pH (75 s). This continues (90 s) and is complete at 100 s.

Each map was calculated from a 16-frame average. This cell was followed continuously for 10 min (9000 frames). Increasing excitation brightness would have decreased the number of frames required for averaging but would have accelerated photobleaching. Although rather simple and highly nonphysiological, this experiment serves to demonstrate the ability to visualize the differential kinetics of two ions.

## B. Depolarization of Pituitary Intermediate Lobe Melanotropes Raises $[Ca^{2+}]_i$ and Lowers $pH_i$

These cells respond to a variety of neurotransmitters and hormones by depolarization and secretion neuropeptide hormones ACTH, $\alpha$-MSH, and $\beta$-endorphin (Kracier et al., 1985; Millington and Chronwall, 1991). This activity is mediated by opening L-type Ca-channels that raise $[Ca^{2+}]_i$ (Cota and Hiriart, 1989; Thomas et al., 1990; Tsien and Tsien, 1990). As preliminary studies for planned investigations of the receptor regulation of secretion in melanotropes, we investigated the effects of depolarization on these cells in primary culture. We used the two-lamp selectable dual-wavelength light source previously described for this and the following study.

Melanotropes were isolated and placed in explant culture (Gehlert et al., 1988), on #00 coverslips, for 14+ days, then double labeled with indo-1-AM and SNARF-1-AM. (The #00 glass plus the #0 bottom of the thermostatted environment chamber give a slightly smaller thickness than a standard #1 coverslip; this increases the working distance of the oil-immersion objectives.) A coverslip was placed in a bath containing 1-ml hepes buffered balanced salt solution and the cells depolarized by adding 1 ml of 130 mM KC1 10 mM Hepes pH 7.3. The emission fluorescence at four wavelengths was recorded as above with a Nikon Fluor DL

FIGURE 6.9

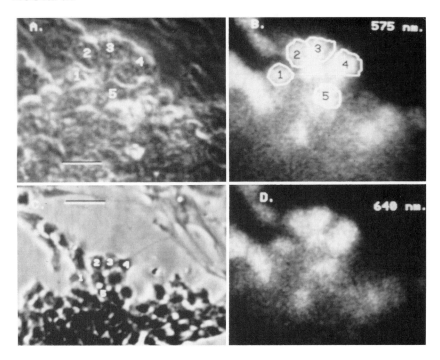

Images of cultured melanotropes. (A) Phase contrast 16-frame average image of the lobule at the beginning of the experiment. (B) 575-nm fluorescence image (16-frame average). (D) 640-nm fluorescence image, taken from the same fluorescence composit as B. (C) Phase-contrast image of the same cells as A, B, and D after fixation, staining, and mounting, as described in the text. This image is offset downward to show the fibroblasts, upper right, which are easily recognized by their morphology and excluded from the recordings and analysis. The cells, numbered 1–5, were chosen for analysis (see text). (Bar = 10 $\mu$m.) (*See Plate 3 for a color version of this figure.*)

phase, $40\times$, 1.3 NA, oil–immersion objective. Figure 6.9 shows several melanotropes in the environment chamber first by (A) phase contrast and (B, D) fluorescence. Panel C shows the same cells imaged after fixation, staining with cresyl violet and mounting in canada balsam (cf. Gehlert *et al.*, 1988, for methodology). A group of melanotropes has migrated out of the lobule (C, lower left) into a monolayer. The fibroblasts (C, upper right) were excluded from the analysis. The five ROIs in (B) are the areas analyzed for $[Ca^{2+}]_i$ and $pH_i$ and graphed in Fig. 6.10 B. Fixation shrinks the tissue; nonetheless, the correspondence between live and fixed cells is easily seen (cf. cells numbered 1–5 in Fig. 6.9 A, B, and C).

The integrated $[Ca^{2+}]_i$ and $pH_i$ were calculated for several cells using the equations previously described. For each cell, two regions of interest (ROIs) of equal size and shape, one for the indo set and one for the SNARF set, were defined. The tapes were played back in real time and a six-frame average was captured. The integrated grey density at both wavelengths for each ROI was written to an ASCII file and the uncorrected ratio displayed on the VGA monitor. This process required a total of 266 ms (eight frames), and was repeated for the length of the experiment. For graphing, four points were averaged and every fourth-averaged point was displayed.

Figure 6.10A shows the graphed integrated intracellular $Ca^{2+}$ and pH for several melanotropes that were followed for several minutes. The pH begins to fall almost immediately. This variable response is also seen in chromaffin cells (Rosario *et al.*, 1991). It may be due to the exciting light. $K^+$ depolarization produces the expected rise in $[Ca^{2+}]_i$ due to activation of L-type Ca-channels. This is coupled to (near) simultaneous increase in the rate of acidification. Note the differential kinetics of the calcium rise in the different cells. To test the responsiveness of the dye, 1 $\mu M$ ionomycin is added at the end of the experiment. Unlike the kidney epithelial cells, which respond with a massive increase in $[Ca^{2+}]_i$, the IL cells show a transient increase in $[Ca^{2+}]_i$, followed by (partial) recovery. The pH shows no response to the second calcium entry. Chromaffin cell $[Ca^{2+}]_i$ also quickly recovers from ionomycin (Roserio *et al.*, 1991). These effects are dependent on entry of extracellular $Ca^{2+}$ rather than direct release

FIGURE 6.10

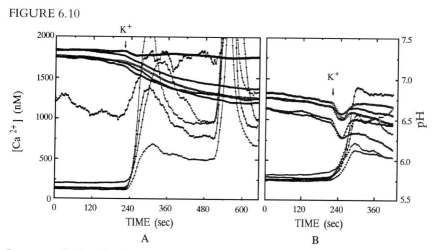

Responses of cultured melanotropes to high-potassium depolarization. (A) Control; (B) in the presence of 100 mM Amiloride. See text for further details.

FIGURE 6.11

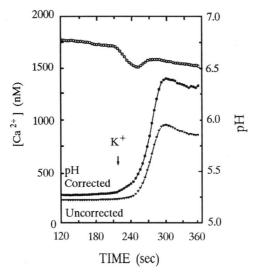

TIME (sec)

One trace from Fig. 6.10B before and after correction of the calcium values for the prevailing pH. See text for further details.

from intracellular stores; no changes in either $Ca^{2+}$ or pH are seen in 1 mM extracellular EGTA (results not shown). As has been shown for chromaffin cells (Rosario *et al.*, 1991), 100 $\mu$M amiloride prolongs the pH drop, presumably by slowing $Na^+–H^+$ exchange (Fig. 6.10B). Comparison of fixed indo $K_d$ and pH 7.00 versus pH corrected results (Figure 6.11) show a slightly elevated $[Ca^{2+}]_i$ and a larger increase in $[Ca^{2+}]_i$ in response to depolarization. In this example, one sees after correction that the fall in pH is coincident with the rise in calcium and does not precede it.

Our results correlate well with those of other labs. Our melanotropes raise their $[Ca^{2+}]_i$ above the 600 nM threshold needed to activate exocytosis (Thomas *et al.*, 1990). Thomas *et al.* (1990) gives a resting $Ca^{2+}$ of 231 ± 23 nM (SEM for 32 cells) in melanotropes versus our values of 202 ± 70 nM (SD for 181 cells) and pH of 7.00 ± 0.26 (181) from our preliminary data.

The imaging approach, besides allowing for individual corrections, can quickly uncover multimodal responses that cannot be seen by fluorimetry. One of the six cells in Fig. 6.10A showed a poor initial response to calcium and did not respond to ionomycin. We find that 15–20% of our IL cells are poor responders. We are also able to reject the fibroblasts, which can be easily identified from the phase-contrast images. This underlines results of other labs that find a high percentage of nonresponders or bimodal responders in their secreting populations (e.g., Malgaroli *et al.*, 1990; Malgaroli and Meldolisi, 1991).

## C. MDCK Cell $Na^+$–$H^+$ Antiporter

Preliminary experiments on intracellular calcium responses to acid loading (Dickens *et al.*, 1989; Jacobsen *et al.*, 1990; Muallem and Loessberg, 1990; Weiner and Hamm, 1990) were studied in MDCK cells (T. B. Wiegmann, L. W. Welling, and S. J. Morris, unpublished). These were grown in RPMI supplemented with 10% FCS on #00 coverslips, double labled with indo–SNARF and placed in $Ca^{2+}$-free medium. An isosmotic exchange for 20-mM $NH_4Cl$-containing medium resulted in a rapid increase in $[Ca^{2+}]_i$, which was closely coupled to the expected increase in $pH_i$ (Fig. 6.12). Calcium levels quickly returned to baseline. When the pH had returned to baseline, subsequent exchange of the $NH_4Cl$ for a low $Na^+$ medium, produced the expected rapid acidification, to which a second increase in $Ca^{2+}$ was coupled. The second calcium peak was biphasic and of lower amplitude than the first. Replacement of the $Na^+$ allows the pH to return to baseline via the $Na^+$–$H^+$ antiporter. The absence of extracellular $Ca^{2+}$ suggests that both calcium elevations were mobilized from intracellular stores. Qualitatively similar results have been reported for N2A neuroblastoma and PC12 pheochromocytoma cells (Dickens *et al.*, 1989) and $GH_4C_1$ anterior pituitary cells (Törnquist and Tashjian, 1992) using microfluorimetry to gather the integrated fluorescence and following either fura-2 or BCECF. Our concurrently recorded results demonstrate the very close coupling of the

FIGURE 6.12

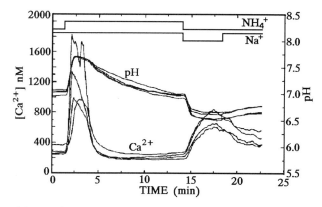

Intracellular calcium and pH responses of MDCK cells to acid loading. Four cells were followed for 25 min. See text for details.

changes. Investigation of the nature of the intracellular calcium store(s) and coupling mechanisms involved are in progress.

## ACKNOWLEDGMENTS

I wish to thank my collaborators, Dr. Bibie M. Chronwall, Ms. Diane M. Beatty, and Mr. Daniel E. Howard at the University of Missouri at Kansas City, and Dr. Thomas B. Wiegmann and Dr. Lawerence W. Welling at the Kansas City VA Medical Center, for their contributions to this project. This work was supported by grants NSF #DIR-9019648; NIH #GM44071; Kansas Affiliates—American Heart Association #KS-90-G17, and the P. S. Astrowe Trust of Menorah Medical Center, Kansas City, Missouri.

## REFERENCES

Aakerlund, L., Gether, U., Fuhlendorff, J., Schwartz, T. W., and Thastrup, O. (1990). Y$_1$ receptors for neuropeprtide Y are coupled to mobilization of intercellular calcium and inhibition of adenylate cylase. *FEBS Lett.* 260, 73–78.

Berk, B. C., Taubman, M. B., Griendling, K. K., Cragoe, E. F. Jr., Fentonand, J. W., and Brock, T. A. (1991). Thrombin-stimulated events in cultured vascular smooth muscle cells. *Biochem. J.* 274, 799–805.

Blackmore, P. F., Waynick, L. E., Blackman, G. E., Graham, C. W., and Sherry, R. S. (1984). α- and β-Adrenergic stimulation of parenchymal cell Ca$^{2+}$ influx; influence of extracellular pH. *J. Biol. Chem.* 259, 12322–12325.

Bookman, R. I. (1990). Characterization of low light level video cameras for fluorescence microscopy. *In* "Optical Microscopy for Biology." (B. Herman and K. Jacobson, eds.), pp. 235–250, Wiley-Liss, New York.

Brecker, P. L., and Fay, E. S. (1987). Photobleaching of fura-2 and its effect on determination of Calcium concentrations. *Am. J. Physiol.* 253, C613–C618.

Bright, G. R. (1992). Multiparameter imaging of cellular function. *In* "Fluorescent probes for biological function of living cells—a practical guide." (W. T. Matson and G. Ref, eds.), Academic Press, New York.

Bright, G. R., and Taylor, D. L. (1986). Imaging at low light level in fluorescence microscopy. *In* "Applications of Fluorescence in the Biomedical Sciences." (D. L. Taylor, A. S. Waggoner, F. Lanni, R. F. Murphy, and R. R. Birge, eds.), Liss, New York.

Bright, G. R., Fisher, G. W., Rogowska, J., and Taylor, D. L. (1987). Fluorescence ratio imaging microscopy; temporal and spatial measurements of cytoplasmic pH. *J. Biol. Chem.* 104, 1019–1033.

Bright, G. R., Fisher, G. W., Rogowska, J., and Taylor, D. L. (1989). Fluorescence ratio imaging microscopy. "Methods in Cell Biology 30." (D. L. Taylor and Y.-L. Wang, eds.), pp. 449–476, Academic Press, New York.

Bright, G. R., Whittaker, J. E., Haugland, R. P., and Taylor, D. L. (1989). Heterogeniety of the changes in cytoplasmic pH upon serum stimulation of quiescent fibroblasts. *J. Cell Physiol.* 141, 410–419.

Caramelo, C., Okada, K., Tsai, P., and Schrier, R. W. (1989). Phorbol esters and arginine vasopressin in vascular smooth muscle cell activation. *Am. J. Physiol.* 256, F875–F881.

Cota, G., and Hiriart, M. (1989). Hormonal and neurotransmitter regulation of Ca channel activity in cultured adenohypophyseal cells. *Soc. Gen. Physiol. Ser. 44,* 143–165.

DeBiasio, R., Bright, G. R., Ernst, L. A., Waggoner, A. S., and Taylor, D. L. (1987). Five parameter fluorescence imaging. *J. Cell Biol. 105,* 1613–1622.

Dickens, C. J., Gillespie, J. I., and Greenwell, J. R. (1989). Interactions between intracellular pH and calcium in single mouse neuroblastoma (N2A) and rat pheochromocytoma cells (PC12). *Q. J. Exp. Physiol. 74,* 671–679.

Diliberto, P. A., Hubert, T., and Herman, B. (1991). Early PDGF-induced alterations in cytosolic free calcium are required for mitogenesis. *Res. Comm. Chem. Path. Pharm. 72,* 3–12.

Foskett, J. K. (1988). Simultaneous nomarski and fluorescence imaging during video microscopy of cells. *Am. J. Physiol. 255,* C566–571.

Gehlert, D. R., Bishop, J. F., Schafner, M. P., and Chronwall, B. M. (1988). Rat intermediate lobe in culture: Dopamine regulation of POMC synthesis and cell proliferation. *Peptides 9,* 161–168.

Gibson, S. F., and Lanni, F. (1990). Measured and analytical point spread functions of the optical microscope for use in 3-D optical serial sectioning. *In* "Optical Microscopy for Biology." (B. Herman, and K. Jacobson, eds.), pp. 119–130, Wiley-Liss, New York (1990).

Giloh, H., and Sedat, J. W. (1982). Fluorescence microscopy: Reduced photobleaching of rhodamine and fluorescein protein conjugates by n-propyl gallate. *Science 217,* 1252–1255.

Grynkiewicz, G., Poenie, M., and Tsien, R. Y. (1985). A new generation of $Ca^{2+}$ indicators with greatly improved fluorescence properties. *J. Biol. Chem. 260,* 3440–3448.

Haugland, R. (1992). "Molecular Probes Handbook of Fluorescent Probes, 1991–1992 Ed." Molecular Probes Corp., Junction City, OR.

Herman, B. (1989). Resonance energy transfer microscopy. *In* "Methods in Cell Biology 30." (D. L. Taylor and Y-L. Wang, eds.), pp. 220–243, Academic Press, New York.

Herman, B., Gore, G. J., Nieminen, A.-L., Kawanishi, T., Harman, A., and Lemasters, J. J. (1990). Calcium and pH in anoxic and toxic injury. *Toxicology 21,* 127–148.

Inoue, S. (1986). "Video Microscopy." Plenum, New York.

Jackson, A. P., Timmerman, M. P. Bagshaw, C. R., and Ashley, C. C. (1987). The kinetics of calcium binding to fura-2 and indo-1. *FEBS Lett. 216,* 35–39.

Jacobsen, C., Mollerup, S., and Sheikh, M. I. (1990). $Ca^{++}$ and pH regulation of $K^+$ channels in membrane vesicles of rabbit proximal tubule. *Am. J. Physiol. 258,* F1634–F1639.

Jericevic, Z., Wiese, B., Bryan, J., and Smith, L. C. (1989). Validation of an imaging system. *In* "Methods in Cell Biology 30." (D. L. Taylor and Y.-L. Wang, eds.), pp. 48–82, Academic Press, New York.

Kao, J. P. Y., and Tsien, R. Y. (1988). $Ca^{2+}$ binding kinetics of fura-2 and azo-1 from temperature-jump relaxation measurements. *Biophys. J. 53,* 635–639.

Khananshvili, D. (1990). Cation antiporters. *Curr. Opin. Cell Biol. 2,* 731–734.

Knight, D. E., von Graffenstein, H., and Athayde, C. M. (1989). Calcium-dependent and calcium-independent exoctyosis. *TINS 12,* 451–458.

Kraicer, J., Gajewsil, T. C., and Moor, B. C. (1985). Release of Pro-opiomelocor-tin-derived peptides from the pars intermedia and pars distalis of the rat pituitary: Effect of corticotrophin-releasing factor and somatostatin. *Neuroendocrinology 41*, 363–373.

Kuijpers, G. A. J., Rosario, L. M., and Ornberg, R. L. (1989). Role of intracellular pH in secretion from adrenal medulla chromaffin cells. *J. Biol. Chem. 264*, 698–705.

Ladoux, A., Krawice, I., Damais, C., and Frelin, C. (1989). Phorbol esters and chemotactic factor induce distinct changes in cytoplasmic $Ca^{2+}$ and pH in granulocyte like HL600 cells. *Biochim. Biophys. Acta. 1013*, 55–59.

Lattanzio, F. A. (1990). The effects of pH and temperature on fluorescent calcium indicators as determined with Chelex-100 and EDTA buffer systems. *Biochem. Biophys. Res. Comm. 171*, 102–108.

Lattanzio, F. A., and Bartschat, D. K. (1991). The effect of pH on rate constants ion selectivity and thermodynamic properties of fluorescent calcium and magnesium indicators. *Biochem. Biophys. Res. Comm. 177*, 184–191.

Lemasters, J. J., Niemann, A. L., Gores, G. J., Dawson, T. L., Wray, B. E., Kawanishi, T., Tanaka, Y., Florine-Casteel, K., Bond, J. M., and Herman, B. (1990). Multiparameter digitized video microscopy (VDIM) of hypoxic cell injury. *In* "Optical Microscopy for Biology." (B. Herman and K. Jacobson, eds.), pp. 523–542, Wiley-Liss, New York.

MacLeod, K. T. (1991). Regulation and interaction of intracellular calcium, sodium and hydrogen ions in cardiac muscle. *Cardioscience 2*, 71–85.

Maly, K., Hochleitner, B. W., and Grunicke, H. (1990). Interrelationship between growth factor induced activation of the $Na^+/H^+$-antiporter and mobilization of intracellular $Ca^{++}$ in NIH3T3-fibroblasts. *Biochem. Biophys. Res. Comm. 167*, 1206–1213.

Marcus, D. A. (1988). High performance optical filters for fluorescence analysis. *Cell Motility and the Cytoskeleton 10*, 62–70.

Malgaroli, A., and Meldolesi, J. (1991). $[Ca^{2+}]_i$ oscillations from internal stores sustain exocytotic secretion from the chromaffin cells of the rat. *FEBS Lett. 283*, 169–172.

Malgaroli, A., Fesce, R., and Meldolesi, J. (1990). Spontaneous $[Ca^{2+}]_i$ fluctuations in rat chromaffin cells do not require inositol 1,4,5-trisphosphate elevations but are generated by a caffine- and rynodine-sensitive intracellular $Ca^{2+}$ store. *J. Biol. Chem. 265*, 3005–3008.

Martinez-Zaguilan, R., Martinez, G. M., Latanzio, F., and Gillies, R. J. (1991). Simultaneous measurements of intracellular pH and $Ca^{2+}$ using the fluorescence of SNARF-1 and fura-2. *Am. J. Physiol. 260*, C297–C307.

Millington, W. R., and Chronwall, B. M. (1991). Dopamine regulation of the intermediate lobe of the pituitary. *In* "Neuroendocrine Perspectives 7." (E. E. Mueler and R. M. MacLeod, eds.), pp. 1–48, Springer, New York.

Miyata, H., Hayashi, H., Suzuki, S., Noda, N., Kobayshi, A., Fujiwake, H., Hirano, M., and Yamazaki, N. (1989). Dual loading of the fluorescent indicator fura-2 and 2,7-biscarboxyethyl-5(6)-carboxyfluorescein (BCECF) in isolated myocytes. *Biochem. Biophys. Res. Comm. 163*, 500–505.

Moolenaar, W. F. (1986). Effects of growth factors on intracellular pH regulation. *Ann. Rev. Physiol. 48*, 363–376.

Morris, S. J. (1990). Real-time multi-wavelength fluorescence imaging of living cells. *BioTechniques 8,* 296–308.

Morris, S. J., Beatty, D. M., Welling, L. W., and Wiegmann, T. B. (1991). Instrumentation for simultaneous kinetic imaging of multiple fluorophores in single living cells. *SPIE Proc. 1428,* 148–158.

Morris, S. J., Sarkar, D. P., Zimmerberg, J., and Blumenthal, R. (1992). Kinetics of cell fusion mediated by viral proteins. *Meth. Enz* (in press).

Muallem, S., and Loessberg, P. A. (1990). Intracellular pH-regulatory mechanisms in pancreatic acinar cells: II. Regulation of $H^+$ and $HCO_3^-$ transporters by $Ca^{2+}$-mobilizing agonists. *J. Biol. Chem. 265,* 12813–12819.

Owen, N. E., Knapik, J., Strebel, F., Tarpley, W. G., and Groman, R. R. (1989). Regulation of $Na^+$-$H^+$ exchange in normal NIH3T3 cells and in NIH3T3 cells expressing the *ras* oncogene. *Am. J. Physiol. 256,* C756–C763.

Rink, T. J., and Sage, S. O. (1990). Calcium signaling in human platelets. *Annu. Rev. Physiol. 52,* 431–449.

Roe, M. W., Hepler, J. R., Harden, T. K., and Herman, B. (1989). Platelet-derived growth factor and angiotensin II cause increases in cytosloic free calcium by different mechanisms in vascular smooth muscle cells. *J. Cell Physiol. 139,* 100–1089.

Roe, M. W., Lemasters, J. J., and Herman, B. (1990). Assessment of fura-2 for measurements of cytosolic free coaclium. *Cell Calcium 11,* 63–73.

Rosario, L. M., Stutzin, A., Cragoe, E. J., and Pollard, H. B. (1991). Modulation of intracellular pH by secretagogues and the $Na^+/H^+$ antiporter in cultured bovine chromaffin cells. *Neuroscience 41,* 269–276.

Sarkar, D. P., Morris, S. J., Eidelman, O., Zimmerberg, J., and Blumenthal, R. (1989). Initial stages of influenza HA-induced cell fusion monitored simultaneously by two fluorescent events: Cytoplasmic continuity and lipid mixing. *J. Cell Biol. 109,* 113–122.

Schulz, I., Thevenod, F., and Dehlinger-Kremer, M. (1989). Modulation of intracellular free $Ca^{2+}$ concentration by IP3-sensitive and IP3-insensitive nonmitochondrial $Ca^{2+}$ pools. *Cell Calcium 10,* 325–336.

Simpson, A. W. M., and Rink, T. J. (1987). Elevation of pH is not an essential step in calcium mubilisation in fura-2-loaded human platelets. *FEBS Lett. 222,* 144–148.

Smallwood, J. I., Waisman, D. M., Lafrinieve, D., and Rasmussen, H. (1983). Evidence that the erythrocyte calcium pump catalyzes a a $Ca^{2+}:nH^+$ exchange. *J. Biol. Chem. 258,* 11092–11097.

Spring, K. R., and Lowy, R. J. (1989). Characteristics of low light level television cameras. *In* "Methods in Cell Biology 29." (Y.-L. Wang and D. L. Taylor, eds.), pp. 270–289, Academic Press, New York.

Takamatsu, T., and Wier, R. G. (1990a). High temporal resolution video imaging of intracellular calcium. *Cell Calcium 11,* 111–120.

Takamatsu, T., and Wier, R. G. (1990b). Calcium waves in mammalian heart: Quantification of origin, magnitude, waveform, and velocity. *FASEB J. 4,* 1519–1525.

Taylor, D. L., Amato, P. A., McNeil, P. L., Luby-Phelps, K., and Tanasugarn, L. (1988). Spatial and temporal dynamics of specific molecules and ions in living cells. *In* "Applications of Fluorescence in the Biomedical Sciences." (D. L.

Taylor, A. S. Waggoner, F. Lanni, R. F. Murphy, and R. R. Birge, eds.), pp. 374–376, Liss, New York.

Thomas, P., Surprenant, A., and Almers, W. (1990). Cytosolic $Ca^{2+}$, exocytosis and endocytosis in single melanotrophs of the rat pituitary. *Neuron 5*, 723–733.

Törnquist, K., and Tashjian, Jr., A. H. (1992). pH homostasis in pituitary $GH_4C_1$ cells: Basal intracellular pH regulated by cytosoloc free $Ca^{2+}$ concentration. *Endocrinology 130*, 717–725.

Tsay, T.-T., Wray, B., Inman, R., Herman, B., and Jacobson, K. (1990). Characterization of low light level cameras for digitized video microscopy. *In* "Optical Microscopy for Biology." (B. Herman and K. Jacobson, eds.), pp. 219–234, Wiley-Liss, New York.

Tsien, R. W., and Tsien, R. Y. (1990). Calcium channels, stores and oscillations. *Ann. Rev. Cell Biol. 6*, 715–760.

Tsien, R. Y. (1989). Fluorescent probes of cell signaling. *Annu. Rev. Neurosci. 12*, 227–53.

Tsien, R. Y., and Harootunian, A. T. (1990). Practical design criteria for a dynamic ratio imaging system. *Cell Calcium 11*, 93–109.

Tsukamoto, Y., Sugimura, K., and Suki, W. N. (1990). Role of $Ca^{2+}/H^+$ antiporter in the kidney. *Kidney Int. Suppl. 33*, S90–94.

Valet, G., Raffael, A., Moroder, L., Wunsch, E., and Rosenstock-Bauer, G. (1981). Fast intracellular pH determination in single cells by flow-cytometry. *Naturwissenschaften 68*, 265–266.

Vicentini, L. M., and Villereal, M. L. (1986). Inositol phosphates turnover, cytosolic $Ca^{++}$ and pH: Putative signals for the control of growth. *Life Sci. 38*, 2269–2276.

Waggoner, A., DeBiasio, R., Conrad, P., Bright, G. R., Ernst, L., Ryan, K., Nederlof, M., and Taylor, D. (1989). Multiple spectral parameter imaging. *In* "Methods in Cell Biology 30." (D. L. Taylor and Y.-L. Wang, eds.), pp. 157–190, Academic Press, New York.

Weiner, I. D., and Hamm, L. L. (1990). Regulation of intracellular pH in the rabbit cortical collecting tubule. *J. Clin. Invest. 85*, 274–281.

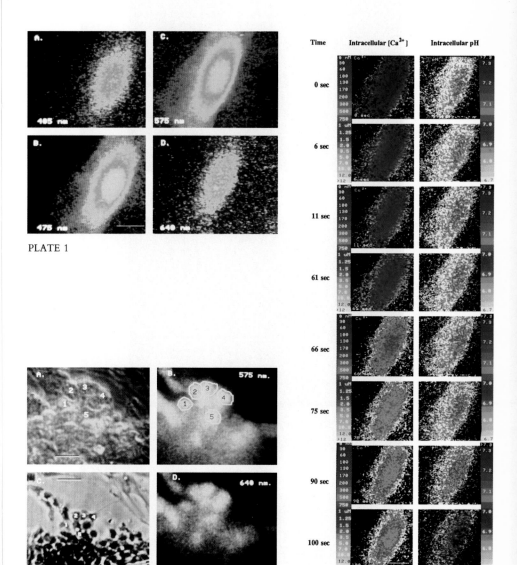

PLATE 1

PLATE 3

Time    Intracellular [Ca²⁺]    Intracellular pH

0 sec

6 sec

11 sec

61 sec

66 sec

75 sec

90 sec

100 sec

PLATE 2

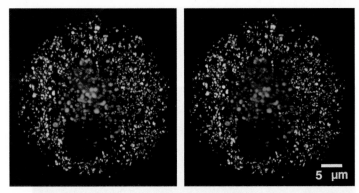

5 µm

PLATE 8

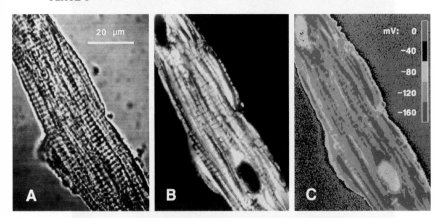

A

B

C

20 µm

mV: 0
-40
-80
-120
-160

PLATE 9

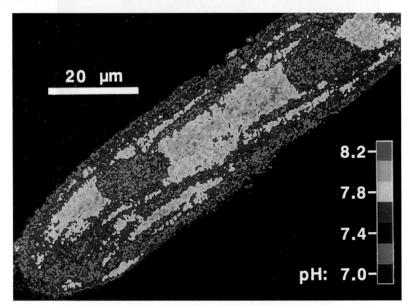

20 µm

8.2
7.8
7.4
pH: 7.0

PLATE 10

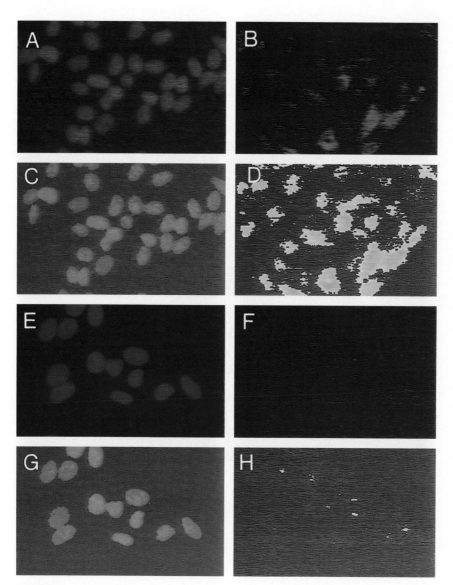

PLATE 11

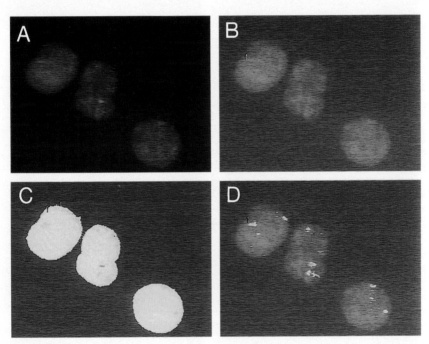

PLATE 12

# 7

# REAL-TIME FLUORESCENCE MICROSCOPY IN LIVING CELLS

## Fluorescence Imaging, Photolysis of Caged Compounds, and Whole-Cell Patch Clamping

### E. Niggli
Department of Physiology
University of Bern
Department of Physiology
University of Maryland School of Medicine

### R. W. Hadley
Department of Physiology
University of Maryland School of Medicine

### M. S. Kirby
Department of Physiology
University of Maryland School of Medicine

### W. J. Lederer
Department of Physiology
Medical Biotechnology Center
University of Maryland School of Medicine

I. INTRODUCTION
II. REAL-TIME IMAGING AND QUANTITATIVE FLUORESCENCE IN LIVING CELLS: GENERAL CONSIDERATIONS
A. Time Resolution
B. Spatial Resolution
C. Detectors
D. Fluorescence Measurements

III. PATCH CLAMP IN WHOLE-CELL MODE: VOLTAGE-CLAMP TECHNOLOGY
A. Scope of Our Work
B. Preparation of Pipettes
C. Fluorescent Calcium Indicator
D. Experiments
E. Cells That Move
F. Cell Lines
G. Recording Data

Optical Microscopy: Emerging Methods and Applications

H. Voltage-Clamp Imaging
   Experiments
IV. PHOTOLYSIS: IMPROVED
   TIME RESOLUTION
A. Example 1:
   Excitation–Contraction
   Coupling Experiments

B. Example 2: Ca-Activated
   Na–Ca Exchanger Currents
V. COMBINED PHOTOLYSIS
   AND FLUORESCENCE
VI. SUMMARY
ACKNOWLEDGMENTS
REFERENCES

## I. INTRODUCTION

An examination of cellular control and signaling mechanisms of many cells requires information on the concentration and/or localization of intracellular molecules. Additionally, living cells must be used because the kinetics of signaling are rapid and processes are not generally seen in fixed tissues. Furthermore, rapid alterations of these intracellular signals may be needed to perturb the system being examined to permit an investigation of the kinetics of the signaling. "Real-time" fluorescence microscopy of living cells provides the basic conditions to make use of a wide range of methods that support such investigations. We have made use of such technology and it employs ultraviolet (UV) and visible light epifluorescence microscopy (Niggli and Lederer, 1990a; Cannell et al., 1987a; Cannell et al., 1987b), confocal microscopy (Niggli and Lederer, 1990a), flash photolysis (Niggli and Lederer, 1991; Niggli and Lederer, 1990b), video imaging and digital image processing (Cannell et al., 1987b; Berlin et al., 1989), and whole-cell patch-clamp methods (Niggli and Lederer, 1991; Niggli and Lederer, 1990b; Bers et al., 1990; Berlin et al., 1989; Cannell et al., 1987a; Wier et al., 1987). While we have been particularly interested in heart cell biology in general and calcium metabolism in particular, the methods we have employed or developed appear to be appropriate for a wide range of cells. In our own work we have made use of these techniques to investigate not only heart cells but also epithelial cell lines (COS-1 and 293) (Kofuji et al., 1992), smooth muscle cells (BC3H1) (Berlin et al., 1990), and insect cells (SF9). Furthermore, the basic approach readily lends itself to adaptation for state-of-the-art techniques, such as two-photon excitation microscopy (Denk et al., 1990) and also fluorescence lifetime imaging (Szmacinski et al., 1992). The goal of this review is to briefly present the methods now used by us and others to investigate the cell biology of cellular and intracellular signaling.

## II. REAL-TIME IMAGING AND QUANTITATIVE FLUORESCENCE IN LIVING CELLS: GENERAL CONSIDERATIONS

### A. Time Resolution

Real-time measurements are achieved when the time-resolution is at least as fast as is needed to acquire the time-resolved data. Unfortunately, we almost never can achieve this goal in any satisfactory fashion. If the temporal resolution is adequate today, our revised theory based on "today's" results will lead to a new understanding or theory that will require greater speed tomorrow. Similarly, what one investigator may feel is "fast enough" for his or her investigations, and is thus "real time," is thoroughly inadequate for another investigator. Thus, in our own experiments, we have attempted to achieve measurements that have both high-temporal and high-spatial resolution but are quite aware of the relativistic nature of our real-time investigation. Furthermore, there is generally a trade-off between high-temporal and high-spatial resolution and thus we may be achieving real-time imaging but producing an image with compromised spatial resolution.

### B. Spatial Resolution

While the optical resolution limits are set by the wavelength of light and the numerical aperture of the object lens in standard light microscopy, improvements in "resolution" or "visibility" or "contrast" or of some combination can be achieved with modern techniques (Somlyo, 1986). Improved resolution may be achieved with confocal microscopy (Pawley, 1990), which also improves contrast in the plane of focus and improved visibility may be achieved with contrast manipulations as used with "video-enhanced microscopy" (Inoue, 1986). With low-light levels that are frequently encountered with epifluorescence microscopy in living cells, there are, however, additional impediments to imaging objects with high spatial and temporal resolution (Taylor et al., 1986).

1. Fluorochromes photobleach. This means that one cannot always adequately improve the signal-by-signal averaging methods. With longer exposure times or greater illumination intensities, the fluorochromes are degraded.

2. The ground states of the fluorochromes may be depleted. This means that the signal cannot be improved by focusing more excitation light onto the cell viewed. Signal does continue to increase from the excited fluorochrome until all of the "ground state" of the fluorochrome is excited. Thus, "ground-state depletion" can limit any increase in signal with greater illumination intensity (Goldman et al., 1991). This may be

true for experiments in which the signals tend to be small or bright lasers are scanned across the target cells. Both of these limitations may apply, for example, to calcium imaging with a laser-scanning confocal microscope or to the tracking of antibody-labeled proteins.

3. Time-dependent changes occur rapidly and/or randomly (in milliseconds or less). The normal voltage–activated calcium transient in striated muscle cells may rise rapidly to a peak level (within milliseconds). Spontaneous calcium transients in cardiac muscle may develop more slowly but occur with random timing and localization and propagate within the cell at an average speed of 1 $\mu$m per 10 ms (i.e., 100 $\mu$m per second) (Berlin *et al.*, 1988; Berlin *et al.*, 1989; Cannell *et al.*, 1987b). The rapidly changing signals lead to smaller time-resolved signals (i.e., the integration period must be reduced to achieve the temporal resolution), and random signal transients cannot be averaged over repeated, stereotyped events. Thus, with faster events, spatial resolution is often sacrificed to achieve a good signal. Obviously, temporal resolution can be sacrificed to improve spatial resolution. Thus, there is a "space-time" bandwidth limitation in real-time fluorescence measurements.

## C. Detectors

Three basic detectors are used in fluorescence imaging: (1) film, (2) solid-state devices, and (3) photomultiplier tubes (PMTs).

### 1. Film

Photographic film is still the most widely used detector for fluorescence imaging. It is inexpensive, adaptable, and easily processed. Virtually every fluorescence microscope can be used with an inexpensive camera back and a roll of film. Film is not only less sensitive than the other two detectors but also cannot be as easily used in quantitative investigations and cannot be enhanced directly with modern digital image processing. Furthermore, the other two detectors can be directly connected to image-processing workstations, permit rapid feedback, and adjustment of conditions to optimize imaging. This feature, along with their greater sensitivity, leads one to favor solid-state devices and photomultiplier tubes for all real-time applications. In none of our applications in living cells do we use film as a detector.

### 2. Solid-State Devices

Charge-coupled devices (CCD) and charge-injection devices (CID) are used in arrays as video detectors and are frequently coupled with amplifiers and readout electronics to meet either video (RS-170) standards or proprietary output standards. Improved sensitivity is often achieved using image intensifiers, such as microchannel plate intensifiers. Decreased background noise may be achieved by cooling to decrease thermal noise in the solid-state detectors. Decreased noise is also achieved by integrating the signal on the detector (CCD) and reading out the data intermittently. This increase in signal-to-noise ratio (SNR) is almost always achieved at a sacrifice of temporal resolution. The solid-state imaging arrays generally operate with a lower dynamic range of signal intensity than one routinely achieves with film. The preponderance of the devices used today operate with 8-bits resolution (1 part in 256), are not cooled, and operate at video rates (i.e., 1 full image per 33 ms). To obtain an adequate signal on the face plate of the solid-state detector array with the kind of low-light signals that are normally available from living cells, an intensifier is almost always needed. The intensifier also adds noise and may decrease resolution. Furthermore, the intensifier image may be altered by the characteristics of the target phosphor on the intensifier. These CCD detectors themselves can be operated with 12 or more bits of dynamic range if the appropriate readout instrumentation is used by the manufacturer. The greater the dynamic range, the slower the readout time if the other elements of the system are the same.

## 3. PMTs

The photomultiplier tube (PMT) is fast, sensitive, and flexible as a detector. The device itself is rather large compared to an element of a CCD array (1–3 cm diameter versus tens of microns). While the PMT offers the greatest sensitivity as a detector, it is a single detector element and thus does not function as an image-acquisition device itself. Relative high-speed processing must be carried on outside the detector since the device itself does not store the signal. The PMT operates quite well, then, to acquire a signal from a spatially averaged source as is illustrated in Fig. 7.1. We use our microscope in this configuration when speed is important and spatially resolved information is not needed. When, however, spatial information is required, the PMT is still an excellent detector but must be used in a different manner.

   The PMT is generally used to acquire images that are produced from raster scanning (as is done with a laser-scanning confocal microscope, see Figs. 7.2 and 7.3). In the case of a cell, a light source is focused to a diffraction-limited spot and scanned across the cell. The entire fluorescence from the "point source" is imaged onto the photocathode of the PMT and the PMT current is measured as a function of time. The image is then reconstructed from knowledge of the position of the spot of light with time. This method is used in the laser-scanning confocal microscope when an aperature is placed in the output-imaging pathway. Limitations of this method include

FIGURE 7.1

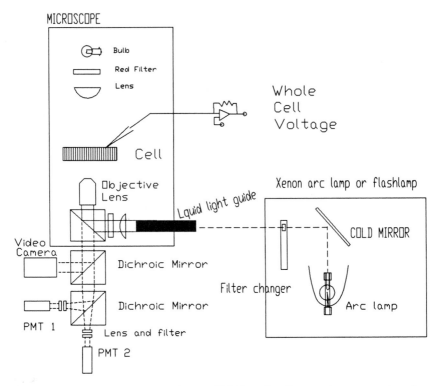

Fluorescence microscope in our experiments. This figure is a diagramatic representation of our experimental equipment. A Nikon Diaphot inverted microscope is used with a high-numerical aperture lens (63× Zeiss Neofluar 1.25 NA). A xenon arc lamp (epifluorescence) or a xenon flash lamp produces light imaged onto the end of a liquid-light guide (Oriel Corporation). The cell of interest is patch clamped (whole-cell mode) and transilluminate with red light. The image of the cell is directed to a video camera (red light only) and to a pair of PMTs or to an intensified CCD camera. Appropriate electrical shielding is needed to minimize electrical noise.

limitations imposed in mechanical scanning (e.g., two mirrors are generally used to produce a full *x-y* scan), high-illumination intensity, and in high-speed processing of the PMT signal in the digital image-storage device. Acustooptic scanning is also possible in two dimensions or in combination with a mirror scanner.

## D. Fluorescence Measurements

Fluorescence measurements in single heart cells require that the fluorescent material be added to the cell. Figure 7.1 shows a diagramatic view of our experimental arrangement. For an intracellular measurement, the material must be injected, loaded through holes in the cell membrane, or loaded as a molecule that can cross the sarcolemma or plasmalemma. Many of the fluorescent agents that have been developed can be made into lipophilic esters. These esters can then be de-esterified by intracellular esterases and thereby returned to their "native" or active form and consequently "trapped" within the cell (Heiple and Taylor, 1982). Acetoxymethyl esters of many fluorescent indicators are now available. These compounds are relatively easy to use. However, they may also produce experimental results that are difficult to interpret. Injection loading is quite valuable for loading the cytoplasmic compartment of single cells because it overcomes many of the problems associated with ester-loading procedures. However, it is the most labor-intensive method and may be associated with other problems. By one means or another, the fluorescent agent must be placed within the cell if intracellular measurements are to be made on living cells.

## 1. Ester-Loading

A typical ester-loading procedure involves incubating the cells of interest with a moderate to low concentration of the fluorochrome-ester. An example would be fluo-3-AM as illustrated in Fig. 7.2A,B. The acetoxymethyl ester of fluo-3 is added to DMSO and 25% pluronic (Tsien, 1981; Tsien, 1989; Niggli and Lederer, 1990a) to make a 2.5 mM solution. This solution is then added to an extracellular solution to make a 10 $\mu$M solution of fluo-3-AM. The cells are incubated in this solution for 15 min to 2 h and then incubated in the same extracellular solution without the fluo-3-AM for 15 min to 2 h. This second incubation permits the intracellular esterases time to convert the fluo-3-AM to fluo-3. If there are many intracellular compartments, it is possible to load these compartments with fluo-3 as well as the cytoplasmic compartment. The fluorescence signal due to fluo-3 will then report the changes in calcium from all compartments loaded with fluo-3. Additionally, partially de-esterified compounds may contribute to the measured signal. When we examined BC3H1 cells with fura-2, loaded using the AM method, we found results that were inconsistent (Berlin $et$ $al.$, 1990). The "calibrated" resting intracellular calcium could be "too high" (hundreds of nanomolar to many micromolar) to "too low" (tens of nanomolar to subnanomolar). Of course, it was found to be "acceptable" sometimes. The problem we faced was that without some sort of gold standard, we did not know which of these many possible answers was "correct." Additionally, we found that conditions that were expected to produce large changes in $[Ca^{2+}]_i$, produced small or no change in the estimated $[Ca^{2+}]_i$ (Berlin $et$ $al.$, 1990). Some investigators have used ester loading to load intracellular organelles to estimate their function, while other investi-

FIGURE 7.2A

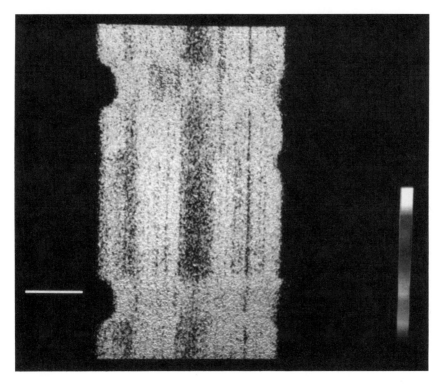

Fluo-3 signal showing nucleus. Fluo-3-AM was used to load a cardiac cell with fluo-3. An extracellular shock was applied to produce a calcium transient and a contraction. A single line scan repeated ever 6 ms along the long axis of the cell through the nucleus produces the video image displayed. The large dark band shows how nuclear fluorescence changes with time. The top of the image shows earlier records while the bottom of the image shows later records (i.e., time runs from top to bottom). There is an increase in fluo-3 fluorescence just before the cell starts to shorten. The increase in nuclear fluorescence lags just behind the increase in cellular fluorescence. We interpret the data to mean that calcium diffuses into the nuclear volume from the cytosol. Part A shows two twitches and the full cell length. Part B is a close-up showing the change in nuclear fluorescence more clearly. This image was taken on a Biorad MCR500 confocal microscope. (Adapted from Niggli and Lederer, 1990; Niggli and Lederer, 1991.) (*See Plates 4 and 5 for color versions of these figures.*)

gators use these same esters to load the cytoplasmic compartment. Thus, the biggest problem with ester loading is knowing whether the measurements are fundamentally useful. However difficult calibration of ionic fluorescent indicators may be when the hydrophilic salt is used (Konishi *et al.*, 1988), the problems are significantly worse when using the ester-loading methods. That warning acknowledged, some very nice qualitative studies

FIGURE 7.2B

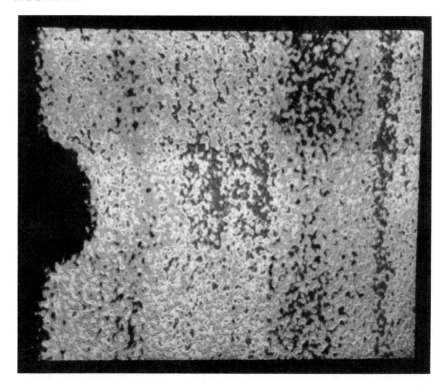

have been produced using ester loading of cells (Cornell Bell *et al.*, 1990). These qualitative results showed the development of waves of elevated intracellular calcium in cultured astrocytes. We too have found that qualitative experiments could be produced in adult heart cells loaded with fluo-3 by the ester-loading procedure. Using a confocal microscope in a single line-scan mode, we were able to measure the fluorescence change seen in heart cells with stimulation (Niggli and Lederer, 1990a). The qualitative result we obtained with these experiments was that the fluorescence signal of the nuclear region lagged behind the increase in fluorescence signal in the bulk of the cell during a calcium transient. The lag appeared to be about 20 ms; this is a finding consistent with the idea that calcium rises in the cytosol and then diffuses into the nucleus. Additional qualitative results have been obtained in our laboratory in cells from the 293 cell line transfected with the human Na–Ca exchanger and loaded with fluo-3 by the ester (Kofuji *et al.*, 1992). Thus, the qualitative use of ester-loaded ion indicators can be readily supported. Quantitative use is certainly possible but is far more limited and must be approached with considerable caution.

## 2. Direct Loading

Lipophilic indicators can be loaded directly. One such agent is shown in Fig. 7.3. In this case hexyl ester of rhodamine B was used to label the sarcoplasmic reticulum (SR) of isolated cardiac muscle cells. The specificity of such loading is always open to question since there are many membranes in a cell and lipophilic indicators will tend to partition into any membrane rather than remain in the aqueous phase. In this case, the known structure of the SR—a 2-$\mu$m repeating pattern—suggests that the loading of the SR membrane was sufficiently specific to provide the contrast needed to make this image. Nevertheless, for critical applications, additional confirmation of the indicators location may be required.

FIGURE 7.3

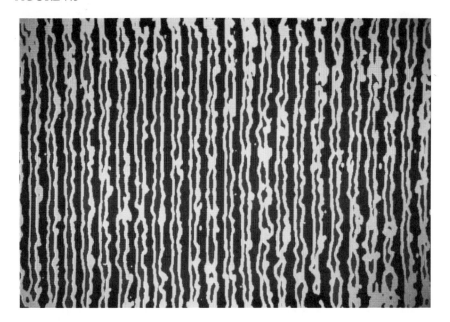

Sarcoplasmic reticulum distribution in heart muscle cells. A confocal optical section of a living heart muscle cell reveals the regular arrangement of the sarcoplasmic reticulum (SR), the subcellular site of calcium-induced calcium release. The 2-$\mu$m periodicity reflects the registration of the SR with the regular organization of the contractile proteins. Hexylester of rhodamine B (R6), a putative SR-specific fluorescent indicator (Molecular Probes, Eugene, OR), was used to obtain this computer-process pseudocolor image. This image was taken on a Biorad MCR500 confocal microscope. (*See Plate 6 for a color version of this figure.*)

3. Injection Loading

Direct injection of the salt of a fluorescent ion indicator into the cytoplasmic compartment of a cell significantly reduces the difficulties associated with quantitative use of the indicators. The problems are not avoided completely, however, and while the quantitative results produced using these methods are much more consistent and "calibratable," they still suffer from alterations of the indicator's properties by intracellular proteins (Konishi et al., 1988). A particularly convenient way to load a cell with a fluorescent ion indicator is through a patch-clamp pipet used in whole-cell clamp mode (see below). This method, while quite good for heart muscle cells, may be poor for very small cells and cells with little buffering capacity for the ion being measured. A problem that can arise comes from the pipet itself. The concentration of the measured substance in the pipet can cominate the intracellular (measured) concentration. We found that this was the case in our measurements of BC3H1 smooth muscle cells (Berlin et al., 1990). Measurements of intracellular calcium that was controlled by the cell were only possible after injection of the calcium indicator, fura-2, and the subsequent removal of the pipet from the cell. Nevertheless, when using appropriate cells and conditions, such measurements are possible as indicated below.

## III. PATCH CLAMP IN WHOLE-CELL MODE: VOLTAGE-CLAMP TECHNOLOGY

### A. Scope of Our Work

Over the last seven years we have been working with the array of fluorescent calcium chelators developed by Roger Tsien and his co-workers (Tsien and Harootunian, 1990; Kao et al., 1989; Minta et al., 1989; Minta and Tsien, 1989; Tsien, 1989; Grynkiewicz et al., 1985; Williams et al., 1985; Tsien, 1981; Tsien, 1980). While we initially began using the esterified indicators to load heart cells (Wier et al., 1987), we came to appreciate the difficulties of this method in our heart cell preparation for quantitative measurements and began using the salts of these indicators exclusively for these measurements (Cannell et al., 1987a; Cannell et al., 1987b; Bers et al., 1990), mainly using fura-2 (for video imaging) and indo-1 (for PMT measurements with greater temporal resolution). In either case, the salt was loaded into the cells by diffusion from a patch-clamp pipet used in whole-cell mode. For us, this approach has the advantage that we could carry out voltage-clamp experiments while measuring inracellular calcium, $[Ca^{2+}]_i$. Since heart cells also contract, we use an independent additional video camera that views the cells transilluminated with red light (Peeters et al., 1987). This transilluminated system permits us to view the cell, its striations, and length throughout the experiment. Appropriately placed dichroic mirrors and interference filters permit us to view the cell while also carrying out all of the experiments describe in this article.

## B. Preparation of Pipettes

In the absence of a fluorescent calcium indicator like indo-1, the glass used to make the patch-clamp pipettes is filamented borosilicate glass tubing. This is thin capillary glass tubing (1.5 mm O.D.) that contains a fine glass filament bonded to the inside of the tube. When pulled to a fine tip, the interior capillary fiber is also pulled and its placement with the tubing facilitates the movement of the filling solution into the tip of the pipette. Generally, we use thin-wall capillary glass (WPI, West Haven, CT), but this is not a critical consideration since both standard wall glass and thick wall glass work. For whole-cell mode, the glass tip is pulled to a resistance between 0.75 and 2 MΩ using a BB-CH pipette puller (Geneva, Switzerland). The tip is not generally fire polished since it is partly "polished" during the pulling process. For use with fluorescent calcium indicators, we always wash the glass in acid. While 1 M nitric acid will do, we often use concentrated chromic acid. With chromic acid, it is necessary to remove heavy metals after the acid wash with at least one EGTA [ethylene glycol-$O,O'$-bis(2-aminoethyl)-$N,N,N'$-tetraacetic acid] postwash. The acid prewash not only removes variable amounts of unwanted calcium and heavy metals but also rids the interior of the glass of processing "junk." We found early in this work that glass pipettes often contained a small amount of material—dust or other particulate matter—that would plug the tip of the pipette. This did not cause severe problems with standard voltage-clamp experiments but did reduce the efficacy of loading calcium indicator via the pipette.

## C. Fluorescent Calcium Indicator

We attempt to use the lowest concentration of indicator possible. We do this largely because we are interested in measuring the "true" $[Ca^{2+}]_i$ signal with a minimal buffering effect of the indicator itself. Thus, we use 35–50 $\mu$M fura-2 or indo-1 in the pipette, along with our "standard" intracellular-filling solution (Cannell et al., 1987a; Bers et al., 1990). It is also important to remember that the solutions used to fill the pipette should be treated to remove heavy metals that could affect the ability of the indicator to report calcium. Cannell and his co-workers have reported remarkably fast and noise-free $[Ca^{2+}]_i$ transients using these methods (Crespo et al., 1990) developed in part by M. B. Cannell (St. George's Hospital Medical School, London, England) and J. R. Berlin (Bockus Research Institute, Graduate Hospital, Philadelphia, PA) (Cannell et al., 1987a).

The fluorescent indicator is made up as a concentrated (2 mM) solution that is added to the pipette-filling solution. Each of these solutions is kept frozen between use. Oxidation of the fluorescent indicator occurs once the salt is added to an aqueous solution and appears to continue even when the solution is frozen although it takes place at a lower rate. Thus, new concentrated stock of fluorescent indicator is made at least once a month.

## D. Experiments

Fluorescence measurements can be made in voltage-clamped cells (using patch-clamp methods in whole-cell mode) quite easily. A low-resistance pipette is filled with an intracellular solution containing the indicator of choice. Let us assume for the moment that we are principally interested in maximum temporal resolution of the $[Ca^{2+}]_i$ signal. PMTs would then be used as detectors and a region of the cell will be imaged onto the photocathode(s) of the PMT(s). If indo-1 is to be used, ratiometric measurements of the fluorescence emission from the cell will be used to estimate $[Ca^{2+}]_i$ and two PMTs will be used. With this arrangement, a single excitation wavelength can be used and continuous output of the two emission signals can be obtained optically.

If greater spatial resolution is desired, we normally would change to fura-2 and replace one of the PMTs with an intensified CCD video camera. The other PMT would not be used. The exictation and emission dichroic mirrors and narrow-band interference filters would be changed to optimize excitation and emission signals. The use of dual video cameras to measure indo-1 emission is possible but requires a pair of video cameras that are matched for position, magnification, gain, and rotation at every pixel. Some of the possible distortion can be corrected after the images are acquired (see Wier and Blatter, 1991). One can circumvent the two-camera difficulties by using fura-2 and one camera. This means that the excitation signal must be rapidly switched between two wavelengths (usually 340 and 380 or 360 and 380) and in a manner synchronized to the video-frame rate. Many brands of automated excitation switchers are on the market today. Video data storage can then be a challenging task, which we will not address in this view but is covered elsewhere in the volume. Suffice it to say that videotape recorders are available, videodisc recorder can be used, and it is possible to digitize the video signal directly for storage on a mass-storage device. Each of these solutions has something to recommend it, but it should be appreciated that approximately 7 Mbytes a second of data must be stored if a 500 by 500 pixel is imaged with 8 bits of intensity information at standard video-frame rates. Similarly, the topic of image processing is an important one that will also not be addressed here but is covered elsewhere in this volume. We use software developed in the Department of Physiology at the University of Maryland by a group of users in the Department of Physiology. Reasonably priced commercial software for use on IBM PCs and Apple Macintosh computers is available as is superb free software for use on Macintosh computers (e.g., NIH "Image").

Whether imaging the cell that is to be voltage clamped or whether using a spatially averaged measurement with greater temporal resolution, a background fluorescence signal is needed before loading the cell with indicator if the signal is to be calibrated. The procedure we use is to place the pipette next to the cell and apply suction on the pipette so that a giga-ohm seal begins to form. After the seal is stable but before "breaking in," a background fluorescence measurement is taken. This background fluorescence measure is subtracted from all subsequent signals to obtain the background-subtracted image or measurement. The background-subtracted data is used to produce the fluorescence ratio measurements. This provides another reason why ester loading of cells may cause problems (see above). It is quite difficult to obtain a background image from ester-loaded cells since the cells are not imaged before the loading procedure. This has led to the use of fluorescence-quenching methods. Thus, after the experiment is completed and while the cell is still in the identical spot in the field of view, an agent (such as Mn) is added to a permeabilized cell. The quenching agent should ideally quench only the fluorescence from the added fluorescent indicator. In this ideal situation, it would then quench all but the native "background" signal and thus provide the background fluorescence measurement or image.

Some cells, such as heart cells, may have a significant amount of native fluorescence. Heart cells have NADH in the mitochondria and the NADH fluoresces measurably when excited with UV light (Eng *et al.*, 1989). Thus, in heart cells loaded with the salt form of the indicator via the pipette, structures that may not be of interest, may provide a significant signal (Niggli and Lederer, 1990a; Lakowicz, 1983).

### E. Cells That Move

Cells that contract like heart cells are particularly difficult to use in quantitative-imaging studies of time-dependent signals. To the extent that they shorten, the background image will be in error since it was obtained when the cell was relaxed (and hence longer). Thus, when motion is severe, an attempt should be made to make the fluorescence signal as large as possible (possibly by using more indicator). The idea is to have a fluorescence signal so large that the background signal becomes insignificant. By regularly making background measurements, an experimenter will know exactly how severe the problem is from one experiment to the next and will also know whether or not it is negligible.

Contracting cells are much less of a problem for data obtained from a large spot within the cell measured with PMTs. This is because a change in background will arise from a change in thickness of the cell, and hence will be relatively small over the region imaged. This problem is smaller because the change in thickness of the cell is not great. Obviously, to make this situation optimal, the region to be imaged should be fixed with respect to the active area of the detector and should not contain a part of the cell that contracts completely out of the field of examination (e.g., the end of the cell). We normally use a region near the pipette since the cell is tethered to the pipette (Berlin *et al.*, 1989).

### F. Cell Lines

The preponderance of cell lines (e.g., BC3H1, COS, CHO, 293, 3T3 cells, etc.) do not move even when their stem cell was of muscle origin. In our experience, their autofluorescence is generally less than heart cells. Thus, these cells can be readily loaded by any method, including by diffusion from the patch-clamp pipette used in whole-cell mode.

### G. Recording Data

As indicated earlier, massive amounts of data can be generated in imaging experiments. We record all video data on videotape using a super VHS instrument and encode the voltage-clamp timing data on the tape. This data can be recorded within the video signal itself and/or on the "hi-fi" audio signal. In order to coordinate the timing of the electrical (i.e., voltage clamp) and optical (imaging) parts of the experiments, two approaches have been used. First, the voltage signal can be timed relative to the video frame using the video timing signal as a marker. This method means that the experimenter can change the voltage of the cell at a precise time with respect to the acquisition of the image. A variety of commercial electronics is available to provide appropriate timing signals (e.g., vertical and horizontal synch signals). Second, the timing of the voltage-clamp signals can simply be recorded on the videotape (on one of the hi-fi audio tracks). Other timing data may also be useful. We, for example, also use an audio track to record the experimenters comments during the experiment. Furthermore, a problem that can vary with the recorder is the linearity of the recording of video data. We often record a gray scale of known voltage levels. This permits us to relinearize any distortion of the video signal that has been introduced by the video recorder. In addition to a video recorder being used for recording the video data, a second video recorder is used to record all of the electrophysiology information. This second video recorder uses a digital-audio interface for data recording now widely available (e.g., Neurocorders made by Neurodata, New York). We generally record up to eight tracts of data at rates between 11 and 44 kHz, and these include (1) voltage, (2) current low gain, (3) current high gain, (4) PMT 1, (5) PMT 2, (6) video-dimension analyzer, (7) miscellaneous data channel (temperature, a timing signal, etc), and (8) equipment signal normalization. The two hi-fi audio tracks are used to record voice information and any additional timing information that is needed.

Direct digital recording of imaging data is almost always preferable if it can be achieved. It achieves high-quality 8-bit data recording (or more) and remains linear. The experimenter must carefully identify the recorded images with respect to the voltage changes produced. The largest two problems with this approach are cost and data-storage capacity and speed.

## H. Voltage-Clamp Imaging Experiments

A need for spatial information leads an experimenter to carry out imaging experiments. When the spatial information changes with membrane voltage, then voltage-clamp imaging experiments may be required. In heart cell experiments, there is one particular circumstance that readily lends itself to this approach: calcium waves (Berlin et al., 1989; Wier et al., 1987; Cannell et al., 1987b; Berlin et al., 1988). Waves of elevated calcium are produced in "calcium overloaded" heart cells. The elevation of $[Ca^{2+}]_i$ that underlies this process leads to increased calcium within the SR of heart muscle cells and altered sensitivity of the calcium-induced calcium release (CICR) process (Niggli and Lederer, 1990b). If $[Ca^{2+}]_i$ rises at some subcellular region, then it may induce local CICR that would propagate as the released calcium diffuses to a neighboring region of SR where the process is repeated. Intracellular $[Ca^{2+}]_i$, the calcium leak, the Na–Ca exchanger, and calcium channels are all quite sensitive to membrane voltage. During the condition of "calcium overload," this sensitivity is manifested by spontaneous releases of calcium and propagated waves of elevated calcium that may be entrained by a "normal" voltage-activated calcium transient (Berlin et al., 1989). This is the kind of study that requires joint imaging and voltage-clamp experiments.

A much more ambitious project requires imaging at a resolution not readily available. The normal calcium-release process originates at the SR and presumably at the Z-line where calcium-release channels in the SR are close to L-type calcium channels in the sarcolemma and T-tubular membranes. Thus, it is thought that calcium is released from this region and diffuses on average 1 $\mu$m (half of a sarcomere) to activate contraction. The close examination of this process, which occurs at the millisecond time scale, cannot be readily imaged at the needed temporal and spatial resolution using standard "wide-field" epifluorescence microscopy. The improved spatial and temporal resolution provided by confocal microscopy may be adequate and represent a current area of investigation.

## IV. PHOTOLYSIS: IMPROVED TIME RESOLUTION

Because we are interested in investigating how cells work, one kind of experiment that is very useful is one that enables us to produce a sudden change in the concentration of a substance of interest, such as calcium. While our discussion will continue to favor changes of calcium, the general issues also apply to changes in magnesium, adenosine-5'-triphosphate (ATP), or other second messengers and substances of greater interest to the reader. The basic idea is that some substance that can be activated by a flash of light is introduced into a cell. On activation, this "caged" compound releases the substance of interest that leads to the step increase in concentration. Since so many proteins are sensitive to calcium directly or indirectly, there is a broad interest in "caged" calcium. We have used two popular caged compounds, DM-nitrophen and nitr-5 (Calbiochem) (Niggli and Lederer, 1991; Niggli and Lederer, 1990b; Hadley and Lederer, 1991). These agents are calcium buffers, which are altered by UV light. Detailed discussions of the physical properties of these chemicals can be found in the original publications (Kaplan and Ellis Davies, 1988; Mulligan *et al.,* 1990; Kao *et al.,* 1989). The important issue is that the affinity of the caged compound for calcium changes when the agent is photolyzed. In the case of DM-nitrophen, the affinity for calcium changes from the nanomolar level to the millimolar level on photolysis. The speed of photolysis can vary but is generally quite fast. DM-nitrophen, for example, produces a step increase in $[Ca^{2+}]_i$ within 200 $\mu$s (Kaplan and Ellis Davies, 1988; Niggli and Lederer, 1991).

Since DM-nitrophen is a buffer whose affinity is changed on photolysis, it is important to provide enough calcium in the pipette so that the buffer can be fully loaded (i.e., about 100 nM free calcium, see above). This is clearly a tricky situation. Providing too little calcium in the pipette will mean that the buffer will not be loaded (diffusing into the pipette) and on photorelease calcium will tend to be bound by the unloaded buffer. If, however, too much calcium is added to the DM-nitrophen pipette solution, the added calcium will exceed the buffering capacity of the DM-nitrophen and it will leak out of the pipette and load the cellular calcium stores and produce "calcium overload." For heart cells, this leads to spontaneous releases of calcium, the development of waves of elevated calcium propagating within the cell, and finally contracture and cell death.

Since DM-nitrophen has a high affinity for calcium and resting calcium is normally around 100 nM, virtually all molecules have calcium bound. However, in order to achieve this, intracellular-free magnesium must be reduced because DM-nitrophen also binds magnesium (Kaplan and Ellis Davies, 1988). With DM-nitrophen largely loaded, it no longer acts as a signficant buffer even when millimolar quantities of this specie of caged calcium are introduced into a cell. Nitr-5 has a lower affinity for calcium and is approximately half-bound half-unbound at physiological free calcium (in heart cells). For that reason, we have carried out many of our experiments using DM-nitrophen.

The DM-nitrophen is introduced into the cell by the patch pipette. Unlike fluorescent indicators, DM-nitrophen is "consumed." Once photolyzed, a molecule of DM-nitrophen can no longer be used as caged calcium. Thus, it is important to load enough DM-nitrophen into a cell so that a series of experiments can be carried out.

The discussion that applies to epifluorescence imaging or PMT measurements while carrying out voltage-clamp experiments largely applies to photolysis experiments. However, the concerns about imaging the fluorescent signal are not relevant here (but see next section). Nevertheless, the caged compounds that are used in these experiments must be loaded into the cell and must reside in the correct compartment when photolysis is activated. For this reason, as well as the absence of DM-nitrophen esters, our photolysis experiments are always done with the caged calcium loaded via a patch-clamp pipette. Furthermore, these experiments are largely used to examine the cell biology of voltage-activated processes.

## A. Example 1: Excitation–Contraction Coupling Experiments

The methods outline above were used to examine how excitation–contraction (EC) coupling is achieved in heart muscle (Niggli and Lederer, 1990b). While CICR has been an established process in both cardiac and skeletal muscle for many years, the actual role of CICR in contraction and the quantitative understanding of CICR has been absent. We examined CICR in heart muscle, the role of voltage changes in EC coupling, and developed a model to understand how the system works (Niggli and Lederer, 1990b). The essence of our findings is that the spatial organization of the SR and the sarcolemma and placement of the L-type calcium channels with respect to the SR calcium-release channels are quite important in mammalian cardiac muscle (Niggli and Lederer, 1990b). This finding has been generally supported by several other investigators (Leblanc and Hume, 1990; Näbauer and Morad, 1990; Lederer *et al.*, 1991; Lederer *et al.*, 1990). Our experiments were carried out using flash photolysis of DM-nitrophen that was coordinated with changes of membrane potential. The supportive experiments used a similar or related protocol or used indo-1 to measure $[Ca^{2+}]_i$.

## B. Example 2: Ca-Activated Na–Ca Exchanger Currents

In its normal operation at the resting potential of a cardiac cell, the Na–Ca exchanger uses the transsarcolemmal sodium gradient to extrude calcium from the cell. We used flash photolysis of caged calcium (DM-nitrophen) to activate the Na–Ca exchanger. A step increase in $[Ca^{2+}]_i$ leads to the activation of the Na–Ca exchanger and the generation of a large inward current. This Na–Ca exchanger current declines as $[Ca^{2+}]_i$ returns to control levels (Niggli and Lederer, 1991; Niggli and Lederer, 1990b). Interestingly, we found that when $[Ca^{2+}]_i$ is increased rapidly, but when the Na–Ca exchanger is blocked with Ni, La, sodium removal, etc., the Na–Ca exchanger current is removed, but an inward current transient remains. We have provided evidence that this rapid current transient (rises in about 200 $\mu$s) comes from the calcium-dependent conformational change of the Na–Ca exchanger proteins (Niggli and Lederer, 1991).

## V. COMBINED PHOTOLYSIS AND FLUORESCENCE

By combining flash photolysis, fluorescence measurements, and voltage-clamp, one has an even more powerful set of tools (see Delaney and Zucker, 1990). If all systems worked ideally, photolysis could produce a rapid increase in a substance of interest and the fluorescence and current measurements would monitor the cellular responses. Our early experiments suggest that there are a number of problems that must be addressed. Since our principal interest is in measuring and controlling $[Ca^{2+}]_i$, we will focus on $[Ca^{2+}]_i$ measurements and photorelease of caged calcium.

To start with, two light sources are needed: one to provide continuous low levels of light for the imaging or PMT fluorescence measurements, another to provide an intense flash of light that is used to produce the photolysis. A flash lamp and a continuous xenon arc lamp can be used together. In principle, it is also possible to use a single lamp that operates at a "simmer" and occasionally is "flashed." While we have not tried this arrangement, calculations suggested that the heat load would be excessive in the lamps available to us.

Because the light used to excite the fluorescent indicator (e.g., indo-1) also leads to photolysis of DM-nitrophen, every effort must be made to minimize the time of UV exposure and also the intensity of the UV light used when exciting indo-1. We have found in preliminary experiments that by opening and closing a shutter, we can largely avoid excessive photorelease. Furthermore, it must be recognized that the flash lamp will tend to produce bleaching of the fluorescent compound. We have found, however, that when such an arrangement is used in combination with whole-cell patch clamping, the fluorescence signal is not reduced excessively. Thus, a side benefit of loading the fluorescent indicator and the caged compounds by the patch-clamp pipette is that the pipette continues to supply the cell with material during the course of the experiment and depletion of these substances is minimized.

Zucker has raised the issue of interactions between the caged compounds and the fluorescent indicators (Zucker, 1992). In cuvette experiments, he shows that the fluorescent signal can be quenched by the caged compound or that the fluorescence spectrum can be altered. This does provide warning that care is needed when carrying out experiments when chemicals with overlapping excitation, absorbance, and emission spectra are used. These agents do not always work independently. Additionally, photolysis and photobleaching products can have biological actions.

## VI. SUMMARY

We have provided a brief overview of the methods used by us when combining electrophysiology experiments with fluorescence and/or photorelease experiments. There are separate concerns about each of the following important issues: temporal resolution, spatial resolution, fluorescence detection, image processing and storage, data coordination, and interaction between fluorescent and photoactive agents. Nevertheless, the experimental approach can provide unique information with resolution appropriate to address many issues. These methods can be readily used in all cells ranging from standard cell lines to challenging contractile cardiac cells.

ACKNOWLEDGMENTS

We would like to thank the Swiss Science Foundation, the NIH, the American Heart Association, the Maryland Heart Association, the DRIF awards from the University of Maryland at Baltimore, and the Medical Biotechnology Center for their support of the work described here. Additionally, we would like to acknowledge the role played by Drs. M. B. Cannell and J. R. Berlin in our early development of fluorescence methods and for their continued advice.

REFERENCES

Berlin, J. R., Cannell, M. B., and Lederer, W. J. (1988). Voltage dependent changes in intracellular calcium in single cardiac myocytes measured with fluorescent calcium indicators. *In* "Biology of adult cardiac myocytes." (Clark, W. A., Decker, R. S. and Borg, T. K., eds.), pp. 366–369. Elsevier Science Publishing Co., Inc., New York.
Berlin, J. R., Cannell, M. B., and Lederer, W. J. (1989). Cellular origins of the transient inward current, $I_{TI}$, in cardiac myocytes: role of fluctuations and waves of elevated intracellular calcium. *Circ. Res. 65*, 115–126.
Berlin, J. R., Wozniak, M. A., Cannell, M. B., Bloch, R. J., and Lederer, W. J.

(1990). Measurement of intracellular $Ca^{2+}$ in BC3H-1 muscle cells with fura-2: Relationship to acetylcholine receptor synthesis. *Cell Calcium 11*, 371–384.

Bers, D. M., Lederer, W. J., and Berlin, J. R. (1990). Intracellular Ca transients in rat cardiac myocytes: Role of Na-Ca exchange in excitation–contraction coupling. *Am. J. Phys. 258*, C944–C954.

Cannell, M. B., Berlin, J. R., and Lederer, W. J. (1987a). Effect of membrane potential changes on the calcium transient in single rat cardiac muscle cells. *Science 238*, 1419–1423.

Cannell, M. B., Berlin, J. R., and Lederer, W. J. (1987b). Intracellular calcium in cardiac myocytes: Calcium transients measured using fluorescence imaging. *In* "In Cell Calcium and the Control of Membrane Transport." (L. J. Mandel and D. C. Eaton, eds.), pp. 202–214. Rockefeller University Press, New York.

Cornell Bell, A. H., Finkbeiner, S. M., Cooper, M. S., and Smith, S. J. (1990). Glutamate induces calcium waves in cultured astrocytes: Long-range glial signaling. *Science 247*, 470–473.

Crespo, L. M., Grantham, C. J., and Cannell, M. B. (1990). Kinetics, stoichiometry and role of the Na-Ca exchange mechanism in isolated cardiac myocytes. *Nature 345*, 618–621.

Delaney, K. R., and Zucker, R. S. (1990). Calcium released by photolysis of DM-nitrophen stimulates transmitter release at squid giant synapse. *J. Phys. 426*, 473–498.

Denk, W., Strickler, J. H., and Webb, W. W. (1990). Two-photon laser scanning fluorescence microscopy. *Science 248*, 73–76.

Eng, J., Lynch, R. M., and Balaban, R. S. (1989). Nicotinamide adenine dinucleotide fluorescence spectroscopy and imaging of isolated cardiac myocytes. *Biophys. J. 55*, 621–630.

Goldman, R. S., Finkbeiner, S. M., and Smith, S. J. (1991). Endothelin induces a sustained rise in intracellular calcium in hippocampal astrocytes. *Neurosci. Lett. 123*, 4–8.

Grynkiewicz, G., Poenie, M., and Tsien, R. Y. (1985). A new generation of $Ca^{2+}$ indicators with greatly improved fluorescence properties. *J. Biol. Chem. 260*, 3440–3450.

Hadley, R. W., and Lederer, W. J. (1991). $Ca^{2+}$ and voltage inactivate $Ca^{2+}$ channels in guinea-pig ventricular myocytes through independent mechanisms. *J. Physiol. 444*, 257–268.

Heiple, J. M., and Taylor, D. L. (1982). An optical technique for measurement of intracellular pH in single living cells. *In* "Intracellular pH: Its Measurement, Regulation, and Utilization in Cellular Functions." (R. Nuccitelli and D. W. Deaner, eds.), pp. 21–54. Alan R. Liss, Inc., New York.

Inoue, S. (1986). "Video Microscopy." Plenum, New York.

Kao, J. P., Harootunian, A. T., and Tsien, R. Y. (1989). Photochemically generated cytosolic calcium pulses and their detection by fluo-3. *J. Biol. Chem. 264*, 8179–8184.

Kaplan, J. M., and Ellis Davies, G. C. (1988). Photolabile chelators for the rapid photorelease of divalent cations. *Proc. Natl. Acad. Sci. U.S.A. 85*, 6571–6575.

Kofuji, P., Lederer, W. J., and Schulze, D. H. (1992). The human cardiac Na-Ca exchanger: Cloning, sequencing and expression. *Biophys. J. 61*, A387.

Konishi, M., Olson, A., Hollingworth, S., and Baylor, S. M. (1988). Myoplasmic binding of fura-2 investigated by steady-state fluorescence and absorbance measurements. *Biophys. J. 54,* 1089–1104.

Lakowicz, J. R. (1983). "Principles of Fluorescence Spectroscopy." Plenum.

Leblanc, N., and Hume, J. R. (1990). Sodium current-induced release of calcium from cardiac sarcoplasmic reticulum. *Science 248,* 372–376.

Lederer, W. J., Niggli, E., and Hadley, R. W. (1990). Sodium-calcium exchange in excitable cells: Fuzzy space [comment]. *Science 248,* 283.

Lederer, W. J., Niggli, E., and Hadley, R. W. (1991). Sodium-calcium exchange (technical comment). *Science 251,* 1370–1371.

Minta, A., and Tsien, R. Y. (1989). Fluorescent indicators for cytosolic sodium. *J. Biol. Chem. 264,* 19449–19457.

Minta, A., Kao, J. P. Y., and Tsien, R. Y. (1989). Fluorescent indicators for cytosolic calcium based on rhodamine and fluorescein chromophores. *J. Biol. Chem. 264,* 8171–8178.

Mulligan, I. P., Adams, S. R., Tsien, R. Y., Potter, J. D., and Ashley, C. C. (1990). Flash photolysis of the caged calcium-chelator, diazo-2, produces rapid relaxation of single skeletal muscle fibres. *Biophys. J. 57.*

Näbauer, M., and Morad, M. (1990). $Ca^{2+}$-induced $Ca^{2+}$ release as examined by photolysis of caged $Ca^{2+}$ in single ventricular myocytes. *Am. J. Phys. 258,* C189–C193.

Niggli, E., and Lederer, W. J. (1990a). Real-time confocal microscopy and calcium measurements in heart muscle cells: Towards the development of a fluorescence microscope with high temporal and spatial resolution. *Cell Calcium 11,* 121–130.

Niggli, E., and Lederer, W. J. (1990b). Voltage-independent calcium release in heart muscle. *Science 250,* 565–568.

Niggli, E., and Lederer, W. J. (1991). Molecular operations of the sodium–calcium exchanger revealed by conformation currents. *Nature 349,* 621–624.

Pawley, J. B. (1990). "Handbook of Biological Confocal Microscopy." Plenum, New York.

Peeters, G. A., Hlady, V., Bridge, J. H. B., and Barry, W. H. (1987). Simultaneous measurement of calcium transients and motion in cultured heart cells. *Am. J. Phys. 253,* H1400–H1408.

Somlyo, A. P. (1986). Recent Advances in Electron and Light Optical Imaging in Biology and Medicine. Ann. N.Y. Acad. Sci. Vol. 483

Szmacinski, H., Nowaczyk, K., Berndt, K., and Lakowicz, J. R. (1992). Fluorescence lifetime imaging. *Biophys. J. 61,* A35.

Taylor, D. L., Waggoner, A. S., Murphy, R. F., Lanni, F., and Birge, R. R. (1986). "Applications of fluorescence in the biomedical sciences." Alan R. Liss, Inc., New York.

Tsien, R. Y. (1980). New calcium indicators and buffers with high selectivity against magnesium and protons: Design, synthesis, and properties of prototype structures. *Biochem. 19,* 2396–2404.

Tsien, R. Y. (1981). A non-disruptive technique for loading calcium buffers and indicators into cells. *Nature 290,* 527–528.

Tsien, R. Y. (1989). Fluorescent probes of cell signaling. *Ann. Rev. Neurosci. 12,* 227–253.

Tsien, R. Y., and Harootunian, A. T. (1990). Practical design criteria for a dynamic ratio imaging system. *Cell Calcium 11*, 93–109.

Wier, W. G., and Blatter, L. A. (1991). $Ca^{2+}$-oscillations and $Ca^{2+}$-waves in mammalian cardiac and vascular smooth muscle cells. *Cell Calcium 12*, 241–254.

Wier, W. G., Cannell, M. B., Berlin, J. R., Marban, E., and Lederer, W. J. (1987). Cellular and subcellular heterogeneity of $[Ca^{2+}]_i$ in single heart cells revealed by Fura-2. *Science 235*, 325–328.

Williams, D. A., Fogarty, K. E., Tsien, R. Y., and Fay, F. S. (1985). Calcium gradients in single smooth muscle cells revealed by the digital imaging microscope using Fura-2. *Nature 318*, 558–561.

Zucker, R. S. (1992). Effects of photolabile calcium chelators on fluorescent calcium indicators. *Cell Calcium 13*, 29–40.

# 8

# Simultaneous Differential Interference Contrast and Quantitative Low-Light Fluorescence Video Imaging of Cell Function

**J. Kevin Foskett**
Division of Cell Biology
Hospital for Sick Children

I. INTRODUCTION
II. CORRELATION OF
FLUORESCENCE WITH CELL
STRUCTURE
A. Background

B. Methods for Combined DIC
and Quantitative Low-Light
Fluorescence Imaging
III. SUMMARY
REFERENCES

## I. INTRODUCTION

Imaging of fluorescence in living cells has become a very powerful tool in cell biology. The coupling of the light microscope to image detectors with sensitivities to very low-light intensities, and their coupling in turn to powerful image-processing computers, has provided in the light microscope an instrument capable of quantitative fluorescence measurements within single cells.

The ability to perform quantitative fluorescence imaging in single cells provides possibilities for gaining new insights into the biochemistry, biophysics, and physiology of the intact cell. For example, a number of fluorescent dyes have been developed that are sensitive to the concentrations of

Optical Microscopy: Emerging Methods and Applications

specific ions or to membrane potential. Quantitation, and in some cases spatial resolution, of the concentrations of $Ca^{2+}$, $H^+$, $Cl^-$, $Na^+$, and $Mg^{2+}$ can be performed with high temporal resolution in single cells during various cell processes, including growth, division, and motility (Haugland and Minta, 1990; Tsien, 1989). When injected into cells, fluorescent analogs and affinity probes of biologically relevant macromolecules permit visualization of dynamic protein behavior in single living cells (Wang, 1989; Maxfield, 1989). These approaches have been especially valuable for elucidating the *in vivo* dynamics and interactions of cytoskeletal proteins (Wang and Sanders, 1990; Sanger and Sanger, 1990). Organellar localization and dynamics can be monitored as well in living cells using fluorescent probes for specific organelles [e.g., endoplasmic reticulum (Terasaki, 1989; Chen and Lee, 1990), golgi apparatus (Pagano, 1989), and mitochondria (Chen, 1989)]. Trafficking kinetics through endosomal and biosynthetic pathways has been determined in intact cells using fluorescent fluid phase and protein markers and lipid analogs (Swanson, 1989; Dunn and Maxfield, 1990; Maxfield and Dunn, 1990). Imaging of fluorescently labeled proteins and lipids embedded in the plasma membrane can provide details regarding molecular interactions and mobilities there (Thomas and Webb, 1990; Edidin, 1989; Angelides, 1989; Kapitza *et al.,* 1985; Ryan *et al.,* 1988). Furthermore, light sensitive probes that are coupled to biologically interesting molecules can be exploited not only to derive information about the intracellular environment but to manipulate it as well (see Chapter 2, this volume). It is clear, therefore, that optical imaging of fluorescence in single living cells is a powerful approach for studying numerous problems of cell biology.

It seems reasonable to expect that the importance of fluorescence imaging for gaining insights into cell biological processes will grow together with improvements in fluorescent probe chemistry. Thus, it is anticipated that an increasing diversity of fluorescence probes will become available for use, which will be specific, minimally perturbing, and easily deliverable to specific environments inside the cell. Such developments will ultimately enable visualization of the spatial and temporal orchestration of molecular events inside the cell that govern cellular physiology. Together with improvements in probe chemistry, attainment of this long-term goal will require the development of techniques to allow many different parameters inside a cell to be visualized simultaneously. The heterogeneity among cells requires that such determinations be made in single cells.

## II. CORRELATION OF FLUORESCENCE WITH CELL STRUCTURE

### A. Background

In addition to fluorescence imaging, other microscopic visualization techniques, including phase, darkfield, and polarization microscopies, nanovid microscopy (DeBrabander *et al.*, 1986) and internal reflection microscopy (Gingell and Todd, 1979) can also be exploited in cell biological studies. Full advantage of the high spatial and temporal resolution provided by low-light-level imaging of fluorescence will only be realized when cell structure is also imaged with high contrast and resolution and with equivalent temporal resolution. In that way, a high correlation between cell structure and fluorescence can be achieved. Differential interference contrast (DIC) imaging permits resolution of structural detail that surpasses other light-microscopic methods. The use of video for DIC imaging has greatly extended the use of resolution and contrast provided by the light microscope (Allen *et al.*, 1981; Inoue, 1981). DIC images have a three-dimensional look, provide a shallow depth of field (out of focus information does not interfere), and have high resolution and contrast. Therefore, DIC imaging allows cell structure to be highly resolved in two or three dimensions. Movements and localization of organelles, identification of specific cell types, and changes in cell shape and behavior can be resolved in DIC. Thus, a combination of low-light-level fluorescence with DIC imaging provides the best means to correlate fluorescence distribution with cell structure in living cells.

Nevertheless, these optical approaches are generally not combined in a microscope designed for quantitative fluorescence determinations, ultimately because of the poor light transmittance of the DIC optical path. Incorporation and illumination of fluorescent molecules inside living cells may not be entirely noninvasive. Excitation of fluorescence, from either intrinsic molecules or extrinsic ones incorporated into the cell, may generate toxic oxygen free-radical species (Martin and Logsdon, 1987; Yonuschot, 1991), buffer intracellular ion levels, or have diverse toxic or metabolic effects. These considerations are especially important in single-cell determinations since the same cell is repeatedly illuminated, unlike the situation in fluorometric measurements of cell suspensions. Two basic strategies have been employed to minimize perturbing effects of fluorescence probes. First, attempts are made to use only the minimal amount of fluorescent probe necessary to make a measurement with an acceptable signal. Second, the excitation light flux is reduced by attenuating the intensity and/or by exposing the cell to only intermittent excitation. Fluorescence intensity is proportional to the amount of probe in the cell, as well as to the level of excitation. Therefore, these strategies to protect cell health require detection systems that are sensitive to very low-light levels. To work at the lowest possible light levels requires that the optical paths within the microscope transmit excitation and emission wavelengths with as little light loss as possible. Therefore, a high numerical aperture objective lens should be used and the number of reflecting surfaces within the microscope should be minimized. Importantly, it is also necessary to eliminate optical elements from the fluorescence light path, particularly from the emission side, which absorb or scatter light or otherwise impede light transmission. Such optical elements, including phase rings and polarizers, which enhance contrast and resolution in transmitted light microscopy, attenuate too much fluorescence emission, necessitating the use of more probe and/or more excitation light flux. For example, an ideal polarizer transmits 50% of incoming light although in practice transmission is generally 30–40%. In a microscope equipped for DIC, the fluorescence emission will pass through a Wollaston prism and a polarizer. If the excitation light also passes through the same elements, the resultant fluorescence intensity will be less than 10% compared with that of the same microscope without these elements. Thus, low-light-level fluorescence imaging is generally associated with elimination of optical elements, which provide high contrast and resolution in transmitted light imaging.

The loss of transmitted light image resolution and contrast resulting from the elimination of these optical elements in low-light-level fluorescence imaging is further compounded by another necessity of video imaging. Because the video detector is operated in a low-light-level mode for the fluorescence imaging, the transmitted light intensity used for bright-field imaging must be greatly attenuated to prevent saturation of the signal as well to protect the camera. This results in an image with poor signal characteristics that is degraded by noise in the camera. Even in digital image-processing systems where it is possible to integrate a number of sequential images, good image resolution cannot be achieved due to insufficient photon flux. Therefore, as a result of the combined effects of eliminating contrast-enhancing optical elements and imaging under photon-poor conditions, low-light-level fluorescence is associated with poor transmitted light imaging. Consequently, dynamic correlations of fluorescence with simultaneous cell structure and function in living cells are difficult. In the last few years, however, two optical systems have been described which solve this problem (Foskett, 1988; Spring, 1990).

## B. Methods for Combined DIC and Quantitative Low-Light Fluorescence Imaging

### 1. Simultaneous Imaging with Two Detectors

*a. Optical Configuration*

A system we developed to simultaneously image cells in low-light-level fluorescence and high-light-level DIC optics, shown in Fig. 8.1, requires two video detectors, the means to separate spectrally the two images to the respective cameras, and a judicious placement of the polarizers in the optical path (Foskett, 1988). The transmitted light optical path incorporates the elements required for DIC imaging. The light from a standard 60-W halo-

FIGURE 8.1

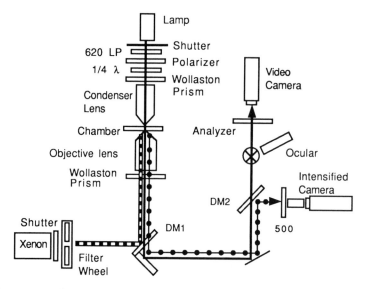

DIC-fluorescence microscope for simultaneous DIC and low-light-level fluorescence imaging. Details of the optical paths are given in the text. 620 LP, 620-nm long pass filter; 500—500-nm interference filter; DM, dichroic mirror. Dashed line, fluorescence excitation; solid line, transmitted light; dotted line, fluorescence emission.

gen lamp is polarized by a calcite prism (Zeta Intl., Prospect, IL) and passes through a ¼-λ plate and a Wollaston prism. The light is focused on the cells by an objective lens, which replaces the standard condenser. With an open-bath perfusion chamber, we generally employ a Zeiss water immersion 40×, 0.75 numerical aperture (NA) lens; for applications that require a longer working distance, a Letiz 32×, 0.40 NA lens is employed as the condenser instead. Fluorescence excitation from a 75-W xenon arc lamp is directed to the preparation by a dichroic mirror (DM1) (Zeiss or Omega Optical) housed in the standard filter cube and focused by an objective lens. We generally employ a Zeiss Neofluor 63×, 1.25 NA lens since it combines high-light-collection capability with a fairly long (0.5 mm) working distance, which is helpful for examining thick specimens and in combined imaging–electrophysiology–microinjection experiments. A Nikon 40×, 1.3 NA objective lens is also very suitable. Fluorescence emission as well as transmitted light that passes through the specimen are collected by the objective lens and pass through a second Wollaston prism associated with it. In the standard Zeiss IM-35 inverted microscope configuration, the light would next pass through DM1 and then a second polarizer (analyzer). However, because this polarizer would be in the fluorescence emission light path, it was repositioned more distally in the optical path, as described later.

In the standard microscope configuration, the light proceeds through the microscope to a set of prisms or mirrors designed to either allow the light to continue to the oculars or to direct it at right angles through a side port to a detector, such as a photomultiplier tube or imaging device. With the use of a trinocular, a second camera is mounted above the oculars and adjusted to be parfocal with the side-port camera. To direct fluorescence emission to the side-port detector and transmitted light to the overhead camera, the prisms are replaced with a dichroic mirror mounted at 45° to the incident light path. The dichroic mirror (DM2) can separate light based on spectral properties; thus, it becomes possible to image the transmitted light and fluorescence separately if they are spectrally distinct. The halogen light is initially filtered to long enough wavelengths (generally near infrared) to pass through DM2. DM1 passes these long wavelengths without interference. DM2 is chosen to be able to pass the long-wavelength halogen light to the overhead camera for transmitted light imaging and reflect the fluorescence emission out the side port for fluorescence imaging. A bandpass interference filter is positioned in the optical path in front of the side-port camera to select the appropriate fluorescence emission wavelengths. The analyzer (calcite prism) that was removed from beneath the nose piece is positioned between DM2 and the transmitted light camera, thereby completing the DIC light path. Thus, the overhead camera views the preparation in high-contrast and high-resolution DIC optics that require high-light levels, whereas the side-port camera views the low-light-level fluorescence. The only DIC optical element in the fluorescence light path is a Wollaston prism. My measurements indicate that this component results in a fluorescence diminution of ~2–4% for most wavelengths.

The overhead camera that views the transmitted light image is coupled to the microscope by a trinocular head, which can direct light to the camera or to the oculars. A $10\times$ ocular and a zoom lens system to adjust the magnification to that of the fluorescence camera are incorporated between the trinocular and the overhead camera. We use accessible space within the trinocular to insert the calcite prism polarizer (analyzer), which completes the DIC light path. The specific optical characteristics of the polarizer that is used in this position determine the precise location of the real image formed by the objective lens. This information needs to be known to enable placement of the ocular in the correct vertical position to achieve parfocality with the side-port camera. Parfocality can be empirically established by viewing the image in the side-view camera and then adjusting the vertical position of the overhead ocular while viewing the object through it, until the object is in the same focal plane as that seen in the side port. The ocular can then be maintained in this position by use of a collar around it to support it on the trinocular head.

The overhead camera will view the preparation in high-light-level DIC optics. Therefore, it should have high resolution and low-noise characteristics, with gain and black-level controls. Suitable cameras, therefore, would include newvicons and video-rate charge-coupled devices (CCD). We currently employ a Dage model 72 video-rate CCD because, in addition to possessing these characteristics, it is extremely small and light.

The side-port camera that views the fluorescence emission is coupled to the microscope using standard manufacturer's coupling lenses incorporated into a tube that holds the dichroic mirror (DM2) in proper position. Our latest version was built by the machine shop of the local Zeiss representative. It incorporates a number of set screws, which allow the position and orientation of the dichroic mirror to be adjusted by the user. In addition, it is designed to allow replacement of the dichroic filter with another. This is of course necessary because of the finite life span of optical filters, but additionally it enhances the flexibility of the system by allowing the user to choose the optical characteristics appropriate for the particular fluorophore in use.

The side-port detector can be either an imaging or photometric device. The imaging device should be a low-light-sensitive detector, such as a cooled slow-scanned CCD or intensified video camera. The reader is directed to excellent recent reviews that evaluate the response characteristics of imaging detectors employed in low-light-level microscopy (Tsay et al., 1990; Bookman, 1990; Spring and Lowy, 1989; Aikens et al., 1989). In our studies with intracellular fluorescent ion-indicator dyes, we have employed a two-stage, inverted-type image intensifier coupled by relay lenses to a newvicon camera (Spring and Smith, 1987) (Videoscope KS-1381). We chose this detector based on its low-light sensitivity, relatively fast response and high spatial resolution, and the "quality" of its noise characterisitics. All automatic features (gain, black level) of this camera are disabled, and a provision for driving the sync signal from an external source is incorporated into it to allow images from the overhead (transmitted light) camera to be inserted into the video field of the intensified camera by use of a video mixer (described below). For imaging applications that require higher spatial resolution of fluorescence, a slow-scanned cooled CCD would be better although at the expense of temporal reolution (Aikens et al., 1989).

Video images contain a tremendous amount of information. Typical video image-processing systems digitize each image to 512 × 480 × 8 bits, or ~250 kbytes. Capture of a video image results in corresponding occupation of digital video memory. When video memory becomes filled, images must be transferred to another storage area, which requires time, thereby decreasing temporal resolution. These problems are exacerbated in the DIC-fluorescence microscope because of the need to record images from two cameras. Image capture to a real-time digital disk avoids these problems, but the expense (although considerably less in recent years) is generally prohibitive. Another solution is to digitize after the experiment from a primary recording device, such as a laser disc recorder (LDR), although this adds another noise component to the data. Furthermore, this approach requires that the video input to the LDR is able to be rapidly switched between the two cameras. Our attempts at this have not been successful because the LDR (Panasonic TQ-2028F) requires too much time to sync to the new signal following a switch from one camera to the other. Our protocol has been to digitize and store images obtained from the low-light-level fluorescence camera during the experiments and capture the transmitted-light DIC images on the LDR (Fig. 8.2). The reason for this strategy is based on the fact that writing images to hard disc is rate limiting

FIGURE 8.2

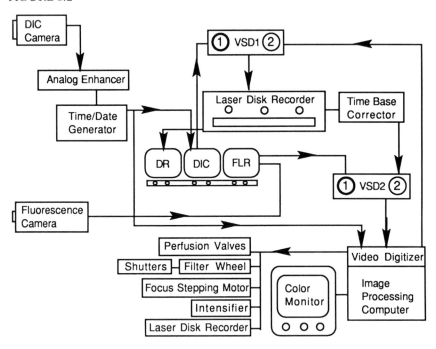

Schematic of the image-processing system that is coupled to the DIC-fluorescence microscope. Details of the flow of video information from the two cameras is given in the text. A tri-pac of monochrome video monitors displays the outputs of the laser disk recorder (DR monitor), the DIC camera (DIC monitor), and the fluorescence camera (FLR monitor). Video switching devices (VSD1, VSD2) are normally set to input position 1. The image-processing computer contains video hardware for real-time video digitization and arithmetic, as well as I/O ports for controlling peripheral devices.

in our system. To improve digital-image throughput, we sacrifice spatial resolution. Fluorescence images are digitized at 256 × 240 resolution instead of 512 × 480, reducing the amount of data by 75% and increasing throughput correspondingly. We have found that this results in relatively little loss of spatial resolution of fluorescence from ion-indicator dyes. However, 256 × 240 digitization significantly and unacceptably degrades the resolution of DIC images. Therefore, the full-resolution DIC images are recorded on the LDR, in order that their capture not impede the data throughput, and hence temporal resolution of the experiments.

The video signal from the DIC camera is analog enhanced (For-A Contour Synthesizer, IV-530), encoded for time and date, and directed to a monochrome monitor (DIC in Fig. 8.2), a video switching device (VSD1) and to one input of the digitizer of the image-processing system (Fig. 8.2). The switching device directs the signal to the LDR. The LDR output is then routed to another monchrome monitor (DR in Fig. 8.2). Thus, the previously captured DIC image can be viewed side by side with the live DIC image (see Fig. 8.4). A second input to VSD1 is the video output of the image-processing system. With this arrangement, digitally processed images can be stored on the LDR.

The video signal from the fluorescence camera proceeds first to a third monochrome monitor (FLR in Fig. 8.2) and then to a second video switching device (VSD2). Normally, VSD2 is set to route the signal to a second video input of the video digitizer in the image-processing computer. A second input to VSD2 is from the LDR, through a time-base corrector, allowing images stored on the LDR to be captured and processed and/or stored digitally. The output of the video image processor is directed to a large color monitor. This output can either be the live DIC or fluorescence images, or images stored in the image processor.

*b. Procedures for Setting Up DIC*

As a result of moving the analyzer from its normal position beneath the nose piece to a more distal position in the trinocular before the DIC camera, the cells cannot be observed in DIC through the oculars. While this is not normally a problem because the DIC image can simply be viewed with the video camera, it does cause a problem when it is necessary to adjust the optical components to achieve proper DIC imaging. This is because alignment of the two Wollaston prisms requires viewing the back focal plane of the objective with the polarizer and analyzer crossed, normally through a telescopic lens placed in one of the ocular positions. Therefore, setting up DIC is accomplished as follows. First, the polarizer is put into a known orientation, which we will refer to as the 0° position. Another polarizer, we normally use a polarizing film, is placed in the normal analyzer position at a polarization orientation of 90° (i.e., the two polarizers are crossed). The Wollaston prisms are adjusted as usual. Having set up proper DIC optics as viewed through the oculars, the next step is to set them up for the overhead camera. This is achieved by removing the overhead camera and zoom-lens assembly from atop the trinocular and peering down the trinocular barrel to the calcite polarizer. With the polarizer above the condenser lens set at 0° and the film polarizer removed, the calcite prism is rotated to achieve maximal extinction. At this point, DIC is set up the same for the overhead camera as it is for the oculars. As long as the orientation of the overhead calcite prism polarizer is fixed, adjusting DIC for the overhead camera simply involves inserting the 90° film polarizer and making the adjustments as normal, viewing through the ocular ports. Final video DIC adjustments are made by rotating the polarizer away from extinction while viewing the preparation on the video monitor.

*c. Performance of the DIC-Fluorescence Microscope*

For low-light-level fluorescence imaging in a microscope with a single camera, DIC optical components are removed from the microscope, and the cells are viewed in bright-field optics under low-light-level conditions. Figure 8.3 demonstrates the advantages of the two-camera system

FIGURE 8.3

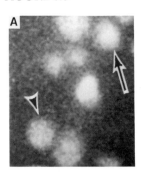 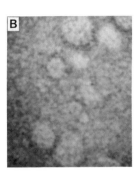 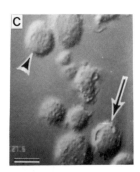

Advantages of the two-camera system for imaging cell structure during low-light-level fluorescence microscopy. (A) Human monocytes in fluorescence, (B) low-light-level bright-field optics, and (C) high-light-level differential interference contrast optics (DIC). Fluorescence is cell autofluorescence (excitation 380 nm). DIC image is inverted (from top to bottom) compared with other images, since it was obtained from a camera mounted at 90° relative to the fluorescence–bright-field camera. Fluorescence image was obtained on one camera while simultaneously imaging in DIC optics on another. Arrows and arrowheads identify the same cells for orientation. See text for details. Bar, 15 um. (From Foskett, 1988.)

FIGURE 8.4

Simultaneous DIC and fluorescence imaging. CHO cells, grown in culture on a glass coverslip, simultaneously imaged in DIC optics (*middle monitor*) and low-light-level fluorescence (*right monitor*). Monitor on left shows DIC image recorded on a laser disk recorder one minute earlier. Arrow points to a cell that had just been injected with the $Ca^{2+}$-indicator dye fura-2. Live DIC image (*middle monitor*) shows an adjacent cell, immediately above the injected cell, in the process of being injected with fura-2. The glass microelectrode can be seen by close examination of the live DIC image. Fura-2 fluorescence in these cells is observed on the right monitor. The cell that is in the process of being injected is not yet filled with dye. The fluorescence image is inverted (from top to bottom) compared to the DIC images, as described in the text.

for imaging cells under low-light-level conditions. Figure 8.3A is a digitized fluorescence image of living human monocytes grown on a coverslip and perfused on the stage of the microscope. · The fluorescence is cellular autofluorescence demonstrating the low-light conditions in which the system is working and emphasizing the sensitivity of the intensified newvicon. Figure 8.3B shows the same field of cells imaged by the same camera at the same gain setting in low-light-level bright-field optics. In Figure 8.3C, the same field of cells is imaged in DIC optics with the second camera while simultaneously imaging the fluorescence with the first camera. Figure 8.3 demonstrates the remarkable gain in transmitted light resolution and contrast obtained in the two-camera system under low-light-level conditions.

In the simplest configuration, the video microscope system consists of two cameras with outputs to monitors. In Fig. 8.4, side-by-side monitors display live images of Chinese hamster ovary (CHO) cells grown in culture on glass coverslips. The field in the middle is imaged in DIC optics, whereas on the right the same field of cells is imaged simultaneously in fluorescence. One cell has been microinjected and an adjacent one is in the process of being injected with the $Ca^{2+}$-indicator dye fura-2. The fluorescence is being excited at 380 nm and observed at 500 ± 20 nm. The left monitor in Fig. 8.4 shows the output of the LDR, which is a DIC image of the same cells captured approximately one minute earlier, immediately after the first cell had been microinjected. Figure 8.5 shows a field of fura-2–loaded human monocytes, simultaneously imaged in fluorescence (left monitor) and DIC (right monitor) optics. In each case, the cells are imaged with high resolution and contrast in DIC while the low-light fluorescence is likewise imaged in good detail. Figures 8.4 and 8.5 demonstrate the parfocality of the fluorescence and DIC images. The inverted nature of the fluorescence images compared with the DIC images results from the deflection of the fluorescence emission at right angles out through the side port of the microscope.

It is possible to make quantitative measurements of fluorescence despite the high-light throughput of the DIC optical path. This can be appreciated by inspection of Fig. 8.6, which demonstrates that the fluorescence intensity is unaffected by the presence of the transmitted light. It is important to note, however, that the degree of spectral contamination of the two light paths will depend on various parameters of the optical configuration, including the degree of spectral separation of the two light paths, the quality of the dichroic filter (DM2) that separates the light to the two cameras, the gain and sensitivity of the fluorescence camera, and the intensities of the fluorescence and transmitted lights.

In image processors that can only digitize from one camera at a time, a modified arrangement allows simultaneous digitization of the DIC and fluorescence images. The DIC image is analog enhanced and time–date coded and then inserted into the video signal of the fluorescence camera using a video splitter–inserter. This results in a video signal containing parts of each

FIGURE 8.5

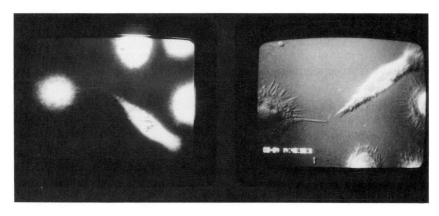

Simultaneous DIC and fluorescence imaging. Human monocytes loaded with fura-2. Left monitor: fluorescence emission. Right monitor: simultaneous DIC image. Monocyte morphology is excellently resolved in DIC, while fluorescence localization is also resolved in good detail. (From Foskett, 1988.)

FIGURE 8.6

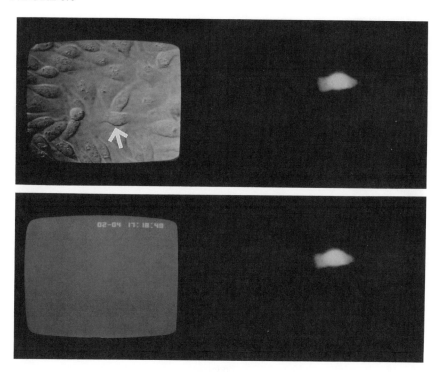

Lack of spectral contamination of the DIC and fluorescence light paths. (*Top*) A field of CHO cells (as in Fig. 8.4) is imaged simultaneously in DIC (*left monitor*) and low-light-level fluorescence (*right monitor*) optics. A single cell had been injected with fura-2, indicated by arrow on the DIC monitor. Fluorescence image is inverted from top to bottom compared to the DIC image. (*Bottom*) Transmitted light shuttered off. There is no effect on fluorescence intensity, demonstrating that the system allows quantitative determinations of fluorescence intensity.

image (Fig. 8.7). Thus, there is no temporal disparity between image capture times from the two cameras.

The fluorescence and DIC images can be viewed simultaneously in real time on side-by-side monitors, or on one monitor if the slitter–inserter is used, or in image-processing systems that are capable of digitizing and combining dual-video signals in real time. However, the fluorescence image is inverted as a result of being reflected through optics at a right angle out of the microscope to the camera. Although this sometimes makes real-time correlation of fluorescence with the DIC image somewhat difficult, it is advantageous for viewing both images on one monitor using the splitter–inserter, as shown in Fig. 8.7. However, there are two ways to flip the image over in real time to facilitate viewing and on-line comparison with the DIC

FIGURE 8.7

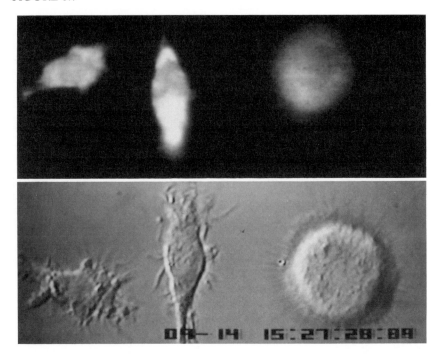

Video splitter–inserter combined DIC and fluorescence images. Cultured human monocytes loaded with fura-2. (*Top*) Fluorescence image. (*Bottom*) Simultaneous DIC image.

image on side-by-side monitors. The first method is an optical approach. Placement of a dove prism in front of the fluorescence camera will invert the image. I have not employed this prism because of some difficulty in properly aligning it. The second method is an electronic approach. As an option on some video cameras, the raster scan can be initiated from the bottom of the field instead of the top.

As mentioned earlier, the fluorescence imaging camera could be replaced by a photomultiplier tube. Thus, in those applications where spatial resolution of fluorescence is not required, for example, whole single-cell ion-indicator dye measurements, fluorescence can be correlated with cell structure and function. This approach requires that an appropriate spatial filter be inserted into an intermediate image plane. This plane is accessible in some microscopes, such as the Zeiss IM–35. The filter limits fluorescence light detection to only the region of interest, increasing the specificity of the signal and reducing noise by eliminating most stray light. This spatial filter is imaged by the DIC camera as well. Thus, only whatever is viewed by the DIC camera will contribute to the fluorescence measured by the photomultiplier tube. Thus, this arrangement allows continuous control of exactly what is being detected by the PMT. For example, in fluorescence photometry of motile cells, the ability to continuously image the cell in DIC optics allows the user to move the stage to ensure that the cell, in spite of its movements, remains in the fluorescence detection light path. This optical arrangement is also especially useful in applications that combine single-cell photometry with electrophysiology or microinjection (Foskett *et al.*, 1989). Thus, imaging the preparation in DIC optics allows specific cells to be carefully chosen based on morphological criteria, and then proper placement of the micro- or patch electrode is facilitated by imaging the electrode tip (for example, as in Fig. 8.4).

*d. Specific Applications*

Our work with the DIC-fluorescence microscope has largely consisted of studies in which ion-indicator dye fluorescence has been imaged in single cells. The ability to simultaneously image the cells in DIC optics is important in our studies because it enables simultaneous determination of cell volume. The relevance of cell-volume determinations can be appreciated based on the following considerations. The water permeability of plasma membranes of most cells is quite high, generally several orders of magnitude greater than ion permeability. The regulation in animal cell membranes of ion "leaks" and active transport pathways confers on the cell an *effective* ion impermeability, which means that the membrane can be regarded as semipermeable. For a perfect osmometer with a semipermeable membrane, from the van't Hoff equation

$$V - b = (RT \sum_i \Phi_i Q_i)/\Pi$$

where $\Pi$ is the osmotic pressure of the medium, $R$ is the gas constant, $T$ is temperature, $\Phi$ is the osmotic coefficient, $V$ is the cell volume, $b$ is the nonosmotically active cell volume, and $Q$ is the amount of the $i$th solute in the cell. Thus, cell volume depends on the osmotic pressure of the medium and on the solute content of the cell. This view of a cell indicates that under conditions when $\Pi$ is not altered, changes in cell volume correspond to parallel changes in cell solute content. Thus, measurements of cell volume during specific ion substitutions and pharmacological manipulations can provide details concerning specific ion-permeation pathways in the plasma membrane, in a noninvasive manner in intact cells. Simultaneous imaging of cell volume and fluorescence from an ion-indicator dye provides the ability, therefore, to correlate in a single cell the activities of specific ion-transport pathways with the activities of putative regulatory ions (e.g.,

FIGURE 8.8

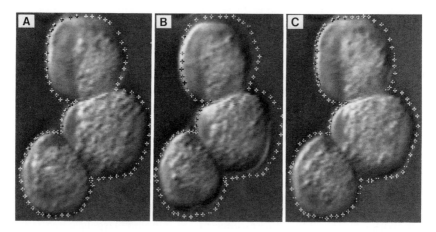

[Ca$^{2+}$]$_i$-induced shrinkage of salivary acinar cells. (A) A triplet of acinar cells imaged in DIC optics; cursor marks delineate the perimeter of the cells. [Ca$^{2+}$]$_i$ had been induced to oscillate by exposing the cells to thapsigargin and caffeine (Foskett and Wong, 1991). [Ca$^{2+}$]$_i$ was simultaneously determined by ratio imaging of fura-2 fluorescence (not shown). Image was recorded at the trough of the [Ca$^{2+}$]$_i$ oscillation. (B) Same cells imaged 30 s later, at the peak of a new [Ca$^{2+}$]$_i$ spike. Cursor marks were transposed from the image in (A). The cells have apparently shrunken. Optical sectioning revealed that changes in cell size in the $x$-$y$ plane were mirrored by changes in the $z$ dimension, allowing volume changes to be computed from the changes in the cross-sectional area of a single optical section. (C) Same cells imaged when [Ca$^{2+}$]$_i$ again returned to a trough before the next rise. Cells have apparently reswelled to a size that is similar to that during the previous [Ca$^{2+}$]$_i$ trough. Thus, [Ca$^{2+}$]$_i$ oscillations drive cell-volume oscillations.

Ca$^{2+}$ and H$^+$) or of specific ions that are being transported by those pathways (e.g., Na$^+$ and Cl$^-$).

In an initial application of this methodology, DIC microscopy and digital imaging of the calcium-indicator dye fura-2 were performed simultaneously in single rat salivary-gland acinar cells to examine the effects of agonist stimulation on cell volume and [Ca$^{2+}$]$_i$ (Foskett and Melvin, 1989). [Ca$^{2+}$]$_i$ was determined by analysis of ratio images of fura-2 constructed from pairs of images obtained at 340-nm and 390-nm excitation. Cell volume was determined by planimetry of single optical sections of images of acinar cells stored on a video disc recorder. By simultaneous optical determinations of cell volume and [Ca$^{2+}$]$_i$ during stimulation of single salivary-gland acinar cells, it was demonstrated that agonist stimulation of fluid secretion is initially associated with a rapid tenfold increase in [Ca$^{2+}$]$_i$ as well as a rapid, substantial cell shrinkage (Fig. 8.8) (Foskett and Melvin, 1989). Close examination of the temporal relationship between the rise of [Ca$^{2+}$]$_i$

and the initiation of the cell shrinkage suggested that elevated $[Ca^{2+}]_i$ triggered the shrinkage. Cell shrinkage under isosomotic conditions suggested that the cell lost solute, presumably $KCl-K^+$ through a basolateral membrane $K^+$ channel and $Cl^-$ through an apical membrane $Cl^-$ channel. Thus, simultaneous imaging of $[Ca^{2+}]_i$ and the cell shrinkage enabled the relationship between the activities of specific ion channels and the level of the intracellular mediator that activates them to be determined in intact cells in a relatively noninvasive fashion. Subsequent to the initial cell shrinkage, cells swelled to a variable extent, and removal of the agonist caused cell volume to return to prestimulation values. Cell swelling implies that the cells gained solute through ion-uptake pathways. Based on these observations, it would appear that ion efflux ($K^+$ and $Cl^-$ channels) and influx mechanisms are both active during agonist stimulation and that the degree of cell shrinkage is in fact a reflection of the secretory activity of the cell (Foskett and Melvin, 1989). Simultaneous $[Ca^{2+}]_i$ determinations revealed that in spite of considerable heterogeneity among individual cells, subsequent changes of cell volume after the initial shrinkage in the continued presence of the agonist were tightly coupled to dynamic levels of $[Ca^{2+}]_i$, even during $[Ca^{2+}]_i$ oscillations (Figs. 8.8 and 8.9). Therefore, dynamic changes in $[Ca^{2+}]_i$ appeared to drive changes in single-cell fluid-secretory rate by regulating the activities of specific transport pathways. This dynamic regulation was reflected in

FIGURE 8.9

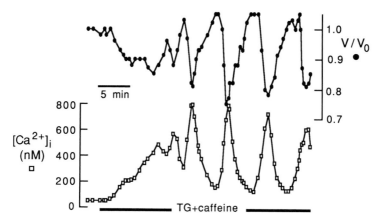

Intracellular $Ca^{2+}$ concentration ($[Ca^{2+}]_i$) (□) and cell volume (●) oscillations in a single parotid acinar cell during exposure to thapsigargin (TG; 2 uM) plus caffeine (10 mM). $[Ca^{2+}]_i$ was determined by quantitative ratio imaging of fura-2. Cell volume, normalized to the initial volume, was determined by planimetry of stored DIC optical sections. (Data from Wong and Foskett, 1991.)

changes in cell-solute content, which could be followed by imaging cell volume.

In a recent study, Ljungström and Chew (1991) also exploited DIC-fluorescence microscopy to explore the relationship between secretagogue elevation of $[Ca^{2+}]_i$ and, in that case, acid secretion in single gastric parietal cells. Acid-secretory responses could be evaluated by DIC imaging of acid-containing vacuoles that changed size during secretion, while $[Ca^{2+}]_i$ was determined by indicator-dye fluorescence.

As indicated earlier, ion-idicator dyes sensitive to $H^+$, $Na^+$, and $Cl^-$ are also available and can be used to probe the intracellular ionic milieu in single cells. The ionic basis of the agonist-induced cell volume changes in salivary acinar cells was explored by simultaneously determining intracellular $Cl^-$ concentration ($[Cl^-]_i$) and cell volume in single acinar cells (Foskett, 1990). $[Cl^-]_i$ was determined by quantitative fluorescence measurements of 6-methoxy-N-[3-sulfopropyl] quinolinium (SPQ), a water soluble fluorescent dye that interacts with halides by a collisional-quenching mechanism. SPQ is particularly useful for biological studies because of its insensitivity other than to $Cl^-$ to anions of biological relevance or to ionic strength and pH (Illsley and Verkman, 1987). The agonist-induced cell shrinkage during the initial 10–30 s of stimulation was associated with a 50% fall of $[Cl^-]_i$. When cell-volume changes were taken into account, this decrease of $[Cl^-]_i$ corresponded to a 60–65% loss of cell $Cl^-$ content. Subsequent cell-volume changes were also tightly coupled to $[Cl^-]_i$. The measured changes in $[Cl^-]_i$ predicted the measured volume changes and vice versa. These experiments provided strong support in a single cell for the idea that cell-solute content in fact determines cell volume under isosmotic conditions.

Determinations of the specific quantitative and temporal relationships between $[Ca^{2+}]_i$ and cell volume or between $[Cl^-]_i$ and cell volume required simultaneous DIC microscopy and fluorescence imaging. Having established these relationships, DIC imaging of cell volume could then be exploited in our studies not only as a direct indicator of cell-solute content but as an indirect indicator of $[Cl^-]_i$ (quantitatively) and $[Ca^{2+}]_i$ (qualitatively) as well. Simultaneous determinations of single-cell volume and fluorescence thereby provide an opportunity to correlate these parameters with other parameters indicated by other fluorescent probes. For example, the relationship between cell volume and $[Na^+]_i$, was examined using a specific fluorescent-indicator dye for $Na^+$ [benzofuran phthalate (SBFI)] (Minta and Tsien, 1989). The relationship between $[Na^+]_i$ and cell volume, determined by simultaneous fluorescence determinations of $[Na^+]_i$ and DIC imaging of cell volume in SBFI-loaded cells in cells in which $[Ca^{2+}]_i$ was induced to oscillate, established the temporal kinetics and relationships of $[Na^+]_i$, cell

FIGURE 8.10

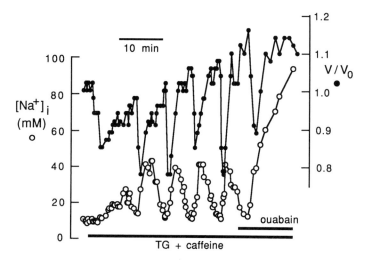

Simultaneous determinations of intracellular Na$^+$ concentration ([Na$^+$]$_i$) (○) and cell volume (●) during [Ca$^{2+}$]$_i$ oscillations induced by thapsigargin (TG; 2 uM) plus caffeine (10 mM). [Na$^+$]$_i$ was determined by quantitative ratio imaging of a Na$^+$ indicator dye, SBFI. Cell volume, normalized to the initial volume, was determined by planimetry of stored DIC images. Cell-volume oscillations indicate that [Ca$^{2+}$]$_i$ is oscillating. Furthermore, since cell volume reflects cell Cl$^-$ content (Foskett, 1990), volume oscillations imply that [Cl$^-$]$_i$ is also oscillating. Phase shift between the volume and [Na$^+$]$_i$ oscillations indicates that [Na$^+$]$_i$ oscillations lag the [Ca$^{2+}$]$_i$, volume, and [Cl$^-$]$_i$ oscillations. Ouabain (1 mM), a specific blocker of the Na$^+$-K$^+$-ATPase, abolished the [Na$^+$]$_i$ oscillations and caused [Na$^+$]$_i$ to increase to levels that saturate the dye, demonstrating that the falling phase of [Na$^+$]$_i$ during oscillations reflects the activity of the Na$^+$ pump. (From Wong and Foskett, 1991.)

volume, cell-solute content, [Ca$^{2+}$]$_i$, and [Cl$^-$]$_i$ during fluid secretion by a single cell (Fig. 8.10). (Wong and Foskett, 1991).

## 2.   Sequential Imaging with One Detector

### a. Background

The optical arrangement of the DIC-fluorescence microscope allows correlation of cell structure with fluorescence in real time. Because the two imaging modalities are not viewed with the same camera, different magnifications and geometric distortions in the two cameras and optical paths do not permit a precise overlay of the fluorescence and DIC images. Nevertheless, for many applications it will be possible to correlate fluorescence with the fine structure of the cell even though it is not possible to strictly superimpose the two images (see Foskett, 1988). Furthermore, in many instances, including the examples previously described, a correlation between fluorescence and cell fine structure is not important although correlation with more macroscopic features (e.g., cell volume, vacuole formation, cell polarity) of the cells may be. At an even somewhat more macroscopic level, this technique provides the ability to accurately identify cell types in a morphologically heterogeneous population for correlation with single-cell fluorescence. Thus, specific cell types or cells exhibiting specific behaviors can be identified and their specific fluorescence analyzed.

Nevertheless, in some instances it will be desirable to precisely correlate fluorescence distribution with cell structure. There are several solutions to this problem. Image-processing techniques can be used to correct registration, magnification, and distortion, allowing the two images to be precisely correlated. Whereas this process was impractical just a few years ago, advancements in computer-processing speed indicate that such an approach should be feasible. Use of slow-scan cooled CCD cameras, which lack geometric distortion, should eliminate the requirement for warping the images, saving considerable time in processing the images.

### b. Optical Configuration

A different solution lies in another method for performing DIC imaging in a fluorescence microscope, recently described by Spring (1990). Unlike the DIC-fluorescence microscope previously described, this technique uses the same camera to image both fluorescence and DIC (Fig. 8.11). Thus, the DIC and fluorescence images are superimposable. This ingenious technique takes advantage of the light-polarizing properties of dichroic mirrors when they are inclined at angles greater than 15°. The dichroic mirror, which reflects fluorescence excitation into the preparation, produces two separate transmission curves depending on the incident beam polarization. Therefore, the dichroic can act as the analyzer in DIC imaging if the wavelength of the transmitted light is properly selected to fall in the narrow spectral region where the $P$ plane is transmitted whereas the $S$ plane is not (Fig. 8.12). Apparently, extinction coefficients are high enough to provide satisfactory DIC. The only significant fluorescent light losses in this system are the same as those in the DIC-fluorescence microscope previously described, namely 3–6% due to the dichroic filter. Absence of an analyzer eliminates the need for a second detector. However, the gain of the intensified camera must be reduced during DIC imaging and gated on during fluorescence imaging. Thus, high-resolution DIC and low-light-level fluorescence imaging cannot be performed simultaneously. Nevertheless, if the gain of the intensifer can be placed under computer control, it should be possible to rapidly sequentially image in DIC and low-light-level fluorescence. The temporal disparity between obtaining the two images will de-

FIGURE 8.11

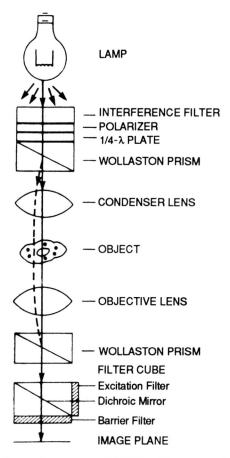

Optical configuration for combined, sequential DIC and fluorescence imaging using a single camera. See text for details. (From Spring, 1990.)

pend on the speed at which the intensifer can be gated, as well as on the lag characteristics of the camera. Use of a computer-controlled intensified video-rate CCD should allow this to be as minimal as 66 ms (Spring, 1990).

### 3. Comparison of Techniques

The significant advantage of the approach described by Spring (1990) (System II) is that a single detector results in precise correlation between the two images. The advantage of the technique described by Foskett (1988) (System I) is that DIC imaging and low-light-level fluorescence can be performed simultaneously. A correlate of this is that the intensified camera can

FIGURE 8.12

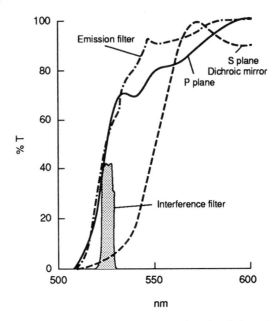

Transmission curves for a dichroic mirror (DM-530) combined with the emission filter selected for the filter cube. The interference filter curve shows the transmission properties of the narrow bandpass filter (centered at 525 nm) chosen to fit between the *S* and *P* polarization plane curves. (From Spring, 1990.)

be replaced with a photomultiplier tube, allowing photometry to be performed simultaneously with DIC imaging. Thus, micromanipulation of cells, including electrophysiology and microinjection, can be performed using high-resolution optics while the cell fluorescence is monitored with high temporal resolution (Foskett *et al.,* 1989; Fig. 8.4). The system can also perform simultaneous dual-emission fluorescence imaging by incorporation of an intensifier in front of the transmitted light camera (see Morris, Chapter 6, this volume). The disadvantages include the expense and complication of operating two cameras and the lack of perfect registration between the DIC and fluorescence images, rendering ambiguous precise correlations between the two images.

In theory, both approaches could be combined in a single microscope that would provide DIC imaging and dual-emission fluorescence in a single microscope. Thus, the camera with gatable gain of System II could be placed at the side port of System I. With another intensified camera atop the trinocular, the side-port camera would view the DIC image generated by the method of System II, as well as shorter wavelength fluorescence emission, while the top camera would view long wavelength emission.

## III. SUMMARY

Multiparameter determinations in single living cells have been possible through use of combined DIC and fluorescence imaging. The applications of this technique previously described have provided insights into the temporal orchestration of intracellular-signaling mechanisms, activities of specific ion-transport pathways, and changes in the intracellular ionic environment, which is the basis of the physiological function of a specific cell type. Although the examples described all employed ion-indicator dyes, simultaneous DIC and fluorescence imaging will be applicable whenever it is necessary or desirable to correlate fluorescence intensity from any probe with specific cell structure, especially when cell structure is a morphological correlate of a cell biological process.

## REFERENCES

Aikens, R. S., Agard, D. A., and Sedat, J. W. (1989). Solid-state imagers for microscopy. *In* "Methods in Cell Biology." (Y.-L. Wang and D. L. Taylor, eds.), Vol. 29, pp. 292–314. Academic Press, San Diego.

Allen, R. D., Allen, N. S., and Travis, J. L. (1981). Video enhanced contrast, differential interference contrast (AVEC-DIC) microscopy: A new method capable of analyzing microtubule-related motility in the reticulopodial network of Allegromia laticollaris. *Cell Motil.* 1, 291–302.

Angelides, K. J. (1989). Fluorescent analogs of toxins. *In* "Methods in Cell Biology." (Y.-L. Wang and D. L. Taylor, eds.), Vol. 29, pp. 29–58. Academic Press, San Diego.

Bookman, R. J. (1990). Temporal response characterization of video cameras. *In* "Optical Microscopy for Biology." (B. Herman and K. Jacobson, eds.), pp. 235–250. Alan R. Liss, Inc., New York.

Chen, L. B. (1989). Fluorescent labeling of mitochrondria. *In* "Methods in Cell Biology." (Y.-L. Wang and D. L. Taylor, eds.), Vol. 29, pp. 103–124. Academic Press, San Diego.

Chen, L. B., and Lee, C. (1990). Probing endoplasmic reticulum in living cells by epifluorescence and digitized video microscopy. *In* "Optical Microscopy for Biology." (B. Herman and K. Jacobson, eds.), pp. 409–418. Alan R. Liss, Inc., New York.

DeBrabander, M., Nuydens, R., Geuens, G., Moeremans, M., and De May, J. (1986). The use of submicroscopic gold particles combined with video contrast enhancement as a simple molecular probe for the living cell. *Cell Motil. Cytoskel.* 6, 105–113.

Dunn, K. W., and Maxfield, F. R. (1990). Use of fluorescence microscopy in the study of receptor-mediated endocytosis. *In* "Non-Invasive Techniques in Cell Biology." (J. K. Foskett and S. Grinstein, eds.), pp. 153–176. Alan R. Liss, Inc., New York.

Edidin, M. (1989). Fluorescent labeling of cell surfaces. *In* "Methods in Cell Biology." (Y.-L. Wang and D. L. Taylor, eds.), Vol. 29, pp. 87–102. Academic Press, San Diego.

Foskett, J. K. (1988). Simultaneous Nomarski and fluorescence imaging during video microscopy of cells. *Am. J. Physiol. 255,* C566–C571.

Foskett, J. K. (1990). [Ca²⁺]ᵢ modulation of Cl⁻ content controls cell volume in single salivary acinar cells during fluid secretion. *Am. J. Physiol. 259,* C998–C1004.

Foskett, J. K., and Melvin, J. E. (1989). Activation of salivary secretion: Coupling of cell volume and [Ca²⁺]ᵢ in single cells. *Science 244,* 1582–1585.

Foskett, J. K., and Wong, D. (1991). Free cytoplasmic Ca²⁺ concentration oscillations in thapsigargin-treated parotid acinar cells are caffeine- and ryanodine-sensitive. *J. Biol. Chem. 266,* 14535–14538.

Foskett, J. K., Gunter-Smith, P. J., Melvin, J. E., and Turner, R. J. (1989). Physiological localization of an agonist-sensitive pool of Ca²⁺ in parotid acinar cells. *Proc. Natl. Acad. Sci. U.S.A. 86,* 167–171.

Gingell, D., and Todd, I. (1979). Interference reflection microscopy. A quantitative theory for image interpretation and its application to cell-substratum separation measurement. *Biophys. J. 26,* 507–526.

Haughland, R. P., and Minta, A. (1990). Design and application of indicator dyes. *In* "Non-Invasive Techniques in Cell Biology." (J. K. Foskett and S. Grinstein, eds.), pp. 1–20. Alan R. Liss, Inc., New York.

Illsley, N. P., and Verkman, A. S. (1987). Membrane chloride transport measured using a chloride-sensitive fluorescent probe. *Biochemistry 26,* 1215–1219.

Inoue, S. (1981). Video image processing greatly enhances contrast, quality and speed in polarization-based microscopy. *J. Cell Biol. 89,* 346–356.

Kapitza, H. G., McGregor, G., and Jacobson, K. A. (1985). Direct measurement of lateral transport in membranes by using time-resolved spatial photometry. *Proc. Natl. Acad. Sci. U.S.A. 82,* 4122–4126.

Ljüngstrom, M., and Chew, C. S. (1991). Calcium oscillations and morphological transformations in single cultured gastric parietal cells. *Am. J. Physiol. 260,* C67–C78.

Martin, J. P., and Logsdon, N. (1987). Oxygen radicals are generated by dye-mediated intracellular photooxidations: A role for superoxide in photodynamic effects. *Arch. Biochem. Biophys. 256,* 39–49.

Maxfield, F. R. (1989). Fluorescent analogs of peptides and hormones. *In* "Methods in Cell Biology." (Y.-L. Wang and D. L. Tayor, eds.), Vol. 29, pp. 13–28. Academic Press, San Diego.

Maxfield, F. R., and Dunn, K. W. (1990). Studies of endocytosis using image intensification fluorescence microscopy and digital image analysis. *In* "Optical Microscopy for Biology." (B. Herman and K. Jacobson, eds.), pp. 357–372. Alan R. Liss, Inc., New York.

Minta, A., and Tsien, R. Y. (1989). Fluorescent indicators for cytosolic sodium. *J. Biol. Chem. 264,* 19449–19457.

Pagano, R. E. (1989). A fluorescent derivative of ceramide: physical properties and use in studying the golgi apparatus of animal cells. *In* "Methods in Cell Biology." (Y.-L. Wang and D. L. Taylor, eds.), Vol. 29, pp. 125–136. Academic Press, San Diego.

Ryan, T. A., Myers, J., Holowka, D., Baird, B., and Webb, W. W. (1988). Molecular crowding on the cell surface. *Science 239,* 61–64.

Sanger, J. W., and Sanger, J. M. (1990). Dynamics of the cytoskeleton and membranous organelles in living cells. *In* "Optical Microscopy for Biology." (B. Herman and K. Jacobson, eds.), pp. 437–448. Alan R. Liss, Inc., New York.

Spring, K. R. (1990). Quantitative imaging at low light levels: Differential interference contrast and fluorescence microscopy without significant light loss. *In* "Optical Microscopy for Biology." (B. Herman and K. Jacobson, eds.), pp. 513–522. Alan R. Liss, Inc., New York.

Spring, K. R., and Lowy, R. J. (1989). Characteristics of low light level television cameras. *In* "Methods in Cell Biology." (Y.-L. Wang and D. L. Taylor eds.), Vol. 29, pp. 270–291. Academic Press, San Diego.

Spring, K. R., and Smith, P. D. (1987). Illumination and detection systems for quantitative fluorescence microscopy. *J. Microsc.* 147, 265–278.

Swanson, J. (1989). Fluorescent labeling of endocytic compartments. *In* "Methods in Cell Biology." (Y.-L. Wang and D. L. Taylor, eds.), Vol. 29, pp. 137–152. Academic Press, San Diego.

Terasaki, M. (1989). Fluorescent labeling of endoplasmic reticulum. *In* "Methods in Cell Biology." (Y.-L. Wang and D. L. Taylor, eds.), Vol. 29, pp. 125–136. Academic Press, San Diego.

Thomas, J., and Webb, W. W. (1990). Fluorescence photobleaching recovery: A probe for membrane dynamics. *In* "Non-Invasive Techniques in Cell Biology." (J. K. Foskett and S. Grinstein, eds.), pp. 129–152. Alan R. Liss, Inc., New York.

Tsay, T.-T., Inman, R., Wray, B. E., Herman, B., and Jacobson, K. (1990). Characterization of low light level video cameras for fluorescence microscopy. *In* "Optical Microscopy for Biology." (B. Herman and K. Jacobson, eds.), pp. 219–234. Alan R. Liss, Inc., New York.

Tsien, R. Y. (1989). Fluorescent indicators of ion concentrations. *In* "Methods in Cell Biology." (D. L. Taylor, and Y.-L. Wang, eds.), Vol. 30, pp. 127–156. Academic Press, San Diego.

Wang, Y.-L. (1989). Fluorescent analog cytochemistry: Tracing functional protein components in living cells. *In* "Methods in Cell Biology." (Y.-L. Wang and D. L. Taylor, eds.), Vol. 29, pp. 1–12. Academic Press, San Diego.

Wang, Y.-L., and Sanders, M. C. (1990). Analysis of cytoskeletal structures by the microinjection of fluorescent probes. *In* "Non-Invasive Techniques in Cell Biology." (J. K. Foskett and S. Grinstein, eds.), pp. 177–212. Alan R. Liss, Inc., New York.

Wong, M. M. Y., and Foskett, J. K. (1991). Oscillations of cytosolic sodium during calcium oscillations in exocrine acinar cells. *Science* 254, 1014–1016.

Yonuschot, G. (1991). Early increase in intracellular calcium during photodynamic permeabilization. *Free Radical Biol. Med.* 11, 307–317.

# 9
# FLUORESCENCE MICROSCOPIC IMAGING OF MEMBRANE DOMAINS

**William Rodgers and Michael Glaser**
Department of Biochemistry
University of Illinois at Urbana-Champaign

I. INTRODUCTION
II. FLUORESCENCE IMAGING OF
MEMBRANE DOMAINS
  A. Calcium- and Protein-Induced
    Domains
  B. Erythrocyte Membrane
    Domains

III. DISCUSSION
  A. Membrane Domains in Cell
    Biology
  B. Summary
ACKNOWLEDGMENTS
REFERENCES

## I. INTRODUCTION

The Fluid Mosaic Model, as proposed 20 years ago, described biological membranes as a mixture of lipid and protein components in a phospholipid bilayer (Singer and Nicolson, 1972). In this model, all membrane components are assumed to have free lateral motion, with no long-range order. Today, the basic tenants of the Fluid Mosaic Model are generally accepted. However, certain refinements have been made since the inception of the model as more data on membrane structure has been collected. These refinements include the asymmetric distribution of phospholipids across the membrane (Op den Kamp, 1979; Devaux, 1991), the role of the cytoskeleton in anchoring transmembrane proteins (Bennett, 1985; Jacobson *et al.*, 1987), and the presence of membrane do-

Optical Microscopy: Emerging Methods and Applications

mains (Pessin and Glaser, 1980; Wolf *et al.*, 1981; Shukla and Hanahan, 1982; Jain, 1983; Thompson and Tillack, 1985; Metcalf *et al.*, 1986; Yechial and Edidin, 1987; Haverstick and Glaser, 1987; Finzi *et al.*, 1989; Davenport *et al.*, 1989; Tocanne *et al.*, 1989; Wolf *et al.*, 1990; Edidin and Strynowski, 1991).

Membrane domains can be described as localized regions within a membrane where the composition differs from that of the bulk membrane. In some cases, domains are morphologically distinct, such as the apical and basolateral domains of polarized epithelial cells, with each region having a different lipid and protein composition (Simons and van Meer, 1988), and here tight junctions appear to be important for maintaining these distinct regions (van Meer and Simons, 1986).

Isolated areas of lipids and proteins could result in a restriction in their lateral diffusion or mobility. Fluorescence recovery after photobleaching has been used to study membrane domains based on the different diffusion rates of labeled membrane lipids and proteins. Fractional recoveries that are less than 100% have been interpreted in terms of populations of fluorophore that are unable to diffuse freely, or are entrapped in membrane domains. Evidence for membrane domains has been obtained for several different biological membranes, including human fibroblast cells and soybean protoplasts (Yechial and Edidin, 1987; Edidin and Strynowski, 1991; Metcalf *et. al.*, 1986).

Measurements, such as fluorescence recovery after photobleaching, however, only indirectly characterize membrane domains. More complete descriptions of membrane domains can be obtained by direct visualization of the domains, in conjunction with experiments designed to characterize the domains in terms of such parameters as composition and stability with time and temperature.

With the growth in microelectronic technology, it is now possible to construct high-quality fluorescence-imaging systems that provide image enhancement and quantitation for fluorescence microscopy (Arndt-Jovin *et al.*, 1985). These fluorescence digital-imaging systems provide a means to describe the lateral distribution of fluorescently labeled membrane components, including measuring the relative composition and enrichment of membrane domains. This article reviews previous studies of membrane lateral topology that have been conducted in both lipid vesicle and labeled biological membranes (Haverstick and Glaser, 1987; Haverstick and Glaser, 1989; Rodgers and Glaser, 1991), and present new data providing a more detailed description of erythrocyte membrane domains using fluorescence digital-imaging microscopy.

## II. FLUORESCENCE IMAGING OF MEMBRANE DOMAINS

### A. Calcium- and Protein-Induced Domains

A variety of agents have been shown to cause the formation of phase separations within membranes. Two common examples of this are divalent cations and basic proteins, most notably $Ca^{2+}$ (Duzgunes and Papahadjopoulos, 1983; Hoekstra, 1982; Ito *et al.*, 1975; Van Dijck *et al.*, 1978; Silvius, 1990) and cytochrome *c* (Gorrissen *et al.*, 1986; de Kruijff and Cullis, 1980; Killian and de Kruijff, 1986; Birrell and Griffith, 1976; Mannella *et al.*, 1987; Mustonen *et al.*, 1987). The interactions of $Ca^{2+}$ and basic proteins with membranes are electrostatic in nature, where the positively charged cations or proteins are interacting selectively with negatively charged lipids in the membrane, such as phosphatidic acid (PA) or phosphatidylserine (PS) (Duzgunes and Papahadjopoulos, 1983).

Earlier work that was done using fluorescence digital-imaging microscopy of large unilamellar vesicles labeled with NBD-PA {1-acyl-2-[N-(4-nitrobenzo-2-oxa-1,3-diazole)-aminocaproyl] phosphatidic acid} showed the formation of $Ca^{2+}$-induced PA-enriched domains (Haverstick and Glaser, 1987). The domains can form quickly, with domains forming within seconds after addition of $Ca^{2+}$ to the vesicles. In these experiments, a $Ca^{2+}$ concentration of 2 mM was routinely used, but concentrations as low as 10 $\mu$M have been found to cause sequestering of NBD-PA into domains. Similar effects are also seen when the liposomes are labeled with NBD-PS {1-acyl-2[N-(4-nitrobenzo-2-oxa-1,3-diazole)-amino caproyl]phosphatidyl serine} rather than NBD-PA.

Erythrocyte membranes labeled with NBD-PA allowed visualization of the formation of $Ca^{2+}$-induced domains. Erythrocytes are well suited for these studies since the cells are relatively large (6–8 $\mu$m in diameter) and contain only the external plasma membrane, thus avoiding the problem of internalization of the label into intracellular membranes. In both intact erythrocytes and erythrocyte ghosts, $Ca^{2+}$-induced domans of NBD-PA were visible after the addition of 10 mM $Ca^{2+}$ (Glaser and Haverstick, 1987). $Ca^{2+}$-induced domains are also apparent in erythrocyte membranes labeled with NBD-PS and incubated with 5 mM $Ca^{2+}$ (Rodgers and Glaser, 1991). Large liposomes composed of extracted erythrocyte lipids showed that the $Ca^{2+}$-induced domains could occur independently of membrane proteins (Haverstick and Glaser, 1987).

Fluorescence imaging of large liposomes was used to study the effects of proteins by following the formation of PA-enriched domains by cytochrome *c* (Haverstick and Glaser, 1989). When 10 $\mu$M cytochrome *c* was added to liposomes labeled with NBD-PA, PA was sequestered into domains. The binding of cytochrome *c* to the PA could be visualized due to the heme absorption of transmitted light by cytochrome *c* at the same location as the NBD-PA domain in the membrane. The size and enrichment of the domains that formed depended on the cytochrome $c$ concentration and the lipid composition of the vesicles.

Overall, these experiments with liposomes demonstrate lipid redistributions that occur when $Ca^{2+}$ or cytochrome $c$ are added to acidic phospholipid containing membrane systems. A critical question is whether domains occur intrinsically in biological membranes.

## B. Erythrocyte Membrane Domains

### 1. Characterization of Lipid Domains

Intrinsic membrane lipid domains have been shown to occur in erythrocytes and erythrocyte ghosts labeled with fluorescent phospholipids (Rodgers and Glaser, 1991). The lipid domains were also evident in intact cells labeled with the fluorescent compound chlorpromazine. The relative distribution of the different phospholipids in the membrane was characterized by double-labeling ghosts with Dansyl-PC {1-acyl-2-[4-[N-[5-(dimethylamino)naptha-lene-1-sulfonyl]amino]butanyl]phosphatidylcholine} and either NBD-PC {1-acyl-2-[N-(4-nitrobenzo-2-oxa-1,3-diazole)aminocaproyl]phosphatidyl-choline}, NBD-PE {1-acyl-2-[N-(4-nitrobenzo-2-oxa-1,3-diazole)-amino-caproyl]phosphatidylethanolamine}, or NBD-PS. Two separate companion images, generated by the selective excitation and emission of the individual fluorescent phospholipids, were collected from each double-labeled membrane. The relative distributions of the two fluorescent phospholipids were compared following normalization of each image to an average gray value of 100 (Rodgers and Glaser, 1991). Differences in the images correspond to different phospholipid distributions in the same membrane. Quantitation of the lipid distributions in terms of absolute concentration in the domains was not possible due to problems inherent in fluorescent microscopy, including photobleaching and out-of-plane fluorescence. Photobleaching was limited by imaging the membranes for only a short period of time (5 s). No anti-bleaching agents or fixatives were used to avoid possible perturbation of the membrane with these reagents. Consequently, measurements of the lipid distributions were made in a relative manner as described using double-labeled ghosts.

Because the domains are heterogeneous with each ghost showing a distinct pattern of domains, it is important to describe the range of domains that are observed in a population of ghosts. The differences in the two-dimensional distributions of radiance values in the companion images generated by the Dansyl-PC and NBD-labeled phospholipids were determined by subtracting the companion images and making a histogram of the distribution of the radiance values after the subtraction (Rodgers and Glaser,

1991). The standard deviation of the histogram was used as an indication of the differences in the two original images. Increasing dissimilarity between the companion images results in a greater range of values in the difference image and therefore, a broader histogram with larger standard deviations. Overall, this method of quantitation allows a convenient means of comparing all of the corresponding pixels in each set of images of the individual double-labeled membranes. As a result, a distribution of the resulting standard deviation values can be made and used to define the differences in the lipid distributions in the membranes for a large population of ghosts.

Based on the distribution of standard deviation values for the different fluorescent phospholipid double-labeling experiments, it was found that the separate phospholipid species are distributed differently in the membrane. In the assessment of the different lipid distributions, the separate experiments were compared to the distribution of standard deviation values for the ghosts double labeled with NBD-PC and Dansyl-PC, where a similar distribution of the two probes occurred in all areas of the ghost due to their identical head groups. The ghosts that showed the most significant differences in their companion images were the ones double labeled with NBD-PS and Dansyl-PC to which $Ca^{2+}$ was added. This arises from the interaction of $Ca^{2+}$ with the PS in the membrane, as described earlier. The experiments using ghosts double labeled with Dansyl-PC and either NBD-PE or NBD-PS had intermediate differences in their phospholipid distributions (Rodgers and Glaser, 1991).

The distribution of standard deviation values reflects different enrichment of the fluorophores in the domains, or different sized domains. To further characterize the amount of lipid enrichment in the domains, the largest radiance value in each area of enrichment was measured. The images also showed areas in the membrane where the fluorophore was less intense than the average radiance of the image, thus representing areas of depletion. These areas were measured using the smallest radiance within the area of depletion. For these measurements, the domains were visually defined using the pseudocolor assignment of pixel radiance. Since most of the pixels in the membrane images are green (centered at a gray value equal to 128) (Fig. 9.3), the domains were visually assigned to the colors above and below green on the pseudocolor scale. This corresponds to radiance values above 155 for the areas of enrichment and less than 100 for the areas of depletion. Also, because the relative concentration of the different phospholipids was being measured, the pixel in the corresponding position in the companion image was measured as well. Thus, each domain in the membrane generated at least two measured radiance values. When both fluorophores showed enrichment in the same domain, but the maximum radiance values of the companion images did not coincide, the domain was measured twice.

Three-dimensional distributions were constructed of the domain radiance values to characterize the relative degree of enrichment or depletion of the different phospholipids. The resulting surface plots showing the distribution of domain radiance values for the separate double–labeling experiments using Dansyl-PC and either NBD-PC, NBD-PE, or NBD-PS are shown in Fig. 9.1. In these plots, each axis represents the radiance values of the pixels in either the Dansyl-PC image ($x$-axis) or the NBD-phospholipid image ($y$-axis). The height at each point ($z$-axis) represents the frequency of the domains with the corresponding Dansyl-PC and NBD-phospholipid pixel radiance values. Both the domains associated with areas of enrichment and the domains associated with the areas of depletion are represented in these plots. An imaginary diagonal passing from the origin across the surface of the $x$-$y$ plane represents the domains where the NBD and the Dansyl [5-(dimethylamino)napthalene-1-sulfonyl] radiance values were equal. Conversely, areas of the plots distributed away from the diagonal represent either NBD or Dansyl domain radiance values that were greater than the corresponding pixel in the companion image. The empty space in the middle of the plots around which the distributions occur represents radiance values that were not associated with maximum or minimum values in the domains.

Importantly, the domain radiance values of the ghosts double labeled with NBD-PC and Dansyl-PC are distributed principally along the $x$-$y$ diagonal (Fig. 9.1A). The equal NBD-PC and Dansyl-PC enrichment and depletion in the domains further substantiate the assumption that the fluorophores are distributed similarly in the membrane because of their identical head group, with no apparent differences due to the separate fluorophores. In the other double-labeling experiments, the values become distributed away from the diagonal (Fig. 9.1B–D), and represent domains with different amounts of enrichment or depletion of the phospholipid species. Some of the domains show enrichment of one fluorophore and only moderate or no enrichment of the second fluorophore. This observation also applies for the domains of depleted regions. The heterogeneity in the distribution of radiance values is most marked for the ghosts that were double labeled with NBD-PS and Dansyl-PC with 10.0 mM $Ca^{2+}$ (Fig. 9.1D).

The surface plots for these experiments also show that the Dansyl-PC enrichment in the domains remains within the same range of radiance values for each of the double-labeling experiments. This demonstrates that the Dansyl-PC distribution in the membrane is occurring independently of the second-labeled phospholipid. The distributions of domain radiance values also demonstrate a similar range of concentrations of NBD-PS (without $Ca^{2+}$), NBD-PE, NBD-PC, and Dansyl-PC in the domains. The NBD-PS in the experiments with $Ca^{2+}$, however, has substantially greater enrichment than what occurs with any of the other lipids (Fig. 9.1D).

Most of the distributions in the surface plots of Fig. 9.1 also show sym-

FIGURE 9.1

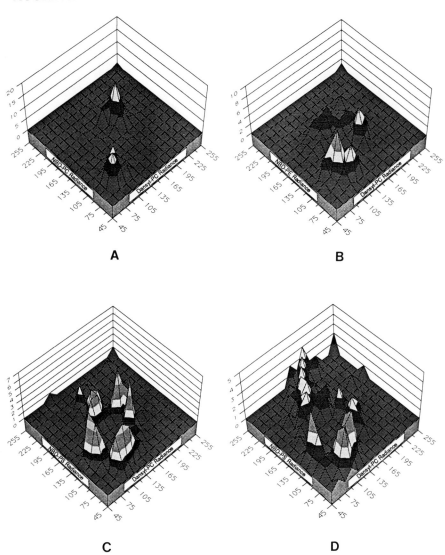

Surface plots showing the distribution of domain radiance values in four separate double-labeling experiments using erythrocyte ghosts. The ghosts were double labeled with Dansyl-PC and either NBD-PC (A), NBD-PE (B), or NBD-PS (C and D). 10 mM $Ca^{2+}$ was added to the ghosts represented in (D). The sample sizes were 88, 96, 89, and 114 for (A), (B), (C), and (D), respectively. The distributions were normalized to a sample size of 114. All images were normalized to a mean gray-level value of 100 ± 4.

FIGURE 9.2

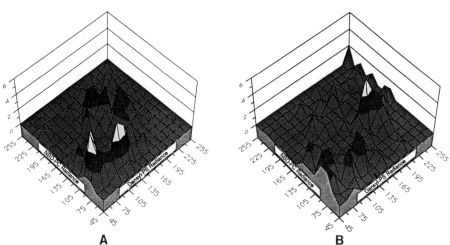

Surface plots showing the distribution of domain radiance values in two separate double-labeling experiments using erythrocyte ghosts. The ghosts were double labeled with Dansyl-PS and NBD-PC. 10 mM $Ca^{2+}$ was added to the ghosts represented in (B). The sample sizes were 127 and 122 for (A) and (B), respectively. All images were normalized to a mean gray-level value of 100 ± 4.

metry relative to the $x$-$y$ diagonal, and this demonstrates an approximately equal number of domains enriched or depleted in either fluorophore of the double-labeling experiments. However, for the ghosts double labeled with NBD-PS and Dansyl-PC with $Ca^{2+}$, the distribution of domain radiance values is not symmetrical relative to the diagonal, with most of the values associated with regions of greater NBD-PS enrichment (Fig. 9.1D). Both the increase in NBD-PS enrichment and the larger number of domains with greater NBD-PS enrichment reflect the specific binding of the $Ca^{2+}$ to the PS in the membrane.

In another experiment, the Dansyl and NBD fluorophores were reversed on PC (phosphatidylcholine) and PS. Figure 9.2 shows the surface plots for the domain radiance values of ghosts that were double labeled with NBD-PC and Dansyl-PS {1-acyl-2-[4-[N-[5-(dimethylamino)napthalene-1-sulfonyl]amino]butanyl]phosphatidylserine}, both without and with $Ca^{2+}$. The changes in the domain radiance value distribution upon the addition of $Ca^{2+}$ are analogous to what was seen in the ghosts double labeled with NBD-PS and Dansyl-PC, with most of the domains now being more enriched in Dansyl-PS. The overall increase in Dansyl-PS radiance values in the areas of enrichment, with little change in the NBD-PC domain radiance values, again reflects the Ca²⁺ binding to PS. Importantly, the range of Dansyl-PS radiance values in both the ghosts without and with Ca²⁺ (Fig. 9.2) is similar to that of the NBD-PS in the analogous experiments using ghosts double labeled with NBD-PS and Dansyl-PC (Figs. 9.1C and 9.1D). The experiment shown in Fig. 9.2 reiterates the observation that the domains are a function of the head group of the phospholipid, and they are not significantly influenced by the fluorophore.

The affect of Ca²⁺ on the Dansyl-PS distribution was also reflected in the distribution of standard deviation values for the images (Fig. 9.3). For example, in the absence of Ca²⁺, the distribution of standard deviation values was centered at 9.0. When Ca²⁺ was added to the sample, the distribution changed so that most of the standard deviation values were now greater than 11.0. The standard deviation value distributions in Fig. 9.3 are also similar to the distributions in the experiments using ghosts that were double labeled with NBD-PS and Dansyl-PC (Rodgers and Glaser, 1991). Both double-labeling experiments with Ca²⁺ also showed the same sample het-

FIGURE 9.3

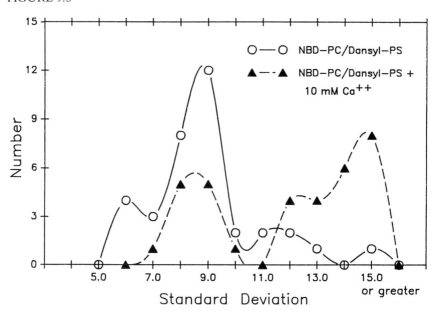

The distribution of standard deviation values for ghosts that were double labeled with NBD-PC and Dansyl-PS. The difference images were generated by subtracting the Dansyl-PS image from the NBD-PC image. The sample sizes were 27 and 35 for the ghosts with and without 10.0 mM Ca²⁺, respectively. The distributions were normalized to a sample size of 35. All images were normalized to a mean gray-level value of 100 ± 4.

erogeneity in the form of a trimodal distribution. This was previously attributed to a combination of ghosts with little $Ca^{2+}$-induced patching with standard deviation values centered at 9.0, ghosts with intermediate amounts of $Ca^{2+}$-induced patching with standard deviation between 11.0–13.0, and ghosts with even greater $Ca^{2+}$-induced patching with standard deviation values of 14.0 and greater (Rodgers and Glaser, 1991).

The lipid distributions from the separate double-labeling experiments were also characterized by measuring the maximum range of radiance values for the two fluorophores in each ghost. These measurements were made by taking the ratio of the largest radiance value over the smallest radiance value within each image of the double-labeled membranes, using the same ghosts that generated the distributions in Figs. 9.1 and 9.2. Overall, the ranges of radiance values averaged between two and three for all of the labeled phospholipids in the absence of $Ca^{2+}$. In the ghosts with $Ca^{2+}$, however, the average range increased to between four and five for the PS-labeled images but remained between two and three for the respective PC-labeled images. The largest range of radiance values for the entire population of ghosts from each experiment was between four and eight for all of the ghosts without $Ca^{2+}$. With $Ca^{2+}$, the largest value for the NBD-PS and Dansyl-PS images increased to between eleven and twelve, while the largest value from the PC-labeled images remained between four and eight. Overall, these ratios are valuable since they indicate the range of concentrations the phospholipids can undergo in the membrane. Using PS as an example, these values show that any membrane component moving from an area of PS depletion to PS enrichment could experience up to a twelvefold change in the PS concentration if $Ca^{2+}$ is present.

The fluorescence labeling of the ghosts was conducted by monomer transfer from donor vesicles (Pagano *et. al.*, 1983). In the double-labeling experiments, the labeling with the two fluorescent phospholipids was done independently of each other since the donor vesicles contained only one fluorescent phospholipid species. For example, in the ghosts represented in Fig. 9.1, the membranes were first labeled with Dansyl-PC by incubation with the Dansyl-PC containing donor vesicles, followed by washing and incubation with the second set of donor vesicles containing the NBD-labeled phospholipid.

The domains do not represent a separate pool of lipids generated by fusion of the donor vesicles to the ghost. This point is demonstrated in Fig. 9.4A with the image of a NBD-PC-labeled ghost that was incubated with DOPC (dioleoylphosphatidylcholine) vesicles containing 1% head group-labeled Dansyl-PE [N-(5-dimethylaminonaphthalene-1-sulfonyl)-dipalmitoyl-L-α-phosphatidylethanolamine], which does not undergo monomer transfer (Struck and Pagano, 1980). If an image of the ghost was obtained

FIGURE 9.4

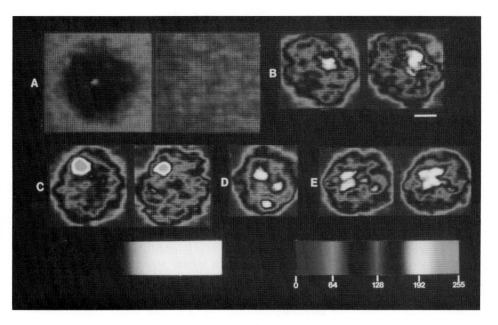

Shown are the images of fluorescently labeled ghosts from five separate experiments using fluorescence-imaging microscopy. (A) The unprocessed NBD image (*left*) and Dansyl image (*right*) of a ghost labeled with NBD-PC and incubated with head-group-labeled Dansyl-PE containing donor vesicles. (B) The initial (*left*) and second image collected 5 min later (*right*) of a Dansyl-PC-labeled ghost. (C) The images of a Dansyl-PC-labeled ghost collected at 4°C (*left*) and 37°C (*right*). (D) The fluorescein image of a ghost with fluorescein-labeled membrane proteins. The fluorescein labeling was done by adding 0.01 mg of fluorescein iodoacetamide from a stock solution in phosphate-buffered saline to 1.0 ml of packed ghosts. The ghosts were stirred in the dark at room temperature for 30–45 min, then washed several times using phosphate-buffered saline. (E) The fluorescein (*left*) and Dansyl-PC (*right*) images of a double-labeled ghost with fluorescein-labeled membrane proteins. The Dansyl image in (A, *right*) was enhanced by multiplying the pixel radiance values by two. The images in (B) through (E) were normalized to a mean gray-level value of 100 ± 4. The white bar equals 2.5 μm. The pseudo-color scheme (*bottom, right*) applied to the images is shown with its equivalent gray values (*bottom, left*). The gray values range from 0 (blue) to 255 (red). The two gray bands in the pseudocolor scale result from the color mixing by the television monitor from which the images are photographed. (*See Plate 7 for a color version of this figure.*)

by Dansyl excitation, it would be from either donor vesicles bound to the surface of the membrane or by labeling through fusion of the head group-labeled Dansyl-PE lipid vesicles to the ghost. However, only the background is visible when the ghost is illuminated with the Dansyl excitation wavelengths (*right*), thus showing no contribution to the image of the ghost by fused or bound donor vesicles. The images in Fig. 9.4A were not processed by background subtraction to avoid the possible removal of a Dansyl image by this operation. The radiance values in the original Dansyl image were multiplied by two to produce the image shown in Fig. 9.4A (*right*). This further illustrates an absence of any image of the ghost by Dansyl fluorescence. Similar observations concerning the absence of any image from the head group-labeled Dansyl-PE vesicles were made with unlabeled whole cells and in both ghosts and whole cells incubated with head group-labeled Dansyl-PE containing 5 mole % DOPS (dioleoylphosphatidylserine) to mimic the NBD-PS donor vesicles. A lipid domain in the NBD-PC image of the ghost is clearly evident in the absence of any image processing. It is also of interest that when $Ca^{2+}$ is added to the double-labeled ghosts, the fluorescent phospholipid enrichment is occurring by diffusion of the fluorophores into the domains from other points within the membrane, rather than having anything to do with the labeling procedure.

Labeled ghosts were examined at different planes of focus to measure what effect this would have on the domain quantitation. This experiment was conducted by first focusing to give the clearest image of the entire ghost, and this was how the images were routinely collected in all of the experiments. After the initial image was collected, two more images were collected where the new planes of focus were 1.0–1.5 $\mu$m above and below the initial plane of focus. In the latter two images, the image of the ghost was usually fuzzy, with only the areas of enrichment being distinguishable. Overall, the sharpest image of the ghost (i.e., the initial image) usually gave the largest value for the domain radiance values. In an image from a plane of focus above or below that used for collecting the initial image, the difference in the enrichment of the domain was always less than 20% of the domain radiance value. The values were calculated using the largest radiance in each domain divided by the average radiance of each image in the separate planes of focus. These small differences in domain radiance values demonstrate that the plane of focus including the entire ghost was adequate for these measurements.

The effect of different concentrations of labeled-phospholipids on the domains was examined by double-labeling ghosts with NBD-PC and Dansyl-PC but using a smaller quantity of NBD-PC donor vesicles for labeling. In Fig. 9.5, the surface plot is shown of the domain radiance values of ghosts that were double labeled with NBD-PC and Dansyl-PC, but where the amount of NBD-PC was approximately 30–40% of that in the ghosts shown in Fig. 9.1A. The concentrations were estimated by measuring the average radiances of the prenormalized images of the ghosts. If the concentration of the fluorophore were to have an affect on the phospholipid enrichment in the domains, then it would be apparent in the distribution of

FIGURE 9.5

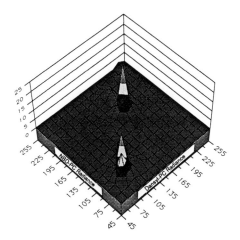

Surface plot showing the distribution of domain radiance values for ghosts that were double labeled with NBD-PC and Dansyl-PC but with a concentration of NBD-PC that was between 30–40% of that in the ghosts in Fig. 9.1A. The sample size was 94. All images were normalized to a mean gray-level value of 100 ± 4.

the NBD-PC domain radiance values relative to that of the Dansyl-PC radiance values. The surface plots in Fig. 9.1A and Fig. 9.5, however, are very similar with both showing that the domain radiance values are distributed almost exclusively along the $x$-$y$ diagonal within the same range of radiance values. The distributions of standard deviation values for the difference images were also similar, with both distributions being relatively narrow and having their peaks centered at 7.0 (Rodgers and Glaser, 1991, and Fig. 9.6). Overall, the similarity between the NBD-PC and Dansyl-PC distributions with the different NBD-PC concentrations illustrates that there is no dependence of the domain enrichment on the concentration of the labeled phospholipids in these experiments.

2. Time and Temperature Studies on the Lipid Domains

The effect of time on the lipid distributions was investigated using ghosts labeled with Dansyl-PC (Figs. 9.4B and 9.6). An initial image was collected, and a second image of the same ghost at the same plane of focus was collected 5 min later. An example of this time-course experiment is shown in Fig. 9.4B where the ghost shows only small changes in the size, shape, and enrichment of the domains.

The changes in the lipid distribution after 5 min were measured for a population of ghosts by subtracting the second image of each ghost 5 min

FIGURE 9.6

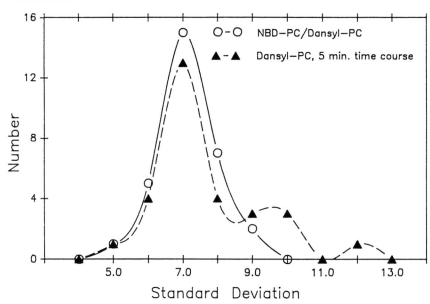

Standard Deviation

The distribution of standard deviation values for ghosts labeled with Dansyl-PC. The differences images were generated by subtracting the second image from the first image. The two images were collected either 5 min or 30 s apart. The distribution of standard deviation values for images collected 30 s apart is taken from Rodgers and Glaser (1991). The sample size for the images collected 5 min apart was 21, and it was normalized to the sample size of the images collected 30 s apart, which was 30. All images were normalized to a mean gray-level value of $100 \pm 4$.

after the initial image. The resulting distribution of standard deviation values from this experiment is shown in Fig. 9.6. The standard deviation value for the ghost shown in Fig. 9.4B is 8.2. Also shown in Fig. 9.6 is the distribution of difference image standard deviation values from the ghosts double labeled with NBD-PC and Dansyl-PC (Rodgers and Glaser, 1991) where both images were collected within 30 s of each other. The similarity of the two standard deviation value distributions in Fig. 9.6 demonstrates that the lipid distributions for the population of ghosts showed little change after 5 min. A small number of ghosts, however, do show large changes after 5 min as indicated by the shoulder in the distribution of standard deviation values, with values greater than 9.0 (Fig. 9.6).

Temperature studies also were conducted on ghosts labeled with Dansyl-PC. Two images of a single ghost are shown in Fig. 9.4C, where the image on the left was collected at 2°C, and the image on the right was collected at 37°C. There were no major changes in the domains after warming the ghost to 37°C. This shows that the domains do not represent gel-phase lipids from a phase transition below 37°C. Some of the ghosts had domains where the lipid enrichment increased with increasing temperature. Domains were still present in the ghosts after warming to as high as 50°C. Also, the time required to warm the sample was about 20–30 min, showing that the domains are stable for longer periods of time relative to what was used for the experiment shown in Fig. 9.4B.

## 3. Distribution of Membrane Proteins

An important consideration in the study of erythrocyte membrane domains is what elements are responsible for forming and stabilizing the lipid domains. One possibility is that proteins are responsible for the heterogeneous distribution of lipids as shown with cytochrome $c$ (Haverstick and Glaser, 1989) and D-$\beta$-hydroxybutyrate dehydrogenase (BDH) (Wang *et al.*, 1988). There are a number of examples of specific interactions between membrane proteins and lipids (Devaux and Seigneuret, 1985; Marsh, 1990). Large domains on the order of microns in size, as observed in the erythrocyte membrane, may also reflect a combination of protein oligomerization and different lipid interactions. In this case, the cell cytoskeleton could also serve as a source of anchoring of the membrane proteins (Jacobson *et al.*, 1987; Bennet, 1985; Edidin and Stroynowski, 1991) and give rise to domains.

The effect of protein aggregation on lipid domain enrichment, the absence of domains in erythrocyte lipid vesicles (Rodgers and Glaser, 1991), and the absence of a phase transition (Fig. 9.4C) implicate membrane proteins in having an important role in forming and stabilizing the lipid domains. To further investigate the lateral distribution of the erythrocyte membrane proteins and their relationship to the lipid domains, fluorescence digital-imaging experiments were conducted on ghost membranes labeled with fluorescein iodoacetamide. Labeling with fluorescein iodoacetamide results in fluorescence labeling of many of the erythrocyte membrane proteins, including both integral membrane proteins, such as Band 3, and the peripheral membrane proteins spectrin, ankyrin and Band 4.1. Lipid extraction (Rodgers and Glaser, 1991) of the fluorescein-labeled ghosts showed all fluorescence to be in the protein disk at the aqueous–organic interface. Thin-layer chromatography of the extracted lipids confirmed that the erythrocyte lipids were not labeled with fluorescein. These experiments, therefore, provide a description of the overall protein distribution in the membrane.

Figure 9.4D shows the fluorescence image of a ghost labeled with fluorescein iodoacetamide. From this image, it is clear that the labeled membrane proteins are distributed into areas of enrichment in a manner analogous to the membrane lipids. In addition, this shows that domains are present in the absence of labeled phospholipids and are not an artifact created by the addition of the labeled phospholipids.

The relationship between the areas of protein enrichment and lipid enrichment in the domains was investigated by fluorescein labeling of the membrane proteins, followed by labeling the ghosts with Dansyl-PC (Fig. 9.4E). In this image, the areas of labeled-protein enrichment largely coincide with the areas of PC enrichment. This correlation in lipid and protein enrichment, thus, further suggests that the erythrocyte membrane proteins have a functional role in forming and maintaining the lipid domains.

The correlation in the areas of protein and lipid enrichment or depletion for a population of ghosts double labeled as described for Fig. 9.4E is shown in the surface plot of the domain radiance values (Fig. 9.7). Importantly, the distribution shows that many of the domain radiance values are distributed along the $x$-$y$ diagonal, thus demonstrating that the domains with Dansyl-PC enrichment frequently have protein enrichment as well. Interestingly, Fig. 9.7 also shows that there are many domains with quite different fluorescein and Dansyl-PC radiance values since there are regions of the distribution that are spread closer to the $x$ or $y$ axis. This heterogeneity may be indicative of interactions between the labeled components and other unlabeled lipids or proteins in the membrane. Overall, Fig. 9.7 is consistent with an important role that proteins have in forming and stabilizing the lipid domains.

FIGURE 9.7

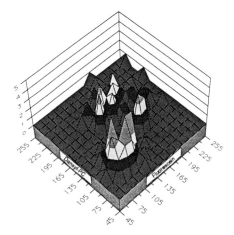

Surface plot showing the distribution of domain radiance values for ghosts that were double labeled with Dansyl-PC and a fluorescein label for proteins. The sample size was 97. All images were normalized to a mean gray-level value of $100 \pm 4$.

## III. DISCUSSION

### A. Membrane Domains in Cell Biology

Membrane domains may represent an important functional as well as structural element in the membrane. For example, domains with different lipid and protein compositions could function as regulatory units by providing an environment favorable for enzyme activation. This type of regulation is conceivable since many membrane-associated proteins, either integral or peripheral, require specific lipids for full activity (Sandermann, 1978; Carruthers and Melchior, 1986). In some cases, the proteins themselves may be responsible for forming the domains, enriching their environment with the lipids that are required for activity. In other cases, the enzyme could partition into or out of a domain depending on the conditions (Haverstick and Glaser, 1989).

One well-characterized example of a membrane protein that requires a specific phospholipid for activity is the mitochondrial membrane protein BDH, which requires PC for enzyme activity (Fleischer et al., 1974). BDH has also been shown to cause a long-range reorganization of membrane phospholipids using resonance energy transfer between tryptophan residues of BDH and Dansyl-labeled phospholipids (Wang et al., 1988). Furthermore, in primary rat hepatocyte cultures where the phospholipid content was manipulated, there was a lag period between the time of the change in the phospholipid content of the membranes and the measured change of the BDH activity (Clancy et al., 1983). Both the energy transfer and the cell culture experiments are consistent with a model of membrane structure where BDH plays a role in organizing the membrane lipids into domains, and the domain in turn influences the activity of the enzyme. The availability of lipids, such as PC, for enrichment into domains may depend on a number of factors, such as competitive interactions with other membrane enzymes.

The important role of lipids in cell signaling and the specific interactions of protein kinase C with membrane lipids (Newton and Koshland, 1990; Bell and Burns, 1991) suggest a role for membrane domains in these events as well. Enrichment of PA into domains by protein kinase C has been shown using self-quenching of NBD-PA in vesicles to which the enzyme was added (Bazzi and Nelsestuen, 1991). Other acidic phospholipids were also found to reorganize in a manner analogous to PA. The enrichment of acidic phospholipids by protein kinase C is thus analogous to what was described with BDH where a membrane enzyme facilitates its own activation by creating a favorable environment in terms of domain formation. The membrane domains also could be necessary for cell signaling.

Membrane domains may also have important functions in other cellular processes, such as protein trafficking, targeting, and endocytosis. For example, the protein oligomerization that is necessary for protein sorting (Hurtley and Helenius, 1989) may represent membrane domains that function to both separate the resident endoplasmic reticulum proteins from the proteins being sorted and also halt the movement of misfolded proteins since they frequently do not oligomerize. Membrane lipids, which are targeted to specific membranes in the cell, may also be incorporated into membrane domains. In this manner, membrane proteins and lipids would be transported as a domain unit to the target membrane (Simons and van Meer, 1988). Further information about membrane domains should provide insight into their functional role of biological membranes. Experiments of this nature will be possible using fluorescence digital-imaging techniques.

## B. Summary

This article describes the use of fluorescence digital-imaging microscopy to characterize domains in both liposomal and biological membranes. This method was originally found to be valuable in characterizing the reorganization of lipids in large unilamellar vesicles upon the addition of either $Ca^{2+}$ or cytochrome $c$. In both cases, large lipid domains were directly visualized. Fluorescence-imaging microscopy was also found to be very useful in characterizing *intrinsic* domains in the erythrocyte membrane. These domains arise by interactions between the proteins and lipids within the membrane (Rodgers and Glaser, 1991).

Further characterization of the erythrocyte membrane domains using fluorescence digital-imaging microscopy demonstrated that the domains are heterogenous with different lipid compositions. $Ca^{2+}$ causes the lipid distribution in the membrane to change, with a marked increase in the PS enrichment in the domains. The lipid distributions that were observed depended on the phospholipid head group and were independent of the nature of the fluorophore.

Fluorescence digital-imaging microscopy of ghosts also showed that the domains were relatively stable with both time (approximately 5 min) and temperature (2–50°C). This demonstrates the absence of a thermotropic phase transition in the ghosts. Labeling the ghosts with fluorescein showed that the proteins were distributed into domains and that interactions between the membrane proteins and lipids were responsible for the formation of the lipid domains.

Experiments using specific labeling of the membrane proteins would provide an opportunity to define specific interactions occurring between membrane proteins and lipids more precisely. The role of these interactions in the membrane lateral organization could then be characterized using fluorescence digital-imaging microscopy.

ACKNOWLEDGMENTS

The authors wish to thank Dr. Gregorio Weber for his continued support and encouragement. W.R. was supported by a Kodak Fellowship.

REFERENCES

Arndt-Jovin, D. J., Robert-Nicoud, M., Kaufman, S. J., and Jovin, T. M. (1985). Fluorescence digital imaging microscopy in cell biology. *Science 230*, 247–256.

Bazzi, M. D., and Nelsestuen, G. L. (1990). Protein kinase C interaction with calcium: A phospholipid-dependent process. *Biochemistry 29*, 7624–7630.

Bazzi, M. D., and Nelsestuen, G. L. (1991). Highly sequential binding of protein kinase C and related proteins to membranes. *Biochemistry 309*, 7970–7977.

Bell, R. M., and Burns, D. J. (1991). Lipid activation of protein kinase c. *J. Biol. Chem. 266*, 4661–4664.

Bennett, V. (1985). The membrane skeleton of human erythrocytes and its implications for more complex cells. *Annu. Rev. Biochem. 54*, 273–304.

Birrell, G. B., and Griffith, O. H. (1976). Cytochrome *c* induced lateral phase separation in a diphosphatidylglycerol-steroid spin-label model membrane. *Biochemistry 15*, 2925–2929.

Carruthers, A., and Melchoir, D. L. (1986). How bilayer lipids affect membrane protein activity. *TIBS 11*, 331–335.

Clancy, R. M., McPherson, L. H., and Glaser, M. (1983). Effect of changes in the phospholipid composition on the enzymatic activity of D-$\beta$-hydroxybutyrate dehydrogenase in rat hepatocytes. *Biochemistry 22*, 2358–2364.

de Kruijff, B., and Cullis, P. R. (1980). Cytochrome *c* specifically induces nonbilayer structures in cardiolipin-containing model membranes. *Biochim. Biophys. Acta 602*, 477–490.

Devaux, P. F. 1991. Static and dynamic lipid asymmetry in cell membranes. *Biochemistry 30*, 1163–1173.

Devaux, P. F., and Seigneuret, M. (1985). Specificity of lipid–protein interactions as determined by spectroscopic techniques. *Biochim. Biophys. Acta 822*, 63–125.

Davenport, L., Knutson, J. R., and Brand, L. (1989). Fluorescence studies of membrane dynamics and heterogeneity. *In* "Subcellular Biochemistry." (J. R. Harris and A. H. Etenadi, eds.), Vol. 14, pp. 145–188. Plenum Press, New York.

Duzgunes, N., and Papahadjopoulous, D. (1983). Ionotropic effects on phospholipid membranes: Calcium/magnesium specificity in binding, fluidity, and fusion. *In* "Membrane Fluidity in Biology." (R. C. Aloia, ed.), Vol. 2, pp. 187–215. Academic Press, New York.

Edidin, M., and Stroynowski, I. (1991). Differences between the lateral organization of conventional and inositol phospholipid-anchored membrane proteins. A further definition of micrometer scale membrane domains. *J. Cell Biol. 112*, 1143–1150.

Finzi, L., Bustamante, C., Garab, G., and Juang, C.-B. (1989). Direct observation of large chiral domains in chloroplast thylakoid membranes by differential polarization microscopy. *Proc. Natl. Acad. Sci. U.S.A. 86,* 8748–8752.

Fleischer, S., Bock, H. G., and Gazzotti, P. (1974). Studies of phospholipid–protein interaction in D-$\beta$-hydroxybutyrate dehydrogenase, a lecithin requiring enzyme. *In* "Membrane Proteins in Transport and Phosphorylation." (G. F. Azzone, M. E. Klingenberg, E. Quagliariello, and N. Siliprandi, eds.), pp. 125–136, North-Holland Publishing Co., Amsterdam.

Gorrissen, H., Marsh, D., Rietveld, A., and de Kruijff, B. (1986). Apocytochrome *c* binding to negatively charged lipid dispersions studied by spin-label electron spin resonance. *Biochemistry 25,* 2904–2910.

Haverstick, D. M., and Glaser, M. (1987). Visualization of Ca$^{2+}$-induced phospholipid domains. *Proc. Natl. Acad. Sci. U.S.A. 84,* 4475–4479.

Haverstick, D. M., and Glaser, M. (1989). Influence of proteins on the reorganization of phospholipid bilayers into large domains. *Biophys. J. 55,* 677–682.

Hoekstra, D. (1982). Fluorescence method for measuring the kinetics of Ca$^{2+}$-induced phase separations in phosphatidylserine-containing lipid vesicles. *Biochemistry 21,* 2833–2840.

Hurtley, S. M., and Helenius, A. (1989). Protein oligomerization in the endoplasmic reticulum. *Ann. Rev. Cell Biol. 5,* 277–307.

Ito, T., Ohnishi, S., Ishinaga, M., and Kito, M. (1975). Synthesis of a new phosphatidylserine spin-label and calcium-induced lateral phase separation in phosphatidylserine-phosphatidtylcholine membranes. *Biochemistry 14,* 3064–3069.

Jacobson, K., Ishihara, A., and Inman, R. (1987). Lateral diffusion of proteins in membranes. *Annu. Rev. Physiol. 49,* 163–175.

Jain, M. K. (1983). Nonrandom lateral organization in bilayers and biomembranes. *In* "Membrane Fluidity in Biology." (R. C. Aloia, ed.), Vol. 1, pp. 1–37. Academic Press, New York.

Killian, J. A., and de Kruijff, B. (1986). The influence of proteins and peptides on the phase properties of lipids. *Chem. Phys. Lipids 40,* 259–284.

Mannella, C. A., Ribeiro, A. J., and Frank, J. (1987). Cytochrome *c* binds to lipid domains in arrays of mitochondrial outer membrane channels. *Biophys. J. 51,* 221–226.

Marsh, D. (1990). Lipid–protein interactions in membranes. *FEBS Letters 268,* 371–375.

Metcalf, T. N., III, Wang, J. L., and Schindler, M. (1986). Lateral diffusion of phospholipids in the plasma membrane of soybean protoplasts: Evidence for membrane lipid domains. *Proc. Natl. Acad. Sci. U.S.A. 83,* 95–99.

Mustonen, P., Virtanen, J. A., Somerharju, P. J., and Kinnunen, P. K. J. (1987). Binding of cytochrome *c* to liposomes as revealed by the quenching of fluorescence from pyrene-labeled phospholipids. *Biochemistry 26,* 2991–2997.

Newton, A. C., and Koshland, D. E. (1990). Phosphatidylserine affects specifity of protein kinase *c* substrate phosphorylation and autophosphorylation. *Biochemistry 29,* 6656–6661.

Op den Kamp, J. A. F. (1979). Lipid asymmetry in membranes. *Ann. Rev. Biochem. 48,* 47–71.

Pagano, R. E., Longmuir, K. J., and Martin, O. C. (1983). Intracellular translocation and metabolism of a fluorescent phosphatidic acid anologue in cultural fibroblasts. *J. Biol. Chem. 258,* 2023–2040.

Pessin, J. E., and Glaser, M. (1980). Budding of Rous Sarcoma virus and Vesicular Stomatitis virus from localized lipid regions in the plasma membrane of chicken embryo fibroblasts. *J. Biol. Chem. 255,* 9044–9050.

Rodgers, W., and Glaser, M. (1991). Characterization of lipid domains in erythrocyte membranes. *Proc. Natl. Acad. Sci. U.S.A. 88,* 1364–1368.

Sandermann, H. (1978). Regulation of membrane enzymes by lipids. *Biochim. Biophys. Acta 515,* 209–237.

Shukla, S. D., and Hanahan, D. J. (1982). Identification of domains of phosphatidylcholine in human erythrocyte plasma membranes. *J. Biol. Chem. 257,* 2908–2911.

Silvius, J. R. (1990). Calcium-induced lipid phase separations and interactions of phosphatidylcholine/anionic phospholipid vesicles. Fluorescence studies using carbazole-labeled and brominated phospholipids. *Biochemistry 29,* 2930–2938.

Simons, M., and van Meer, G. (1988). Lipid sorting in epithelial cells. *Biochemistry 27,* 6197–6202.

Singer, S. J., and Nicolson, G. L. (1972). The fluid mosaic model of the structure of cell membranes. *Science 175,* 720–731.

Struck, D. K., and Pagano, R. E. (1980). Insertion of fluorescent phospholipids into the plasma membrane of a mammalian cell. *J. Biol. Chem. 255,* 5402–5410.

Thompson, T. E., and Tillack, T. W. (1985). Organization of glycosphingolipids in bilayers and plasma membranes of cells. *Ann. Rev. Biophys. Biophys. Chem. 14,* 361–386.

Tocanne. J.-F., Dupou-Cezanne, L., Lopez, A., and Tournier, J.-F. (1989). Lipid lateral diffusion and membrane organization. *FEBS Letters 257,* 10–16.

Treistman, S. N., Moynihan, M. M., and Wolf, D. E. (1987). Influence of alcohols, temperature, and region on the mobility of lipids in neuronal membrane. *Biochim. Biophys. Acta 898,* 109–120.

Van Dijck, P. W. M., de Kruijff, B., Verkleij, A. J., Van Deenen, L., and De Gier, L. M. J. (1978). Comparative studies on the effects of pH and $Ca^{2+}$ on bilayers of various negatively charged phospholipids and their mixtures with phosphatidylcholine. *Biochim. Biophys. Acta 512,* 84–96.

Van Meer, G., and Simons, K. (1986). The function of tight junctions in maintaining differences in lipid composition between the apical and basolateral cell surface domains of MDCK cells. *Embo. J. 5,* 1455–1464.

Wang, S., Martin, E., Cimino, J., Omann, G., and Glaser, M. (1988). Distribution of phospholipids around gramicidin and D-$\beta$-hydroxybutyrate dehydrogenase as measured by resonance energy transfer. *Biochemistry 27,* 2033–2039.

Wolf, D. E., Kinsey, W., Lennarz, W., and Edidin, M. (1981). Changes in the organization of the sea urchin egg plasma membrane upon fertilization: Indications from the lateral diffusion rates of lipid-soluble fluorescent dyes. *Dev. Biol. 81,* 133–138.

Wolf, D. E., Maynard, C. A., McKinnon, C. A., and Melchior, D. L. (1990). Lipid domains in the ram sperm plasma membrane demonstrated by differential scanning calorimetry. *Proc. Natl. Acad. Sci. U.S.A. 87,* 6893–6896.

Yechiel, E., and Edidin, M. (1987). Micrometer-scale domains in fibroblast plasma membranes. *J. Cell Biol. 105,* 755–760.

# 10
# MEASUREMENT OF MEMBRANE GLYCOPROTEIN MOVEMENT BY SINGLE-PARTICLE TRACKING

**Michael P. Sheetz**
Department of Cell Biology
Duke University Medical School

**Elliot L. Elson**
Department of Biochemistry and Molecular Biophysics
Division of Biology and Biomedical Sciences
Washington University Medical School

I. INTRODUCTION
II. SINGLE-PARTICLE TRACKING
III. COMPARISON OF SPT WITH FPR
  A. Video-Microscopy Techniques
IV. CELLULAR REQUIREMENTS AND LIMITATIONS
  A. Digital Video Analysis

V. STATISTICAL ANALYSIS
VI. APPLICATIONS
  A. Factors Contributing to $D$ for Membrane Proteins
  B. Barriers to Lateral Diffusion
VII. SUMMARY
REFERENCES

## I. INTRODUCTION

The single-particle tracking (SPT) method of diffusion measurements uses recent advances in video technology to track with high-precision individual microscopic particles attached to relevant molecules. The resulting trajectories provide useful information about the mechanisms and forces that drive and constrain the particles' motion. Tracking individual particles has the great advantage of detecting behaviors shown by only a small fraction of the population at any given time but which may be important. These behaviors could not be seen in measurements of a large population of particles [e.g., in fluorescence photobleaching recovery (FPR) measurements]. SPT also allows analysis of the lateral mobility of particles confined to small domains. This SPT approach has proved especially useful in characterizing the motions of small particles on cell surfaces attached to a small number of membrane glycoproteins. We will describe the strengths and limitations of SPT and the related nanovid technology in terms of the parameters that are reliably measured and questions reliably answered.

It is evident that a cell membrane is not simply a homogenous solution of proteins embedded in a fluid lipid lamina. An increasing body of evidence suggests that the lateral distribution of membrane glycoproteins is directed by interactions with the cytoskeleton, which can both drive their systematic motions and restrain their passive diffusion. In this paper, we outline the basic experimental details of the SPT technique, which can provide routine measurements of the active movements and the barriers to diffusion responsible for the heterogeneous distribution of membrane proteins.

To account for the nonuniform distribution and hindered diffusion of membrane proteins, the fluid mosaic model of membranes has been modified in recent years by invoking the existence of specialized lipid or protein domains. The structure and function of some of these, for example, the neuromuscular endplate, have been extensively described (Bloch and Pumplin, 1988). Others have been characterized only indirectly (Yechiel and Edidin, 1987). From the analysis of these systems, the domain structure of plasma membrane components has a spatial scale in the range of $\mu$m. The SPT method makes it possible to observe particle motion over such small distances. Although the resolution limit of the light microscope does not permit structures separated by less than 0.2 $\mu$m to be distinguished, it is possible to measure movements of particles on the nm scale.

Systematic transport of particles forward or backward over the cell surface has been related to cellular locomotion (e.g., Bray and White 1988; Kucik et al., 1991). Under some conditions, such a small fraction of the particles participates in this behavior that it could not be detected by conventional methods, such as FPR. Using SPT, however, this systematic transport can be readily analyzed since motions of individual particles are observed without interference from the other particles in the population (e.g., Kucik et al., 1989). This approach has helped to distinguish between the dynamic behaviors of particles in the domains defined by the lamellar ectoplasm and the endoplasmic region of the cell that surrounds the nucleus.

## II. SINGLE-PARTICLE TRACKING

In SPT measurements, the positions of a particle determined at a sequence of times provide its trajectory for the period of observation. Frequently, simple inspection of the particle trajectory yields useful qualitative information about the mechanism of motion. In addition, the trajectory can be conveniently analyzed quantitatively by calculating the mean squared displacement of the particle, $\rho(\delta\tau)$, as a function of time (Sheetz et al., 1989). If $r_{\tau+\delta\tau}$ and $r_{\tau}$ are the positions of the particle at times $\tau + \delta\tau$ and $\tau$, then $\rho(\delta\tau)$ is simply the average of the values $(r_{\tau+\delta\tau} - r_{\tau})^2$ at various values of $\tau$ along the trajectory and as a function of $\delta\tau$. For simple diffusion in a plane, $\rho(\delta\tau) = 4D\delta\tau$. For systematic transport at constant velocity $v$, $\rho(\delta\tau) = (v\delta\tau)^2$. If a particle is both diffusing and systematically drifting, $\rho(\delta\tau) = 4D\delta\tau + (v\delta\tau)^2$. Therefore, $D$ and $v$ can be determined from a plot of $\rho(\delta\tau)$ versus $\delta\tau$ Moreover, constraint of the diffusion to a small domain will cause the plot of $\rho(\delta\tau)$ to curve downward (Sheetz et al., 1989; Qian et al., 1991).

The major advantages of the SPT are the possibilities of observing and characterizing motions, which appear in only a small fraction of the particles and of characterizing motions of particles confined to very small domains. The major disadvantage of the technique is that it is difficult to determine or control the number of ligands that bind the particle to the membrane glycoproteins. One way of minimizing the polyvalency of the particles is to allow the protein molecules that act as ligands for membrane glycoproteins to compete for the binding sites on the particles with protein molecules that do not bind the membrane glycoproteins. Conditions can be arranged so that on average a particle is likely to have only one or no ligand molecules bound to its surface (Lee et al., 1991). Even without this precaution, however, it has been observed that there was only a small difference between the diffusion coefficients measured by SPT with antibody-coated gold particles and those measured by FPR with fluorescence-labeled Fab fragments for the same antibody (Edidin et al., 1991). The comparison of multivalent particle diffusion with monovalent Fab diffusion is an important control for the potential effects of crosslinking membrane proteins by the particles.

## III. COMPARISON OF SPT WITH FPR

For the past 15 years, FPR has most often been used to characterize the lateral diffusion of cell surface components. FPR and related methods measure the time course of fluorescence emitted from a population of molecules within an observation region typically a micrometer or more in diameter. Mainly because of interference from background fluorescence, one typically requires approximately 1000 fluorophores in this region to obtain a minimally adequate signal. Then, even if there were five fluorophores per membrane protein, the minimum number of proteins detectable would be 200. With so large a population it would be very difficult, if not impossible, to detect behavior displayed by only a few members of the population. For example, it would be difficult or impossible using FPR to detect the systematic transport of only one or two molecules within a population of 200 other comparably labeled molecules. Thus, it is difficult to observe intermittent, relatively infrequent active events over the background of simple diffusive behavior with FPR. Furthermore, the minimum distance of travel detected by FPR is the range of 200–1000 nm and that by SPT in the range of 10–100 nm. Therefore, SPT can observe motion over a spatial range at least an order of magnitude smaller than FPR can. Thus, SPT may be especially useful for studying behavior within small membrane domains.

SPT and FPR measurements of diffusion also differ in their statistics. Diffusion coefficients must be determined either by measuring the motion of a single particle over a long period of time or of a population over a shorter time. Therefore, FPR provides a statistically valid measurement of a diffusion coefficient (within the limits of the measurement accuracy) from a single fluorescence recovery transient (lasting a time period on the order of $w^2/4D$, where $w$ is the radius of the bleached spot). In contrast, the motion of a single particle must be continued for a much longer time interval for a correspondingly accurate estimate of the diffusion coefficient (Qian et al., 1991).

## A. Video-Microscopy Techniques

There are several methods for imaging particles in the video microscope that have distinct advantages for certain situations. The first application of bright-field microscopy to the imaging of gold particles, called "nanovid microscopy" (DeBrabander et al., 1988; Geerts et al., 1987), has the advantage of enhancing the contrast of the particles over other structures in the cells. Using this approach, it is possible to follow small particles (10–40 nm) moving both on the surface and within cells. To visualize the smaller particles requires signal averaging with a corresponding decrease in time resolution. For example, to observe a 10-nm particle requires averaging for about 1 s to obtain a sufficient signal over the noise to permit analysis. Although the bright-field image of a cell has very low contrast, there are often detectable cellular features that could be confused for gold particles.

We have generally perferred differential interference contrast (DIC) optics because the narrow depth of focus (less than 0.5 $\mu$m) allows a precise determination of the vertical location of the image within the cell. On the other hand, the greater depth of focus of the bright-field microscope allows tracking of particles over a greater vertical distance than does the DIC optics used in SPT. Furthermore, the contrast of the gold particles is greater than with bright-field microsocopy, and latex particles as small as 80 nm are readily imaged by video-enhanced DIC but not by bright field.

A third alternative is to use fluorescence microscopy to observe fluorescent particles (Gross and Webb, 1988). Fluorescence microscopy is especially useful for detecting particles on a cell that has many small organelles (e.g., intracellular vesicles, which could be mistaken for gold or latex particles under bright-field or DIC optics) (Jay and Elson 1992).

## IV. CELLULAR REQUIREMENTS AND LIMITATIONS

Most of the studies of particle diffusion have been performed on highly spread cells. The peripheral lamellar regions of these cells are frequently free of cytoplasmic organelles and so provide a clean featureless background on which the gold particles are readily identified. In thicker cells or tissues, the narrow depth of focus of video-enhanced DIC occasionally allows the observation of particle movements on the upper surface. In many cases, however, the presence of microvilli and other surface features, such as membrane folds, make it virtually impossible to follow two-dimensional movements of membrane glycoproteins. In these instances, tracking measurements can still be carried out using the greater contrast and depth of focus provided by fluorescense microscopy.

### A. Digital Video Analysis

To carry out an SPT measurement it is necessary to acquire, process, and store the video images from which the particle trajectories are determined. The underlying principles and the methods for acquiring video-enhanced DIC images have been well described (Allen, 1985). To minimize photon noise these images are obtained using high-light intensity. This necessitates subtraction of a background image from each frame to restore contrast. The video-enhanced DIC images can be stored on a S-VHS tape and redigitized as previously described (Gelles et al., 1988).

For simple diffusion in an unlimited space neither the spatial nor the temporal resolution of the measurement is crucial for the accuracy of the result. It is possible simply to continue the measurement for a long time so that the particle moves a great enough distance to assure both adequate spatial and temporal resolution. If the particle is confined to a small area, high spatial and temporal resolution are necessary to allow determination of the diffusion coefficient within the available domain. Frame-by-frame analysis yields approximately 33 ms time resolution for each position measurement. This is complicated by the interlacing of fields to form a frame that combines signal gathered over a 50 ms period in any given region of the image. By using noninterlaced and other nonstandard formats, it is possible to increase the temporal resolution within a limited portion of a field by 7–10-fold (4.6–3.3 ms). A customized camera for this purpose has been constructed from which the images can be stored in standard video (RS170) format using an Imaging Technologies VSI (variable scan interface) board (Gelles *et al.*, unpublished results). High-speed cameras are appearing on the market, which may be adaptable to SPT measurements.

The particles can be located simply by calculating the centroid of the appropriately thresholded particle-intensity distribution in the image for either nanovid or SPT. For DIC images, however, that have both a light and a dark portion, about half of the information is discarded in the centroid calculation. Greater precision and reliability can be obtained using a cross correlation algorithm to compute the region of an experimental image that best fits the known image of the particle.

Actual measurements of position for 40-nm gold particles are typically accurate to within ±5 nm. The locations of higher contrast 150-nm latex spheres can be determined to within ±1.5 nm (Gelles *et al.*, 1988) at a frequency of 30 Hz. There is clearly some positional averaging because of the rapid diffusion rate for some glycoproteins and most lipids. (A protein with a diffusion coefficient of $3 \times 10^{-9}$ cm$^2$ s$^{-1}$ would move on the average approximately 100 nm during a frame time). Therefore, with the temporal resolution now available there is no advantage to increased spatial resolution of fast-moving particles because the range of their diffusional motion during a frame period is greater than the currently available measurement precision. Similarly, calculations for the frequency of vibration of the gold particle with respect to other protein ligands to which it is attached show that averaging of position requires less than milliseconds. Hence, only the position of the particle averaged over the range of vibration would be seen. In addition to potential effects on the precision of measurement, these diffusional movements will decrease the apparent contrast of the particle and will make automated analysis of position difficult.

## V. STATISTICAL ANALYSIS

A detailed analysis of the theoretical basis of SPT, a comparison of SPT and FPR, and a discussion of errors in SPT measurements have recently been published (Qian *et al.*, 1991). The analysis of errors must take into account that diffusion is a stochastic process. Thus, even with infinitely precise position measurements, the diffusion coefficients and flow velocities determined from plots of the mean square displacement will have variances that depend on the number of data points acquired and that can be derived theoretically. The variance due to this stochastic character of diffusion diminishes as the number of position measurements increases. Although the detailed analysis of this subject is complex (Qian *et al.*, 1991), the result can be described in simple terms for free diffusion. The experimental measurements of the positions of a particle are obtained at intervals of $\Delta T$. Over an interval of time $n\Delta T$ the mean square displacement of the particle will be $\rho(n\Delta T) = 4Dn\Delta T$. To obtain this value of $\rho(n\Delta T)$, $N$ measurements of position are included in the calculation of the mean square displacement. Then the expected variance of this estimate $\rho(n\Delta T)$ is $\{(2n^2 + 1)/3n(N - n + 1)\}(\rho(n\Delta T))^2$. To obtain an accurate estimate of $\rho(n\Delta T)$, it is necessary that $N >> n$. Then a useful figure of merit, $\Phi$, for the accuracy of a measurement is provided by the square root of the ratio of the variance of $\rho(n\Delta T)$ divided by the square of $\rho(n\Delta T)$ [i.e., $\Phi \sim (2n/3N)^{1/2}$]. The accuracy of the estimate increases as $\Phi$ decreases. Hence, the accuracy increases as $N^{1/2}$ and diminishes as $n$ increases. The latter is due to the fact that as $n$ increases, the number of statisically independent samples of displacement within $N\Delta T$ decreases.

## VI. APPLICATIONS

We have used SPT to observe the reversible transitions of particles on cell surfaces between states of random diffusion and systematic transport (Sheetz *et al.*, 1989; Kucik *et al.*, 1990). These relatively rare events could not have been observed and the rates of systematic transport could not have been obtained by FPR measurements of a large population of molecules. We observed not only systematic rearward (centripetal) transport, which had often been seen before, but also a previously unreported and much rarer forward (centrifugal) transport. Furthermore, we were able to demonstrate characteristic differences between both the rates of transport forward (approximately 1 $\mu$m/s) and rearward (approximately 0.2 $\mu$m/s) and between the relative probabilities of diffusion and forward and rearward systematic transport (Kucik *et al.*, 1989). The distinct advantages for the SPT method, which made these studies possible, include the abilities both to observe relatively rare dynamic events even in the presence of a large number of particles that do not exhibit that behavior and to analyze quantitatively the diffusion rates and transport velocities of particles in both the common and rare dynamic states. It is notable that attachment of the glycoproteins to the cytoskeletal motor, which drives their systematic transport, reduces their diffusion coefficient by about 100-fold. The rapid kinetics of the reversible attachment to the cytoskeletal motor are also seen from the fact that there is no long period of reduced diffusion coefficient before the particles experience systematic transport.

### A. Factors Contributing to $D$ for Membrane Proteins

The basis for the inhibition of lateral movement of the same membrane glycoproteins in biological membranes as opposed to lipid bilayers has been of concern for some time. Recent studies using expression of altered membrane glycoproteins have shed some light on the different determinants for specific proteins that contribute to the inhibition of diffusion (Zhang et al., 1991). The surprising finding is that the contributions of the different regions of the molecule are simply additive in inhibiting diffusion. The major contributions are at or near the membrane surface whereas our evidence from the movement of gold particles indicates that the water phase contributes very little to the inhibition of diffusion. This is consistent with the concept that the water phase is of a much lower viscosity than the membrane surface. The multivalency of the gold particles could contribute to the apparent diffusion coefficient, particularly in the case of lipids where the diffusion coefficients are very high (i.e., the resistance to diffusion is low). In the case of lipids, the cross-linking of lipids can contribute to the drag and decrease the apparent diffusion coefficient. In the case of membrane glycoproteins where there is already a significant drag due to the membrane viscosity, the contribution of the cross-linking and the water phase appear less. Cross-linking of glycoprotein receptors can activate cellular responses (e.g. the secretory response of basophilic leukemia cells in response to IgE) (Metzger et al., 1986). Thus, it is important for many studies to use monovalent ligands for tagging although in the absence of activation of cellular processes we have not seen a significant decrease in the diffusion coefficient with particle size or extent of cross-linking.

## B. Barriers to Lateral Diffusion

There has been evidence for a long time that barriers to the lateral diffusion of membrane glycoproteins contribute significantly to the reduced diffusion coefficients of proteins in membranes. FPR measurements with different bleaching areas (Yechiel and Edidin, 1987) and SPT measurements at long times (Sheetz et al., 1989) all have indicated that there are barriers on the scale of $0.2-2.0\ \mu$m that restrict the lateral movement of membrane glycoproteins. Using SPT labeling of specific glycoproteins and the laser tweezers to manipulate the particles, it has been possible to measure directly the distance between membrane barriers (Edidin et al., 1991). From FPR studies it was evident that the diffusing fraction of an MHC-1 (major histocompatibility complex 1) protein with a transmembrane tail decreased as the area of the bleached spot increased whereas a lipid-linked MHC-1 in the same cell showed no change. This was interpreted to suggest that there were barriers to membrane diffusion. With gold particles attached to the same two MHC-1s, it was found recently that the particles could be dragged three times further when they were bound to the lipid-linked MHC-1 than when bound to the transmembrane anchored protein (see Table 10.1). A surprising feature of these measurements of the barrier-free pathlength (BFP) in

TABLE 10.1
BARRIER-FREE PATHLENGTHS OF LASER-TRAPPED MHC MOLECULES[a]

|        | H-2D[b]          | Qa2             |
|--------|------------------|-----------------|
| 23°C   | 0.6 ± 0.1 (68)   | 1.7 ± 0.2 (53)  |
| 34°C   | 3.5 ± 0.6 (29)   | 8.5 ± 0.8 (50)  |

[a] Values are arithmetic mean ± standard error of the mean (sample size).

the membrane is that a 10° rise in temperature caused a 5-fold increase in the BFP. Thus, at physiological temperatures the transmembrane protein could be moved over a distance of on average 3.0 $\mu$m before it would encounter a barrier that would pull it out of the trap. This seems like a long distance between the barriers and there is a much higher density of membrane associated skeletal proteins in most cells. The dramatic increase in BFP with temperature has been interpreted as a result of an increased dynamics in the membrane skeleton. Much more work is needed in this area for a more complete understanding of the basis for the BFP measurements.

## VII. SUMMARY

SPT provides another important tool for understanding the structure of biological membranes. It is particularly useful in the analysis of the behavior of individual or small numbers of glycoproteins; events that would be missed in the FPR or similar types of measurements. The advent of computer analysis of movements makes it possible to analyze a large number of particle movements to define the average behavior. New technologies, such as the laser tweezers, enable even more specific studies of membrane properties and provide many new dimensions for the future.

REFERENCES

Allen, R. D. (1985). New observations on cell architecture and dynamics by video-enhanced contrast optical microscopy. *Ann. Rev. Biophys. Biophys. Chem. 14*, 265–290.

Bloch, R. J., and Pumplin, D. W. (1988). Molecular events in synaptogenesis: Nerve-muscle adhesion and postsynaptic differentiation. *Am. J. Physiol. 254*, C345–C364.

Bray, D., and White, J. G. (1988). Cortical flow in animal cells. *Science 239*, 883–887.

DeBrabander, M., Nuydens, R., Geerts, H., and Hopkins, C. R. (1988). Dynamic behavior of the transferrin receptor followed in living epidermoid carcinoma (A431) cells with nanovid microscopy. *Cell Motil. Cytoskel. 9*, 30–47.

Edidin, M., Kuo, S., and Sheetz, M. P. (1991). Lateral movements of membrane glycoproteins restricted by dynamic cytoplasmic barriers. *Science 254*, 1379–1382.

Geerts, H., de Brabander, M., Nuydens, R., Geuens, S., Moeremans, M., deMey, J., and Hollenbeck, P. (1987). Nanovid tracking: A new automatic method for the study of mobility in living cells based on colloidal gold and video microscopy. *Biophy. J. 52*, 775–782.

Gelles, J., Schnapp, B. J., and Sheetz, M. P. (1988). Tracking kinesin-driven movements with nanometer-scale precision. *Nature 331*, 450–453.

Gross, D. J., and Webb, W. W. (1988). Cell surface clustering and mobility of the liganded LDL receptor measured by digital fluorescence microscopy. *In* "Spectroscopic Membrane Probes." (L. M. Leow, ed.), pp. 19–48, (II). CRC Press Inc., Boca Raton, FL.

Jay, P. Y., and Elson, E. L. (1992). Surface particle transport mechanism independent of myosin II in Dictyostelium. *Nature 356*, 438–440.

Kucik, D. F., Elson, E. L., and Sheetz, M. P. (1989). Forward transport of glycoproteins on leading lamellipodia in locomoting cells. *Nature 340*, 315–317.

Kucik, D. F., Elson, E. L., and Sheetz, M. P. (1990). Cell migration does not produce membrane flow. *J. Cell Biol. 111*, 1617–1622.

Kucik, D. F., Kuo, S. C., Elson, E. L., and Sheetz, M. P. (1991). Preferential attachment of membrane glycoproteins to the cytoskeleton at the leading edge of lamella. *J. Cell Biol. 115*, 1029–1036.

Lee, G. M., Ishihara, A., and Jacobson, K. A. (1991). Direct observation of Brownian motion of lipids in a membrane. *Proc. Nat. Acad. Sci. USA 88*, 6274–6278.

Metzger, H., Alcaraz, G., Hohman, R., Kinet, J.-P., Pribluda, V., and Quarto, R. (1986). The receptor with high affinity for immunoglobulin E. *Ann. Rev. Immunol. 4*, 419–470.

Qian, H., Sheetz, M. P., and Elson, E. (1991). Single Particle Tracking. Analysis of diffusion and flow in two-dimensional systems. *Biophys. J. 60*, 910–921.

Sheetz, M. P., Turney, S., Qian, H., and Elson, E. L. (1989). Nanometer-level analysis demonstrates that lipid flow does not drive membrane glycoprotein movements. *Nature 340*, 284–288.

Yechiel, E., and Edidin, M. (1987). Micrometer-scale domains in fibroblast plasma membranes. *J. Cell Biol. 105*, 753–760.

Zhang, F., Crise, B., Su, B., Hou, Y., Rose, J. K. Bothwell, A., and Jacobson, K. (1991). Lateral diffusion of membrane-spanning and glycosylphosphatidylinositol-linked proteins: Toward establishing rules governing the lateral mobility of membrane proteins. *J. Cell Biol. 115*, 75–84.

# 11
# TOTAL INTERNAL REFLECTANCE FLUORESCENCE MICROSCOPY

**Lukas K. Tamm**
Department of Physiology
University of Virginia
School of Medicine

I. INTRODUCTION
II. THEORY
  A. Principle
  B. Polarized Evanescent Wave
     Intensities
  C. Fluorescence Emission at an
     Interface
  D. Intermediate Layers
III. INSTRUMENTATION
  A. Laser and Microscope
  B. Optical Geometries for
     Fluorescence Excitation by TIR
  C. Sample Compartments
  D. Fluorescence Detection
     (Photometry and Imaging)
IV. APPLICATIONS
  A. Imaging of Cell-Substrate
     Contacts
  B. Binding Equilibria and Kinetics
     at Solid–Liquid Interfaces
  C. Binding Equilibria at Supported
     Membranes
  D. Cell-Supported Membrane
     Contacts
  E. Surface Diffusion by
     TIR–FRAP
  F. Polarized TIR: Fluorophore
     Orientation and Rotational
     Diffusion
  G. TIR Resonance Energy Transfer
V. PROSPECTUS
ACKNOWLEDGMENTS
REFERENCES

## I. INTRODUCTION

Total internal reflectance fluorescence microscopy is a relatively recent addition to the still expanding repertoire of new techniques of light microscopy. Although the phenomenon of total internal reflection has been known since the early days of optics and although the evanescent field that is associated with total internally reflected electromagnetic radiation has long been explored in many spectroscopies and over a wide range of the electromagnetic spectrum, it is only about a decade ago that total internal reflection has been introduced into light microscopy. In total internal reflectance fluorescence microscopy (TIRFM), the region near an interface between two media with different refractive indices is illuminated by an evanescent wave. The evanescent wave penetrates only a short distance into the medium of lower refractive index and selectively excites fluorescence in that region. Therefore, TIRFM combines subwavelength resolution in the direction perpendicular to the plane of the interface with the high sensitivity of fluorescence. Microscopic observation further allows one to image structures at or near the interface and to fluorometrically probe these structures with a spatial resolution in the $\mu$m (to sub-$\mu$m) range.

Many different scientific problems have now been approached with TIRFM. They include the imaging of live cells near a substrate, specific cell-substrate interactions, the physical chemistry of the interactions of macromolecules with solid surfaces, testing and screening for biocompatibility of solid surfaces, interactions of macromolecules with membranes, ligand-membrane receptor interactions, and interactions between live cells and reconstituted model membranes.

In this chapter, I will first summarize briefly the relevant theory for total internal reflection or evanescent wave fluorescence microscopy. In the second section, the practical implementation of total internal reflectance (TIR) illumination on standard microscopes will be discussed. Several optical setups have been developed and successfully used in different laboratories, and they will be briefly described with their respective advantages and limitations. The third section is devoted to a summary of the major applications of TIRFM. In addition to cell-substrate imaging, this section also contains information on some more quantitative microspectroscopic uses of TIRFM in order to study molecular interactions at interfaces with applications in many different areas of the biomedical sciences. Specifically, it is shown how TIRFM can be combined with other well-known techniques of fluorescence microscopy, such as fluorescence recovery after photobleaching (FRAP), polarization microscopy, and resonance energy-transfer microscopy. Finally, in the last section, some future potential developments and applications to biomedical problems of total internal reflectance fluorescence microscopy will be briefly presented and discussed.

## II. THEORY

### A. Principle

When a light beam propagating through a medium of high-refractive index encounters an interface with a medium of lower refractive index, it is either reflected or refracted according to Snell's Law

$$n_1 \sin \theta_1 = n_2 \sin \theta_2 \tag{1}$$

where $n_1$ and $n_2$ are the refractive indices of the high- and low-refractive index media, and $\theta_1$ and $\theta_2$ are the angles of incidence relative to the normal to the interface. When $n_1 > n_2$ and $\theta_1$ is greater than the critical angle, $\theta_c$, total internal reflection occurs in medium 1. The critical angle is

$$\theta_c = \sin^{-1}(n_2/n_1) = \sin^{-1} n_{21} \tag{2}$$

Although the incident light beam is totally internally reflected under these conditions, an electromagnetic field penetrates a small distance into medium 2. The intensity of this field decays exponentially with the distance $z$ from the interface

$$I(z) = I_o e^{-z/d_P} \tag{3}$$

with a characteristic penetration depth

$$d_P = \frac{\lambda_o/n_1}{4\pi\sqrt{\sin^2 \theta - n_{21}^2}} \tag{4}$$

This field is often called the "evanescent wave." As the evanescent wave is set up in medium 2, a standing wave is set up in medium 1. The standing and evanescent waves are depicted schematically in Fig. 11.1. For a finite width beam, it can also be shown that the electromagnetic wave travels some small distance in medium 2 before it reenters into medium 1. This distance of propagation along the surface is called the Goos-Hänchen shift.

### B. Polarized Evanescent Wave Intensities

The light intensity at the interface ($I_o$ in Eq. 3) is a function of the angle of incidence ($\theta = \theta_1$) and of the polarization of the incident light. A coordinate

FIGURE 11.1

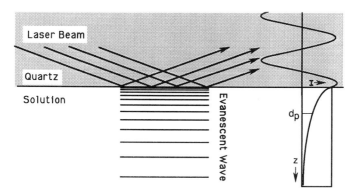

Schematic drawing of the evanescent wave produced by a totally internally reflected laser beam at a quartz-buffer interface. On the right-hand side, the standing wave in the solid medium and the evanescent wave in the liquid medium are shown. When 488-nm laser light is incident at an angle of 72° on a quartz-buffer interface, the penetration depth $d_p$ of the evanescent wave is about 97 nm.

system is defined such that the plane of incidence is the $x$-$z$ plane, with $z$ normal to the interface. The $y$ direction is parallel to the quartz surface and perpendicular to the $x$-$z$ plane. The incident light is either $p$-polarized with the electric field vector parallel to the $x$-$z$ plane or $s$-polarized with the electric field vector perpendicular ("senkrecht") to the $x$-$z$ plane. The intensities in the different directions of the evanescent field, at $z = 0$, and for unit incident intensities are given by (Harrick, 1967)

$$I_x = \frac{4 \cos^2\theta \ (\sin^2\theta - n_{21}^2)}{(1 - n_{21}^2) \ [(1 + n_{21}^2) \sin^2\theta - n_{21}^2]}$$

$$I_y = \frac{4 \cos^2\theta}{1 - n_{21}^2} \tag{5}$$

$$I_z = \frac{4 \cos^2\theta \ \sin^2\theta}{(1 - n_{21}^2 \ [(1 + n_{21}^2) \sin^2\theta - n_{21}^2]}$$

The total evanescent intensities for $p$-polarized and $s$-polarized incident light are

$$I_\parallel = I_x + I_z$$

and $$\tag{6}$$

$$I_\perp = I_y$$

respectively. The evanescent intensities $I_\parallel$ and $I_\perp$ are plotted in Fig. 11.2 as a function of the incident angle $\theta$ for $n_1 = 1.46$ (fused quartz) and $n_2 = 1.33$ (water or aqueous buffer). One of the most salient features of these curves is that the evanescent intensities near the interface and within 10° of the critical angle are up to five times larger than the intensities of the incident plane waves.

## C. Fluorescence Emission at an Interface

Having excited fluorophores in medium 2 near the interface to medium 1, peculiarities of their fluorescence emission require some further discussion. Unlike in bulk solution, fluorophores near interfaces do not emit light isotropically but rather in complex spatial patterns (Hellen and Axelrod, 1987). Two independent factors can contribute to the highly anisotropic fluorescence emission: first, the different dielectric properties of the media 1 and 2 impose different propagation speeds and angles on the light that is emitted from the fluorophores; and second, in most (if not all) cases of practical interest for TIRFM, the fluorophores (and therefore, their emission transi-

FIGURE 11.2

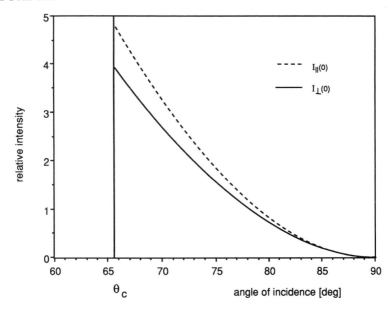

Evanescent intensities $I_\parallel$ (*broken line*) and $I_\perp$ (*solid line*) for *p*-polarized and *s*-polarized incident light ($\lambda_o = 488$ nm) at the quartz-buffer interface ($z = 0$) as a function of the angle of incidence, $\theta$. The critical angle $\theta_c$ is 65.7° for an interface between quartz ($n = 1.46$) and aqueous buffer ($N = 1.33$).

tion dipoles) are oriented with respect to the interface for durations that are at least as long as their lifetimes.

Hellen and Axelrod (1987) modeled fluorophores near an interface as constant *power* (rather than constant amplitude) oscillators. They determined the radiated intensity in all directions, $\mathbf{r}$ as a function of the distance $z'$ of the transition dipole $\boldsymbol{\mu}$ from the interface and as a function of the orientational distribution (described by $\theta'$ and $\phi'$) of the dipole with respect to the interface. This intensity is proportional to

$$S(\mathbf{r}, z', \theta', \phi') = \frac{cn_i |\mathbf{E}(\mathbf{r}, z', \theta', \phi')|^2}{8\pi P_\mathrm{T}(z')} \tag{7}$$

where $c$ is the speed of light in vacuum, $n_i$ is the respective refractive index, $\mathbf{E}(\mathbf{r}, z', \theta', \phi')$ is the electric field amplitude of the oscillator, and the total power released by the dipole is

$$P_\mathrm{T}(z') = (\omega/2)\mathrm{Re}(\boldsymbol{\mu}\cdot\mathbf{E}) \tag{8}$$

with $\omega$ being the angular frequency of the light.

Explicit calculations of Eq. 7 and 8 are rather tedious, but the graphical results of two specific examples are given in Fig. 11.3.

In both cases, the transition dipole is located in water at a distance $z' = 80$ nm from a quartz surface and the fluorophores are assumed to be azimuthally unoriented. However, the dipole is assumed to be parallel to the surface ($\theta' = 90°$) in Fig. 11.3A, and perpendicular ($\theta' = 0°$) to the surface in Fig. 11.3B. Highly nonuniform emission distribution patterns are expressed in both cases. Since in most practical applications the fluorescence is collected by an objective that is located on the water side of the interface, reasonable to good collection efficiencies can be expected even for low aperture objectives. If, however, fluorescence collection is attempted from the glass side with fluorescent dipoles, which are primarily oriented perpendicular to the surface, only very high-numerical aperture objectives can be used. Apart from the fluorescence intensity, the emission polarization and the fluorescence lifetimes are also affected by the presence of a nearby interface. Due to the complicated dependence of the fluorescence intensity on the distance $z'$ and the orientation of the fluorophores, accurate measurements of distances of fluorophores from the interface and of their orientation distributions are rather difficult to obtain from TIRFM data.

## D. Intermediate Layers

Several useful experimental situations exist where the interface is not a simple interface between two media but rather consisting of several in-

FIGURE 11.3

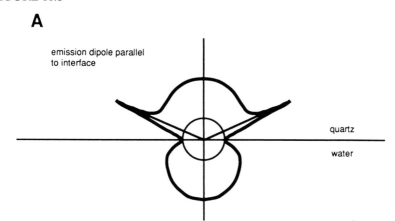

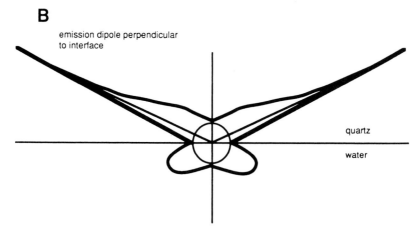

Anisotropic fluorescence emission of a fluorophor near a quartz–buffer interface. The fluorophor is assumed to be located in the aqueous phase 80 nm from the interface and oriented either parallel (A) or perpendicular (B) to the interface. (Adapted from Hellen and Axelrod, 1987.)

terfaces between one or more thin layers of materials of refractive index $n_i$ and sandwiched between the bulk materials of refractive indices $n_1$ and $n_2$. These intermediate layers may be model membranes, protein layers, or the membranes of whole cells. A situation that has some particularly interesting consequences on the optical properties near the interface is a thin film of metal between the solid and liquid bulk phases. Several combinations of refractive indices can be envisaged with intermediate thin films:

1. For $n_1 > n_i > n_2$, total internal reflection occurs at the $n_1/n_i$ interface, when $\theta > \theta_{c1} = \sin^{-1}(n_i/n_1)$, or at the $n_i/n_2$ interface, when $\theta_{c1} > \theta > \theta_{c2} = \sin^{-1}(n_2/n_i)$.
2. For $n_i > n_1 > n_2$, TIR occurs at the $n_i/n_2$ interface, when $\theta > \theta_{c2}$.
3. For $n_1 > n_2 > n_i$, TIR occurs at the $n_1/n_i$ interface, when $\theta > \theta_{c1}$.

However, in Case 3 the wave is evanescent in medium 2 only, when $\theta > \theta_{c2}$. (For $\theta_{c1} < \theta < \theta_{c2}$ in Case 3 the wave is evanescent in the film but propagates as a normal plane wave in medium 2.)

The theory for thin, nonmetallic films becomes particularly simple when the film is much thinner than the penetration depth $(d_p)$ of the evanescent wave [i.e., when the thickness, $d$, of the film is smaller than $\lambda_o/(10\pi n_i) << d_p$] and when the absorption of the thin film is small [i.e., when the absorption coefficient $\alpha = 2.303\varepsilon c < \pi n_i/(2\lambda_o)$, [where $\varepsilon$ is the decadic molar extinction coefficient and $c$ is the concentration in moles/liter]. Under these conditions, the polarized evanescent intensities next to the thin film are well approximated by Eq. 5 when $I_z$ is multiplied by the thin film correction factor $(n_2/n_i)^4$ (Harrick, 1965, 1966). The distance dependence of the evanescent wave is independent of the presence of an intermediate layer and is still accurately described by Eqs. 3 and 4, but with changed subscripts of $n$, depending on which interface is the TIR-supporting interface. The nonabsorbing thin film conditions are normally fulfilled for visible light TIRFM of thin films of biological materials, such as cell membranes or model membranes, which are adsorbed to the surface of glass or quartz microscope slides.

When the intermediate layer is absorbing, such as for the case of a thin film of metal, and when its thickness is of the order of $d_p$ or smaller, the complex refractive index, $\hat{n}_i = n_i(1 - i\kappa_i)$, has to be considered, where $\kappa_i = \alpha\lambda_o/4\pi n_i$ is the attenuation index of the intermediate layer. The relevant Fresnel coefficients for transmission and reflection that describe this situation are rather complicated but have been derived by Hellen et al. (1988).

The most salient features of the results of these Fresnel calculations are the dramatic increase of $I_\parallel$ for $p$-polarized incident light and the virtually vanishing intensity $I_\perp$ for $s$-polarized incident light (Fig. 11.4). The maximal intensity $I_\parallel$ is no longer found at the critical angle $\theta_c$ but rather at the "surface plasmon angle" $\theta_p$, which is greater than $\theta_c$. This evanescent component can become an order of magnitude (or more) larger than the intensity of the incident light. The reason for this resonance-like effect is due to the excitation of a surface plasmon mode at the metal–water interface when the incident light matches some electronic transition in the metal. In addition to strongly exciting fluorophores with highly polarized light at the

FIGURE 11.4

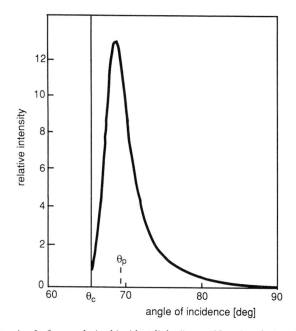

Evanescent intensity $I_\parallel$ for $p$-polarized incident light ($\lambda_o$ = 488 nm) at the interface between a quartz substrate ($n$ = 1.46) that has been coated with a 20-nm-thick film of aluminum ($\hat{n}$ = 0.74 + $i$ 5.75) and water ($n$ = 1.33). The critical angle $\theta_c$ is 65.7°, and the surface plasmon angle $\theta_p$ is 69°. (Adapted from Hellen *et al.*, 1988.)

surface plasmon angle, a thin film of metal (or of a semiconductor) has the further effect of efficiently quenching fluorescence within approximately 10 nm from the surface. This latter effect can be experimentally exploited to distinguish between adsorbed layers (e.g., of a cell) that are very close (<10 nm) or farther away (10–200 nm) from the solid surface.

## III. INSTRUMENTATION

### A. Laser and Microscope

Although, in principle, it is possible to use a conventional light source (e.g., a mercury lamp) to excite fluorescence on a microscope under TIR illumination, lasers are generally preferred in TIRFM for two reasons. First, laser light is well collimated, it can be easily directed into the TIR prism, and if properly alligned, it illuminates areas of defined geometries at the TIR-supporting interface, which is positioned in the object plane of the microscope. Second, the high-light intensity of ion lasers is useful if the microscope is also intended to be used for fluorescence recovery after

FIGURE 11.5

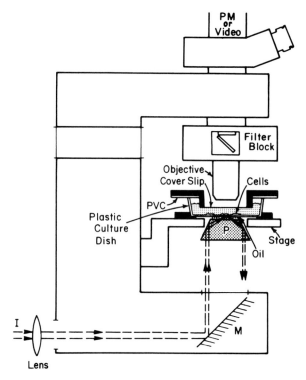

Optical configuration for TIRFM adapted to an upright microscope. (From Axelrod et al., 1984, with permission.)

photobleaching (FRAP) studies in either the TIR or the conventional epi-illumination mode. Argon ion lasers are used most often because they have strong lines at 488 and 514 nm [i.e., two lines that are adequate to excite the most commonly used fluorophores, such as fluorescein, 4-nitrobenzo-2-oxa-1,3-diazole (NBD), and rhodamines]. For some fluorophores, which absorb in the green-to-red region of the electromagnetic spectrum, krypton ion lasers may be preferable. A further advantage of lasers over conventional light sources is that they produce coherent and polarized light. Coherence is important for setting up interference fringes for fringe pattern photobleaching (see Section IV.E), and some interesting experiments using polarized light are described in Section IV.F.

Upright or inverted microscopes can be used for TIRFM. Two examples for TIR optical configurations adapted to an upright and an inverted microscope are shown in Figs. 11.5 and 11.6, respectively. The choice for a particular kind of microscope depends on the application. An upright microscope might be preferred for viewing cell-substrate contacts of cells that are growing in culture as a monolayer on the bottom of a plastic petri dish. (Plastics of standard petri dishes also support TIR illumination. However, their autofluorescence may obscure some of the contrast in weakly fluorescent samples.) Inverted microscopes are generally preferred for studies of *specific* molecular contacts of cells or soluble macromolecules with substrates or supported membranes because the bound molecules are "hanging" from the upper inner surface of the measuring cell. In contrast, cells that are not bound to the substrate, as well as cellular debris or contaminating protein, aggregate sediment to the bottom of the cell without obscuring the probed surface in an inverted microscope configuration. Inverted microscopes have the further advantage that they generally leave more working space for inserting optical and other parts (e.g., the TIR coupling prism) in close proximity to the sample.

Focusing is achieved in some microscopes by moving the stage, in

FIGURE 11.6

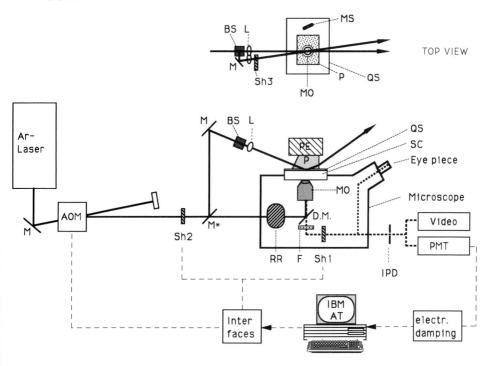

Optical and electronic configuration of an inverted laser fluorescence microscope designed to perform TIRFM, epiillumination FRAP, and TIR–fringe FRAP experiments. (From Kalb *et al.*, 1992, with permission.)

others by moving the objective lens. TIR is more difficult (although not impossible) to implement on moving stage microscopes because any adjustment of the focus requires a realignment of the totally internally reflected laser beam into the focal plane and the optical axis of the microscope. Another general consideration for choosing a good microscope for TIRFM should be its mechanical stability. Especially for the more sophisticated techniques of TIRFM, it is very important to guarantee a mechanically stable evanescent illumination with a pointing stability of the focused laser beam on the sample within a couple of $\mu$m over several hours. This can only be achieved by fixing in a stable fashion the laser, the microscope, and *all* optical parts in between the laser and the microscope on a vibration isolation table.

## B. Optical Geometries for Fluorescence Excitation by TIR

Several designs of prisms have been used in different laboratories in order to couple the laser beam into the TIR-supporting solid substrate. In some designs, it is sufficient to use fused quartz ($n = 1.46$) or crown glass ($n = 1.53$), whereas in others, flint glass ($n = 1.62$) or other high-refractive index materials may be needed. In all designs, the prisms need to be optically coupled to the solid substrate with a small amount of glycerol or immersion oil. In the following, some common designs are listed..

1. In the upright microscope configuration shown in Fig. 11.5, a flint glass prism is installed in place of the microscope condenser. The prism is made by truncating and polishing the top of a commercial equilateral triangle prism.
2. Figure 11.7 shows the optical configuration for a glass or quartz cube prism (1 cm$^3$) on an inverted microscope (Axelrod, 1981). The cube is fixed on the optical axis ($z$) of the microscope. It can be translated along this axis, which allows one to bring it into contact with the sample chamber after the latter has been set up on the stage of the microscope. Since the prism is fixed relative to the entering beam, the TIR beam remains aligned even when the microscope stage is translated laterally to view different areas of the sample. A focusing lens (typically, about 50–100-mm focal length) is used to focus the beam to a small illuminated area on the sample. Changing the focus of this lens allows one to change the size of the illuminated area. The laser beam is refracted at the vertical face of the cube at different angles across the beam profile. This gives rise to a geometric aberration, and the elliptical illuminated area at the TIR interface is elongated compared to the elliptical area that is expected for parallel incident light with a Gaussian beam profile. The flat top surface permits an easy switch between TIR

FIGURE 11.7

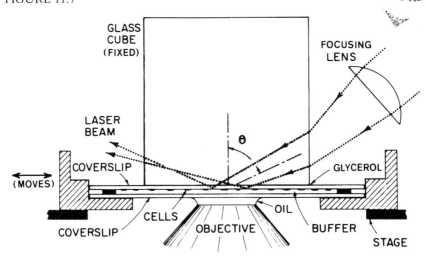

Sample compartment and TIR coupling with a cube prism for an inverted microscope. (From Axelrod, 1981, with permission.)

and conventional illumination systems, the latter consisting of a lamp and a condenser, which are normally located above the prism.

3.  A truncated hemicylindrical quartz prism has been used by Kalb *et al.* (1990). If the beam width is small relative to the radius of curvature of the prism, the rays enter this prism more normally than a cube prism at any angle of incidence and, as a consequence, the aberration of the beam profile is smaller. A still more perfect elliptical TIR illumination profile is obtained with a trapezoidal prism, such as the one depicted in Fig. 11.6. The entering surface of this prism is normal to the director of the incident beam. This geometry has been found to be very useful for setting up clean interference fringes between two intersecting beams in the TIR mode (Kalb *et al.*, 1992). Obviously, the prism surface is only normal to the incident beam at one angle of incidence, chosen to be 72° in this particular setup.

4.  Interference fringes have also been produced by coupling two laser beams with a trapezoidal prism that is located at one end of a 1-mm-thick quartz microscope slide and off the center of the optical axis of the microscope (Weis *et al.*, 1982). The beams undergo multiple internal reflections in the quartz slide before they superimpose and interfere in the focal plane above the objective. A second identical prism is used as output coupler on the other end of the slide. This geometry permits, without realignment of the laser beams, only one-dimensional sample translation normal to the incoming laser beam.

5.  A still different method for producing TIR interference fringes is the use of a parabolic mirror placed on top of the stage of an inverted microscope with its focal point superimposed with the focal point of the microscope objective (Hellen *et al.*, 1988). The two components of a split laser beam hit the parabolic mirror at identical angles, pass through a hemisphere prism, and meet and interfere at the TIR surface above the microscope objective. Very high-intersection angles that produce interference fringes of very high-spatial frequencies can be produced by this method.

## C. Sample Compartments

Different applications require different sample compartments. For high-resolution TIRFM imaging on an inverted microscope, a very thin sample chamber is preferred, such as the one shown in Fig. 11.7. This permits the use of high-numerical aperture objectives with short working distances. In the example of Fig. 11.7, the cells adhere to the upper glass coverslip, which is spaced from the lower coverslip with thin spacers (e.g., double stick tape) by less than 0.1 mm.

Measurements of binding equilibria or binding kinetics on surfaces (e.g., on supported membranes) require more sophisticated sample compartments, which permit the experimentalist to perform several operations without readjusting any of the optical alignments: (i) two access holes to the inside of the cell allow the cell to be perfused with different solutions, namely for titrations with soluble ligands in equilibrium or kinetic-binding experiments; (ii) the inner vertical dimension should be about 1–2 mm (i.e., just large enough to permit stirring with a thin magnetic stir bar in order to attain a quick homogeneous distribution of solutes in titration experiments but still small enough to permit focusing to the upper inner surface with a microscope objective of intermediate NA; for most experiments, we use a Zeiss 40×, 0.75 NA, water immersion objective with a working distance of about 2 mm); and (iii), especially when TIRFM is used in conjunction with supported membranes, the design must be absolutely tight (not leaking buffer) because air bubbles will irreversibly destroy the integrity and extended bilayer configuration of supported planar membranes on the surface of the TIR-supporting substrate. The design of the cell that we are currently using in our TIRFM-binding experiments on supported planar membranes is depicted in Fig. 11.8. It is made with two solid frames of aluminum, which are screwed together and tightly press, from bottom to top, the bottom glass coverslip window (25 × 40 mm), the spacer piece made from Teflon (the spacing between the lower and upper internal surfaces is approximately 1 mm), and the quartz slide (25 × 40 × 1 mm) with or without a supported membrane on its lower surface. This

FIGURE 11.8

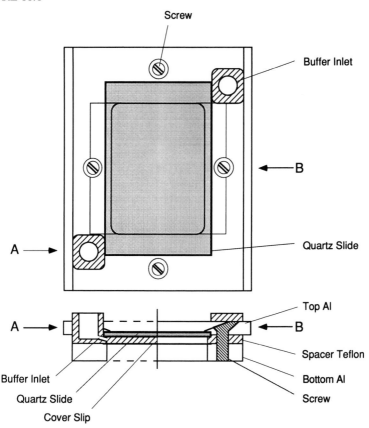

Design of a TIRFM sample compartment for measuring the binding of macromolecules, membranes, or whole cells to surfaces. The upper inner surface of the compartment that is probed by the evanescent wave may be hydrophilic or hydrophobic, or, for studying membrane receptor–ligand interactions, it may have attached a single supported planar membrane.

cell fits tightly into a specially made stage that is adapted to the $xy$-translation stage of the microscope.

## D. Fluorescence Detection (Photometry and Imaging)

### 1. Imaging

The most straightforward method to qualitatively record images of cell-surface contacts and other interface structures in TIRFM is to use standard microphotography on high-speed photographic films. However, often only relatively few fluorescently labeled molecules are present in the small volume illuminated by the evanescent wave, and therefore, TIR fluorescence might often be quite dim. In these cases, and especially when repeated images of the same fields are to be recorded, low-light-level video or charge-coupled device (CCD) cameras are preferred. There is extensive literature available on low speed and video rate electronic recording and digital processing of low-light-level microscope images (see e.g., Inoué, 1986; Aikens *et al.*, 1989; Arndt-Jovin and Jovin, 1989; Spring and Lowy, 1989).

## 2. Photometry

Photomultiplier tubes (PMTs) are commonly used for the spatially unresolved but fast and sensitive quantitation of the fluorescence intensities, which arise from the field that is illuminated by the evanescent wave. Parts of this field may be selected with a variable aperature or image-plane diaphragm (IPD), which is located in an image plane of the observation path of the microscope. Photon counting and analog PMTs have been used in TIR microspectrofluorometry. Especially when long sampling intervals with low-light intensities are needed, photon counting is superior to analog detection. Although sampling times are often relatively short (typically 1–40 ms), photoncounting PMTs are normally used in microspectrofluorometry. However, in our setup we used an analog PMT with an analog electronic noise reduction system. A PMT with extended red characteristics (EMI 9658A) has been selected. With this system, we can measure, with a good signal-to-noise ratio (SNR), fluorescence intensity changes that are caused by the binding or unbinding of about $10^4$ fluorescein or about $10^5$ NBD molecules. Therefore, changes in surface concentration of about 1 to 10 molecules per $\mu m^2$ are easily detected by this instrument on supported planar membranes where we typically select an observation area of about 4000 $\mu m^2$.

For special purposes, a monochromator may be intercalated between the image plane diaphragm and the detector (Hlady *et al.*, 1985; Watts *et al.*, 1986). This configuration allows one to obtain spectral information on the adsorbed molecules.

In order to compare fluorescence intensities measured on different days (even over several months), it is important to carefully control the excitation light intensity. This can be done by measuring the excitation light intensity with a photodiode immediately before it enters the sample. With this procedure, one can correct for differences in the laser intensity, the performance of the acoustooptic modulator (AOM), and the alignment of the beam before the sample.

For many applications of quantitative TIRFM, a microprocessor-controlled regulation of the incident light and of the data acquisition is very useful. We use a standard PC with software that is written in ASYST (Asyst Technologies, Rochester, NY) to control the voltage of the AOM and the two electronic shutters in the system and to store all raw data during acquisition. The electronic shutters are important to avoid unnecessary photobleaching by the observation beam when data is collected over long periods of time (e.g. in slow-binding kinetics) and to avoid damage of the PMT by the intense laser beam during the bleach pulses in FRAP experiments.

## IV. APPLICATIONS

### A. Imaging of Cell-Substrate Contacts

Because of the short characteristic penetration depth of the evanescent wave, the contact region between the cells and a solid substrate to which they adhere can be imaged selectively and with high contrast by TIRFM. Fluorescence that might arise from other parts of the cell, including cellular autofluorescence, is highly suppressed by this imaging technique. Imaging of cell substrate contacts was first demonstrated by fibroblasts that were labeled with the lipid dye, diI, and for primary culture rat myotubes that were labeled with tetramethyl-rhodamine-α-bungaratoxin (Axelrod, 1981). By varying the angle of incidence, evanescent wave penetration depths between 100 and 400 nm were achieved, which resulted in quite different TIRFM images, indicating that the fibroblasts adhered to the quartz substrate along several longitudinal folds and that the depth of invagination of these folds was of the order of a few hundred nanometers. Since bungaratoxin binds to acetylcholine receptors on the surface of myotubes, it was possible to discern by TIRFM acetylcholine receptors that clustered in fairly large patches in the contact region with the quartz substrate. Much smaller and fewer clusters were found on the membrane that was not in contact with the substrate.

The contacts between chick heart explant cells and glass (Gingell et al., 1985) and between Dictyostelium amoebae and glass (Todd et al., 1988) were investigated by a negative contrast TIRFM technique. In this technique, the cells were surrounded by a solution that contained a nonreactive fluorescent volume marker, such as a fluorescein-labeled dextran. The focal contacts of the cell with the glass excluded the marker and therefore appeared dark under TIR illumination. When viewed with this technique, the cell contacts of the amoebae to polylysine-treated glass were uniform and tight over the whole contact area.

Lanni et al. (1985) studied the structural organization of interphase 3T3 fibroblasts near a glass substrate interface by TIRFM. Fluorescent markers for several subcellular compartments and organelles were used in this study.

Mitochondria and nuclei were shown to be distant from the contact regions. However, areas of high-fluorescence intensities indicative of cell-substrate contact areas were observed with the cytoplasmic membrane marker C18-diI and the cytoplasmic markers fluorescein-dextran and carboxyfluorescein. The distribution of fluorescein-labeled actin was investigated in adherent and mobile cells. Stress fibers that terminated in the focal contact zones were visible in adherent cells, whereas motile cells exhibited, as visualized by TIRFM, a relatively even actin distribution with no significant accumulation in the close contact zones.

In another study, the distribution of microfilaments and actin-associated proteins at the sites membrane-substrate attachment of rat myotubes was determined (Bloch *et al.*, 1989). Detergent-permeabilized samples were observed by TIRFM and epifluorescence microscopy. Fluorescent phalloidin derivatives and antibodies to vinculin, $\alpha$-actinin, filamin, and talin preferentially labeled domains of tight cell-substrate contact. This suggested that bundles of microfilaments associated with the membrane at the sites of myotube-substrate attachment. Interdigitated with these close contact domains, there were found less closely attached membrane domains that were enriched in acetylcholine receptors, a 43K acetylcholine receptor-linked protein, and unbundled actin.

The distribution and lateral mobility of the angiotensin converting enzyme and another specific endothelial surface protein were assessed by TIRFM using fluorescein-conjugated monoclonal antibodies against these two proteins on endothelial cells grown on quartz slides that were coated with extracellular matrix components (Nakache *et al.*, 1986). The extracellular matrix and particularly, the basal membrane, which is built up from extracellular matrix proteins, are known to organize and induce the development of various tissues already at very early stages. When the endothelial cells were grown on the extracellular matrix-coated substrate, they were polarized and the two membrane proteins whose distribution was investigated by TIRFM were restricted to the apical membranes of the cells, independent on whether the cells were grown at a sparse density or at confluence (Fig. 11.9). However, when the cells were brought into suspension, the fluorescein-conjugated antibody was found to be distributed over the entire cell surface.

Red blood cells were adsorbed to polylysine-coated glass slides and subsequently hemolyzed by hypoosmotic shock (Axelrod *et al.*, 1986). It was concluded that the resulting erythrocyte ghosts flattened down onto the substrate because both membranes could be viewed by the evanescent wave, which has a very low intensity at distances greater than about 200 nm. The "ghosts" appeared with a "crescent" and a "bite," and the crescent could be identified by labeling with rhoadmine-labeled wheat germ agglutinin to

FIGURE 11.9

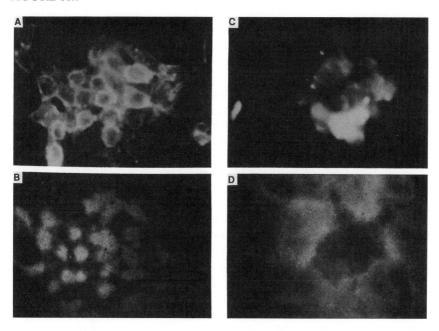

Endothelial cells grown on extracellular matrix (ECM)-coated quartz slides and preincubated with (A, B) the fluorescent lipid marker NBD-phosphatidycholine (NBD-PC) or (C, D) with fluorescein-labeled antiangiotensin-converting enzyme antibodies. (A) and (C) are epifluorescence, (B) and (D) are total internal reflectance fluorescence micrographs. The ECM-facing basolateral membrane is well stained with NBD-PC but, relative to the apical membrane, appears to be depleted of the angiotensin-converting enzyme. (From Nakache *et al.,* 1986, with permission.)

represent the outer surface of the outer membrane. In contrast, the bite represented the cytoplasmic surface of the substrate-attached membrane.

## B. Binding Equilibria and Kinetics at Solid–Liquid Interfaces

The binding of proteins to solid surfaces has been studied by nonmicroscopic TIRF much earlier than by TIRFM. For instance, early applications included binding studies of fluorescein to glass (Hirschfeld, 1965), serum albumin to glass (Harrick and Loeb, 1973; Watkins and Robertson, 1977; Burghardt and Axelrod, 1983), serum proteins to various biocompatible materials (Lowe *et al.,* 1986), antibodies to antigen-coated glass or polydimethylsiloxane (Kronick and Little, 1975; Darst *et al.,* 1988), and insulin to insulin receptors in supported membranes (Sui *et al.,* 1988). Also, intrinsic tryptophan fluorescence of proteins has been used to study protein adsorption to various solid–liquid interfaces by TIRF (Hlady et al., 1985; Hlady et al., 1986).

Microscopic TIRF has several advantages over nonmicroscopic TIRF configurations: (i) binding experiments can be localized to small areas, such as, for example, flattened cells or membrane fragments; (ii) the visual selection of "good" areas on a nonuniformly coated substrate can significantly increase the difference between specific and nonspecific binding; and (iii) TIRFM can be combined with two other fluorescence techniques, namely FRAP and fluorescence correlation spectroscopy (FSC), which require the spatial resolution of a microscope.

Surface desorption rates can be observed without a perturbation of the chemical equilibrium by combining TIR with fluorescence recovery after photobleaching or fluorescence correlation spectroscopy (Thompson et al., 1981). In TIR–FRAP, adsorbed molecules are bleached by a flash of the focused and totally internally reflected laser beam; subsequent fluorescence recovery is monitored by an attenuated evanescent intensity as bleached molecules exchange with unbleached ones from the solution or the surrounding nonbleached areas of the surface (Fig. 11.10).

In TIR–FCS, a large area of the interface is illuminated with a relatively dim evanescent wave, and fluorescence fluctuations due to individual molecules entering and leaving a small and well-defined portion of the evanescent field are autocorrelated. In general, the shape of the theoretical TIR–FRAP and TIR–FCS curves depends in a complex manner on the bulk and surface diffusion coefficients, the size of the illuminated or observed region, and the kinetic rate constants for adsorption and desorption. However, the experimental conditions can be varied such that the rate constants or the surface diffusion coefficients can be readily obtained. To further discuss measurements of surface kinetics by TIR–FRAP and TIR–FCS, the following surface-binding reaction is considered

$$A + B \underset{k_{-1}}{\overset{k_1}{\rightleftharpoons}} AB$$

where $A$ is a solute that can bind to free surface sites $B$ to form occupied sites $AB$. $k_1$ and $k_{-1}$ are surface adsorption and desorption rate constants and $\bar{A}$, $\bar{B}$, and $\overline{AB}$ are the equilibrium concentration values of $A$, $B$, and $AB$. An interesting question is whether this reaction is limited by bulk diffusion or by the reaction rate constants. This can be decided by changing $\bar{A}$ (and therefore, the ratio $\bar{A}/\overline{AB}$) in TIR–FRAP experiments. If the recovery curves change their shape, bulk diffusion contributes to the measured signal. If they do not change shape, the reaction is reaction limited and the rate

FIGURE 11.10

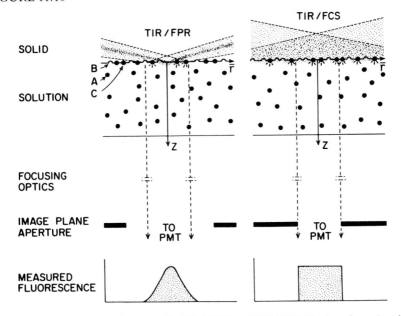

Schematic drawing of optical systems for TIR/FRAP and TIR/FCS. The laser beam is either focused (TIR/FRAP) or unfocused (TIR/FCS) at the solid–liquid interface, and fluorescence is collected from a small area defined by either the extension of the illuminated field (TIR/FRAP) or an image plane diaphragm (TIR/FCS). (From Thompson *et al.*, 1981, with permission.)

constants are easier to extract from TIR–FRAP or TIR–FCS experiments. Within certain experimental limitations, the recovery and autocorrelation curves can be forced toward the reaction limit by decreasing the size of the observation region and/or increasing $\bar{A}$. For a reaction-limited system, the following equations are useful to determine the rates $k_1$, $k_{-1}$, and the lateral diffusion coefficient on the surface, $D_L$. For TIR–FRAP with a linear Gaussian beam profile of a $1/e^2$-width $w$,

$$\frac{F_\infty - F(t)}{F_{pre} - F(0)} = (1 + \frac{4D_L}{w^2} t)^{-1/2} e^{-k_{-1}t} \tag{9}$$

and for TIR–FCS with a circular Gaussian observation profile with a $1/e^2$-radius $w$,

$$\frac{G(t)}{G(0)} = (1 + \frac{4D_L}{w^2} t)^{-1} e^{-(k_1\bar{A}+k_{-1})t} \tag{10}$$

In Eq. 9, $F_{pre}$, $F(0)$, $F(t)$, and $F_\infty$ are the fluorescence intensities before the bleach pulse, immediately after the bleach pulse, at times $t$, and a very long time after the bleach pulse, respectively. In Eq. 10, $G(t)$ is the autocorrelation function that has a magnitude of $G(0)$ at $t = 0$. To obtain both rate constants, the equilibrium binding constant, $K = k_1/k_{-1}$, must be known from independent experiments, such as, for example, a Langmuir-binding isotherm obtained by TIRFM.

TIR–FRAP has been applied to study the surface dynamics of the adsorption of bovine serum albumin (BSA) to quartz (Burghardt and Axelrod, 1981). The experimental FRAP curves were at least double exponential and revealed a minimum of three classes of adsorbed protein layers: an irreversibly bound fraction, a slow reversibly bound fraction, and a fast reversibly bound fraction. The relative proportions of these fractions changed with increasing protein concentration in the bulk phase. The fast reversible off-rate, $k_{-1}$, was $0.17 \pm 0.01$ s$^{-1}$ for a wide beam and $0.26 \pm 0.02$ s$^{-1}$ for a narrow beam. This dependency of $k_{-1}$ on the beam width suggested the presence of surface diffusion for this fraction, and an approximate diffusion coefficient $D_L = (5 \pm 1) \times 10^{-9}$ cm$^2$/s could be calculated from Eq. 9. The slow reversible desorption rate, $k_{-2}$, was $(4.8 \pm 1.0) \times 10^{-3}$ s$^{-1}$, and was independent of the beam size.

TIR–FRAP was also used to study the adsorption of lysozyme and BSA to alkylated and plain quartz slides (Schmidt *et al.,* 1990; Zimmermann *et al.,* 1990). Photobleaching experiments were carried out in order to determine the relative contributions of surface versus bulk fluorescence. This enabled these workers to calibrate the absolute amount of adsorbed protein. Three classes of adsorbed lysozyme could be distinguished on alkylated silicon dioxide: approximately 65% were quasi-reversibly bound, about 25% exchanged on a time scale of several hours, and about 10% of the protein were in a fast chemical equilibrium (of the order of fractions of a second) with the bulk solution.

The TIR–FCS combination was applied by Thompson and Axelrod (1983) to investigate the surface-binding kinetics of anti–DNP-immunoglobulins at a quartz surface that had been coated with an irreversibly bound layer of BSA or DNP-BSA. The binding of the antibodies to the DNP-BSA coated substrate was virtually irreversible. When BSA-coated quartz was used as the substrate, reversible and irreversible binding components were found for rhodamine-labeled antibodies and insulin. The adsorption rate constant, $k_1$, for the reversibly bound fraction of antibodies on BSA-coated quartz was estimated to be in the range between $0.8 \times 10^8$ and $4 \times 10^8$ M$^{-1}$ s$^{-1}$, and the correlation function was found to be at or near the bulk diffusion limit.

## C. Binding Equilibria at Supported Membranes

One important step forward for the application of TIRFM to binding phenomena came with the development of supported phospholipid monolayers (von Tscharner and McConnell, 1981) and supported phospholipid bilayers (Tamm and McConnell, 1985). It already became apparent then that the phosphatidylcholine head groups of oriented phospholipids were among the best surfaces to largely prevent *nonspecific* surface binding of many proteins (L. Tamm, unpublished results). Nonspecific surface binding had been a major problem in many of the earlier TIR-binding studies. These findings strongly supported confidence that it should be possible to measure *specific* protein binding to well-defined binding sites in supported planar membranes by TIRFM.

Suggestions to measure ligand binding to membrane receptors by TIRFM on supported planar membranes had been made in earlier reviews (McConnell *et al.*, 1986; Thompson *et al.*, 1988), but it was only in 1990 when this approach was first used in actual studies of protein binding to membranes (Kalb *et al.*, 1990; Poglitsch and Thompson, 1990). Here, I briefly review the presently available experimental studies that have used TIRFM to measure the binding of proteins to specific binding sites (receptors) on supported membranes. Some of the advantages and limitations of this relatively recent method for measuring ligand-membrane receptor binding are also discussed.

Due to the spatial extension of the evanescent field, the total measured fluorescence intensity is the sum of the fluorescence of molecules that are bound to the membrane and the fluorescence of molecules that are present in the illuminated volume but not bound to the membrane. The ratio of the fluorescence of bound ($F_b$) to unbound ($F_f$) ligands is

$$\frac{F_b}{F_f} = \frac{c_b \exp(-z/d_p)}{c_f d} \tag{11}$$

where $d_p$ is the $1/e$ penetration depth of the evanescent field (Eq. 4), $z$ is the mean distance of the bound fluorophores from the interface, and $c_b$ and $c_f$ are the concentrations of the bound and free ligands, respectively. In this simple, first-order approximation, the volume of the supported membrane itself is not excluded for free fluorescent ligands. Therefore, the ratios $F_b/F_f$ as calculated from Eq. 11 are lower limits for the actual values. Some useful limiting concentrations as calculated by Eq. 11 are tabulated in Table 11.1. In these calculations, $z$ was taken to be 20 nm and $d_p$ was 93 nm. It is evident from this table that the concentrations that give rise to equal fluorescence from bound and free ligands are shifted to lower free ligand concentrations when the receptor surface concentrations are decreased. For instance, when the binding-site concentration is 20,000 molec./$\mu$m$^2$, which corresponds to one receptor/50 nm$^2$, binding constants down to $10^4$ M$^{-1}$ can be determined (Kalb *et al.*, 1990). The upper limit for measurable binding constants is theoretically limited by the sen-

TABLE 11.1

Estimate of Useful Concentration Ranges of the TIRFM Method[a]

| | | $F_b/F_f$ | | |
|---|---|---|---|---|
| | $c_{receptor}$ (molec. $\mu m^{-2}$): | 20,000 | 5,000 | 500 |
| $c_{ligand}$ (M) | area/receptor (nm²): | 50 | 200 | 2,000 |
| $10^{-3}$ | | 0.27 | 0.068 | 0.0068 |
| $10^{-4}$ | | 2.7 | 0.68 | 0.068 |
| $10^{-5}$ | | 27.2 | 6.8 | 0.68 |
| $10^{-6}$ | | 272. | 68 | 6.8 |

[a] The TIRFM signal ratios of bound ($F_b$) and unbound ($F_f$) ligands, which are present in the volume illuminated by the evanescent field, are tabulated at various receptor densities. A limiting line is drawn between ligand concentrations above which the fluorescence intensity arising from the free ligand gets larger than the intensity from the bound ligand. The $F_b/F_f$ ratios are calculated for full saturation of the binding sites.

sitivity of the detection system, but in practice, unspecific binding often becomes the limiting factor.

## 1. Monoclonal Antibody-Lipid Hapten Binding

The binding of a monoclonal antibody to lipid haptens in a supported membrane has been used as a test system to critically evaluate TIRFM for measuring binding constants to membrane receptors (Kalb et al., 1990). The binding of monoclonal antibodies to monovalent haptens in solution is a biochemically and biophysically well-defined process. However, although immunochemically very important, the binding of antibodies to membrane-bound target sites is theoretically much more complex and still not fully understood. Nevertheless, the binding reaction of the bivalent monoclonal antibody GK14-1 to a trinitrophenol (TNP)-lipid hapten, which has a TNP group covalently linked to a phospholipid head group via a six methylene segment linker, has been studied in considerable detail and at many different surface concentrations of the lipid hapten in vesicles by a fluorescence resonance energy-transfer technique (Tamm and Bartoldus, 1988). Therefore, this system could serve as an almost ideal reference system for establishing and evaluating the usefulness of TIRFM for quantitative binding studies of membrane receptor-ligand interactions.

Figure 11.11 shows the binding of the monoclonal antibody GK14-1 to supported phospholipid bilayers at three different surface concentrations of the TNP-lipid hapten (Kalb et al., 1990). With increasing antibody concentration, the fluorescence intensities on the surface, as measured with TIRFM, increased up to different saturating values. These saturating values increased when the lipid hapten concentration in the membrane was increased from 0.2 to 1 to 5 mol%. Assuming an occupied area of 60 nm² per antibody on the membrane (and a lipid cross-sectional area of 0.7 nm²), one would expect the entire surface to be covered with antibody at and above 2.3 mol% lipid hapten. In agreement with these considerations, the experimentally determined saturating fluorescence intensities increased by a factor of about 5 when the lipid hapten concentration was increased from 0.2 to 1 mol% but only by a factor of 2 to 3 for the increase from 1 to 5 mol% lipid hapten.

The numerical values of the binding constants that were obtained with the TIRFM method (Kalb *et al.*, 1990) were in good agreement with those obtained with the resonance energy-transfer assay in suspension (Tamm and Bartoldus, 1988) although light-scattering problems limit the application of the latter technique to high lipid hapten concentrations. The overall (apparent) binding constant for the binding of the *bivalent* antibody GK14-1 to the membrane bound hapten was $1.5 \times 10^7 \, M^{-1}$. This value is much lower than the expected value if the measured binding energies for *two soluble* haptens were additive [i.e., $(5 \times 10^7)^2 \, M^{-1} = 2.5 \times 10^{15} \, M^{-1}$]. We do not believe that the relatively low-binding constant on membranes is primarily due to monovalent antibody binding to the lipid haptens because experiments with $F_{ab}$ fragments resulted in still lower binding constants (L. Tamm, unpublished results). Rather, a reduced accessibility of the hapten on the membrane surface and unfavorable entropy effects (orientation)

FIGURE 11.11

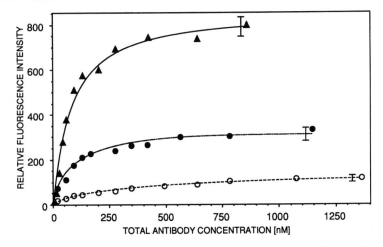

Background-corrected binding isotherms of FITC-labeled antibody to 0.2 (○), 1.0 (●), or 5.0 mol% (▲) lipid hapten in supported planar bilayers. The error bars indicate the uncertainty (±5%) in the fluorescence intensity at saturation, which was estimated from repeated experiments with 1 mol% hapten. The lines represent the best fits of the data to simple Langmuir adsorption isotherms. (From Kalb *et al.*, 1990, with permission.)

FIGURE 11.12

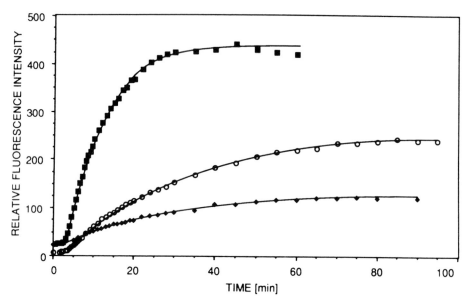

Association kinetics of the fluorescein-labeled monoclonal antibody GK14-1 to lipid haptens in a supported bilayer as measured by TIRFM. The increases of the relative fluorescence intensity at 1 mol% lipid hapten after addition of 75 nM (◆) and 150 nM (○) antibody, and at 5 mol% lipid hapten after addition of 150 nM antibody (■) are shown. (From Kalb et al., 1990, with permission.)

could account for the observed discrepancy. It is also possible to measure the kinetics of antibody binding to lipid haptens in supported phospholipid bilayers by TIRFM. An example is shown in Fig. 11.12. Again, the kinetic data obtained by TIRFM were in good agreement with those obtained by the resonance energy-transfer technique.

The binding of another monoclonal antibody, the antinitroxide antibody ANO2, to supported membranes, which contained a nitroxide spin-label hapten, has also been studied by TIRFM (Pisarchick and Thompson, 1990). The affinity of ANO2 for the spin-label-lipid hapten is much lower than that of GK14-1 for the TNP lipid hapten, and therefore, specific binding to supported lipid monolayers was only observable at binding site concentrations greater than 5 mol% (Timbs et al., 1991). The requirement of this antibody for such high-binding site concentrations in the membrane (i.e., a site density that is larger than the highest possible antibody packing density) can cause serious problems when the data are evaluated by simple Langmuir adsorption isotherms (for a discussion of the "large ligand" effect, see Tamm and Bartoldus, 1988). However, when compared on a qualitative, empirical basis, the apparent binding constant for bivalent antibody-ligand binding in solution ($K = 2.7 \times 10^6$ M$^{-1}$) is similar to the one for the binding of the bivalent antibody to membrane-bound target sites ($K = 3 \times 10^6$ M$^{-1}$), as it has also been found for the higher affinity antibody GK14-1. Part of this effect could be due to bivalency of the antibody because the binding of $F_{ab}$ fragments to the lipid hapten was decreased by an order of magnitude or more. Timbs *et al.* (1991) have further tried to measure the relative fractions of mono- and bivalently bound antibodies by competitive binding studies in which membrane-bound bivalent antibodies were displaced with soluble hapten. These data suggested that from 50 to 90% of the antibody was bivalently bound to the membrane although a clear decision could not be made. The antibody surface concentration at saturation was estimated to be 6000 molecules/$\mu$m$^2$ by taking the ratio of the fluorescence intensities that were obtained in the presence and absence, respectively, of 25 mol% lipid hapten in the supported membranes. This value translates into one antibody per 170 nm$^2$, which is in approximate agreement with the loose packing of antibodies on monolayer-coated EM grids (Uzgiris and Kornberg, 1983) but much larger than the actual cross-sectional area (about 60 nm$^2$) of IgGs as observed by electron microscopy (Coleman *et al.*, 1976).

## 2. Antibody-$F_c$-Receptor Binding

The first TIRFM binding study in which the membrane binding sites were integral membrane proteins was reported by Poglitsch and Thompson (1990). The supported membrane was prepared by directly fusing vesicles that were prepared from plasma membrane fragments of a macrophage-related cell line to hydrophilic quartz slides. Among other proteins, these membranes contained an $F_{c\gamma}$-receptor, specific for binding the $F_c$ region of $\gamma$-immunoglobulins. The receptor concentration in these membranes was measured to be about 50 molecules/$\mu$m$^2$, which was too low to study directly the binding of the antibodies via their $F_c$ domains but adequate for studying the binding of a monoclonal anti-$F_c$-receptor antibody. However, $F_c$–$F_{c\gamma}$-receptor interactions could be measured with an indirect method in which the inhibition of the binding of the fluorescently labeled monoclonal anti-$F_c$-receptor antibody by the $F_c$-domain-containing, unlabeled antibodies was followed by TIRFM. A binding constant of about $10^5$ M$^{-1}$ was obtained.

Following up on this work, Poglitsch *et al.* (1991) then purified the $F_{c\gamma}$-receptor and reconstituted it into liposomes and finally into supported planar bilayers. By this method, they were able to increase the surface concentration of receptors to about 800 to 1700 molecules/$\mu$m$^2$ and to directly measure the binding of labeled IgGs by TIRFM. Binding constants of $1 \times 10^5$ and $9 \times 10^5$ M$^{-1}$ were determined depending on whether polyclonal mouse IgG or the monoclonal ANO2 antibody was used. The binding con-

FIGURE 11.13

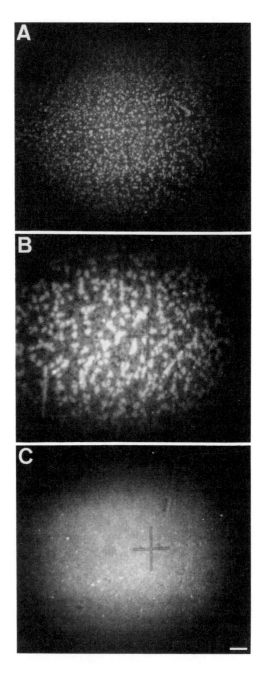

stant of the monoclonal anti-$F_c$-receptor antibody was more than 10-fold smaller when measured with purified, reconstituted $F_c$-receptors instead of membranes prepared from native membrane fragments. The reasons for this difference are not yet well understood.

## 3. Laminin-Sulfatide Binding

Based on conventional solid phase radio-immune assays, it has been concluded that laminin, a large extracellular matrix glycoprotein, binds specifically to sulfatide lipids (e.g., 3-$SO_4$-galactosylceramide) with a binding constant of the order of $10^9$ $M^{-1}$ (Roberts *et al.*, 1985). TIRFM on supported phospholipid bilayers (SPBs) that contain the sulfatide should be a system to characterize these interactions in more detail and, more important, in a well-defined membrane environment that is not guaranteed by the undefined lipid deposition in plastic wells as is customary in conventional solid phase assays.

When fluorescein-labeled laminin was bound to sulfatide-containing SPBs no saturation could be detected (Kalb and Engel, 1991). Also, relatively high fluorescence intensities were observed when laminin was bound in the absence of bovine serum albumin (BSA) to lipid bilayers that contained no sulfatides. The binding levels were similar for all control lipid systems tested, and even for membranes containing up to 30 mol% sulfatides. In the presence of BSA, no binding was detectable to membranes that contained low amounts of sulfatides. The kinetics of the binding and the dissociation of bound laminin after the addition of ethylenediamine-$N,N,N',N'$-tetraacetic acid (EDTA) were comparable to the kinetics of the calcium-dependent self-assembly of laminin in solution. When the membrane surface was observed directly by TIRFM, the bound laminin displayed an inhomogeneous fluorescence distribution (Figs. 11.13A and 11.13B). Distinct areas of high fluorescence intensity could be distinguished from the background fluorescence and the size of these areas depended on

---

TIRFM images showing the aggregation of the laminin–nidogen complex at the surface of lipid bilayers. Different amounts of fluorescein-labeled laminin were bound to supported bilayers, which contained 100% sulfatide in the outer leaflet. A grainy appearance of the fluorescence indicates the aggregation of the protein on the surface. Larger aggregates are visualized as bright spots. Their average size increases with increasing protein concentration. (A) and (B) are typical fields at 10 nM and 40 nM laminin, respectively. In (C), 40 nM laminin was bound and subsequently treated with a 2 M excess of EDTA over $Ca^{2+}$. The brightness of the photographs cannot be taken as a measure of the absolute fluorescence intensity, which was much lower in (C) than in (B). The bar corresponds to 10 $\mu$m. (From Kalb and Engel, 1991, with permission.)

the lipid composition and the laminin surface concentration. Upon the addition of EDTA, these spots almost disappeared and a relatively homogeneous fluorescence was observed (Fig. 11.13C). These observations led to the conclusion that the binding of laminin to lipid membranes is closely linked with the (membrane-independent) laminin self-association, and, therefore, that the binding constant of laminin cannot be correctly determined by either the solid phase or the TIRFM assays. However, this example further demonstrates that TIRFM is a technique that can reveal various aspects of very complex interactions on membrane surfaces, including equilibrium binding, binding kinetics, and surface-induced self-aggregation.

## 4. Binding of Prothrombin to Negatively Charged Lipid Bilayers

In the blood coagulation cascade, prothrombin is converted to thrombin by factors $X_a$ and $V_a$. This conversion physiologically takes place on the surface of platelet membranes, but it can also be catalysed *in vitro* on the surface of negatively charged vesicles. The binding of prothrombin and two fragments of prothrombin, fragment 1 and prethrombin 1, to negatively charged lipid surfaces was recently demonstrated by TIRFM on SPBs (Tendian *et al.*, 1991). Binding of prothrombin and its fragments to bilayers of palmitoyloleoyl-phosphatidylcholine (POPC) was absent or low. When the membranes were composed of 70 mol% POPC and 30 mol% negatively charged bovine brain phosphatidylserine (PS), the specific binding of prothrombin depended markedly on the presence of $Ca^{2+}$. The apparent binding constants were measured to be $2.5 \times 10^6 \, M^{-1}$ in the presence of $Ca^{2+}$, and (3.3–7.7) $\times 10^4 \, M^{-1}$ in the absence of $Ca^{2+}$. The $Ca^{2+}$-independent binding site(s) could further be attributed to the carboxy-terminal two thirds of prothrombin (i.e., to prethrombin 1) because this fragment bound to POPC/PS (7:3) membranes in the presence *or* absence of $Ca^{2+}$ with an apparent binding constant of (6–11) $\times 10^4 \, M^{-1}$. In contrast the $Ca^{2+}$-specific binding site(s) could be attributed to prothrombin fragment 1, because this fragment was found to specifically bind to POPC/PS (7:3) only in the presence of 5 mM $Ca^{2+}$ (apparent binding constant, $K = 1 \times 10^6 \, M^{-1}$).

## 5. Formation of Supported Phospholipid Bilayers by Fusion of Vesicles to Supported Phospholipid Monolayers

Recently, we have employed TIRFM to study the adsorption and fusion of small unilamellar vesicles to supported phospholipid monolayers (Kalb *et al.* 1992). The goal of this study was to develop a new technique for the preparation of supported phospholipid bilayers. TIRFM is an ideal tool to follow in real time the formation of supported membranes, and TIR–FRAP can be used concomittantly to characterize the physical properties (lateral diffusion) of the bilayers during their construction. Figure 11.14 shows the kinetics of the adsorption of small unilamellar vesicles of phosphatidylcholine (POPC) to preformed supported phospholipid monolayers of POPC on quartz. The vesicles were labeled with 0.2 mol% NBD-phosphatidyl-ethanolamine. The rates of adsorption increased with increasing vesicle concentration, and for all concentrations, a biphasic time course was observed. The switch from the fast phase to the slow phase occurred at about 1500 fluorescence units and independent of the vesicle concentration. This critical fluorescence level exactly corresponded to the fluorescence level that was expected (from Langmuir–Blodgett experiments) for a single completed phospholipid bilayer. Therefore, the first phase of vesicle adsorption in Fig. 11.14A was attributed to the formation of a single supported phospholipid bilayer by vesicle fusion, and the second phase was attributed to the adsorption without fusion of further vesicles on the supported bilayer. The fast adsorption–fusion process was irreversible and the slow adsorption process was reversible because the latter vesicles could be washed off the membrane by extensive perfusion with buffer (Fig. 11.14B). Evidence for the fusion process on the supported monolayer came from TIR–FRAP experiments (see Section IV.E below).

FIGURE 11.14

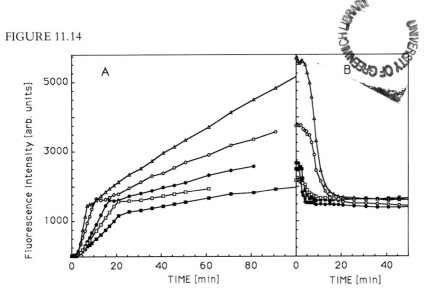

Formation of supported planar bilayers by fusion of POPC vesicles to supported phospholipid monolayers. (A) Time courses of the TIR fluorescence intensity increases at the interface between a quartz-supported planar phospholipid (POPC) monolayer and buffer after the addition of 35 $\mu$M ($\blacksquare$), 50 $\mu$M ($\square$), 75 $\mu$M ($\bullet$), 100 $\mu$M ($\circ$), and 200 $\mu$M ($\Delta$) phospholipid (POPC) vesicles containing 0.2 mol% NBD-egg PE. (B) Wash kinetics showing the removal of (reversibly) adsorbed phospholipid vesicles from the completed and stably bound supported phospholipid bilayer. (From Kalb *et al.*, 1992, with permission.)

## D. Cell-Supported Membrane Contacts

TIRFM and supported membranes have also successfully been used to image the contacts between single cells and reconstituted membranes. Cell–cell and membrane–membrane contacts are very important in many different biological systems. For instance, cell adhesion molecules of different kinds mediate cellular contacts between like and nonlike cells in development; different classes of lymphocytes cooperate at various stages in several immune responses by engaging in physical contacts of their respective cell surfaces (for instance, T-cell receptors interact with antigens and histocompatibility antigens on target cells); macrophages engulf foreign cells by antibody-$F_c$-receptor mediated phagocytosis; and enveloped viruses as well as exocytotic vesicles fuse with plasma membranes of eucaryotic cells after specific contacts between the two membranes have been formed. The supported membrane-attached cell configuration in conjunction with TIRFM offers a unique opportunity to directly visualize these cell-membrane contacts and to follow by specific labeling the fate of the molecules of interest upon contact formation. Since the evanescent wave penetrates about 100 to 200 nm into the liquid medium, cytoplasmic rearrangements near the cell-membrane contact regions may also be followed by this technique.

The first example in which the contact of a cell with a supported membrane was studied was the attachment of rat basophil leukemia cells to a haptenated supported phospholipid monolayer via a monoclonal antidinitrophenyl (DNP) IgE antibody (Weis *et al.*, 1982). When the DNP-specific antibody was labeled with fluorescein, a punctuate distribution of labeled domains within the contact region of each cell with the supported membrane was observed, which provided evidence for the clustering of the IgE and the bound IgE receptor in the cell membrane in these areas (Fig. 11.15A).

FIGURE 11.15

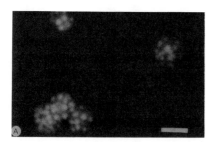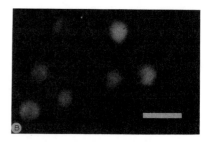

TIRFM images of basophil leukemia cells bound to dipalmitoylphosphatidylcholine monolayers containing 2 mol% dinitrophenyl lipid hapten. Before binding to the supported monolayers, the cells were incubated with either (A) fluorescein-labeled monoclonal antidinitrophenyl IgE or (B) 1:1 antidinitrophenyl IgE:fluorescein-labeled rat myeloma IgE. The scale bars are 25 μm. (From McConnell *et al.*, 1986, with permission.)

When, however, the cell was bound with unlabeled specific antibodies, but at the same time incubated with a fluorescein-labeled nonspecific IgE antibody, the fluorescence distribution in the contact region was uniform (Fig. 11.15B). This experiment indicated that only those IgE receptors that were linked via a specific antibody to the target membrane clustered into large domains at the interface between the cell and the artificial model membrane.

Other experiments involving cell-supported membrane contacts will be discussed below in Section IV.G, TIR Resonance Energy Transfer.

### E. Surface Diffusion by TIR–FRAP

Surface diffusion can be measured by TIR–FRAP in the presence of relatively large amounts of fluorescence in solution. Two techniques are available, namely spot photobleaching and interference fringe photobleaching. Spot photobleaching has already been discussed in Section IV.B on the surface binding kinetics of macromolecules. Fringe-pattern photobleaching uses interference fringes produced by two intersected totally internally reflected laser beams at the solid–liquid interface (Weis *et. al.*, 1982). If the interface region in which lateral diffusion is to be measured is not confined to a very small area (i.e., smaller than a single cell), fringe-pattern photobleaching is preferable over spot photobleaching because the measurements can be performed on larger areas and, therefore, with a better signal-to-noise ratio and because the fluorescence recoveries follow a single exponential function from which the lateral diffusion coefficient, $D_L$, can easily be extracted (Davoust *et al.*, 1982).

$$F(t) = F_\infty - (F_\infty - F_o) \, e^{-D_L a^2 t} \tag{12}$$

In Eq. 12, $F_o$, $F(t)$, and $F_\infty$ are fluorescence intensities at times zero, $t$, and a long time after the bleach pulse, and $a = 2\pi/p$ ($p$ = period) is the spatial frequency of the interference fringe pattern. Unfortunately, it is more difficult to determine the fraction of mobile molecules by fringe-pattern FRAP than it is, for instance, in normal pattern photobleaching with a Ronchi ruling. Therefore, TIR–FRAP is the preferred technique only when there is a need for surface selectivity [i.e., in the presence of significant amounts of fluorophores in solution (or in parts of a cell, which are far from the TIR interface)]. In all other instances, epiillumination FRAP yields much larger signal-to-noise ratios. The measurement of the mobile fraction depends on the depth of photobleaching and the maximal fluorescence recovery that can be expected for a 100% mobile fraction is 22%. (For theoretical curves of the dependency of the mobile fraction on the depth of photobleaching, see Tamm and Kalb, 1992).

The TIR–fringe FRAP technique was a valuable tool to follow the process of vesicle fusion to supported phospholipid monolayers (Kalb *et al.*, 1992). Lateral diffusion experiments were performed during vesicle fusion (see Fig. 11.14) with supported monolayers in 1 min time intervals. When the vesicles were labeled with a fluorescent phospholipid analog, no long-range lateral diffusion was observed during the fast kinetic phase of Fig. 11.14. At the breakpoint between the fast and slow kinetic phase of Fig. 11.14, the lipids suddenly started to diffuse over long distances with a lateral diffusion coefficient of $(4.0 \pm 0.5) \times 10^{-8}$ cm$^2$/s, which is expected for lipid diffusion in a phospholipid bilayer. The mobile fraction was high, about 90%, at this point. When more vesicles adsorbed during the slow kinetic phase, the diffusion coefficient remained constant and the mobile fraction decreased roughly inversely proportional to the fluorescence intensity. This was interpreted as evidence against further fusion of vesicles into large planar bilayers after the first bilayer was completed. The lateral diffusion coefficients in the first (Langmuir–Blodgett-deposited) leaflet of the bilayer were also measured during vesicle fusion by either TIR–FRAP or epiillumination FRAP. This layer was always connected and showed fast lateral diffusion with high mobile fractions at all times.

The lateral diffusion coefficient of an NBD-labeled phosphatidylethanolamine (NBD-PE) was measured by TIR–fringe FRAP in a supported monolayer of dipalmitoyl-phosphatidylcholine (Weis *et al.*, 1982). At room temperature, this diffusion coefficient was $2 \times 10^{-10}$ cm$^2$/s, indicative of a phospholipid monolayer in the "gel" state. A similar, or still smaller, lateral diffusion coefficient was found when fluorescein-labeled monoclonal IgE antibodies were bound to lipid haptens in the same type of supported membranes.

Lateral diffusion coefficients of adsorbed bovine serum albumin on polydimethylsiloxane were also measured by TIR–fringe FRAP (Tilton *et al.*, 1990a,b). When the surface concentration was increased from 0.1 to 0.7 fractional occupation of the total area by the protein, the diffusion coefficients decreased from about $5 \times 10^{-8}$ cm$^2$/s to about $6 \times 10^{-9}$ cm$^2$/s. Independent of the protein surface concentration, about 40% of all adsorbed BSA molecules were found to be mobile. The authors concluded that protein–protein lateral interactions significantly hindered the rate of surface diffusion, but a percolation threshold for diffusion was not found in the examined concentration range.

### F. Polarized TIR: Fluorophore Orientation and Rotational Diffusion

Some general theoretical aspects of polarized fluorescence excitation and emission near solid–liquid interfaces have been discussed in Sections II.B. and II.C. The two major applications of polarized light in microspectrofluorometry are determinations of orientation distributions of fluorophores at interfaces and measurements of rotational diffusion coefficients of fluorescently labeled molecules in membranes.

Orientation distributions are often conveniently expressed in terms of order parameters. For distributions that are axially symmetric around a principal director axis, it is sufficient to consider a single order parameter

$$S = \frac{3 < \cos^2\theta > - 1}{2} \tag{13}$$

where $\theta$ denotes the angle between the transition dipole moment of the fluorophore and the director, and the angular brackets are a time and ensemble average over the possible fluctuations and distributions of the fluorophore in its environment. $S$ can vary between $-0.5$ (for a rigid alignment perpendicular to the director) and $+1.0$ (for a rigid alignment parallel to the director). Intermediate values of $S$ occur for other orientations and when the molecules undergo rapid orientation fluctuations. For the case of supported membranes, the director is oriented parallel to the membrane normal (i.e., normal to the TIR-supporting solid–liquid interface).

Orientation distributions of head group-labeled NBD-dipalmitoyl phosphatidylethanolamine (DPPE), chain-labeled NBD(C12)-PE, and a chain-labeled lipid–peptide were measured by polarized TIR fluorescence in monolayers of dipalmitoyl phosphatidylcholine (DPPC) on alkylated microscope slides as a function of the monolayer transfer pressure (Thompson *et al.*, 1984). The order was high at all pressures for chain-labeled lipids with and without the conjugated peptide and decreased monotonically for the head group-labeled peptide when the coating pressure was increased from 10 to 40 mN/m. The same type of experiment was performed on dioctadecyl-indocarbocyanine (diI) in monolayers, which were composed of DSPC:DNP-DOPE (7:3) and supported on alkylated microscope slides (Timbs and Thompson, 1990). An order parameter of $-0.3$ was measured, which again suggested a preferential orientation distribution (weighted as defined by Eq. 13) of the absorption dipole moments near the plane of the membrane. This order parameter did not change when antibodies were bound to the surface of the membranes.

Rotational diffusion coefficients of macromolecules near a solid–liquid interface may be measured by polarized fluorescence photobleaching recovery (Smith *et al.*, 1981; Velez and Axelrod, 1988). This technique has been used so far only in transmission (epi-) fluorescence microscopy, but it should also be possible to combine it with TIR illumination. As with nonpolarized FRAP for the determination of lateral diffusion, polarized TIR–FRAP may only be advantageous over polarized epifluorescence for the determination of rotational diffusion, when fluorescent molecules far away from the interface obscure the desired signal from the interface. The

theory of polarized photobleaching for measurements of rotational diffusion coefficients has been worked out by Velez and Axelrod (1988). The
final expressions for the fluorescence recoveries of parallel and perpendicular
polarized light. $\Delta F_{\parallel, \perp}$, and for the time-resolved photobleaching anisotropy, $r_b(t)$, are, for the case of two-dimensional rotational diffusion,

$$\Delta F_{\parallel, \perp}(t) = a \pm b\, e^{-4D_Rt} - c\, e^{-16D_Rt} \tag{14}$$

and

$$r_b(t) = \frac{2b\, e^{-4D_Rt}}{3a - b\, e^{-4D_Rt} - 3c\, e^{-16D_Rt}} \tag{15}$$

respectively, where $D_R$ is the rotational diffusion coefficient and $a$, $b$, and $c$
are complicated functions of the duration and intensity of the bleach pulse
and the order parameter of the fluorophore. Expressions that are similar to
Eqs. 14 and 15 have been derived for three-dimensional rotational diffusion
(Velez and Axelrod, 1988). Corrections for high-aperture observation can
usually be neglected when working at $\leq 0.75$ NA and when the order parameter is significantly less than 1.

## G. TIR Resonance Energy Transfer

When two fluorophores with sufficient overlap between the donor emission
and the acceptor absorption spectra are in close ($<100$ Å) proximity, non-
radiative transfer of energy can occur between the excited states of the two
molecules (Förster, 1948). This resonance energy transfer manifests itself in
a decreased quantum yield of the fluorescence emission of the donor and an
increased fluorescence emission of the acceptor.

$$\frac{Q}{Q_o} = \frac{\tau_F}{\tau_o} = 1 - E \tag{16}$$

The transfer efficiency, $E$, in Eq. 16 can range between $0 \leq E \leq 1$ and is
measured from either the ratio of the quantum yields $Q$ or from the fluorescence lifetimes $\tau_F$ of the donor in the presence and absence of acceptor,
respectively. It has been shown that $E$ is a function of the sixth power of the
distance $r$ between the donor and acceptor (Förster, 1948):

$$E = \frac{R_o^6}{R_o^6 + r^6} \tag{17}$$

$$R_o = 9.8 \times 10^3\, (\kappa^2 n^{-4} Q_o\, J)^{1/6} \tag{18}$$

$$J = \int_0^\infty q(\lambda)\, \varepsilon(\lambda)\, \lambda^4\, d\lambda \tag{19}$$

Where $R_o$ is the critical distance (in Å), $\kappa^2$ is a factor that describes the
orientation of the emission transition dipole moment of the donor relative

to the absorption transition dipole moment of the acceptor (usually assumed to be 2/3), $n$ is the index of refraction of the medium between the donor and the acceptor, $J$ is the spectral overlap integral, $q(\lambda)$ is the spectral quantum yield distribution function of the donor in each interval $\lambda$ to $\lambda + d\lambda$, and $\varepsilon(\lambda)$ is the corresponding decadic extinction coefficient of the acceptor. Equations 17–19 describe molecular properties of the sample and are independent of the system that is used for fluorescence detection. Since relative ratios are compared in Eq. 16, this equation is also valid for the measurement of $E$ under high-aperture conditions on a fluorescence microscope (Herman, 1989) and for the measurement of $E$ under TIR illumination. Because of the sharp dependence on distance of Eq. 17, resonance energy transfer is extremely useful to measure distance changes in the neighborhood of $R_o$ (i.e., at around 40 Å for many donor-acceptor pairs). Molecular rearrangements that occur at distances larger than about $2R_o$ are unlikely to be detectable unless there is a marked change in the orientation factor $\kappa^2$. The theory of resonance energy transfer has also been worked out for two-dimensional systems and other specialized geometries of donor–acceptor distributions (Fung and Stryer, 1978; Shaklai *et al.*, 1977), but no closed form solution, such as the one for the three-dimensional fixed distance case (Eqs. 17–19), is found for these situations.

To date, there are only two reports in the literature in which resonance energy-transfer microscopy has been combined with TIR. In both cases, interactions of molecules of the immune system were studied in supported membranes (Watts *et al.*, 1986; Watts and McConnell, 1986). In these experiments, supported bilayers were prepared that contained the purified and Texas Red–labeled class II histocompatibility antigen, I-A$^d$. A fluorescein-labeled peptide was added and excited with an evanescent wave at 488 nm. No Texas Red fluorescence emission was observed under these conditions. Significant resonance energy transfer and Texas Red emission occurred only when peptide-specific and I-A$^d$–restricted T-lymphocytes were specifically bound to the supported planar membrane. This effect could be reversed with an excess of unlabeled peptide, with an antibody against I-A$^d$, or with irrelevant helper–inducer T-lymphocytes. These experiments demonstrated that a specific ternary complex of the histocompatibility antigen, the peptide antigen, and the T-cell receptor had been formed at the interface between the cell and the supported membrane and that a stable association of the peptide antigen with the histocompatibility antigen only occurred in the presence of a T-cell receptor that was at the same time specific for the antigen and the histocompatibility antigen.

## V. PROSPECTUS

Total internal reflectance fluorescence microscopy has now come to a stage where a practically unlimited number of important biochemical and cell biological experiments can be planned and carried out successfully. The relevant basic theory is well developed and the practical implementation to studying cell-substrate contacts by imaging and surface-binding equilibria by microspectrofluorometry has been shown to be experimentally feasible in numerous cases. A particularly promising area of TIRFM is its application to studying the binding of specific ligands to receptors in supported membranes. Receptor interactions of macromolecular or low molecular weight ligands can be studied, provided they can be fluorescently labeled without an interference of the label with the binding properties of the ligand. Alternatively, binding may also be studied indirectly with labeled competitive inhibitors, or, if quartz optics are available, by intrinsic tryptophane fluorescence. The major advantages of TIRFM over other techniques for a quantitative and thermodynamically meaningful characterization of ligand–receptor interactions in membranes are (i) the receptors are kept in their native lipid bilayer environment in supported membranes, which is not the case in the very popular solid phase assays, such as radioimmune assays (RIA) and enzyme-linked immunosorbant assays (ELISA); (ii) depending on the receptor density in the membrane, binding constants in the range from about $10^4$ $M^{-1}$ to about $10^8$ $M^{-1}$ are expected to be measurable by TIRFM on supported membranes; (iii) light-scattering artifacts, which often become a serious problem in many other spectroscopic techniques, are virtually nonexistent in TIRFM; and (iv) dynamic processes, such as lateral diffusion and receptor clustering before and after ligand binding, are addressable by (epifluorescence or TIR) FRAP and/or direct imaging in the supported membrane-TIRFM system. Also, fluorescence resonance energy transfer has found so far only limited application in TIRFM (Watts et al., 1986), but it will doubtlessly become a powerful additional tool for studying molecular interactions in membranes or other biological structures near solid–liquid interfaces.

One common problem of TIRFM binding studies is the proper calibration of absolute fluorescence intensities in the sample. Even if the excitation intensity is carefully controlled (e.g., by a photodiode before the TIR coupling prism), fluorescence intensities are difficult to convert into surface concentrations. Two possible approaches to this problem are (i) to calibrate the system with a (membrane) surface that has a well-defined concentration of binding sites that can be saturated at high ligand concentrations, such as the antibody-lipid hapten system (Kalb et al., 1990); or (ii) to measure the fluorescence intensity of a known concentration of fluorescent ligand in the evanescent wave in the absence of specific binding sites on the surface (Pisarchick and Thompson, 1990). Although these two methods work pretty well, they are still not entirely satisfactory for routine applications. Better calibration systems are needed in the future.

Another very important future development for applications of TIRFM to membrane studies will be to improve the currently available methods for reconstituting large integral membrane proteins into supported planar bilayers. Although the present reconstitution techniques can produce membranes with receptors that maintain their specific binding functions, it has not yet been possible to reconstitute these proteins into planar membranes in a laterally mobile form. However, the lateral mobility of receptors in cell membranes is believed to be a key element of many signal transduction pathways. Therefore, once membrane receptors can be reconstituted into supported bilayers in a laterally mobile form, several postbinding steps of signal transduction may also be studied in this system. We have recently undertaken some steps towards this goal and have developed a new method to prepare supported phospholipid bilayers. The technique consists of a fusion of vesicles to the hydrophobic surface of a supported (Langmuir–Blodgett-deposited) phospholipid monolayer (Kalb *et al.*, 1992). Initial experiments on the reconstitution of the small membrane protein cytochrome $b_5$ by the monolayer-fusion technique have yielded a significant fraction (about 30 to 40%) of laterally mobile cytochrome $b_5$ in supported planar bilayers (Kalb and Tamm, 1992).

In conclusion, TIRFM on supported membranes and other substrates appears to be a very powerful new technique to quantitatively study solute–interface interactions, which are ubiquitous in so many biological recognition processes.

## ACKNOWLEDGMENTS

Supported by grants from the Swiss National Science Foundation and the U.S. Public Health Service.

## REFERENCES

Aikens, R. S., Agard, D. A., and Sedat, J. W. (1989). Solid-state imagers for microscopy. *Meth. Cell Biol.* **29**, 291–313.

Arndt-Jovin, D. J., and Jovin, T. (1989). Fluorescence labeling and microscopy of DNA. *Meth. Cell Biol.* **30**, 417–447.

Axelrod, D. (1981). Cell-substrate contacts illuminated by total internal reflection fluorescence. *J. Cell Biol.* **89**, 14–145.

Axelrod, D., Burghardt, T. P., and Thompson, N. L. (1984). Total internal reflection fluorescence. *Ann. Rev. Biophys. Bioeng.* **13**, 247–268.

Axelrod, D., Fulbright, R. M., and Hellen, E. H. (1986). Adsorption kinetics on biological membranes: Measurement by total internal reflection fluorescence. *In* "Applications of Fluorescence in the Biomedical Sciences." (D. L. Taylor, A. S. Waggoner, R. E. Murphy, F. Lanni, and R. R. Birge, eds.), pp. 461–476, Alan R. Liss, Inc., New York, NY.

Bloch, R. J., Velez, M., Krikorian, J. G., and Axelrod, D. (1989). Microfilaments and actin-associated proteins at sites of membrane-substrate attachment within acetylcholine receptor clusters. *Exp. Cell Res. 182,* 593–596.

Burghardt, T. P., and Axelrod, D. (1981). Total internal reflection/fluorescence photobleaching recovery study of serum-albumin adsorption dynamics. *Biophys. J. 33,* 455–467.

Burghardt, T. P., and Axelrod, D. (1983). Total internal reflection fluorescence study of energy transfer in surface-adsorbed and dissolved bovine serum albumin. *Biochemistry 22,* 979–985.

Coleman, P. M., Deisenhofer, J., and Huber, R. (1976). Structure of the human antibody molecule Kol (Immunoglobulin G1): An electron density map at 5Å resolution. *J. Mol. Biol. 100,* 257–282.

Darst, S. A., Robertson, C. R., and Berzofsky, J. A. (1988). Adsorption of the protein antigen myoglobin affects the binding of conformation-specific monoclonal antibodies. *Biophys. J. 53.*

Davoust, J., Devaux, P. F., and Leger, L. (1982). Fringe pattern photobleaching, a new method for the measurement of transport coefficients of biological macromolecules. *EMBO J. 1,* 1233–1238.

Förster, T. (1948). Intermolecular energy migration and fluorescence. *Ann. Phys. (Leipzig) 2,* 55–75.

Fung, B. K., and Stryer, L. (1978). Surface density determination in membranes by fluorescence energy transfer. *Biochemistry 17.*

Gingell, D., Todd, I., and Bailey, J. (1985). Topography of cell-glass apposition revealed by total internal reflection fluorescence of volume markers. *J. Cell Biol. 100,* 1334–1338.

Harrick, N. J. (1965). Electric field strengths at totally reflecting interfaces. *J. Opt. Soc. Am. 55,* 851–857.

Harrick, N. J. (1967). "Internal Reflection Spectroscopy." Second Edition, Harrick Scientific Corporation, Ossining, NY.

Harrick, N. J., and du Pré, F. K. (1966). Effective thickness of bulk materials and of thin films for internal reflection spectroscopy. *Appl. Opt. 5,* 1739–1743.

Harrick, N. J., and Loeb, G. I. (1973). Multiple internal reflection fluorescence spectrometry. *Anal. Chem. 45,* 687–691.

Hellen, E. H., and Axelrod, D. (1987). Fluorescence emission at dielectric and metal-film interfaces. *J. Opt. Soc. Am. 4,* 337–350.

Hellen, E. H., Fulbright, R. M., and Axelrod, D. (1988). Total internal reflection fluorescence: Theory and application at biosurfaces. *In* "Spectroscopic Membrane Probes, Vol. II." (L. M. Loew, ed.), pp. 47–79, Press Inc., Boca Raton, Fla.

Herman, B. (1989). Resonance energy transfer microscopy. *Methods in Cell Biology 30,* 219–243.

Hirschfeld, T. (1965). Total reflection fluorescence (TRF). *Can. Spectroscopy 10,* 128.

Hlady, V., Reinecke, D. R., and Andrade, J. D. (1985). Total internal reflection intrinsic fluorescence (TIRIF) spectroscopy applied to protein adsorption. *In* "Surface and Interfacial Properties of Biomedical Polymers, Vol. 2: Protein Adsorption." (J. D. Andrade, ed.), pp. 81–119, Plenum Press, Inc., New York, NY.

Hlady, V., Reinecke, D. R., and Andrade, J. D. (1986). Fluorescence of adsorbed protein layers: Quantitation of total internal reflection fluorescence. *J. Colloid Interface Sci. 3,* 555–569.

Inoue, S. (1986). "Video Microscopy." Plenum Press, New York, NY and London.

Kalb, E., and Engel, J. (1991). Binding and calcium-induced aggregation of laminin onto solid lipid bilayers. *J. Biol. Chem. 267,* 19047–19052.

Kalb, E., and Tamm, L. K. (1992). Incorporation of cytochrome $b_5$ into supported phospholipid bilayers by vesicle fusion to supported monlayers. *Thin Solid Films 210/211,* 763–765.

Kalb, E., Engel, J., and Tamm, L. K. (1990). Binding of proteins to specific target sites in membranes measured by total internal reflection fluorescence microscopy. *Biochemistry 29,* 1607–1613.

Kalb, E., Frey, S., and Tamm, L. K. (1992). Formation of supported planar bilayers by fusion of vesicles to supported phospholipid monolayers. *Biochim. Biophys. Acta 1103,* 307–316.

Kronick, M. N., and Little, W. A. (1975). A new immunoassay based on fluorescence excitation by internal reflectance spectroscopy. *J. Immunol. Meth. 8,* 235–240.

Lanni, F., Waggoner, A. S., and Taylor, D. L. (1985). Structural organization of interphase 3T3 fibroblasts studied by total internal reflection fluorescence microscopy. *J. Cell Biol. 100,* 1091–1102.

Lowe, R., Hlady, V., Andrade, J. D., and van Wagenen, R. A. (1986). Human haptoglobin adsorption by a total internal reflectance fluorescence method. *Biomaterials 7,* 41–44.

McConnell, H. M., Watts, T. H., Weis, R. M., and Brian, A. A. (1986). Supported planar membranes in studies of cell–cell recognition in the immune system. *Biochim. Biophys. Acta. 864,* 95–106.

Nakache, M., Gaub, H. E., Schreiber, A. B., and McConnell, H. M. (1986). Topological and modulated distribution of surface markers on endothelial cells. *Proc. Natl. Acad. Sci. U.S.A. 83,* 2874–2878.

Pisarchick, M. L., and Thompson, N. L. (1990). Binding of a monoclonal antibody and its $F_{ab}$ fragment to supported phospholipid monolayers measured by total internal reflection fluorescence microscopy. *Biophys. J. 58,* 1235–1249.

Poglitsch, C. L., and Thompson, N. L. (1990). Interaction of antibodies with $F_c$ receptors in substrate-supported planar membranes measured by total internal reflection fluorescence microscopy. *Biochemistry 29,* 248–254.

Poglitsch, C. L., Sumner, M. T., and Thompson, N. L. (1991). Binding of IgG to MoFcγRII purified and reconstituted into supported planar membranes as measured by total internal reflectance fluorescence microscopy. *Biochemistry 30,* 6662–6671.

Roberts, D. D., Rao, C. N., Magnani, J. L., Spitalnik, S. L., Liotta, L. A., and Ginsberg, V. (1985). Laminin binds specifically to sulfated glycolipids. *Proc. Natl. Acad. Sci. U.S.A. 82,* 1306–1310.

Schmidt, C. F., Zimmerman, R. M., and Gaub, H. E. (1990). Multilayer adsorption of lysozyme on a hydrophobic substrate. *Biophys. J. 57,* 577–588.

Shaklai, N., Yguerabide, J., and Ranney, H. M. (1977). Interaction of hemoglobin with red blood cell membranes as shown by fluorescent chromophore. *Biochemistry 16*, 5585–5592.

Smith, L. M., Weis, R. M., and McConnell, H. M. (1981). Measurement of rotational motion in membranes using fluorescence recovery after photobleaching. *Biophys. J. 36*, 73–91.

Spring, K. R., and Lowy, R. J. (1989). Characteristics of low-light level television cameras. *Meth. Cell Biol. 29*, 269–289.

Sui, S., Urumow, T., and Sackmann, E. (1988). Interaction of insulin receptors with lipid bilayers and specific and non-specific binding of insulin to supported membranes. *Biochemistry 27*, 7463–7469.

Tamm, L. K. (1988). Lateral diffusion and fluorescence microscope studies of a monoclonal antibody specifically bound to supported phospholipid bilayers. *Biochemistry 27*, 1450–1457.

Tamm, L. K., and Bartoldus, I. (1988). Antibody binding to lipid model membranes: The large ligand effect. *Biochemistry 27*, 7453–7458.

Tamm, L. K., and Kalb, E. (1992). Microspectrofluorometry on supported planar membranes. *In* "Molecular Luminescence Spectroscopy: Methods and Applications, Part Three." (S. G. Shulman, ed.), John Wiley & Sons, Inc., New York, NY (forthcoming).

Tamm, L. K., and McConnell, H. M. (1985). Supported phospholipid bilayers. *Biophys. J. 47*, 105–113.

Tendian, S. W., Lentz, B. R., and Thompson, N. L. (1991). Evidence from total internal reflection fluorescence microscopy for calcium-independent binding of prothrombin to negatively charged planar phospholipid membranes. *Biochemistry 30*, 10991–10999.

Thompson, N. L., and Axelrod, D. (1983). Immunoglobulin surface-binding kinetics studied by total internal reflection with fluorescence correlation spectroscopy. *Biophys. J. 43*, 103–114.

Thompson, N. I., Burghardt, T. P., and Axelrod, D. (1981). Measuring surface dynamics of biomolecules by total internal reflection fluorescence with photobleaching recovery or correlation spectroscopy. *Biophys. J. 33*, 433–454.

Thompson, N. L., McConnell, H. M., and Burghardt, T. P. (1984). Order in supported phospholipid monolayers detected by the dichroism of fluorescence excited with polarized evanescent illumination. *Biophys. J. 46*, 739–747.

Thompson, N. L., Palmer III, A. G., Wright, L. L., and Scarborough, P. E. (1988). Fluorescence techniques for supported planar model membranes. *Comments Mol. Cell Biophys. 5*, 109–131.

Tilton, R. D., Robertson, C. R., and Gast, A. P. (1990a). Lateral diffusion of bovine serum albumin adsorbed at the solid–liquid interface. *J. Colloid Interface Sci. 137*, 192–203.

Tilton, R. D., Gast, A. P., and Robertson, C. R. (1990b). Surface diffusion of the interacting proteins. Effect of concentration on the lateral mobility of adsorbed bovine serum albumin. *Biophys. J. 58*, 1321–1326.

Timbs, M. M., and Thompson, N. L. (1990). Slow rotational mobilities of antibodies and lipids associated with substrate-supported phospholipid monolayers as measured by polarized photobleaching recovery. *Biophys. J. 58*, 413–428.

Timbs, M. M., Poglitsch, C. L., Pisarchick, M. L., Sumner, M. T., and Thompson, N. L. (1991). Binding and mobility of antidinitrophenyl monoclonol antibodies on fluid-like, Langmuir–Blodgett phospholipid monolayers containing dinitrophenyl-conjugated phospholipids. *Biochim. Biophys. Acta 1064*, 219–228.

Todd, I., Melior, J. S., and Gingell, D. (1988). Mapping cell-glass contacts of dictyostelium amoebae by total internal reflection aqueous fluorescence overcomes a basic ambiguity of interference reflection microscopy. *J. Cell Science 89*, 107–114.

Uzgiris, E. E., and Kornberg, R. D. (1983). Two-dimensional crystallization technique for imaging macromolecules with application to antigen–antibody–complement complexes. *Nature 301*, 125–129.

Velez, M., and Axelrod, D. (1988). Polarized fluorescence photobleaching recovery for measuring rotational diffusion in solutions and membranes. *Biophys. J. 53*, 575–591.

von Tscharner, V., and McConnell, H. M. (1981). Physical properties of lipid monolayers on alkylated planar glass surfaces. *Biophys. J. 36*, 421–427.

Watkins, I., and Robertson, C. R. (1977). A total internal-reflectance technique for the examination of protein adsorption. *J. Biomed. Mater. Res. 11*, 915–938.

Watts, T. H., and McConnell, H. M. (1986). High-affinity fluorescent peptide binding to I-A$^d$ in lipid membranes. *Proc. Natl. Acad. Sci. U.S.A. 83*, 9660–9664.

Watts, T. H., Gaub, H., and McConnell, H. M. (1986). T-cell-mediated association of peptide antigen and major histocompatibility complex protein detected by energy transfer in an evanescent wave field. *Nature 320*, 179–181.

Weis, R. M., Balakrishnan, K., Smith, B. A., and McConnell, H. M. (1982). Stimulation of fluorescence in a small contact region between rat basophil leukemia cells and planar lipid membrane targets by coherent evanescent radiation. *J. Biol. Chem. 257*, 6440–6445.

Zimmerman, R. M., Schmidt, C. F., and Gaub, H. E. (1990). Absolute quantities and equilibrium kinetics of macromolecular adsorption measured by fluorescence photobleaching in total internal reflection. *J. Colloid Interfac. Sci. 139*, 268–280.

# 12
# LASER SCANNING CONFOCAL MICROSCOPY OF LIVING CELLS

**John J. Lemasters, Enrique Chacon, George Zahrebelski, Jeffrey M. Reece, and Anna-Liisa Nieminen**
Department of Cell Biology & Anatomy
University of North Carolina at Chapel Hill

I. INTRODUCTION
II. OPTICAL PRINCIPLES OF CONFOCAL MICROSCOPY
III. PINHOLE SIZE AND CONFOCALITY
IV. COMPARISON OF CONVENTIONAL AND CONFOCAL MICROSCOPE IMAGES
V. PRACTICAL CONSIDERATIONS FOR VIEWING LIVING CELLS BY CONFOCAL MICROSCOPY

VI. IMAGING CELL VOLUMES AND SURFACES
VII. VISUALIZATION OF ORGANELLES
VIII. ELECTRICAL GRADIENTS IN LIVING CELLS
IX. ION IMAGING
X. CONCLUSION
ACKNOWLEDGMENTS
REFERENCES

## I. INTRODUCTION

Conventional light microscopes create images with a depth of field at high power of 2–3 $\mu$m. Since the resolving power of optical microscopy is 0.2 $\mu$m, superimposition of detail within this thick plane of focus obscures structural detail that might otherwise be resolved. For specimens more than a few microns thick, light from out-of-focus planes also degrades the image. This is especially noticeable in fluo-

Optical Microscopy: Emerging Methods and Applications

rescence microscopy where out-of-focus fluorescence creates a diffuse halo around objects under study. One way to circumvent these problems is to use thin specimens, such as 1-$\mu$m plastic sections of fixed tissues. However, for study of living cells and tissues, this approach is not feasible.

The advent of confocal microscopy solves the dilemma of studying thick specimens by optical microscopy. Confocal microscopy creates images whose depth of field is less than 1 $\mu$m thick. In addition, confocal microscopes reject light from out-of-focus planes. Together smaller depth of field and rejection of out-of-focus light combine to produce images that are remarkably detailed. In many respects, a confocal microscope is computerized axial tomography (CAT) for cells. The improvement in resolution of useful information by confocal microscopy over conventional microscopy is comparable to that by CAT scanning over conventional radiographs. Increasingly, confocal microscopy is becoming an essential analytical tool for studying the structure and physiology of living cells.

## II. OPTICAL PRINCIPLES OF CONFOCAL MICROSCOPY

In confocal microscopy, a demagnified light beam is projected on the specimen (Fig. 12.1) (Minsky, 1961). The crossover of the conical beam of light is typically about 0.2 $\mu$m in diameter for a high numerical aperture objective lens. Light reflected or fluoresced by the specimen is focused on a pinhole aperture by the objective lens. Only light from a narrow plane of focus traverses the pinhole to strike a light detector beyond. Light originating from above or below the image plane strikes the walls of the pinhole aperture and is not transmitted to the detector. In this optical arrangement, one point in the image plane is observed at a time. Therefore, to generate a two-dimensional image, the light beam is scanned across the specimen. This is accomplished either by using a rotating disk (Nipkow disk) containing pinholes (Petran et al., 1968) or a laser beam reflected off of vibrating mirrors (White et al., 1987; Wilson, 1990). Using a Nipkow disk in what is called a tandem-scanning confocal microscope, the confocal image may be viewed directly and recorded by photographic film. In laser scanning confocal microscopy, a cathode ray tube (CRT) recreates the image. Much as in scanning electron microscopy, detector signal modulates CRT intensity, and CRT scan rate is synchronized to the laser scan rate. Typically, a computer memory also stores the confocal images. By collecting additional scans after moving the plane of focus up and down, subsequent image processing can reconstruct a three-dimensional image.

A new variant of confocal microscopy is slit-scanning confocal microscopy (Wilson and Hewlett, 1990). In this approach, a vibrating mirror scans a narrow slit of light across the specimen. Reflected or fluoresced light is descanned by the same mirror and passed through a variable detection slit.

FIGURE 12.1

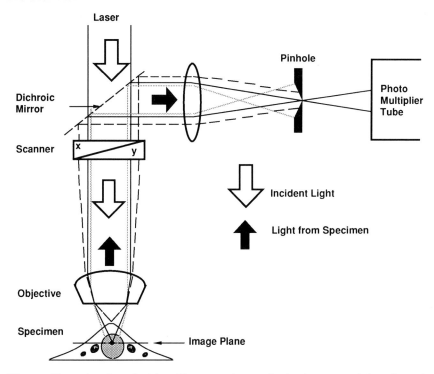

Diagram illustrating the principle of laser scanning confocal microscopy. A laser beam is passed through an x-y scanner, collimated to a small spot by the objective lens, and scanned across the specimen. Fluoresced light is collected by the objective, descanned, and directed by a dichroic mirror to a pinhole aperture placed in the conjugate image plane. Light originating from the specimen plane of focus is passed through to a photomultiplier detector. Light from above or below the specimen focal plane strikes the walls of the aperture and is not transmitted. Contrary to what is depicted, for a typical confocal system the dichroic mirror reflects laser light and transmits fluorescence, although the confocal imaging principle is the same.

Light transmitting the detection slit is scanned again, and an image is viewed through oculars by photographic film or by a video camera. Because a slit aperture rather than a pinhole aperture is employed, the degree of confocality is less with a slit-scanning system. Nevertheless, confocal slices approaching 1 $\mu$m in thickness are achievable.

## III. PINHOLE SIZE AND CONFOCALITY

In essence, a confocal microscope produces slices through a specimen of defined thickness. For a high numerical aperture oil-immersion lens, thickness of each confocal slice can be less than 1 $\mu$m. Thickness of confocal

FIGURE 12.2

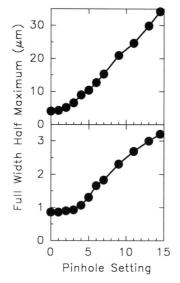

Pinhole diameter and thickness of confocal slices. The full width half-maximal thickness of confocal slices was determined for Nikon (*upper panel*) 10×, 0.5 NA Fluor and (*lower panel*) 60×, 1.4 NA planapochromat lenses as the pinhole micrometer setting was varied. A Biorad MRC-600 laser scanning confocal microscope was operated in the reflectance mode using a front-reflecting mirror tilted at a slight angle and illuminated with the 488-nm line of an argon laser. Full width half-maximal depth of field was determined from the width of the reflected band of light in the confocal image and the angle of the mirror. Pinhole micrometer settings of 0 and 14.5 correspond to pinhole diameters of 0.5 and 15 mm, respectively.

slices increases as the diameter of the pinhole aperture increases (Fig. 12.2). For large pinholes, confocality is essentially lost. Conversely, slice thickness decreases as the pinhole becomes smaller. However, below a minimum pinhole size, confocality no longer improves appreciably as the pinhole diameter decreases although brightness continues to decline. Thus, for instruments with variable pinholes, an optimal diameter should be determined empirically for the best combination of slice thickness and brightness. For the Biorad MRC-600 laser scanning confocal microscope, maximum brightness at the smallest depth of field occurs at a pinhole setting of about 4 (Fig. 12.2).

## IV. COMPARISON OF CONVENTIONAL AND CONFOCAL MICROSCOPE IMAGES

The improvement of resolution and clarity provided by confocal microscopy is readily apparent when a photomicrograph obtained with conventional microscope optics is compared to the corresponding confocal image. Figure 12.3 illustrates cultured rat hepatocytes stained with rhodamine 123, a cationic fluorophore that accumulates electrophoretically into mitochondria in response to the mitochondrial membrane potential (Johnson *et al.,* 1981; Emaus *et al.,* 1986). Figure 12.3A is an image obtained using conventional epillumination fluorescence microscopy. Individual mitochondria are easily identified at the periphery where the hepatocyte is thinnest, but in thicker central portions of the cell, superimposition and out-of-focus fluorescence obscure mitochondrial outlines. Figure 12.3B is a confocal image of a hepatocyte labeled under similar conditions. In the confocal image,

FIGURE 12.3

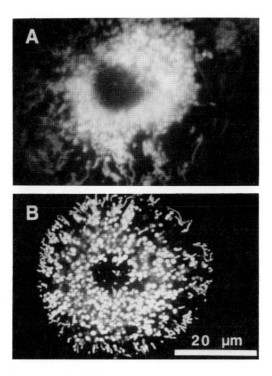

Comparison of rhodamine-123-loaded hepatocytes viewed by conventional and confocal fluorescence microscopy. (A) Conventional micrograph of rhodamine-123-labeled mitochondria in a cultured hepatocyte. The image was collected with an MTI-Dage Model 66 ISIT video camera using a 1.25 NA oil-immersion lens on a Zeiss IM35 inverted microscope. The image is the average of 256 video frames collected in 8.5 s. (B) Confocal image of a rhodamine-123-labeled hepatocyte loaded under similar conditions. The confocal image was obtained in 9 s from six scans using a Biorad MRC-500 laser scanning confocal microscope. Magnification of the two images is approximately the same.

mitochondria are sharply defined spheres and filaments. The improvement of resolution is immediately obvious.

## V. PRACTICAL CONSIDERATIONS FOR VIEWING LIVING CELLS BY CONFOCAL MICROSCOPY

Confocal microscopy intrinsically requires greater illumination than non-confocal microscopy. This is because confocal microscopy collects light from only a fraction of the illuminated volume. As a consequence, photo-damage and photobleaching are greater considerations in confocal micros-copy, especially for the study of living cells where repeated measurements over time may be required.

To produce confocal images with the least amount of photodamage, the first step is to select a confocal microscope with maximal detector sensitivity and optimal light-transmitting characteristics. In particular, a highly efficient light path from the specimen to the detector is impor-tant. In general, photomultiplier circuits should be operated at the high-est gain for maximal sensitivity. Laser power usually greatly exceeds that required for imaging. For most applications, laser intensity can be at-tenuated by 100–1000-fold using neutral density filters without loss of image quality. Casual viewing should be avoided. A typical beginner's mistake is to scan the specimen continuously at high laser power while making minor adjustments or previewing fields of interest. This leads to significant photobleaching and phototoxicity before an experiment can even begin.

Some fluorophores, for example fluorescein and acridine orange, are particularly prone to photobleaching and/or phototoxicity. They should be avoided. Other fluorophores, such as Texas Red and rhodamine dyes, are quite stable. The examples of Figs. 12.3–12.7 illustrate living cells loaded with calcein, rhodamine 123, rhodamine-dextran, tetramethylrhodamine methyl ester, and SNARF-1. By using maximal photomultiplier gain and minimal laser intensity, hundreds of images were obtained with these fluo-rophores with negligible photobleaching and photodamage.

To improve signal-to-noise ratios, it is usually necessary to average sev-eral scans, but this can be overdone. In cultured hepatocytes and myocytes, adequate signal-to-noise ratios can be achieved from only four or five scans. Additional scanning provides little improvement relative to the increased exposure to the laser beam.

Since not all applications require the thinnest possible confocal slice, sensitivity can be increased by opening the pinhole aperture. Doubling the diameter of the pinhole quadruples sensitivity but only doubles the thick-ness of the optical section (Fig. 12.2). Many laser scanning confocal micro-scopes offer a variable aperture or choice of pinholes for this reason. Thus, for light-sensitive specimens, a large pinhole setting should be considered so that laser power can be attenuated to an acceptable level.

## VI. IMAGING CELL VOLUMES AND SURFACES

Basic aspects of cell structure include size, shape, and surface topography. Confocal microscopy combined with three-dimensional digital reconstruction techniques can provide this structural information for single living cells. The experimental strategy is to load the cytoplasm with fluorescent dye and collect confocal images through the entire thickness of the cell. Subsequently, digital image processing reconstructs the cell volume in three dimensions. Calcein, whose fluorescence is unaffected by physiologic changes of pH and ion concentration, is a useful fluorophore for this purpose (Zahrebelski *et al.*, 1991; Chacon *et al.*, 1992). Calcein is loaded by incubating cells with calcein acetoxymethyl ester. Intracellular esterases cleave the ester bonds to trap the free acid form of fluorophore inside the cytoplasm. Cultured hepatocytes and myocytes load well when incubated with about 5 $\mu$M calcein-acetoxymethyl ester for 30–60 min at 37°C.

Once loaded, confocal images are collected through the entire thickness of the cell. Many confocal systems will do this automatically by using stepper motors attached to the focus knobs of the microscope. After images are collected, they are transferred to a high-speed imaging workstation for reconstruction or "rendering" of the cell volume in three dimensions. Presently, we use a Silicon Graphics computer and software developed by Vital Images for this purpose.

Figure 12.4 illustrates the results of such a volume rendering. A cultured adult rabbit cardiac myocyte was loaded with calcein and exposed to the metabolic inhibitors, NaCN and 2-deoxyglucose, a model of "chemical hypoxia" that mimics the adenosine-5'-triphosphate (ATP) depletion and reductive stress of anoxia (Bond *et al.*, 1991). After 24 min before any major structural changes had occurred, confocal images were collected, and cell volume was rendered on an imaging workstation. The top panels show a stereo pair of the volume rendering, which can be viewed in three dimensions using stereoscopic glasses. In addition, the volume rendering is shaded to give a sense of depth and surface contour.

The reconstruction of the cultured myocyte reveals a wealth of surface detail. In particular, regularly spaced riblike structures are the dominant surface feature. These structures represent impressions of underlying sarcomeres on the sarcolemma. Interestingly, these structures are not discernable in single confocal images but only become evident in a three-dimensional reconstruction of a through-focus series. Volume renderings obtained in this way rival scanning electron micrographs in detail and clarity, but unlike scanning electron micrographs, they are three-dimensional images of

FIGURE 12.4

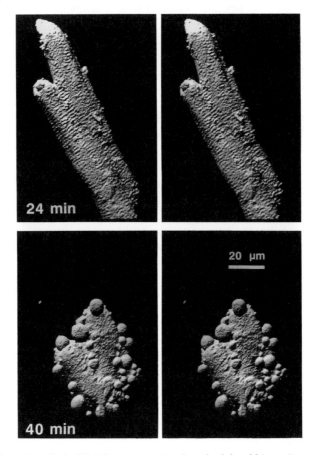

Reconstruction of a calcein-labeled myocyte. A cultured adult rabbit cardiac myocyte was loaded with calcein and exposed to 2.5 mM NaCN and 20 mM 2-deoxyglucose. After 24 and 40 min, confocal image slices were collected through the thickness of the cell in 1-µm increments. Three-dimensional distribution of calcein fluorescence was reconstructed using a Silicon Graphics IRIS 4D/310VGX imaging workstation (Mountain View, CA) operating VoxelView software by Vital Images (Fairfield, IA). Pairs of images rotated in the y-axis are presented for stereoscopic viewing. Software-generated shading enhances perception of depth and surface detail.

living cells and can be collected repeatedly. The lower panels of Fig. 12.4 show the same myocyte at a later time of metabolic inhibition. Shape and topography of the cell have changed dramatically. The myocyte has hyper-contracted and developed surface protrusions called blebs. This cell went on to die as indicated by abrupt loss of trapped calcein. Also at the onset of cell death, nuclear uptake of membrane impermeant dyes, such as trypan blue and propidium iodide, occurs (Lemasters *et al.*, 1987; Herman *et al.*, 1988).

In particular, the red nuclear fluorescence of propidium iodide is readily imaged by confocal microscopy.

## VII. VISUALIZATION OF ORGANELLES

Several fluorophores are available to label specific organelles of living cells. By exploiting spectral differences between individual probes, two or more of these probes may be imaged simultaneously. Multiline argon and argon-krypton lasers provide a selection of different excitation wavelengths. Dichroic and barrier filters permit separation of emitted fluorescence into longer and shorter wavelengths, directing this light through confocal apertures to different light detectors.

Rhodamine 123 and rhodamine-dextran (or Texas Red-dextran) are convenient labels of mitochondria and lysosomes, respectively, which can be used together (Gores *et al.*, 1989). Simple incubation of cells with 0.1–1 $\mu$M rhodamine 123 for 15–30 min provides excellent mitochondrial loading. Lysosomal loading requires overnight incubation of cultures with 1 $\mu$g/ml rhodamine-dextran or intraperitoneal injection of 20 mg rhodamine-dextran/100 g body weight 1 day prior to hepatocyte isolation. In cells loaded with both fluorophores, mitochondria are identified by green fluorescence and lysosomes are identified by red fluorescence. Using dual detectors, both mitochondria and lysosomes can be imaged simultaneously in the confocal microscope.

After collecting serial confocal images of rhodamine 123 and rhodamine-dextran through the cell thickness, the three-dimensional distribution of mitochondria and lysosomes may then be reconstructed. Figure 12.5 shows a stereo pair illustrating the distribution of mitochondria and lysosomes in a cultured hepatocyte. Lysosomes occupy a predominantly supranuclear region of the cell, whereas mitochondria are distributed in more basal regions closer to the adhering substrate. Indeed, it was only through confocal microscopy that we recognized these different distributions of mitochondria and lysosomes within cultured hepatocytes. In addition, many other organelle-specific fluorophores have been described that are adaptable to confocal microscopy, for example BODIPY-ceramide derivatives for Golgi (Pagano *et al.*, 1991) and carbocyanine dyes for endoplasmic reticulum (Lee and Chen, 1988).

## VIII. ELECTRICAL GRADIENTS IN LIVING CELLS

Cationic fluorophores, such as rhodamine 123, distribute into negatively charged cellular compartments in accordance with the Nernst equation:

$$\Delta \psi = -60 \log F_{in}/F_{out} \tag{1}$$

FIGURE 12.5

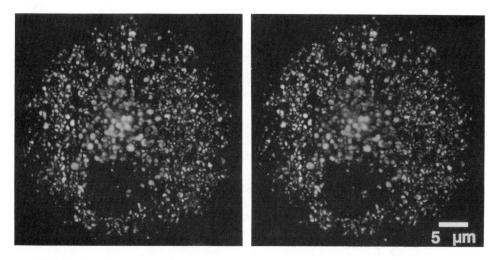

Distribution of lysosomes and mitochondria in a cultured hepatocyte. A cultured hepatocyte was loaded with rhodamine 123 to label mitochondria (green) and rhodamine-dextran to label lysosomes (red). Confocal images were collected using 488-nm (rhodamine 123) and 568-nm (rhodamine-dextran) excitation light from an argon-krypton laser. Rhodamine 123 fluorescence was imaged through a 522 ± 15-nm bandpass filter, and rhodamine-dextran fluorescence was collected through a 585-nm long pass filter. The stereo pairs are volume renderings reconstructed as described for Fig. 12.4. (*See Plate 8 for a color version of this figure.*)

where $F_{in}$ and $F_{out}$ are fluorophore concentrations inside and outside the compartment of interest, and $\Delta\psi$ is the electrical potential difference inside the compartment relative to outside. Accordingly, a 10 to 1 uptake ratio corresponds to a $-60$-mV gradient, a 100 to 1 gradient to a $-120$-mV gradient, and so forth. For excitable cells like neurons and myocytes, a plasma membrane potential of $-90$ mV and a mitochondrial potential of up to $-150$ mV are representative values. These potentials are additive, and thus mitochondria inside an excitable cell are 240 mV more negative than the extracellular space. This corresponds to a huge 10,000 to 1 gradient of fluorophore concentration inside mitochondria to that outside the cell. With 8-bit frame memories storing only 256 levels of light intensity for each picture element (pixel), measurement of such large gradients using a linear scale is impossible. So-called gamma circuits, long used in scanning electron microscopy, must be used instead. Gamma circuits apply a logarithmic function to the output of the light detector, effectively condensing a very large signal range into the available 256 gray levels of video memory.

Figure 12.6 illustrates this approach for the measurement of electrical potentials in a cultured adult rabbit cardiac myocyte loaded with 100 nM tetramethylrhodamine methyl ester. Figure 12.6A is a nonconfocal bright-field image of the myocyte showing prominent striations of the contractile apparatus. Mitochondria, although abundant, are not readily discerned. Figure 12.6B shows an unprocessed confocal image of tetramethylrhodamine methyl ester fluorescence using a gamma circuit. Mitochondria are obvious as bright spheres and rods. In collecting this image, we first set the black level (or dark current) to zero while focusing within the coverslip under the cell. This avoids the necessity of a background subtraction, which is not easily performed with gamma circuits. In Figure 12.6C, the image was pseudocolored to represent electrical potential. Pixel values were converted to a linear fluorescence scale and divided by average extracellular fluorescence. The ratio values for each pixel were then converted to $\Delta\psi$ using the Nernst equation (Eq. 1) and displayed using different colors to represent different electrical potentials. In areas just under the sarcolemma, in the nucleus and in a few open spaces between mitochondria, pseudocoloring showed an electrical potential averaging about $-80$ mV. Since potential of the extracellular space was zero, sarcolemmal $\Delta\psi$ was $-80$ mV, very close to the expected value. Mitochondria showed heterogeneity of electrical potential, ranging between $-120$ and $-160$ mV. This heterogeneity was due to the small size of mitochondria in relation to the thickness of the confocal slice. Hence, only a few mitochondria extended

FIGURE 12.6

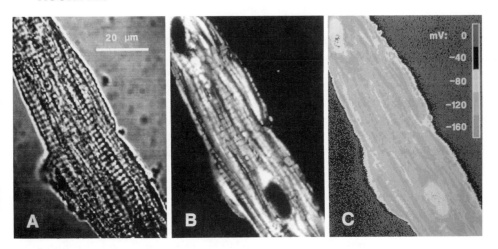

Distribution of electrical potential in cultured cardiac myocyte. (A) Nonconfocal bright-field image of a cultured adult rabbit cardiac myocyte loaded with tetramethylrhodamine methyl ester. (B) Confocal gray-scale fluorescence image obtained at a gamma setting of $+4$ using a Biorad MRC-600 laser scanning confocal microscope and 568-nm excitation light from an argon-krypton laser. (C) The image is pseudocolored to show electrical potential as described in the text. (*See Plate 9 for a color version of this figure.*)

through the plane of focus of the confocal slice. As a result, dye uptake for most mitochondria was underestimated since observed pixel intensity represents a weighted average of mitochondrial and cytosolic fluorophore concentration. In this experiment, apparent mitochondrial potentials were as great as $-160$ mV relative to the outside. Since cytosolic potential was $-80$ mV, the difference, $-80$ mV, represented a minimal value for mitochondrial $\Delta\psi$.

The validity of this approach for measuring intracellular electrical potentials depends on ideal behavior of the cationic fluorophore. Many widely used fluorescent mitochondrial dyes, such as rhodamine 123, do not behave ideally. They quench when taken up into mitochondria and show nonspecific binding unrelated to electrical potential (Emaus et al., 1986). Methyl and ethyl esters of tetramethylrhodamine seem to lack these undesirable qualities. Thus, they are well suited for estimating membrane potential by confocal imaging (Ehrenberg et al., 1988; Farkas et al., 1989).

## IX. ION IMAGING

The development by Roger Tsien and his collaborators of the $Ca^{2+}$-indicating fluorescent probes, quin-2, fura-2, and indo-1, has revolutionized the study of ion homeostasis in living cells (Tsien et al., 1984; Grynkiewicz et al., 1985). Using optical microscopy, the concentration and distribution of free-ionic $Ca^{2+}$ were imaged inside cells for the first time (reviewed in Roe et al., 1990). Additional probes for other cations, including $H^+$, $Na^+$, $K^+$, and $Mg^{2+}$, have also become available (Minta and Tsien, 1989; Bright et al., 1987; Raju et al., 1989; Jesek et al., 1990; Bassnett et al., 1990). For such ion measurements in single cells, the general approach is to load ion-reporting probes as their lipid-soluble ester derivatives, allowing endogenous esterases to liberate and trap free-acid forms within the cytosol. Inside the cell, probe fluorescence becomes dependent on the concentration of the ion it measures. However, fluorescence will also depend on probe content in the microscopic light path and will vary due to differences of cell thickness, presence of organelles excluding the probe, and variable probe loading. To correct for these variations, a ratioing procedure is used, which takes advantage of ion-induced shifts of the excitation or emission spectra of many probes. For example, the fluorescence of fura-2 increases with increasing $Ca^{2+}$ when excited at 340 nm but decreases when excited at 380 nm. Thus, the ratio of fluorescence at 340 nm to that at 380 nm is proportional to free $Ca^{2+}$ but is independent of probe concentration. Other $Ca^{2+}$-indicating fluorophores, such as fluo-3 (Minta et al., 1989), cannot be used for ratio imaging. However, because fluo-3 is readily excited by the 488-nm line of argon lasers, it is a useful probe for confocal $Ca^{2+}$ imaging.

Because wavelengths available from lasers are limited, probes that shift their emission spectrum are most useful for confocal microscopy. Such probes include indo-1 for $Ca^{2+}$ and SNARF-1 for $H^+$. Figure 12.7 illustrates the measurement of intracellular pH in a cultured myocyte by ratio imaging of SNARF-1. SNARF-1 is desirable in comparison to other pH-sensitive fluorophores, such as BCECF, because it is less susceptible to photobleaching and because it has a pH-dependent shift of its fluorescence emission spectrum. The image in Fig. 12.7 was obtained by exciting at 568 nm and measuring emitted fluorescence simultaneously with two detectors, one collecting 584 ± 5-nm fluorescence and the other collecting >620-nm fluorescence. After background subtraction, the images were divided into one another on a pixel-by-pixel basis. The resulting ratios were converted to pH values based on standard curves of SNARF-1 fluorescence as a function of pH. Notably, pH values from SNARF-1 fluorescence showed marked heterogeneity within the myocyte. Subsarcolemmal and nuclear areas had a pH of about 7.2, but for regions corresponding to the

FIGURE 12.7

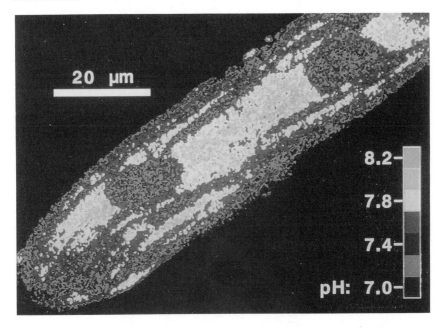

Intracellular pH in a cultured cardiac myocyte. A cultured adult rabbit cardiac myocyte was loaded with SNARF-1. Using 568-nm excitation light, fluorescence at 584 ± 5 nm was divided by fluorescence at >620 nm. Using an *in situ* calibration, the confocal image was pseudo-colored to represent distribution of pH. (*See Plate 10 for a color version of this figure.*)

distribution of mitochondria (see Fig. 12.6), pH was 7.8–8.2. Dual-labeling studies with rhodamine 123 confirmed that these regions of higher pH were mitochondria. Thus, ΔpH across mitochondrial membranes was as high as 1 pH unit.

The mitochondrial protonmotive force $\Delta p$ is the driving force for ATP formation and other energy-requiring mitochondrial functions, where

$$\Delta p = \Delta \psi + 60 \, \Delta pH \qquad (2)$$

Using tetramethylrhodamine methyl ester to measure $\Delta \psi$, and SNARF-1 to measure $\Delta pH$, both components of $\Delta p$ can now be evaluated within mitochondria of individual living cells. In the examples of Figs. 12.5 and 12.6, we can infer a mitochondrial $\Delta p$ of at least $-140 \, \mathrm{mV}$ in resting adult cardiac myocytes.

This example illustrates some of the advantages of confocal microscopy for imaging ion-sensitive probes. First, ion-sensitive fluorophores often do not load exclusively into the cytosol, as is often assumed. In particular, mitochondria possess esterases and may load heavily (Steinberg *et al.*, 1987; Gunter *et al.*, 1988; Lemasters *et al.*, 1990). Mitochondrial loading may not be evident by conventional fluorescence optics because of the relatively large depth of field. Moreover, in thick cells, such as primary cultured hepatocytes and myocytes, out-of-focus fluorescence contributes substantially to conventional fluorescence images to obscure punctate fluorescence of mitochondria and degrade two-dimensional spatial resolution. Confocal microscopy largely avoids these problems by allowing mitochondrial and cytosolic ions to be measured separately.

## X. CONCLUSION

The examples presented here illustrate the principal advantages of confocal microscopy: (1) thin optical "slices" through thick cells, (2) rejection of out-of-focus light from other focal planes, and (3) resolution in all three dimensions from multiple optical slices. Using stable probes and precautions against excess excitation energy, confocal images of parameter-specific fluorophores in single cells can be collected almost indefinitely. This new technology promises to furnish significant new insights into the workings of living cells in health and disease.

ACKNOWLEDGMENTS

This work was supported, in part, by Grants AG07218 and DK30874 from the National Institutes of Health and Grant N00014-89-J-1433 from the Office of Naval Research. E.C. is the recipient of a National Research Service Award through Grant

ES07126 to the Curriculum of Toxicology from the National Institute of Environmental Health Sciences.

## REFERENCES

Bassnett, S., Reinisch, L., and Beebe, D. C. (1990). Intracellular pH measurement using single excitation-dual emission fluorescence ratios. *Am. J. Physiol. 258,* C171–C178.

Bond, J. M., Herman, B., and Lemasters, J. J. (1991). Recovery of cultured rat neonatal myocytes from hypercontracture after chemical hypoxia. *Res. Commun. Chem. Pathol. Pharmacol. 71,* 195–208.

Bright, G. R., Fisher, G. W., Rogowska, J., and Taylor, D. L. (1987). Fluorescence ratio imaging microscopy: Temporal and spatial measurements of cytoplasmic pH. *J. Cell Biol. 104,* 1019–1033.

Chacon, E., Bond, J. M., Reece, J. M., Zahrebelski, G., Nieminen, A.-L., Herman, B., and Lemasters, J. J. (1992). Laser scanning confocal microscopy (LSCM) of mitochondrial membrane potential, cell swelling, lysosomal disruption, and plasma membrane integrity during chemical hypoxia in cultured cardiac myocytes. *Toxicologist 12,* 371.

Ehrenberg, B., Montana, V., Wei, M.-D., Wuskell, J. P., and Loew, L. M. (1988). Membrane potential can be determined in individual cells from the Nernstian distribution of cationic dyes. *Biophys. J. 53,* 785–794.

Emaus, R. K., Grunwald, R., and Lemasters, J. J. (1986). Rhodamine 123 as a probe of transmembrane potential in isolated rat liver mitochondria: Spectral and metabolic properties. *Biochem. Biophys. Acta. 850,* 436–448.

Farkas, D. L., Wei, M.-D., Febbroriello, P., Carson, J. H., and Loew, L. M. (1989). Simultaneous imaging of cell and mitochondrial membrane potentials. *Biophys. J. 56,* 1053–1069.

Gores, G. J., Nieminen, A.-L., Wray, B. E., Herman, B., and Lemasters, J. J. (1989). Intracellular pH during "chemical hypoxia" in cultured rat hepatocytes: Protection by intracellular acidosis against the onset of cell death. *J. Clin. Invest. 83,* 386–396.

Grynkiewicz, G., Poenie, M., and Tsien, R. Y. (1985). A new generation of $Ca^{2+}$ indicators with greatly improved fluorescence properties. *J. Biol. Chem. 260,* 3440–3450.

Gunter, T. E., Restroepo, D., and Gunter, K. K. (1988). Conversion of esterified Fura-2 and Indo-1 to $Ca^{2+}$-sensitive forms by mitochondria. *Am. J. Physiol. 255,* C304–C310.

Herman, B., Nieminen, A.-L., Gores, G. J., and Lemasters, J. J. (1988). Irreversible injury in anoxic hepatocytes precipitated by an abrupt increase in plasma membrane permeability. *FASEB J. 2,* 146–151.

Jesek, P., Mahdi, F., and Garlid, K. D. (1990). Reconstitution of the beef heart and rat liver mitochondrial $K^+/H^+$ ($Na^+/H^+$) antiporter. Quantitation of $K^+$ transport with the novel fluorescent probe, PBFI. *J. Biol. Chem. 265,* 10522–10526.

Johnson, L. V., Walsh, M. L., Bockus, B. J., and Chen, L. B. (1981). Monitoring of relative mitochondrial membrane potential in living cells by fluorescence microscopy. *J. Cell Biol. 88,* 526–535.

Lee, C., and Chen, L. B. (1988). Dynamic behavior of endoplasmic reticulum in living cells. *Cell 54*, 37–46.

Lemasters, J. J., DiGuiseppi, J., Nieminen, A.-L., and Herman, B. (1987). Blebbing, free $Ca^{2+}$ and mitochondrial membrane potential preceding cell death in hepatocytes. *Nature 325*, 78–81.

Lemasters, J. J., Nieminen, A.-L., Gores, G. J., Dawson, T. L., Wray, B. E., Kawanishi, T., Tanaka, Y., Florine-Casteel, K., Bond, J. M., and Herman, B. (1990). Multiparameter digitized video microscopy (MDVM) of hypoxic cell injury. *In* "Optical Microscopy for Biology." (B. Herman and K. A. Jacobson, eds.), pp. 523–542, Alan R. Liss, Inc., New York.

Minsky, M., (1961). Microscopy apparatus. United States Patent 3,013,467, Dec. 19, 1961 (Filed Nov. 7, 1957).

Minta, A., and Tsien, R. Y. (1989), Fluorescent indicators for cytosolic sodium. *J. Biol. Chem. 264*, 19449–19457.

Minta, A., Kao, J. P. Y., and Tsien, R. Y. (1989). Fluorescent indicators for cytosolic calcium based on rhodamine and fluorescein chromophores. *J. Biol. Chem. 264*, 8171–8178.

Pagano, R. E., Martin, O. C., Kang, H. C., and Haugland, R. P. (1991). A novel fluorescent ceramide analogue for studying membrane traffic in animal cells: Accumulation at the Golgi apparatus results in altered spectral properties of the sphingolipid precursor. *J. Cell Biol. 113*, 1267–1279.

Petran, M., Hadravsky, M., Egger, M. D., and Galambos, R. (1968). Tandemscanning reflected-light microscopy. *J. Opt. Soc. Am. 58*, 661–664.

Raju, B., Murphy, E., Levy, L. A., Hall, R. D., and London, R. E. (1989). A fluorescent indicator for measuring cytosolic free magnesium. *Am. J. Physiol. 256*, C540–C548.

Roe, M. W., Lemasters, J. J., and Herman, B. (1990). Assessment of Fura-2 for measurements of cytosolic free calcium. *Cell Calcium 11*, 63–73.

Steinberg, S. F., Bilezikian, J. P., and Al-Awqati, Q. (1987). Fura-2 fluorescence is localized to mitochondria in endothelial cells. *Am. J. Physiol. 253*, C744–C747.

Tsien, R. Y., Pozzan, T., and Rink, T. J. (1984). Measuring and manipulating cytosolic free $Ca^{2+}$ with trapped indicators. *Trends Biochem. Sci. 9*, 263–266.

White, J. G., Amos, W. B., and Fordham, M. (1987). An evaluation of confocal versus conventional imaging of biological structures by fluorescence light microscopy. *J. Cell Biol. 105*, 41–48.

Wilson, T. (1990). "Confocal Microscopy." Academic Press, London.

Wilson, T., and Hewlett, S. J. (1990). Imaging in scanning microscopes with slit-shaped detectors. *J. Microsc. 160*, 115–139.

Zahrebelski, G., Owens, K., Nieminen, A.-L., Reece, J. M., Herman, B., and Lemasters, J. J. (1991). Confocal microscopy shows synchronization of mitochondrial depolarization and lysosomal disintegration preceding cell death. *Hepatology 14*, 131A.

# 13
# INTRAVITAL MICROSCOPY

**Robert S. McCuskey**

Department of Anatomy, College of Medicine, University of Arizona

I. INTRODUCTION
II. *IN VIVO* MICROSCOPY OF
   ORGANS
   A. Basic Methods
   B. Transillumination Methods

C. Epiillumination Methods
D. Chronic Preparations
III. CONCLUSION
REFERENCES

## I. INTRODUCTION

Improvements during the past several decades in the optical microscopic methods used to examine living tissues and their organs have resulted in a better understanding of their morphology as related to function during health and disease; this has been particularly true for the microvasculature in these sites. Since the microcirculation is dynamic, it is best studied using *intravital* microscopic methods, which permit dynamic events in health and disease to be directly visualized, evaluated, quantified, and recorded continuously in life. Such studies have utilized mostly small rodents, but other species also have been used, including animals ranging from amphibians to man. The vast majority of intravital studies of the microcirculation has been conducted on thin tissues that are relatively easy to image (e.g., mesentery, hamster cheek pouch, selected skeletal muscle preparations). However, most internal organs also have been studied, but to a much lesser extent. As a result, this chapter focuses on some of the techniques used for the study of organs using the liver, spleen, kidney and bone marrow as examples.

## II. *IN VIVO* MICROSCOPY OF ORGANS

### A. Basic Methods

The basic methods that have been used to study living organs with the light microscope include (1) examination of surgically exposed organs *in situ* or as isolated, perfused preparations, (2) examination of organs *in situ* through windows implanted in the body wall, and (3) examination of grafts of organs contained within chambers implanted in ectopic sites. Each method has advantages as well as limitations (McCuskey, 1981).

Microscopic study of surgically exposed organs *in situ* usually permits high-resolution examination of the microvasculature with its supplying vessels and nerve supplies intact. However, these preparations are subject to movements induced by the heart, respiratory, and gastrointestinal systems; sometimes such movements can be troublesome and preclude critical microscopic study of the organ. Another limitation is the requirement to use anesthesia, which limits the duration of the study to relatively short periods of time (2–12 h). In spite of these limitations, the use of such methods has resulted in a better understanding of the structure and function of the microvasculature in a number of organs since the rate, duration, magnitude, and direction of histologic, pathologic, and pharmacologic events can be monitored in life as a function of time in the same preparation (Bloch, 1965; McCuskey, 1981, 1986). Most tissues and organs are amenable to such study, but only a few, such as the liver and, to a lesser extent, the spleen, have been studied in any detail (for the liver; Bloch, 1955, 1970; Bloch *et al.*, 1972, 1975; Cilento *et al.*, 1981; Eguchi *et al.*, 1991; Knisely *et al.*, 1957; Koo *et al.*, 1975, 1976a,b, 1977; Koo and Liang, 1977, 1979a,b,c,d; McCuskey, R. S., 1966, 1968, 1971; McCuskey *et al.*, 1979, 1982, 1983a,b, 1984a,b, 1987; Oda *et al.*, 1983; Rappaport 1973, 1977; Reilly and Dimlich, 1984, 1985; Reilly *et al.*, 1981, 1982, 1983, 1984, 1988; Unger and Reilly, 1986a,b; for the spleen; Cilento *et al.*, 1980; Fleming and Parpart, 1958; Knisely, 1936; Groom *et al.*, 1991; MacDonald *et al.*, 1987; MacKenzie *et al.*, 1941; McCuskey and McCuskey, 1985; McCuskey and Meineke, 1977a,b; McCuskey *et al.*, 1972a,b, 1973; Reilly and McCuskey, 1977a,b; Schmidt *et al.*, 1990; Stock *et al.*, 1983). While most such studies have been of intact organs *in situ*, recent results also indicate that isolated, perfused organs (e.g., liver and spleen) also are suitable candidates for high-resolution *in vivo* microscopy (Burkart *et al.*, 1982; Cilento *et al.*, 1980; Stock *et al.*, 1983, 1989; Suematsu *et al.*, 1991). While such preparations are not subject to induced movements and permit critical control of pressure and flow, they usually are of very short duration since structural and functional deterioration frequently becomes evident after 1 to 2 h.

Most organs can be studied by light microscopy in a variety of small laboratory animals, including rats, mice, hamsters, rabbits, guinea pigs, etc., by transillumination of relatively thin (3–5 mm) areas of the organ.

Thicker areas of the organs in these species as well as the thicker organs of larger animals, such as cats, dogs, and monkeys, can be examined only by epiillumination. It should be noted, however, that the resolution obtainable using epiillumination usually is inferior to that realized with transillumination (Bloch, 1973; McCuskey, 1966, 1981).

## B. Transillumination Methods

Two basic methods of transillumination have been used for light microscopy of organs. These include the use of quartz, glass, or plastic light rods or fiber-optic light guides (Bloch, 1955; Branemark, 1959; Knisely, 1936, 1967; McCuskey, 1981), which generally are not focusable; alternatively, a focusable condenser contained on a modified compound microscope is used.

### 1. Quartz-Rod Method

Using the quartz-rod method, light is conducted to the organ by a fused, hollow quartz-rod, which is bent so that its tip can be slipped underneath the organ (Knisely, 1936, 1967; Bloch, 1955). The organ floats on the tip of the rod, which, in addition to light, also conducts Ringer's solution to the upper surface of the organ to maintain homeostasis. Constant irrigation of Ringer's solution also is maintained over the upper surface of the organ. Both sources of Ringer's solution are temperature controlled at the animal's body temperature. Then, the transilluminated organ is examined with a Leitz stereo binocular or monocular compound microscope at magnifications of 50–700×.

While the quartz-rod method permits microscopic examination of the organ in its normal anatomical position, examination at high magnifications with good resolution rarely is possible. This is due to difficulties in controlling movements of the organ induced by the heart, respiratory, and gastrointestinal systems as well as vibrations transmitted by the light-guide assembly and microscopy suspension system both of which usually are not rigidly fixed. In addition, critical focusing of the light upon the site of study is not possible.

To overcome the above difficulties, most in vivo microscopic studies of the microvasculature of organs during the past 30 years have used a modified compound trinocular microscope (Bloch, 1970; McCuskey, 1966, 1981, 1986). Figure 13.1 is a block diagram of the equipment currently in use in our laboratory.

### 2. Compound Microscopic Method

The compound trinocular microscope (Leitz) normally is equipped for both transillumination as well as epiillumination of the organ to be studied. After

FIGURE 13.1

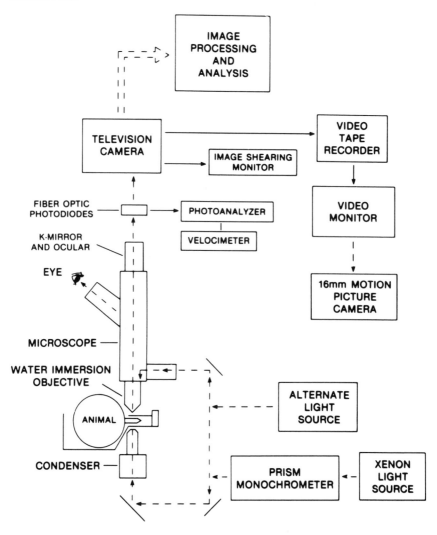

Diagram of optical and electronic equipment used for high-resolution, *in vivo* microscopic studies of tissues and organs.

the animal is anesthetized, the organ to be studied is exteriorized through an appropriate incision in the body wall and positioned over a window of optical grade mica or glass in a specially designed tray mounted on the microscope stage. The tray has provision for the drainage of irrigating fluids; and, the window overlies a long working distance condenser. The organ is covered with a piece of Saran or Mylar film, which sometimes is cemented to a movable U-shaped metal or plastic frame. The Saran or Mylar holds the organ in position and limits movement induced by the heart, respiration, and intestines, yet it is flexible enough to avoid compression of the underlying microvasculature in the organ being studied. In addition, the plastic film helps to maintain homeostasis by limiting exposure of the surface of the organ to the external environment. Homeostasis is further insured by constant suffusion of the organ with Ringer's solution, which is maintained at body temperature by proportional regulating heaters electronically clamped to rectal temperature.

Once the organ is exteriorized and positioned on the window overlying the substage condenser, it is transilluminated with white light obtained from a xenon or quartz-halogen lamp or, more commonly, with selected wavelengths of monochromatic light between 400–800 nm obtained by placing interference filters or a Leitz prism monochromater in the light path between the lamp and the substage condenser. The microscopic images of the microvasculature and its surrounding tissue are secured at magnifications up to 1500 × using both dry- and water-immersion objectives (Leitz). Even at the lower magnifications, the use of water-immersion objectives is preferred in order to provide high-quality images; their use also precludes fogging of the lens by the underlying warm, moist tissue preparation. However, the use of dry, long working distance objectives of limited numerical aperture sometimes is necessary, such as when micropipets must be introduced into the microscopic field. Such is the case when drugs are to be applied topically or when measurements are to be secured of microvascular pressures or tissue and cellular pH or $PO_2$. To date, the use of phase contrast and differential interference contrast optics has not been useful in the study of most transilluminated organs due to the thickness of the tissue. However, preliminary studies using confocal microscopy suggest new possibilities in this regard.

The resulting optical images are either studied and photographed directly or are televised through a projection ocular (1.6–5.0×) by a vidicon television system. A silicon or intensified silicon (SIT or ISIT) vidicon camera is used depending on the sensitivity and resolution required as well as the wavelengths of light to be imaged. The resulting video images are either taped (Sony) onto ¾-in. U-matic video cassettes or kine-recorded on 16-mm motion picture film using an Arriflex 16-S camera whose motor is synchronized with the framing of the video images.

Transillumination with white light results in a microscopic image in which the blood circulating in the microvasculature has a normal red color and, thus, it contrasts highly against the clear, usually unpigmented tissue surrounding the vessels. Only limited cellular detail is visualized, however, when using white light. In contrast, the use of specific wavelengths of monochromatic light enhances definition of cellular detail through the selective absorption or transmission of these wavelengths by specific tissue and cellular components. When such monochromatic, microscopic images are televised, the contrast between tissue and cellular components can be enhanced further by readjustments of the brightness and contrast controls on the video monitor. Thus, the images of a particular structure(s) can be enhanced or suppressed depending on the wavelength of light selected and the adjustments of the television system. For example, the use of wavelengths of light that are selectively absorbed by hemoglobin contained in the circulating erythrocytes aids in the study of patterns of blood flow or the overall morphological organization of a microvascular bed in an organ, particularly at low or moderate magnifications using 5–40× objectives. However, for studying highly vascular organs, such as liver or spleen, at high magnifications (McCuskey, 1966, 1981, 1986; McCuskey and Mc-Cuskey, 1985), it is useful to transilluminate the organ at wavelengths of light between 575 and 750 nm to eliminate the absorption of light by the hemoglobin contained in the numerous erythrocytes flowing or sequestered in the microvasculature. This not only increases the amount of light transmitted through the organ, but it also enhances the definition of the endothelium and other cellular components. When such images are televised using a silicon vidicon having a peak spectral response between 600 and 800

FIGURE 13.2

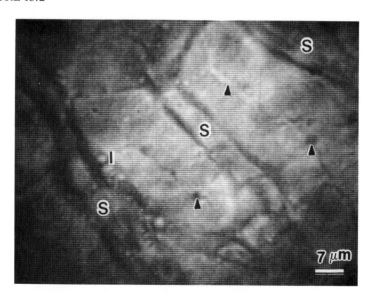

*In vivo* photomicrograph of the rat liver: S, sinusoids; I, fat-storing cell of Ito; arrows, bile canaliculi.

FIGURE 13.3

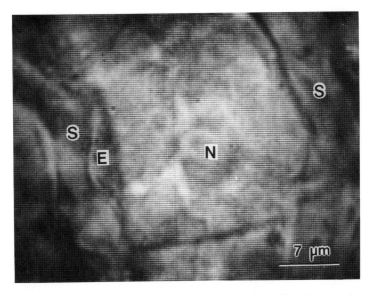

*In vivo* photomicrograph of the rat liver: S, sinusoids; N, nucleus of hepatic parenchymal cell; E, endothelial cell.

nm, the following can be observed in most organs: differentiation of the microvasculature into arterioles, capillaries or sinusoids, and venules; patterns of blood flow in these vessels; the shape and deformation of individual blood cells; differentiation of the endothelium and smooth muscle of most vessels; identification of most cells contiguous with the microvasculature; and some cytoplasmic and nuclear details (e.g., mitochondria, lysosomes, fat droplets, secretory granules, nucleoli, in all of the above cell types). Under optimal conditions, the measured resolution is 0.3–0.5 $\mu$m when using 80–100× water-immersion objectives. Figures 13.2 and 13.3 illustrates some in vivo microscopic images obtained with this system.

## C. Epiillumination Methods

While most organs and tissues are amenable to one of the above techniques of transillumination, a few can only be imaged using epiillumination through the objective lens using appropriate optics (e.g., brain) (Rosenblum, 1963; McCuskey and McCuskey, 1984). In such situations, Leitz Ultrapak objectives (both dry and water immersion) are useful since they have a coaxial condenser surrounding the objective for conducting and focusing the light onto the surface of the organ. Figure 13.4 illustrates an *in vivo*

FIGURE 13.4

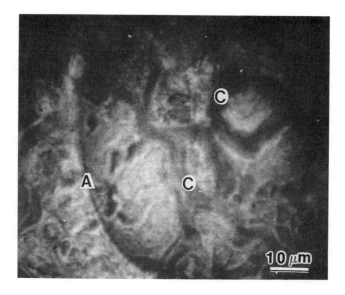

*In vivo* photomicrograph of the rat lung: C, alveolar capillaries; A, border of an alveolus.

microscope image of the lung using epiillumination. As indicated above, the resolution obtained using this method of illumination is considerably less than that obtained by transillumination. Epiillumination also is particularly useful for studying the patterns and distribution of fluorescent probes of microvascular and cellular function. In the liver, for example, fluorescent probes have been used to study the phagocytic and endocytotic properties of Kupffer cells under a variety of conditions, the transport of material from the sinusoid into parenchymal cells and from parenchymal cells into bile canaliculi, the patterns of flow and entrapment of leukocytes and tumor cells, etc. (Bagge *et al.*, 1983; Burkart *et al.*, 1982; McCuskey *et al.*, 1983b, 1984a,b; Stock *et al.*, 1989; Wisse and McCuskey, 1986). for intensely fluorescing materials, epi- and transillumination can be combined to provide improved definition of the cellular localization of a variety of fluorescent probes. Alternatively, weakly fluorescing probes may first be imaged and recorded by epiillumination and their localization subsequently identified by transillumination. In many cases, the use of intensified [SIT, ISIT, or cooled charge-coupled device (CCD)] video cameras coupled with digital image processing and/or filtering techniques is necessary to obtain images of reasonable quality and for extraction of the desired information, especially if this information is to be quantified (Schmidt, 1990; Stock *et al.*, 1989; Suematsu *et al.*, 1991).

Finally, the use of confocal microscopy (see Chapter 12 by Lemasters *et al.*, this volume) has the potential in providing "optical sections" of living tissue that can be digitally reconstructed to yield three-dimensional images as well as provide more precise localization of fluorescent probes within cells. To be useful for imaging rapidly moving systems, however, it has to operate at television framing rates (25–30 fps) and with limited illumination intensity so as not to produce cellular injury. This frequently precludes the use of laser-scanning systems whose high-intensity illumination may produce tissue damage and whose framing rates are not fast enough to prevent blurred images if movement is present. Confocal systems using scanning disks to achieve "real-time" images to date suffer from poor efficiency in light emission and, thus, require the use of image intensifiers. Nevertheless, future improvements are on the horizon with the result that such techniques are just beginning to be used in studying organs by *in vivo* microscopy (Andrews *et al.*, 1991).

## D. Chronic Preparations

Organs visualized through windows in the body wall or organs grafted into chambers in ectopic sites can be studied chronically, frequently for several months, and often without the necessity for anesthesia (Algire and Merwin, 1955; Branemark *et al.*, 1964; Feleppa *et al.*, 1971; Greenblatt *et al.*, 1969a,b; Levasseur *et al.*, 1975; McCuskey and McCuskey, 1984; McCuskey *et al.*, 1971, 1975; Nims and Irwin, 1973; Oestermeyer and Bloch, 1977; Papenfuss *et al.*, 1979; Wagner, 1969). However, it must be remembered that, like isolated, perfused organs, grafts in ectopic sites are not in their normal anatomical position or environment and are deprived of their normal vascular and nerve supply. Another limitation of these preparations is that the microscopic images of organs contained in chambers or visualized through windows in the body wall usually are not of as high definition as those obtained in acute preparations *in situ*. Details are often partially obscured by the presence of inflammatory exudate and granulation or connective tissue between the window and the underlying organ or graft. For certain types of studies, however, these latter preparations are extremely useful.

A number of chambers have been devised for chronic implantation in various sites. Examples are the mouse and rat-back chamber (Algire and Merwin, 1955; Feleppa, *et al.*, 1971; Papenfuss, *et al.*, 1979), the hamster cheek-pouch chamber (Greenblatt, *et al.*, 1969a,b; McCuskey, *et al.*, 1975; Oestermeyer and Bloch, 1977), and the rabbit-ear chamber (Williams, 1967; Nims and Irwin, 1973). Such chambers permit the study of their contained tissue or grafts of organs for considerable periods of time. In addition, chambers or windows have been implanted for the chronic study of some organs [e.g., brain (Minard, *et al.,* 1954; Levasseur, *et al.,* 1975), bone and bone marrow (Albrektsson, 1987; Branemark, *et al.,* 1964; McCuskey, *et al.,* 1971), and lung (Krahl, 1967; Wagner, 1969)]. The useful life of such preparations varies from 4 weeks (Hamster cheek-pouch chamber) to over a year (rabbit-ear chamber and bone-marrow chamber).

Chambers or windows have been implanted to chronically study organs in their normal anatomical sites. The rabbit tibial bone marrow is such an example (Albrektsson, 1987; Branemark *et al.,* 1964; McCuskey, *et al.,* 1971). The chamber is designed for chronic, transverse installation in the proximal marrow cavity of the tibia. In our laboratory (McCuskey, *et al.,* 1971; McClugage, *et al.,* 1971), it is constructed of stainless steel or tantalum tubing encasing a glass coverslip window and a glass rod for light transmission (Fig. 13.5). The distance between the glass rod and the window is adjusted to provide a gap of 0.2 to 2 mm into which bone marrow can regenerate from preexisting marrow surrounding the chamber. Microscopic examination of the marrow within the gap of the chamber is made at magnifications of 32× to 500× using Leitz UMK 10, 20, 32, and 50× objectives with appropriate oculars. Basically the same electrooptical system described previously is used to provide video images and motion picture records. Such chambers have provided good cellular detail of the microvasculature and the hemopoietic elements for periods in excess of one year.

The hamster cheek-pouch chamber is an example of a chronically implanted chamber that permits grafted organs to be studied for periods up to 4–5 weeks. A number of organs have been grafted successfully into this

FIGURE 13.5

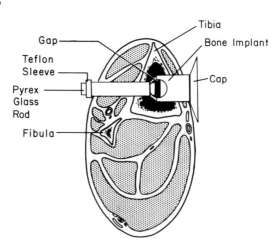

Diagram of a bone-marrow chamber installed in the proximal tibia of the rabbit.

FIGURE 13.6

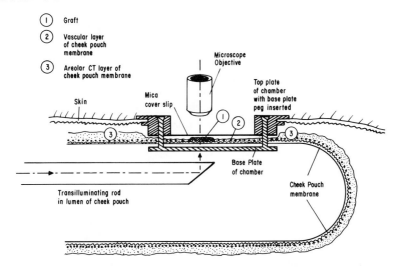

Diagram of a cheek-pouch chamber containing a graft after installation in the hamster.

immunologically privileged site; examples are the kidney, bone marrow, spleen, pituitary, heart, lung, and liver. In addition to providing a method for chronic *in vivo* microscopy, the use of the hamster cheek-pouch chamber also has permitted study of structures not easily visualized in acute preparations (e.g., mammalian renal glomeruli and bone marrow) (Greenblatt, *et al.*, 1969a,b; McCuskey, *et al.*, 1975; Oestermeyer and Bloch, 1977).

The chamber consists of a base plate with two pegs and a top plate, which contains a transparent central mica coverslip. During installation the cheek-pouch membrane is exposed and its mucoareolar layer is removed; the tissue to be grafted is placed on this site. When in place, the cheek-pouch membrane containing the graft is sandwiched between the base and top plates, which are inserted through the skin overlying the cheek (Fig. 13.6).

Five to seven days after implantation, the grafts are observed microscopically using a modified Leitz Panphot microscope equipped with Leitz UMK objectives (10, 20, 32, 50×) and 6× or 10× oculars. The stage of the modified microscope contains a holder for securing the chamber for examination and also has a removable fiber optic or plexiglass rod that serves as a light conduit. This is inserted into the lumen of the cheek pouch to transilluminate the tissue contained within the chamber. Both white and monochromatic light sources can be used. Optical images are studied and photographed directly; or, the optical images are televised and the resulting video images recorded on tape or are kine recorded on 16-mm film as previously described. Figure 13.7 illustrates an *in vivo*

FIGURE 13.7

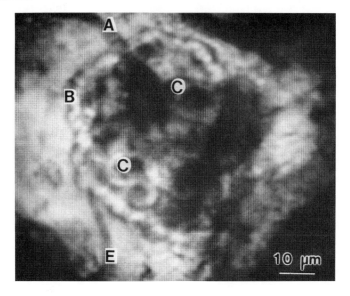

*In vivo* photomicrograph of a renal glomerulus in a kidney graft contained within a hamster cheek-pouch chamber: A, afferent arteriole; E, efferent arteriole; B, Bowman's capsule; C, glomerular capillaries.

microscopic image of a renal glomerulus within a kidney graft contained in a hamster cheek-pouch chamber.

## III. CONCLUSION

In conclusion, high-resolution electrooptical techniques now permit quantitative evaluation of intact, living organs at the cellular levels *in situ*. Responses to pharmacodynamic substances, physiologic stimuli, and pathological conditions can be evaluated in a quantitative manner, as can biological processes, such as transvascular exchange of solutes (Stock *et al.*, 1989; Schmidt and Buscher, 1991), cellular secretory phenomena (McCuskey and Chapman, 1969; Watanabe *et al.*, 1991), endocytotic mechanisms (Wisse and McCuskey, 1986), and cellular injury (Suematsu *et al.*, 1991). While many of the above are just now being studied in intact living organs, similar studies in isolated cells and tissues have been ongoing for a number of years. In short, many of the techniques developed for investigations in cell biology of isolated cells are adaptable for *in vivo* microscopic studies of such cells in intact organs where their normal microenvironment is preserved. Thus, the possibilities of obtaining new information using these tools are both exciting and challenging.

REFERENCES

Albrektsson, T. (1987). Implantable devices for long-term vital microscopy of bone tissue. *CRC Critical Rev. Bio-compatibility 3*, 25–51.

Algire, G. H., and Merwin, R. M. (1955). Vascular patterns in tissues and grafts within transparent chambers in mice. *Angiology 6*, 311–318.

Andrews, P. M., Petroll, W. M., Cavanagh, H. D., and Jester, J. V. (1991). Tandem scanning confocal microscopy (TSCM) of normal and ischemic living kidneys. *Amer. J. Anat. 191*, 95–102.

Bagge, V., Skolnik, G., and Ericson, L. E. (1983). The arrest of circulating tumor cells in the liver microcirculation. *J. Cancer Res. Clin. Oncol. 105*, 134–140.

Blankenship, L. L., Jr., Cilento, E. V., and Reilly, F. D. (1991). Hepatic microvascular regulatory mechanisms. XI. Effects of serotonin on intralobular perfusion and volumetric flowrates at the inlet of periportal and outlet of centrilobular sinusoids. *Microcirc. Endoth. Lymphat. 7*, 57–75.

Bloch, E. H. (1955). The *in vivo* microscopic vascular anatomy and physiology of the liver as determined with the quartz-rod method of transillumination. *Angiology 6*, 340–349.

Bloch, E. H. (1965). The dynamic histology of organs *in situ*. II. The lung, liver and kidney. *Anat Rec. 151*, 498.

Bloch, E. H. (1970). The termination of hepatic arterioles and the functional unit of the liver as determined by microscopy of the living organ. *Ann. N.Y. Acad. Sci. 170*, 78–87.

Bloch, E. H. (1973). The illumination, sensing and recording of the living microvascular system. *Biblio. Anat. 11*, 185–210.

Bloch, E. H., Abdel Wahab, M. F., and Warren, K. S. (1972). *In vivo* microscopic observations of the pathogenesis of hepatosplenic schistosomiasis in the mouse liver. *Amer J. Trop. Med. Hyg. 21*, 546–557.

Bloch, E. H., Warren, K. S., and Rosenthal, M. S. (1975). *In vivo* microscopic observations of the pathogenesis of acute mouse viral hepatitis. *Brit. J. Path. 56*, 256–264.

Branemark, P. I., (1959). Vital microscopy of bone marrow in rabbit. *Scand. J. Clin. Lab. Invest. 38* (Supplement), 1–82.

Branemark, P.-I., Breine, V., Johansson, B., Roylance, P. J., Rockert, H., and Yoffey, J. M. (1964). Regeneration of bone marrow. *Acta Anat. 59*, 1–46.

Burkart, V., Suss, R., Urbaschek, R., and Friedrich, E. A. (1982). *In vivo* microscopy of cell to cell interactions in the perfused liver. *In* "Sinusoidal Liver Cells." (D. L. Knook and E. Wisse, eds.), pp. 455–456, Elsevier, Amsterdam.

Cilento, E. V., McCuskey, R. S., Reilly, F. D., and Meineke, H. A. (1980). A compartmental analysis of the circulation of erythrocytes through the spleens of rats. *Amer. J. Physiol. 239*, H272–277.

Cilento, E. V., Reilly, F. D., and McCuskey, R. S. (1981). Quantification of volumetric flow within segments of the hepatic microvasculature following norepinephrine administration. *Microvas. Res. 21,* 239.

Eguchi, H., McCuskey, P. A., and McCuskey, R. S. (1991). Kupffer cell activity and hepatic microvascular events after acute ethanol ingestion in mice. *Hepatology 13,* 751–757.

Feleppa, A. E., Meineke, H. A., and McCuskey, R. S. (1971). Studies of transplanted hemopoietic tissue utilizing a modified Algire back chamber. I. Vasoproliferation following erythropoietin stimulation. *Scand. J. Hematol. 8,* 86–91.

Fleming, W. W., and Parpart, A. K., (1958). Effect of topically applied epinephrine, norepinephrine, acetycholine, and histamine on the intermediate circulation of the mouse spleen. *Angiology 9,* 294–302.

Greenblatt, M., Choudari, D. V. R., Sanders, A. G., and Shubik P. (1969a). Mammalian microcirculation in the living animal. *Amer. J. Path. 56,* 317.

Greenblatt, M., Choudari, K. V. R., Sanders, A. G., and Shubik P. (1969b). Mammalian microcirculation in the living animal: Methodologic considerations. *Microvas. Res. 1,* 420–432.

Groom, A. C., Schmidt, E. E., and MacDonald, I. C. (1991). Microcirculatory pathways and blood flow in spleen; new insights from washout kinetics, corrosion casts, and quantitative intravital microscopy. *Scanning Microscopy 5,* 159–174.

Knisely, M. H. (1936). Spleen studies. I. Microscopic observations of the circulatory system of living unstimulated mammalian spleens. *Anat. Rec. 65,* 25.

Knisely, M. H. (1967). Fused quartz rod living tissue illuminators. *In* "*In Vivo* Techniques in Histology." (G. H. Bourne, ed.), pp. 137–148, Williams and Wilkins, Baltimore.

Knisely, M. H., Harding, F., and Debacker, H. (1957). Hepatic sphincters. *Science 125,* 1023–1026.

Kinosita, R., and Ohno, S. (1961). Studies on bone marrow biodynamics observations on microcirculation in rabbit bone marrow, "*in situ.*" *Bibl. Anat. 1,* 106–109.

Koo, A., and Liang, Y. Y. S. (1977). Blood flow in hepatic sinusoids in experimental hemmorhagic shock in the rat. *Microvasc. Res. 13,* 315–325.

Koo, A., and Liang, I. Y. S. (1979a). Vagus-mediated vasodilator tone in the rat terminal liver microcirculation. *Microvasc. Res. 18,* 413–420.

Koo, A., and Liang, I. Y. S. (1979b). Microvascular filling pattern in the rat liver sinusoids during vagal stimulation. *J. Physiol. 295,* 191–199.

Koo, A., and Liang, Y. Y. S. (1979c). Stimulation and blockade of cholinergic receptors in terminal liver microcirculation in rats. *Am. J. Physiol. 236,* E728–E732.

Koo, A., and Liang, Y. Y. S. (1979d). Parasympathetic cholinergic vasodilator mechanism in the terminal liver microcirculation in rats. *Quart. J. Exp. Physiol. 64,* 149–159.

Koo, A., Liang, Y. Y. S., and Cheng, K. K. (1975). The terminal hepatic microcirculation in the rat. *Quart. J. Exp. Physiol. 60,* 261–266.

Koo, A., Liang, I. Y. S., and Cheng, K. (1976a). Intrahepatic microvascular changes in carbon tetrachloride-induced cirrhotic livers in the rat. *Austral. J. Exp. Biol. Med. Sci. 54,* 277–286.

Koo, A., Liang, I. Y. S., and Cheng, K. (1976b). Effect of the ligation of hepatic artery on the microcirculation of the cirrhotic liver in the rat. *Austral. J. Exp. Biol. Med. Sci. 54,* 287–295.

Koo, A., Liang, I. Y. S., and Cheng, K. K. (1977). Adrenergic mechanisms in the hepatic microcirculation in the rat. *Quart. J. Exp. Physiol. 62,* 199–208.

Krahl, V. (1967). Thoracic windows for the study of living pulmonary histology and physiology. In *"In Vivo* Techniques in Histology." (G. H. Bourne, ed.), pp. 222–238, Williams and Wilkins, Baltimore.

Levasseur, J. E., Wei, E., Raper, R. J., Kontos, H. A., and Patterson, J. L. (1975). Detailed description of a cranial window technique for acute and chronic experiments. *Stroke 6,* 308–317.

MacDonald, I. C., Ragan, D. M., Schmidt, E. E., and Groom, A. C. (1987). Kinetics of red blood cell passage through interendothelial slits into venous sinuses in rat spleen, analyzed by *in vivo* microscopy. *Microvas. Res. 33,* 118–134.

MacKenzie, D. W., Whipple, A. O., and Wintersteinner, M. P. (1941). Studies on the microscopic anatomy and physiology of living transilluminated mammalian spleens. *Amer. J. Anat 68,* 397–456.

McClugage, S. G., McCuskey, R. S., and Meineke, H. A., (1971). Microscopy of living bone marrow *in situ.* II. Influence of the microenvironment of hemopoiesis, *Blood 38,* 96–107.

McCuskey, R. S. (1966). A dynamic and static study of hepatic arterioles and hepatic sphincters. *Amer. J. Anat. 119,* 455–487.

McCuskey, R. S. (1967a). Dynamic microscopic anatomy of the fetal liver. I. Microcirculation. *Angiology 18,* 648–653.

McCuskey, R. S. (1967b). Dynamic microscopic anatomy of the fetal liver. II. Effect of pharmacodynamic substances on the microcirculation. *Bibl. Anat. 9,* 71–75.

McCuskey, R. S. (1967c). Erythropoietin: Effect on the living fetal hepatic microvascular system *in situ. Life Sci. 6,* 2129–2133.

McCuskey, R. S. (1968). Dynamic anatomy of the fetal liver. III. Erythropoiesis, *Anat. Rec. 161,* 267–280.

McCuskey, R. S. (1971). Sphincters in the microvascular system. *Microvas. Res. 2,* 428–433.

McCuskey, R. S. (1981). *"In vivo"* microscopy of internal organs. *Prog. Clin. Biol. Res. 59,* 79–87.

McCuskey, R. S., (1986). *"In vivo"* microscopy of organs. In "The Science of Biological Specimen Preparation." (M. Muller, R. Becker, A. Boyde, and J. J. Wolosewick, eds.), pp. 73–77, SEM, Inc., Chicago.

McCuskey, R. S., and Chapman, T. M. (1969). Microscopy of the living pancreas *in situ. Amer. J. Anat. 126,* 395–408.

McCuskey, P. A., and McCuskey, R. S. (1984). *In vivo* and electron microscopic study of the development of diabetic cerebral microangiopathy. *Microcir., Endothel. Lymphatics 1,* 221–224.

McCuskey, R. S., and McCuskey, P. A. (1985). *In vivo* and electron microscopic studies of the splenic microvasculature in mice. *Experientia 41,* 179–187.

McCuskey, R. S., and Meineke, H. A., (1973). Studies of the hemopoietic micro-environment. III. Differences in the splenic microvascular system and stroma between S1/S1^d and W/W^v anemic mice. *Amer. J. Anat. 137,* 187–198.

McCuskey, R. S., and Meineke, H. A., (1977a). Studies of the hemopoietic micro-environment. V. Erythropoietin-induced release of vasoactive substance(s) from erythropoietin-responsive stem cells. *Proc. Soc. Exper. Biol. Med. 156,* 181–185.

McCuskey, R. S., and Meineke, H. A., (1977b). Erythropoietin and the hemopoietic microenvironment. In *"Kidney Hormones,* Vol. II." (J. W. Fisher, ed.), pp. 311–327, Academic Press, New York.

McCuskey, R. S., McClugage, S. G., and Younker, W. J. (1971). Microscopy of living bone marrow *in situ. Blood 38,* 87–95.

McCuskey, R. S., Meineke, H. A., and Townsend, S. F., (1972a). Studies of the hemopoietic microenvironment. I. Changes in the microvascular system and stroma during erythropoietin regeneration and suppression in the spleens of CF₁ mice. *Blood 39,* 697–712.

McCuskey, R. S., Meineke, H. A., and Kaplan, S. M. (1972b). Studies of the hemopoietic microenvironment. II. Effect of erythropoietin on the splenic microvasculature of polycythemic CF₁ mice. *Blood 39,* 809–813.

McCuskey, R. S., Meineke, H. A., Kaplan, S. M., and Reed, P. A., (1973). Mechanism of ESF-induced vasodilatation in erythropoietic tissue. *Microvas. Res. 6,* 124–125.

McCuskey, P. A., McCuskey, R. S., and Meineke, H. A. (1975). Studies of the hemopoietic microenvironment. IV. *"In vivo"* microscopic and histochemical study of allografts of bone marrow in the hamster cheek pouch chamber. *Exp. Hematol. 3,* 297–308.

McCuskey, R. S., Reilly, F. D., McCuskey, P. A., and Dimlich, R. V. W. (1979). *In Vivo* microscopic studies of the hepatic microvascular system. *Biblio. Anat. 18,* 73–76.

McCuskey, R. S., Urbaschek, R., McCuskey, P. A., and Urbaschek, B. (1982). *In vivo* microscopic responses of the liver to endotoxins. *Klin. Wochenschr. 60,* 749–751.

McCuskey, R. S., Urbaschek, R., McCuskey, P. A., and Urbaschek, B. (1983a). *In vivo* microscopic observations of the responses of Kupffer cells and the hepatic microcirculation to Mycobacterium bovis BCG alone and in combination with endotoxin. *Infect. Immunity 42,* 362–367.

McCuskey, R. S., Vonnahme, F. J., and Grun, M. (1983b). *In vivo* and electron microscopic observations of the rat hepatic microvasculature after portacaval anastomoses. *Hepatology 3,* 96–104.

McCuskey, R. S., McCuskey, P. A., Urbaschek, R., and Urbaschek, B. (1984a). Species differences in Kupffer cells and endotoxin sensitivity. *Infect. Immunity 45,* 278–280.

McCuskey, R. S., Urbaschek, R., McCuskey, P. A., Sacco, N., Stauber, W. T., Pinkstaff, C. A., and Urbaschek, B. (1984b). Deficient hepatic phagocytosis and lysosomal enzymes in the low endotoxin-responder, C3H/HeJ mouse. *J. Leukocyte Biol. 36,* 591–600.

McCuskey, R. S., McCuskey, P. A., Urbaschek, R., and Urbaschek, B. (1987). Kupffer cell function in host defense. *Rev. Infect. Dis. 5*, S616–S619.

Minard, D., Osserman E. F., and Howell, S. R. (1954). The lucite calvarium for direct observation of the brain in monkeys. *Anat. Rec. 120*, 317.

Nims, J. C., and Irwin, J. W. (1973). Chamber techniques to study the microvasculature. *Microvas. Res. 5*, 105–118.

Oda, M., Nakamura, M., Watanabe, N., Ohya, Y., Sikizuka, E., Tsukada, N., Yonei, Y., Komatsu, H., Nagata, H., and Tsuchiya, M. (1983). Some dynamic aspects of the hepatic microcirculation demonstration of sinusoidal endothelial fenestrae as a possible regulatory factor. *In* "Intravital Observation of Organ Microcirculation." (M. Tsuchiya, H. Wayland, M. Oda, and I. Okazaki, eds.), pp. 105–138, Excerpta Med., Amsterdam.

Oestermeyer, C. F., and Bloch, E. H. (1977). *In Vivo* microscopy of hamster renal allografts. *Microvas. Res. 13*, 153–180.

Papenfuss, H. D., Gross, J. F., Intaglietta, M., and Treese, F. A. (1979). A transparent access chamber for the rat dorsal skin fold. *Microvas. Res. 18*, 311–318.

Paulo, L. G., McCuskey, R. S., Fink, G. D., Roh, B. L., and Fisher, J. W., (1973). Effect of posterior hypothalamic stimulation on reticulocyte release and bone marrow microcirculation. *Proc. Soc. Exper. Biol. Med. 143*, 986–990.

Rappaport, A. M. (1973). The microcirculatory hepatic unit. *Microvas. Res. 6*, 212–228.

Rappaport, A. M. (1977). Microcirculatory units in the mammalian liver. *Biblio. Anat. 16*, 116–120.

Reilly, F. D., and Dimlich, R. V. W. (1984). Hepatic microvascular regulatory mechanisms. IV. Effect of Lodoxamide tromethane and arterenol-HC1 on vascular responses evoked by compound 48/80. *Microcir. Endothel. and Lymph. 1*, 87–106.

Reilly, F. D., and Dimlich, R. V. W. (1985). Hepatic microvascular regulatory mechanisms. VI. Effects of lodoxamide tromelthane or phentolamine HC1 in the early hemodynamic and glucotropic responses evoked by endotoxin. *Microcir. Endothel. Lymph. 2*, 271–292.

Reilly, F. D., and McCuskey, R. S. (1977a). Studies of the hemopoietic microenvironment. VI. Humoral regulatory mechanisms in the splenic microvascular system of mice. *Microvas. Res. 13*, 79–90.

Reilly, F. D., and McCuskey, R. S. (1977b). Studies of the hemopoietic microenvironment. VII. Neural mechanisms in splenic microvascular regulation in mice. *Microvas. Res. 14*, 293–302.

Reilly, F. D., McCuskey, R. S., and Cilento, E. V. (1981). Hepatic microvascular regulatory mechanisms. I. Adrenergic mechanisms. *Microvas. Res. 21*, 103–116.

Reilly, F. D., Dimlich, R. V. W., Cilento, E. V., and McCuskey, R. S. (1982). Hepatic microvascular regulatory mechanisms. II. Cholinergic mechanism. *Hepatology 2*, 230–235.

Reilly, F. D., Dimlich, R. V. W., Cilento, E. V., and McCuskey, R. S. (1983). Hepatic microvascular regulatory mechanisms. III. Aminergic mechanisms as related to mast cells. *Microcircul. Clin. Exper. 2*, 61–73.

Reilly, F. D., Cilento, E. V., and McCuskey, R. S. (1984). Hepatic microvascular regulatory mechanisms. V. Effects of lodoxamide tromethamine or phentolamine-Hcl on the vascular responses elicited by serotonin. *Microir. Endoth. Lymphat. 1,* 671–689.

Reilly, F. D., McCuskey, R. S., and Cilento, E. V. (1988). Hepatic microvascular regulatory mechanisms. X. Effects of alpha-one or two adrenoceptor blockade on glucoregulation in normotensive endotoxic rats with optimal perfusion and flowrates. *Microirc. Endoth. Lymphat. 4,* 293–309.

Rosenblum, W. I., and Zweifach, B. W. (1963). Cerebral microcirculation in the mouse brain. *Arch. Neurol. 9,* 414.

Schmidt, R., and Buscher, H.-P. (1991). Hepatic uptake of fluorescein, investigated by video fluorescence microscopy and digital image analysis. *J. Hepatol. 13,* 208–212.

Schmidt, E. E., MacDonald, I. C., and Groom, A. C. (1990). Interactions of leukocytes with vessel walls and with other blood cells, studied by high resolution intravital microscopy of spleen. *Microvas. Res. 40,* 99–117.

Stock, R. J., Cilento, E. V., Reilly, F. D., and McCuskey, R. S. (1983). A compartment analysis of the splenic circulation in the rat spleen, *Amer. J. Physiol. 245,* H17–21.

Stock, R. J., Cilento, E. V., and McCuskey, R. S. (1989). A quantitative study of FITC-dextran transport in the microcirculation of the isolated perfused rat liver. *Hepatology 9,* 75–82.

Suematsu, M., Kato, S., Ishii, H., Asako, H., Yanagisawa, T., Suzuki, H., Oshio, C., and Tsuchiya, M. (1991). Intralobular heterogeneity of carbon Tetrachloride-induced oxidative stress in perfused rat liver visualized by digital imaging fluorescence microscopy. *Lab. Invest. 64,* 167–173.

Unger, L. S., and Reilly, F. D. (1986a). Hepatic microvascular regulatory mechanisms. VII. Effects of endoportally-infused endotoxin on microcirculation and mast cells in rats. *Microirc. Endoth. Lymphat. 3,* 47–74.

Unger, L. S., and Reilly, F. D. (1986b). Hepatic microvascular regulatory mechanisms. IX. Effects of compound 48/80 on endotoxin-induced vascular responses in rats. *Microirc. Endoth. Lymphat. 3,* 109–128.

Wagner, W. W. (1969). Pulmonary microcirculatory observations *in vivo* under physiological conditions. *J. Appl. Physiol 26,* 375–377.

Watanabe, N., Tsukada, N., Smith, C. R., and Phillips, M. J. (1991). Motility of bile canaliculi in the living animal: Implications for bile flow. *J. Cell Biol. 113,* 1069–1080.

Williams, R. G. (1967). The transparent chamber method as applied to rabbit ears. *In* "In Vivo Techniques in Histology." G. H. Bourne, ed. pp. 113–136, Williams and Wilkins, Baltimore.

Wisse, E., and McCuskey, R. S. (1986). The application of *in vivo* microscopy to liver research. *In* "The Science of Biological Specimen Preparation: 1985." (M. Muller, R. Becker, A. Boyde, and J. J. Wolosewick, eds.), pp. 73–77, SEM, Inc., Chicago.

# 14
# Time-Resolved Fluorescence Lifetime Imaging

**M. vandeVen**
ISS Inc.
Champaign, Illinois

**E. Gratton**
Laboratory for Fluorescence Dynamics
Department of Physics
University of Illinois at Urbana-Champaign

I. INTRODUCTION
II. RECENT TECHNOLOGICAL ADVANCES IN SPATIAL AND TEMPORAL IMAGING: THE EMERGING FIELD
III. INSTRUMENTATION USED FOR TIME-RESOLVED FLUORESCENCE IMAGING
  A. Modulation of the Light Intensity
  B. Fast Detectors for Microscopy
IV. APPROACHES TO TIME-RESOLVED IMAGING
V. THEORY FOR PHASE AND MODULATION SENSITIVE DETECTION
  A. Heterodyne Detection
  B. Homodyne Detection
  C. Contrast Enhancement
VI. TIME-RESOLVED IMAGING WITH CCD CAMERA: SUBFRAME SCANNING AND GATING

VII. MULTICHANNEL PHOTON COUNTING
VIII. IMAGING PHOTON DETECTORS
IX. MICROCHANNEL PLATE MULTIPLIERS WITH POSITION SENSITIVE DETECTORS.(PSD)
  A. One-Dimensional Resistive Anode Detector
  B. Two-Dimensional Resistive Anode Detector
X. SYNCHROSCAN STREAK CAMERA
XI. FRAMING CAMERA
XII. TWO-PHOTON TIME-RESOLVED IMAGING
XIII. CONCLUSION
REFERENCES

## I. INTRODUCTION

In this chapter, we present some of the basic motivations for developing lifetime detection capabilities for a microscope, the various approaches proposed for fluorescence lifetime imaging, the light sources and the detectors appropriate for obtaining lifetime resolution, and some examples of implementation of lifetime-resolved microscopes.

Fluorescence microscopy, initially introduced as a method to enhance the image content by labeling specific elements of cells, has become a highly sophisticated tool for the study of cell structure and dynamics. The microscopy methods have become more quantitative to the level that it is not unrealistic to expect to perform spectroscopic experiments directly in the living cell. Although a living cell contains an enormous number of various molecular components, by selective labeling it is frequently possible to detect only a few of them. Since the microscope field of view is becoming the "cuvette," it is apparent that to investigate such complex sample systems

FIGURE 14.1

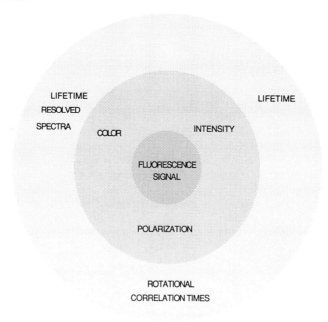

Complexity of steady-state (*inner circle*) and time-resolved (*outer circle*) fluorescence emission phenomena.

using microscopy, it is desirable to use all of the sophisticated methods that have been successfully used in fluorescence spectroscopy performed in cuvettes. The levels of complexity can be illustrated as shown in Fig. 14.1; the inner circle represents steady-state observations and the external circle represents time-resolved measurements. For fluorescence experiments in a cuvette, the simplest approach for studying a fluorescent system consists of measuring steady-state properties. When these studies are coupled with perturbation methods, such as quenching by external agents, changes in temperature and/or viscosity of the solvent, and other variables (i.e., pH, pressure, ionic strength) can provide detailed information on the dynamics of the system. However, steady-state methods are of little help when the system under investigation is complex and a number of different molecular species contribute to the overall observed signal. To a great extent, unscrambling the signal from the various components can be obtained using time-resolved methods. In many practical situations, time-resolved methods are the only possibility for studying systems that are intrinsically heterogeneous.

Apart from the high cost, one of the major obstacles for the general applicability of fluorescence time-resolved methods is the reduced sensitivity that most of those methods introduce. For most applications in fluorescence microscopy, a reduction in the light-collection efficiency cannot be tolerated. This is one of the major reasons why time-resolved methods are very rare in microscopy. However, as we will discuss in this chapter, technological revolutions in light sources and detectors reduce this intrinsic limitation to a reasonable amount.

## II. RECENT TECHNOLOGICAL ADVANCES IN SPATIAL AND TEMPORAL IMAGING: THE EMERGING FIELD

There is a great deal of interest in obtaining information about the dynamic properties of cell physiology and biology. The potential of fluorescence microscopy was recognized early, for a recent review, see Weber (1986). This interest has been further stimulated by rapid advances in several areas. Until recently, spatially resolved detection of the two- and three-dimensional fluorescence emission intensity pattern was feasible only under a conventional microscope. Detection was often hampered by difficulties that arose from autoluminescence of the intrinsic cellular components, scattered light and glare from inside the microscope, weak light sources, the bleaching of extrinsic fluorophores, and the limited capabilities of photomultiplier or cumbersome photographic plate detection. (For a review see, for example, Jovin and Arndt-Jovin, 1989.) On the other hand, time-resolved information could usually be obtained only in a cuvette or for samples on a solid surface. At this time, several successful schemes have been developed to retrieve the time-resolved emission and spatial information to study the dynamic properties of living tissues. This development has been possible due to the availability of stable and bright nonlaser (V. Chen, 1989) and laser light sources (Gratton and vandeVen, 1989), the latter operating in continuous wave (CW) or pulsed mode, and often with wavelength tunability. Also responsible for breakthroughs in such studies are confocal–scanning microscopes with excellent three-dimensional resolution and background rejection (Pawley, 1989; Robert-Nicoud *et al.*, 1989a,b; Stelzer *et al.*, 1989; Schormann *et al.*, 1989; Hiraoka *et al.*, 1990); digital-imaging detectors, with far better signal-to-noise ratio (SNR), dynamic range, readout speed, and linearity (Lansing-Taylor *et al.*, 1986; Art, 1989; Jovin and Arndt-Jovin, 1989); fast, affordable computers for automation and data collection; and new data analysis algorithms (Agard *et al.*, 1989; H. Chen *et al.*, 1989). Certainly, the success is also due to the development of new fluorophores less susceptible to bleaching (Grinvald *et al.*, 1982; Benson *et al.*, 1985; Tsien and Waggoner, 1989) with good quantum yield and engineered in such a way that they can be incorporated in a site-specific way while exhibiting excellent sensitivity for local intracellular variations of ion concentration (Krapf *et al.*, 1988; Linderman *et al.*, 1990; Moore *et al.*, 1990; Niggli and Lederer, 1990; Ryan *et al.*, 1990; Takamatsu and Wier, 1990; Tsien and Harootunian, 1990), local fluidity (Eisinger, 1989; Dix and Verkman, 1990b), oxygen concentration (Rumsey *et al.*, 1988; Wilson *et al.*, 1988; Carrero and Gratton, 1991; Carrero *et al.*, 1992), pH (Tsien, 1989), voltage (Grinvald *et al.*, 1982, 1983; Gross *et al.*, 1986; Kinosita *et al.*, 1988a), cell locomotion (Jacobson *et al.*, 1989), etc. All of these developments have resulted in the ability to not only rapidly obtain quantitative information about the two- and three-dimensional structure of a whole cell via steady-state fluorescence images but also to gather time-resolved information about the local environs in the various cell compartments on millisecond to second time scales. This time-resolved spectroscopy exploits the wealth of extra information present by adding the time dimension, and new dynamic cell processes can be studied.

In a fluorescence measurement, the basic photophysical parameters that can be recovered are quantum yield, excitation and emission spectra, polarization, fluorescence, and phosphorescence lifetime. Confocal microscopy adds three-dimensional images. Optical sectioning reduces background by two orders of magnitude, as compared to conventional microscopy of the cell contents, as does two-photon fluorescence detection (Denk *et al.*, 1990; Sandison and Webb, 1991; Piston *et al.*, 1992). This latter technique provides the best approach for reduction of background noise.

Three-dimensional images of the time-resolved two-photon fluorescence decay of living cells stained with an indicator dye can be rapidly recorded with a diffraction limited resolution in just a few (1–10) seconds.

Three images are taken at 120° phase shift for a particular modulation frequency. Short integration times of 18 $\mu$s/pixel are feasible with the outlined time-resolved imaging laser scanning confocal microscope (Biorad MRC6000). Promising examples of phase and modulation lifetime images of cell nuclei stained with DNA indicator dye have recently been presented by Piston et al. (1992). They obtained excellent spatial resolution of 0.3 $\mu$m/pixel although the signal-to-noise characteristics of the images still has to be improved.

Time-resolved imaging of polarization properties has not yet been extensively explored. However, Eisinger (1989) proposed its use to study membrane fluidity. Beach et al. (1988) studied the differential polarization images of hemoglobin S in red blood sickle cells. Fushimi et al. (1990) and also Dix and Verkman (1990a) used fluorescence anisotropy imaging to study cell membrane fluidity. Single-pixel detection was used by Keating and Wensel (1990, 1991) for the study of polarization properties of a single cell; while Velez and Axelrod (1988) observed polarized fluorescence photobleaching recovery to rotational diffusion in membranes in a similar way.

A few reports have appeared regarding time-resolved imaging of the entire field of view of objects not positioned under a microscope; for example: fingerprints (Menzel, 1989; Mitchell and Menzel, 1989). Several avenues have been followed in the past to obtain more information on dynamic cell processes. In general, lifetime changes across an emission spectrum can be expected to occur for heterogeneous ground state systems, excited state reactions, dipolar relaxation processes, etc. Extensive reviews are given in Arndt-Jovin et al. (1985) and Jovin and Arndt-Jovin (1989).

There are a number of situations where the time resolution of the fluorescence emission can add a new contrast parameter in a microscope image. For example, some dyes bind to DNA with different lifetime values depending on the base on which they are interacting. Also, several fluorescent probes have different lifetime values when bound to particular receptors with respect to the unbound case. In general, fluorescence lifetime values are more sensitive to the nature of the surrounding of the probe than the average spectral emission. Therefore, a selection of the locations of the probe with a particular lifetime value should also provide the location of the particular receptor. What is of interest for the purpose of imaging and contrast is the capability of selecting different lifetime values rather than the knowledge or the accurate measurement of the value of the lifetime. In this regard, the best approach would be some form of "gating" that can select the time of the emission from different fluorophores. From the imaging viewpoint, this selection should be equivalent to one based on the color or other spectroscopic properties. Most of the approaches suggested so far are based on this principle.

## III. INSTRUMENTATION USED FOR TIME-RESOLVED FLUORESCENCE IMAGING

Early attempts mainly used $N_2$-lasers or $N_2$-laser pumped dye lasers coupled to a microscope and photomultiplier detectors to study the feasibility of obtaining time-resolved properties with adequate spatial and temporal resolution (Kinosita *et al.*, 1976; Bottiroli *et al.*, 1979; Docchio *et al.*, 1984; Rodgers and Firey, 1985; Murray *et al.*, 1986; Borst *et al.*, 1987; Vigo *et al.*, 1987). Combined simultaneous spectral and temporal data acquisition were obtained by Ramponi and Rodgers (1987) with an intensified silicon intensified target (ISIT) camera coupled to an optical multichannel analyzer. Photon-counting electronics and phase-sensitive detection were primarily used to observe time-resolved luminescence on a pixel basis since the single-channel photomultiplier detection made two-dimensional data collection rather cumbersome.

The general layout of an imaging instrument is shown in Fig. 14.2 and has not significantly changed over time. Recently, several reviews have appeared describing the various components along with their opto-, mechanical, and electronic characteristics, advantages, and drawbacks (Pawley, 1989; Jovin and Arndt-Jovin, 1989). Microscope components, (confocal) microscopes, laser and noncoherent lamp light sources, and detectors were reviewed. An area that shows phenomenal growth is semiconductor laser technology. These small-sized, powerful diode lasers of various wavelength and modulation capabilities, along with fiber-optic coupling systems, will certainly make inroads into the time-resolved imaging area. To add the capability to time resolve the fluorescence emission, the light source must be modulated at a frequency comparable to the rate of the process to be measured and the detector must be able to respond to this frequency.

### A. Modulation of the Light Intensity

For amplitude modulation of a CW excitation source in the low-frequency range, mechanical choppers are placed in the optical path as an economical means to separate the various temporal components (Menzel, 1989; Mitchell and Menzel, 1989; Jovin *et al.*, 1989; Jovin and Arndt-Jovin, 1989; Clegg *et al.*, 1990, 1991, 1992a,b; Marriott *et al.*, 1991, 1992). Electrooptic (Pockels cell) modulators (Goldstein, 1986) and acoustooptic modulators (Koppel, 1988; Piston *et al.*, 1989; Marriott *et al.*, 1991; Clegg *et al.*, 1992b) are often utilized in the megahertz frequency range. Mode-locked and synchronously pumped, cavity-dumped laser systems emit a train of very short, subpicosecond to 100 ps, full width at half maximum (FWHM), optical pulses with repetition rates between several hundred kilohertz to hundreds of megahertz. These light sources are very well suited for single-photon

FIGURE 14.2

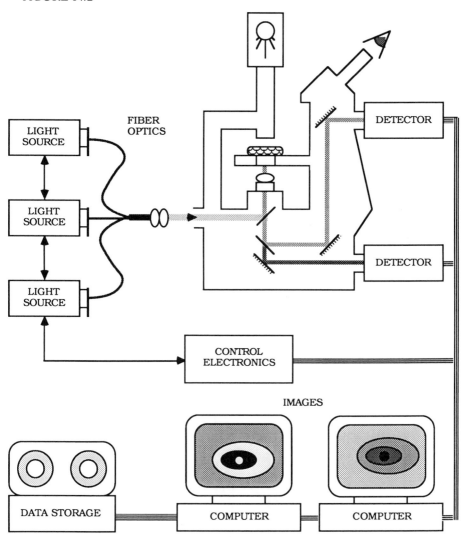

General layout of an imaging microscopy instrument. Several lamp and laser light sources for alignment, calibration, and illumination are coupled via shuttered fiber-optic bundles to the microscope. Various two-dimensional detectors simultaneously image different spectral information selected with dichroic beam splitters and optical filters. Real-time images are displayed on color monitors and stored for further processing.

counting applications but also for multifrequency phase-sensitive detection: the Fourier transform of the pulse train contains a large number of harmonics of the base frequency.

## B. Fast Detectors for Microscopy

Two-dimensional detectors, like photographic plates in the past and, more recently, video and charge-coupled device (CCD) or ISIT cameras, are frequently used to record the displayed two-dimensional information (Burle, 1987; Jovin and Arndt-Jovin, 1989). Also one-dimensional detectors, such as diode arrays, can be used (Clegg *et al.*, 1992b; Feddersen *et al.*, 1989a,b; 1991; Marriott *et al.*, 1991). Of prime importance for time-resolved applications is the availability of a fast (5 ns) gateable microchannel plate (MCP) image intensifier, which can be placed in front of a slow detector. The gate allows for discrimination of prompt and delayed luminescence in photon-counting applications and, with a small modification for gain modulation of an intensifier (Fig. 14.3), in time-resolved phase and modulation spectroscopy applications (Clegg *et al.*, 1992b; Feddersen *et al.*, 1989a). In this last application, the 1-ms frequency response of the imaging phosphor screen might, at first sight, form a formidable obstacle for the detection of any frequency above 1 kHz. It is possible, however, to circumvent this

FIGURE 14.3

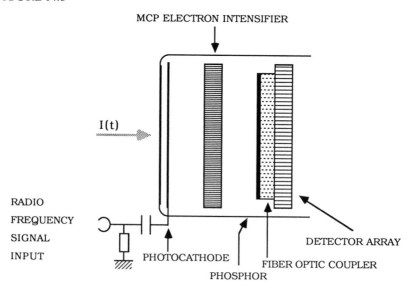

Gain-modulated, proximity-focused microchannel plate detector with modified 5-ns fast-gate circuitry.

bottleneck by the application of the cross-correlation principle (Spencer and Weber, 1969; Gratton and Limkeman, 1983; Alcala *et al.*, 1985). The information contained in the high-frequency signal is down converted to lower frequencies of a few to several tens of hertz in the image intensifier. The intensifier works like an electronic mixer. The cross-correlation frequency must be lower than the readout frequency of the camera. Some detectors have very slow reading frames of seconds or longer; in this case, application of a second down conversion step to even lower frequencies is feasible. In order to position the gate in the linear response region, a small bias voltage of about 50 V is added (Feddersen *et al.*, 1989a). No reference detector is needed when a scattering reference component with known luminescence lifetime can be imaged at the same time (Gratton and Limkeman, 1983).

Having discussed the essential system components, let us now focus our attention to the various ways in which imaging detectors can be used to obtain a time-resolved fluorescence image.

## IV. APPROACHES TO TIME-RESOLVED IMAGING

Several methods are currently being used to measure time-resolved emission:

*In the time domain*—time-correlated single-photon counting (TCSPC) coupled with pulsed light sources (Cundall and Dale, 1983; O'Connor and Phillips, 1984; Chang *et al.*, 1984).

*In the frequency domain*—multifrequency, cross-correlated phase and modulation fluorometry (MPF) with both modulated CW and pulsed light sources (Gratton and Limkeman, 1983; Lakowicz, 1983; Cundall and Dale, 1983; Alcala *et al.*, 1985).

Often, microchannel plate detectors have been incorporated into the instruments. Since they are essentially two dimensional, a lifetime instrument can readily be converted to a time-resolved imaging device by exchanging the sample compartment for a microscope or by using bifurcated fiber optics coupled directly to the detector.

Since the developments in our laboratory have been carried out in the frequency domain with multifrequency phase and modulation fluorometry (MPF), we will discuss in detail the operating principles of this technique. It is our opinion that the frequency-domain technique provides better potential in the imaging area as compared with correlated single-photon counting. The frequency-domain method was developed into a practical technique by the introduction of the cross-correlation principle by Spencer and Weber (1969) and by the addition of the multifrequency capability by Gratton and Limkeman (1983). Recently, digital acquisition methods have been introduced (Feddersen *et al.*, 1989b; Gratton *et al.*, 1990), which further enhance the speed and accuracy of the method while reducing a number of systematic errors. Similar advantages also hold for quantitative digital-imaging techniques.

## V. THEORY FOR PHASE AND MODULATION SENSITIVE DETECTION

### A. Heterodyne Detection

Sinusoidally amplitude-modulated, excitation light with angular frequency $\omega = 2\pi f$ is absorbed in cell tissues. The emitted fluorescence is shifted in time (phase shift) and decreased in amplitude (demodulation) with respect to the excitation light. The relations between phase shift, $\phi$, demodulation ratio, $M$, and lifetime, $\tau$, are defined by

$$\tan(\phi) = \omega\tau \tag{1}$$

and

$$M = M_{\text{fluor.}}/M_{\text{exc.}} = [1 + (\omega\tau)^2]^{-1/2} \tag{2}$$

The fluorescence emission can be described as

$$F(t) = F_0 [1 + M_{\text{fluor.}} \sin(\omega t + \phi_{\text{fluor.}})] \tag{3}$$

where $F_0$ is the average emitted fluorescence intensity and $\phi$ a phase delay. The cross-correlation method modulates the gain, $G(t)$, of the detector, a photomultiplier, or the intensifier section of an array detector with a frequency slightly offset from the excitation modulation frequency by an amount $\Delta\omega$. $\Delta\omega = (\omega_{\text{det.}} - \omega)$, with $\omega$ and $\omega_{\text{det.}}$ ranging from several hundred kilohertz to hundreds of megahertz and with $\Delta\omega$, tens of hertz:

$$G(t) = G_0 [1 + M_{\text{det.}} \sin(\omega_{\text{det.}} t + \phi_{\text{det.}})] \tag{4}$$

The detector functions as a mixer that down converts the high-frequency (MHz) information via multiplication of Eq. 3 with Eq. 4. As outlined in Gratton and Limkeman (1983), the final result is the detection via a low pass filter (a phosphor screen, for example) of the phase shift and demodulation in this low-frequency cross-correlation signal:

$$S(t) = S_0 \left[1 + \frac{M_{\text{fluor.}}}{2} M_{\text{det.}} \cos(\Delta\omega\, t + \Delta\phi)\right] \tag{5}$$

with $\Delta\phi = (\phi_{\text{fluor}} - \phi_{\text{det}})$. The time average signal obtained from a pixel (picture element) of an array detector is then given by

$$S_{\text{p}} = S_0 \left[1 + \frac{M_{\text{fluor.}}}{2} M_{\text{det.}} \cos(\Delta\phi)\right] \tag{6}$$

Equation 6 is summed over the number of emitting species if several of them are present. The method described so far is also called *heterodyne detection*. Using an array detector and the heterodyning technique, readout of the detector has to occur in a time less than $\Delta\omega^{-1}$. This requirement often cannot be met due to constraints in the available detector electronics, computer speed, memory size, etc.

## B. Homodyne Detection

An alternate approach to heterodyne detection is to impose an additional modulation that has a frequency of exactly $\Delta\omega$ on top of the excitation light. This is done using an optical chopper or other type of modulator, which is coupled to the rest of the instrument via a phase-locked loop. As previously described, the down-converted heterodyne signal is modulated at the frequency difference, $\Delta\omega$, between excitation and emission detector modulation. By varying the value of the variable phase delay between optical chopper and the rest of the instrument, and by sampling consecutive windows in a "boxcar detector-like" fashion, the complete low-frequency sinewave signal can be reconstructed for each pixel of the detector array (Clegg *et al.*, 1990, 1992a,b; Feddersen *et al.*, 1991; Marriott *et al.*, 1991). Averaging of the signal can continue until a sufficient signal-to-noise ratio is obtained. Calibration of the phase delay is done with standard single lifetime reference components (Gratton and Limkeman, 1983; Lakowicz, 1983).

## C. Contrast Enhancement

As discussed in Gratton and Jameson (1985) and Lakowicz (1983), the observed phase delay in a mixture of luminescent species is the sum of all phase delays together. By choosing the detected phase angle, the signal from a certain component can be made to disappear at the detector output. This constitutes a very elegant way to increase contrast. Detector-gating techniques cannot provide the same contrast when the long component of a two species luminescent sample has to be eliminated. As can be seen in Eq. 6, for a given luminescent species one can eliminate its contribution to an image by choosing the phase difference of ($\phi_{fluor.}$ − $\phi_{det.}$) to be 90°. The constant DC term in Eq. 6 can be measured by turning the modulating signal off. Subtraction of both measurements eliminates one component. Sometimes lifetimes are very close together; in that case, one still could change the modulating frequency and, often, the excitation wavelength to enhance the observable phase differences. Several additional contrast-enhancing procedures are described in Lakowicz and Berndt (1991). To our knowledge, however, these techniques have not yet been applied to biological samples apart from some studies on DNA material and several cell types (Clegg *et al.*, 1990, 1991; Feddersen *et al.*, 1991; Marriott *et al.*, 1991). Figure 14.4 shows the general layout of their heterodyne–homodyne instrumentation.

Fluorescence lifetime imaging microscopy (FLIM) on free and protein-bound NADH and calcium concentrations were presented by Szmacinski *et al.* (1992) and discussed in detail by Lakowicz *et al.* (1992). Lakowicz (1992) shows the contrast enhancement potential with phase-resolved imaging for calcium images of quin-2-loaded COS cells. Certainly, new dyes have to be developed to better explore the capabilities of FLIM. The great potential for increased resolution by the extra temporal resolution, elimination of auto fluorescence, and higher sensitivity for particular quenching concentrations is evident from Marriott *et al.* (1991). They are able to resolve the very weak long-lived luminescence (delayed fluorescence and phosphorescence) from several organic chromophores attached to anoxic biological materials like chromosomes and 3T3 cells. An example of the increased contrast that can be obtained with time-resolved imaging of prompt fluorescence from 2H3 tumor cells is presented by Clegg *et al.* (1992b). When compared with other techniques these heterodyne and heterodyne–homodyne techniques of time-resolved imaging under a microscope are at present clearly the most advanced and promising ones for a rapid development of applications to living biological materials.

## VI. TIME-RESOLVED IMAGING WITH CCD CAMERA: SUBFRAME SCANNING AND GATING

ISIT cameras have a wide variety of drawbacks generally associated with TV-type cameras, such as limited dynamic range, high noise level, and nonlinear response. Nevertheless, they have been extensively used, for example, to study the spatial variability of photobleaching in cells by Benson *et al.* (1987). Video cameras were also used by Ryan *et al.* (1990) and Takamatsu and Wier (1990), to mention a few. Presently, an ever-increasing number of

FIGURE 14.4

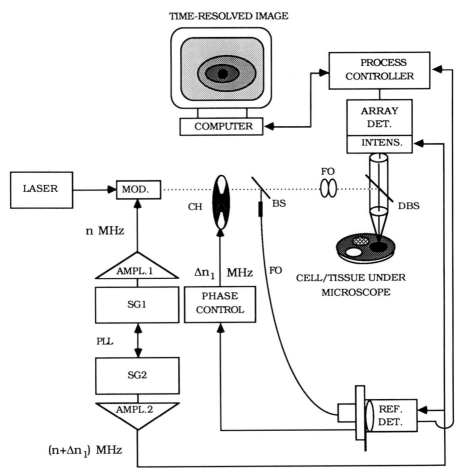

General layout of imaging microscope with heterodyne–homodyne detection. Ampl.1,2: radio frequency amplifiers; ARRAY DET. and INTENS.: two-dimensional array detector with gain-modulated intensifier section; BS: beam splitter; DBS: dichroic beam splitter; FO: fiber optics; REF. DET.: one-dimensional photomultiplier detector (for reference phase and modulation detection); heterodyne modulation frequency $n$ MHz; cross-correlation frequency $\Delta n_1$ MHz; PLL: phase-locked loop; SG1,2: radio frequency signal generators.

CCD cameras are being used; for example, by Linderman *et al.* (1990), Niggli and Lederer (1990), and Tsien and Harootunian (1990). Three-dimensional images with a time resolution of 1 ms of calcium movement in single smooth muscle cells were recently reported by Etter *et al.* (1992) and Tuft *et al.* (1992). They used a laser-light source and a masked CCD camera attached to a microscope with fast piezo-electric-driven focus adjustment of

0.1 $\mu$m/ms over a total range of 20 $\mu$m. Loew et al. (1992) obtained with this system similar images of the membrane potential of single mitochondria in a living cell. The advantages of using cameras in quantitative optical microscopy have been extensively reviewed in Hiraoka et al. (1987), Art (1990), and Moore et al. (1990) and, with an emphasis on applications in luminescence digital-imaging microscopy (LDIM), by Jovin et al. (1989), Jovin and Arndt-Jovin (1989), and Jovin (1991). The main advantages of a CCD camera are high resolution, high sensitivity, wide dynamic range, and low dark noise. The cameras currently used often have an approximate 512 × 512 pixel frame size, but the trend is toward larger arrays of 1024 × 1024 elements or even 2048 × 2048. Naturally, the demands on readout speed, computer memory, and data-processing speed also increase and may become prohibitive for the largest area sizes. However, since the complete image is not usually of importance, certain smaller image sections can be selected in a subframe mode (see for example Hansen et al., 1988; Jovin and Arndt-Jovin, 1989). The introduction of these digital cameras made reliable, quantitative, time-resolved luminescence imaging possible for the first time. Although some fast-gated CCD cameras are appearing on the market, most researchers still rely on the standard 30 frame/s readout rate. A large number of publications have appeared about $Ca^{2+}$ or other ion monitoring in a cell or tissue. For example, Vergara et al. (1991) were able to follow $Ca^{2+}$ transients in skeletal muscle fibers with rastering at 16.7 ms. The rising slope of the signal is barely resolved, but the use of subframe monitoring reduced the scan time to 4 ms. Subframe scanning or random pixel access may also be necessary to circumvent image blurring from cellular motion (Hansen et al., 1988). At a 500-kHz clock rate, only two complete 512 × 512 frames could be read out per second, while 256 × 256 and 128 × 128 subframes could be dumped at a rate of 2.6 and 6.3 frames/s, respectively. The sequences of images obtained show a change in intensity on the time scales investigated. Lasser-Ross et al. (1990) used a frame-transfer CCD camera with increased clock speed to obtain 50 × 50 and 20 × 20 subframe images of calcium transients in Purkinje cells and leech neurons every 25 and 10 ms, respectively.

As discussed in Jovin et al. (1989), the fluorescence, especially from weakly emitting molecules or from bright emitters, present in low concentration can be completely obscured by autofluorescence from intrinsic cellular components but also by scattered light and glare inside the microscope. Light-emitting species not only emit prompt, nanosecond fluorescence but also delayed luminescence from singlet-state fluorescence and triplet-state phosphorescence, see Fig. 14.5. These components possess different temporal behavior. Delayed luminescence can be observed by combining pulsed laser excitation with gated detection via mechanical (Menzel,

FIGURE 14.5

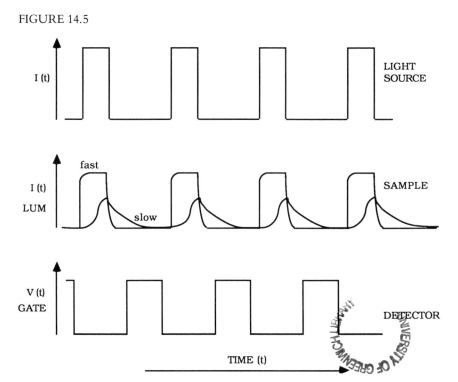

Principle of operation of instrument with pulsed light source and gated detector. Sample luminescence with two-component luminescence emission. Gated detector only observes the slow decay.

1989; Mitchell and Menzel, 1989) or acoustooptic modulator-controlled beam interruption. One chopper interrupts the light source. A second chopper may be added in front of the detector (Jovin *et al.*, 1989). The second chopper is adjustable to pass the proper emission and is phase locked to the first one. The phase delay between the two choppers and the chopper speeds can be varied. In this way, prompt, time-resolved fluorescence and luminescence images can be obtained. The addition of an MCP intensifier stage equipped with a typically 5 ns fast gate to the CCD camera makes the system suited for recording time-resolved images of the prompt fluorescence, which occurs on a nanosecond time scale; however, pulsed light sources are essential for this application.

Two-dimensional time-resolved images of a binary mixture of standard fluorophores deposited on a microscope slide were collected by Wang *et al.* (1991). Integration times of 1 s were used. A fast-gated image intensifier was coupled to their CCD camera. Other system features were multigate detection and data processing algorithms with short calculation times. Their measurement speed and accuracy could be improved by replacing the slow repetition rate $N_2$ laser with its inherent pulse-to-pulse instability and by improving the A/D converter used. Furthermore, the system should be coupled to a microscope to evaluate its capabilities for measuring time-resolved images on living tissues.

## VII. MULTICHANNEL PHOTON COUNTING

Wang *et al.* (1990) describe a multichannel, photon-counting time-resolved microscope equipped with photomultiplier tubes. Triggered by a pulse from an $N_2$ laser, all fluorescence photons emitted are sampled simultaneously and stored in a 64-channel memory. After a number of pulses, sufficient counts have accumulated to reconstruct the decay of the emission for one single spot. Scanning the *x-y* stage enables the construction of images. Results show the feasibility of the technique for several dyes with different lifetimes deposited on a thin glass surface. The main limitation in speed is caused by the slow repetition rate of the $N_2$ laser and the detector type used, which might be replaced by two-dimensional image intensified detectors.

## VIII. IMAGING PHOTON DETECTORS

This type of microchannel plate electron multiplier used in time-resolved fluorescence detection has recently been reviewed by Knutson (1988). It consists of a transparent photocathode (Rees *et al.,* 1981). Specially shaped resistive anodes or wire grids are used to obtain position information of the incoming photon. Knutson (1988) proposed the addition of an extra semi-transparent grid to obtain timing information, which can also be retrieved from the capacitive surge of the MCP. Also, multianode MCPs are capable of producing spatially resolved images. However, the electronics associated with the data retrieval from the anodes become prohibitively expensive when their number reaches a hundred or more. An image dissection tube is another variant (Fig. 14.6). By varying the horizontal and vertical deflector voltages, electrons are focused on the hole in the image plate and further amplified. Its efficiency is low and many electrons are lost; but it possesses very high speed. Beach *et al.* (1988) used this type of camera combined with a photoelastic light modulator and lock-in detection to study the hemoglobin S alignment inside red blood cells via differential polarization imaging.

Wang *et al.* (1989) show a two-dimensional time-resolved image of a mixture of two standard fluorophores deposited on a glass slide. To scan the object they used an image-dissector tube and heterodyne detection with

FIGURE 14.6

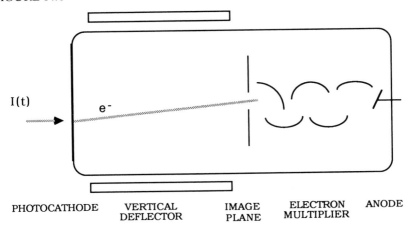

| PHOTOCATHODE | VERTICAL<br>DEFLECTOR | IMAGE<br>PLANE | ELECTRON<br>MULTIPLIER | ANODE |

Principle of operation of image dissector tube. Vertical and horizontal (*not shown*) deflectors and focusing elements scan and direct successive electron beams to the opening in the image plane with subsequent amplification.

gain modulation of the blanking electrode. The frequency response of the device reaches 1 GHz, but the sensitivity of this method may not be sufficient to routinely obtain time-resolved images of living tissues under a microscope.

## IX. MICROCHANNEL PLATE MULTIPLIERS WITH POSITION SENSITIVE DETECTORS (PSD)

### A. One-Dimensional Resistive Anode Detector

A one-dimensional position sensitive device, based on a three-stage MCP, has been described in Matsuura *et al.* (1987). A position analyzer, attached to two preamplifiers located at each end of the resistive anode chain, is able to determine the position of an incoming electron pulse, Fig. 14.7. To our knowledge, this type of one-dimensional position sensitive detector has not been used in time-resolved fluorescence imaging since similar two-dimensional devices are available.

### B. Two-Dimensional Resistive Anode Detector

The two-dimensional resistive anode detector consists of an MCP electron multiplier with resistive anode (MCP-RA) coupled to a position retrieval system. Every photon–generated charge pulse can be individually located

FIGURE 14.7

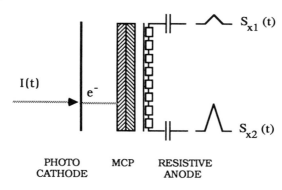

I(t)

$S_{x1}(t)$

$e^-$

$S_{x2}(t)$

PHOTO     MCP     RESISTIVE
CATHODE           ANODE

Principle of operation of one-dimensional position sensitive microchannel plate with resistive anode. Signals $S_{x1}(t)$ and $S_{x2}(t)$ determine the position of the incoming photon.

on the anode. The dark current is very low, less than 1 count per hour per anode element since the associated photocathode area is very small (Mc-Mullan *et al.*, 1987). Morgan *et al.* (1990) used an IPD (imaging photon detector) based on an MCP-RA with modified electronics. After photons strike the photocathode, photoelectrons are generated in a proximity focused MCP electron multiplier. The resulting electron pulse strikes a shaped anode, equipped at each corner with a charge sensitive preamplifier. In total, four preamplifiers were installed. Processing electronics calculate positional information and correct for noise and cosmic ray counts. The image accumulates in a frame store. Fast-timing pulses, from the rear of the MCP stack, trigger a constant fraction discriminator. Its timing pulse output is correlated with radio frequency signals, and the emission signal is modulated by a Pockels cell electrooptic modulator for the excitation. The detection method is a combination of photon counting and phase techniques. So far, the feasibility of the technique has only been demonstrated on crystals of standard fluorophores and not on samples of biological interest.

## X. SYNCHROSCAN STREAK CAMERA

Presently, the photoelectronic streak camera is very well suited for measuring very short, picosecond phenomena. Temporal information of a fast optical pulse is converted into an image on a phosphor screen. In general, radiation emitted on a time scale ranging from $10^{-7}$ to $10^{-13}$ s can be measured. The wavelength range is primarily determined by the spectral sensitivity of the photocathode.

Fluorescence photons emitted by the object under study are converted into photoelectrons on the photocathode of the streak camera (Fig. 14.8) after passing through a spectrograph. These photoelectrons are accelerated towards the phosphor screen and deflected with a high (kV) sweep pulse, which is properly synchronized with the passage of an electron pulse between the deflector plates: the *synchroscan* mode of operation. The electron beam is swept by a sinusoidal voltage at a predetermined rate over a given distance. Therefore, accurate knowledge of the temporal information is contained in the image displayed on the phosphor screen. Usually, an MCP image intensifier amplifier stage is placed in close proximity to the phosphor screen. Finally, the phosphor screen converts the electrons back into photons. With spectral information, for example, displayed along the horizontal and temporal (spatial) information along the vertical, simultaneous measurement of the emission wavelengths versus time is possible, while the intensity forms the third axis (Fig. 14.9). This is one of the biggest advantages of a streak camera system. Information about commercial streak camera systems can be found in manufacturer documentation (see, for example, Hamamatsu, 1989, 1990).

A concise summary and an extensive discussion of synchroscan streak camera systems, their merits, and optical and electronic components can be found in Nordlund (1988) who also discusses time resolution ranging from 0.5 to 10 ps for modern cameras, dynamic range (still mostly based on empirical evaluation), signal-to-noise ratio, sensitivity to polarization (none for normal incident light), detectability of fluorescence quantum yields, lifetimes, and polarization data. Kusumi *et al.* (1988) mention the use of a synchroscan camera to investigate the structure and dynamics of cell membranes.

In order to convert the synchroscan image converter camera system into

FIGURE 14.8

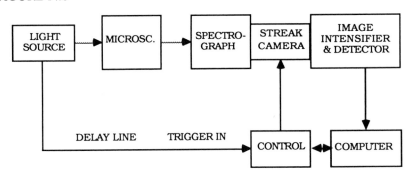

Layout of a fluorescence lifetime imaging instrument equipped with a synchroscan image converter.

FIGURE 14.9

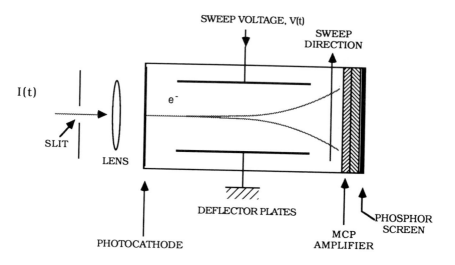

IMAGE CONVERTER TUBE

Principle of operation of a synchroscan image converter (streak camera) tube.

a fluorescence lifetime imaging microscope (FLIM), also called a picosecond fluorescence microscope (PFM) by Hamamatsu (1989), the fluorescence lifetime and intensity can be mapped for the object under study by scanning the x-y stage of the microscope (Hamamatsu, 1989). This is, of course, not a very fast method and was demonstrated on solid-state samples. Kusumi *et al.* (1988) also used a synchroscan camera for fluorescence decay measurements but only mentions its possible use as a FLIM.

## XI. FRAMING CAMERA

An early application of a framing camera in time-resolved fluorescence can be found in Young *et al.* (1986). Itoh *et al.* (1990) used a framing camera to overcome the limitations of a single-image time-resolved microfluorometer as mentioned in Kinosita *et al.* (1988b). The framing camera is capable of projecting several images (in this case, four; see Fig. 14.10) next to each other on the phosphor screen (Hamamatsu, 1990). The position of the image on the phosphor screen can be changed by applying a properly shaped, step voltage sequence to the horizontal and vertical deflection plates. An ultrafast gated MCP serves as a shutter. A voltage pulse of about 1 kV turns on the MCP. The width of the pulse determines the exposure time on the phosphor screen. Under highly intense, continuous light from a mercury lamp, or CW argon ion laser, exposure times for each individual image can be as short as 100 ns (Itoh *et al.*, 1990). Time intervals between sequential images can be varied between 300 ns and a few seconds. The instrument is capable of capturing four 10-ns images at intervals of about 10 $\mu$s. In this example, a quadruple pulse, Q-switched Nd-YAG laser was used for illumination.

## XII. TWO-PHOTON TIME-RESOLVED IMAGING

In the two-photon methods, the fluorescence intensity is proportional to the square of the excitation light intensity. This has the important consequence that the bleaching effect is limited to the roughly diamond-shaped volume covered by the overlapping laser beams (Denk *et al.*, 1990; Piston and Webb, 1991; Piston *et al.*, 1992). Furthermore, two paired, red photons show far less severe bleaching compared to one single UV photon. Other advantages are optical sectioning without the need for a pinhole, used as a spatial filter in confocal laser-scanning microscopy, and the circumvention of chromatic aberration in the objective lens. Femtosecond laser pulses, 100-fs pulse width (FWHM), are produced by a colliding pulse, mode-locked dye laser. Images were recorded in 10 s for a laser power at the focal plane of 3 mW. Of course, this is not exceedingly fast to study cellular dynamics and still many questions remain regarding the applicability of this technique in physiological samples.

## XIII. CONCLUSION

The present status of time-resolved imaging instrumentation shows quite an assembly of varying instrumentation. Microchannel plate electron multipliers and array detectors, like diode arrays and CCD cameras equipped with fast gateable intensifiers, provide almost ideal detectors with good linearity, sensitivity, and signal-to-noise ratio. The latest addition of gain-modulated one- and two-dimensional arrays has the possibility of increasing contrast by separating the fluorescence of spectrally overlapping but temporally separated species and of generating phase-resolved luminescence images with reduced background fluorescence. In particular, we have been

FIGURE 14.10

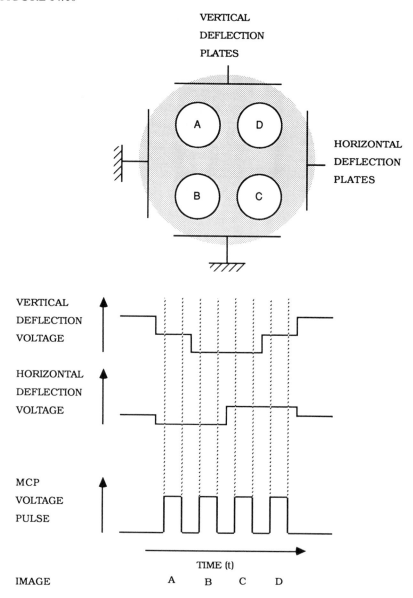

Principle of operation of a framing camera. Images are produced on the phosphor screen in positions A, B, C, and D via well-chosen vertical and horizontal deflection voltages. Exposure time is determined by the on time of the gated MCP.

involved in the development of frequency-domain techniques to obtain time-resolved images of the entire field of view. Among the major advantages of these methods are the relatively high sensitivity and the fact that they can be used with little modification of the standard digital-imaging setup based on a CCD detector. Similar developments are to be expected for two-photon, time-resolved imaging since the technique is capable of generating excellent three-dimensional images without some of the constraints of confocal microscopy, such as an intensity-limiting pinhole. Red photons also cause less bleaching than ultraviolet photons. The fluorescence from surrounding regions is effectively cut away because the two-photon intensity depends on the square of the incoming excitation intensity. The aforementioned techniques should also be very beneficial in the medical field since they would increase the resolution of fluorescent images obtained from dye-stained tumor cells and tissues. The entire field of time-resolved imaging is in a state of flux and, in the coming years, we will certainly see many new and exciting applications.

## REFERENCES

Alcala, J. R., Gratton, E., and Jameson, D. M. (1985). A multifrequency phase fluorometer using the harmonic content of a mode-locked laser. *Anal. Biochem. 14*, 225–250.

Agard, D. A., Hiraoka, Y., Shaw, P., and Sedat, J. W. (1989). Microscopy in three dimensions. *Meth. Cell Biol. 30*, 353–377.

Arndt-Jovin, D. J., Robert-Nicoud, M., Kaufman, S. J., and Jovin, T. M. (1985). Fluorescence digital imaging microscopy in cell biology. *Science 230*, 247–256.

Art, J. (1989). Photon detectors for confocal microscopy. *In* "Handbook of Biological Confocal Microscopy." (J. Pawley, ed.), pp. 127–139, Plenum Press, New York.

Beach, D. A., Bustamante, C., Wells, S., and Foucar, K. M. (1988). Differential polarization imaging III theory confirmation. Patterns of polymerization of hemoglobin S in red blood sickle cells. *Biophys. J. 53*, 449–456.

Benson, D. M., Bryan, J., Plant, A. L., Gotto Jr., A. M., and Smith, L. C. (1985). Digital imaging fluorescence microscopy: Spatial heterogeneity of photobleaching rate constants in individual cells. *J. Cell. Biol. 100*, 1309–1323.

Borst, W. L., Gangopadhyay, S., and Pleil, M. W. (1987). Fast analog technique for determining fluorescence lifetimes of multicomponent materials by pulsed laser. *In* "Fluorescence Detection." (E. R. Menzel, ed.) Vol. 743, pp. 15–23, *SPIE Proceedings, Los Angeles, CA*.

Bottiroli, G., Prenna, G., Andreoni, A., Sacchi, C. A., and Svelto, O. (1979). Fluorescence of complexes of quinacrine mustard with DNA. I. Influence of the DNA base composition on the decay time in bacteria. *Photochem. Photobiol. 29*, 23–28.

Burle (1987). "Electro-Optics Handbook." Burle Industries, Inc., Tube Products Division, Lancaster, PA.

Carrero, J., and Gratton, E. (1991). Oxygen imaging in tissue. *Biophys. J. 59,* 167a.

Carrero, J., French, T., and Gratton, E. (1992). Oxygen imaging in tissues. *Biophys. J. 61,* A177.

Chang, M. C., Courtney, S. H., Cross, A. J., Gulotty, R. J., Petrich, J. W., and Fleming, G. R. (1984). Time-correlated single photon counting with microchannel plate detectors. *Anal. Instrum. 14,* 433–464.

Chen, V. (1989). Non-laser illumination for confocal microscopy. *In* "Handbook of Biological Confocal Microscopy." (J. Pawley, ed.), pp. 53–67, Plenum Press, New York.

Chen, H., Sedat, J. W., and Agard, D. A. (1989). Manipulation, display, and analysis of three-dimensional biological images. *In* "Handbook of Biological Confocal Microscopy." (J. Pawley, ed.), pp. 141–150, Plenum Press, New York.

Clegg, R. M., Marriott, G., Feddersen, B. A., Gratton, E., and Jovin, T. M. (1990). Sensitive and rapid determinations of fluorescence lifetimes in the frequency-domain in a light microscope. *Biophys. J. 57,* 375a.

Clegg, R. M., Marriott, G., Arndt-Jovin, D. J., and Jovin, T. M. (1991). Time-resolved fluorescence and phosphorescence image microscopy. *Photochem. Photobiol. 53,* 76S.

Clegg, R. M., Marriott, G., Feddersen, B., Gratton, E., Arndt-Jovin, D. J., and Jovin, T. M. (1992a). Time-resolved image microscopy. *Biophys. J. 61,* A415.

Clegg, R. M., Feddersen, B., Gratton, E., and Jovin, T. (1992b). Time resolved imaging fluorescence microscopy. *In* "Time-Resolved Laser Spectroscopy in Biochemistry III" (J. Lakowicz, ed.), SPIE Proceedings, Los Angeles, Ca., Vol. 1640, 448–460.

Cundall, R. B., and Dale, R. E. (1983). "Time-Resolved Fluorescence Spectroscopy in Biochemistry and Biology." NATO ASI Series A: Life Sciences. Vol. 69, Plenum Press, New York.

Denk, W., Strickler, J. H., and Webb, W. W. (1990). Two-photon laser scanning fluorescence microscopy. *Science 248,* 73–76.

Dix, J. A., and Verkman, A. S. (1990a). Mapping of fluorescence anisotropy in living cells by ratio imaging. Application to cytoplasmic viscosity. *Biophys. J. 57,* 231–240.

Dix, J. A., and Verkman, A. S. (1990b). Pyrene eximer mapping in cultured fibroblasts by ratio imaging and time-resolved microscopy. *Biochem. 29,* 1949–1953.

Docchio, F., Ramponi, R., Sacchi, C. A., Bottiroli, G., and Freitas, I. (1984). An automatic pulsed laser microfluorometer with high spatial and temporal resolution. *J. Microsc. 134,* 151–160.

Eisinger, J. (1989). Membrane fluidity and diffusive transport. *In* "Fluorescent Biomolecules: Methodologies and Applications." (D. M. Jameson and G. D. Reinhart, eds.), pp. 151–171, Plenum Press, New York.

Etter, E. F., Kuhn, M. A., Tuft, R. A., Bowman, D. S., and Fay, F. S. (1992). High speed imaging of local cytoplasmic $[Ca^{2+}]$. *Biophys. J.* A159.

Feddersen, B., vandeVen, M., and Gratton, E. (1989a). Parallel wavelength acquisition of fluorescence decay with picoseconds resolution using an optical multichannel analyzer. *Biophys. J. 55,* 190a.

Feddersen, B. A., Piston, D. W., and Gratton, E. (1989b). Digital parallel acquisition in frequency-domain fluorometry. *Rev. Sci. Instrum. 60,* 2929–2936.

Feddersen, B. A., Gratton, E., Clegg, R. M., and Jovin, T. (1991). An optical and electronic heterodyning technique for use with CCD cameras and array detectors for time-resolved fluorescence with subnanosecond resolution. *Biophys. J. 59,* 155a.

Fushimi, K., Dix. J. A., and Verkman, A. S. (1990). Cell membrane fluidity in the intact kidney proximal tubule measured by orientation-independent fluorescence anisotropy imaging. *Biophys. J. 57,* 241–254.

Goldstein, R. (1986). Electro-optic devices in review: The linear electro-optic (Pockels) effect forms the basis for a family of active devices. *Lasers and Applic. 5,* 67–73.

Gratton, E., and Limkeman, M. (1983). A continuously variable frequency cross-correlation phase fluorometer with picosecond resolution. *Biophys. J. 44,* 315–423.

Gratton, E., and Jameson, D. M. (1985). New approach to phase and modulation resolved spectra. *Anal. Chem. 57,* 1694–1697.

Gratton, E., and vandeVen, M. (1989). Laser sources for confocal microscopy. *In* "Handbook of Biological Confocal Microscopy." (J. Pawley, ed.), pp. 53–67, Plenum Press, New York.

Gratton, E., Feddersen, B., and vandeVen, M. (1990). Parallel acquisition of fluorescence decay using array detectors. *In* "Time-resolved Laser Spectroscopy in Biochemistry II." (J. R. Lakowicz, ed.), pp. 21–25, Vol. 1204, SPIE Proceedings, Los Angeles, CA.

Grinvald, A., Hildesheim, R., Farber, I. C., and Anglister L. (1982). Improved fluorescent probes for the measurement of rapid changes in membrane potential. *Biophys. J. 39,* 301–308.

Grinvald, A., Fine, A., Farber, I. C., and Hildesheim, R. (1983). Fluorescence monitoring of electrical responses from small neurons and their processes. *Biophys. J. 42,* 195–198.

Gross, D., Loew, L. M., and Webb, W. W. (1986). Optical imaging of cell membrane potential changes by applied electric fields. *Biophys. J. 50,* 339–348.

Hamamatsu Photonics K. K. (1989). Picosecond fluorescence microscope system. The 2-d measurement and analysis system of photoluminescence lifetime and intensity. Catalog No. ETV-188, Dec./89, CR-3000. Hamamatsu City, Japan.

Hamamatsu Photonics K. K. (1990/). Streak tubes. Technical data No. T-121-02. Hamamatsu City, Japan.

Hansen, E. W., Zelten, J. P., and Wiseman, B. A. (1988). Laser scanning fluorescence microscope. *In* "Time-Resolved Laser Spectroscopy in Biochemistry." (J. Lakowicz, ed.), pp. 304–311, Vol. 909, SPIE Proceedings, Los Angeles, CA.

Hiraoka, Y., Sedat, J. W., and Agard, D. A. (1987). The use of a charge-coupled device for quantitative optical microscopy of biological structures. *Science 238,* 36–41.

Hiraoka, Y., Sedat, J. W., and Agard, D. A. (1990). Determination of three-dimensional imaging properties of a light microscope system. Partial confocal behavior in epifluorescence microscopy. *Biophys. J. 57,* 325–333.

Itoh, H., Hibino, M., Shigemori, M., Koishi, M., Takahashi, A., Hayakawa, T., and Kinosita, K. (1990). Multi-shot pulsed laser fluorescence microscope system. *In* "Time-Resolved Laser Spectroscopy in Biochemistry II." (J. Lakowicz, ed.), pp. 49–53, Vol. 1204, SPIE Proceedings, Los Angeles, CA.

Jacobson, K., Ishihara, A., Holifield, B., and DiGuiseppi, J. (1989). Digitized fluorescence microscopy in cell biology applications. *In* "Fluorescent Biomolecules: Methodologies and Applications" (D. M. Jameson and G. D. Reinhart, eds.), pp. 139–149, Plenum Press, New York.

Jovin, T. M. (1991). Digital imaging microscopy of DNA. *Photochem. Photobiol.* 53, 76S–77S.

Jovin, T. M., and Arndt-Jovin, D. J. (1989). Luminescence digital imaging microscopy. *Ann. Res. Biophys. Chem.* 18, 271–308.

Jovin, T. M., Marriott, G., Clegg, R. M., and Arndt-Jovin, D. J. (1989). Photophysical processes exploited in digital imaging microscopy: Fluorescence resonance energy transfer and delayed luminescence. *Ber. Bunsengesellschaft Phys. Chem.* 93, 387–391.

Keating, S. M., and Wensel, T. G. (1990). Nanosecond fluorescence microscopy of single cells. *In* "Time-Resolved Laser Spectroscopy in Biochemistry II." (J. Lakowicz, ed.), pp. 42–48, Vol. 1204, SPIE Proceedings, Los Angeles, CA.

Keating, S. M., and Wensel, T. G. (1991). Nanosecond fluorescence microscopy: Emission kinetics of Fura-2 in single cells. *Biophys. J.* 59, 186–202.

Kinosita, K., Jr., Mitaku, S., Ikegami, A., Ohbo, N., and Kunii, T. L. (1976). Construction of a nanosecond fluorometric system for applications to biological samples at cell or tissue levels. *Japan. J. Appl. Phys.* 15, 2433–2440.

Kinosita, K., Jr., Ashikawa, I., Saita, N., Yoshimura, H., Itoh, H., Nagayama, K., and Ikegami, A. (1988a). Electroporation of cell membrane visualized under a pulsed-laser fluorescence microscopy. *Biophys. J.* 53, 1015–1019.

Konosita, K., Jr., Ashikawa, I., Hibino, M., Shigemori, M., Yoshimura, H., Itoh, H., Nagayama, K., and Ikegami, A. (1988b). Submicrosecond imaging under a pulsed-laser fluorescence microscope. *In* "Time-Resolved Laser Spectroscopy in Biochemistry" (J. Lakowicz, ed.), pp. 271–277, Vol. 909, SPIE Proceedings, Los Angeles, CA.

Knutson, J. R. (1988). Fluorescence detection: schemes to combine speed, sensitivity and spatial resolution. *In* "Time-Resolved Laser Spectroscopy in Biochemistry." (J. Lakowicz, ed.) pp. 51–60, Vol. 909, SPIE Proceedings, Los Angeles, CA.

Koppel, A. (1988). "Acousto Optics." Marcel Dekker Inc., New York.

Krapf, R., Berry, C. A., and Verkman, A. S. (1988). Estimation of intracellular chloride activity in isolated perfused rabbit proximal convoluted tubules using a fluorescent indicator. *Biophys. J.* 53, 955–962.

Kusumi, A., Tsuji, A., Murata, M., Sako, Y., Yoshizawa, A. C., Hayakawa, T., and Ohnishi, S.-I (1988). Development of a time-resolved microfluorimeter with a synchroscan streak camera and its application to studies of cell membranes. *In* "Time-Resolved Laser Spectroscopy in Biochemistry." (J. Lakowicz, ed.), pp. 350–351, Vol. 909, SPIE Proceedings, Los Angeles, CA.

Lakowicz, J. R. (1983). "Principles of Fluorescence Spectroscopy." Plenum Press, New York.

Lakowicz, J. R. (1992). Fluorescence lifetime sensing. *Laser Focus World.* May, 60–80.

Lakowicz, J. R., and Berndt, K. (1991). Lifetime-selective fluorescence imaging using an rf phase-sensitive camera. *Rev. Sci. Instrum. 62*, 1727–1734.

Lakowicz, J. R., Szmacinski, H., Nowaczyk, K., and Johnson, M. L. (1992). Fluorescence lifetime imaging of $Ca^{2+}$ using visible wavelength excitation and emission. *In* "Time-Resolved Laser Spectroscopy in Biochemistry III" (J. Lakowicz, ed.), *SPIE Proceedings, Los Angeles, Ca.*, Vol. 1640, 390–404.

Lansing-Taylor, D., Waggoner, A. S., Lanni, F., Murphy, R. F., and Birge, R. R. (1986). Part II: Imaging and quantitative fluorescence microscopy. *In* "Applications of Fluorescence in the Biomedical Sciences" (D. Lansing Taylor, A. S. Waggoner, F. Lanni, R. F. Murphy, and R. R. Birge, eds.), pp. 257–521, Alan R. Riss Inc., New York.

Lasser-Ross, N., Miyakawa, H., Lev-Ram, V., Young, S., and Ross, W. N. (1990). High time-resolution fluorescence imaging with a CCD camera. *Biophys. J. 57*, 376a.

Linderman, J. J., Harris, L. J., Slakey, L. L., and Gross, D. J. (1990). Charge-coupled-device imaging of rapid calcium transients in cultured arterial smooth muscle cells. *Cell Calcium 11*, 131–144.

Loew, L. M., Wei, M.-d., Carrington, W., Tuft, R. A., and Fay, F. S. (1992). 3D imaging of membrane potential in individual mitochondria within a neurite. *Biophys. J.* A229.

Marriott, G., Clegg, R. M., Arndt-Jovin, D. J., and Jovin, T. M. (1991). Time-resolved imaging microscopy: Phosphorescence and delayed fluorescence imaging. *Biophys. J. 60*, 1374–1387.

Matsuura, S., Inuzuka, E., Koshikawa, T., Katagawa, T., Uchiyama, T., Kikuchi, R., and Suzuki, T. (1987). Position-sensitive detection system for wide and high resolution angular distribution measurement in medium energy ion scattering. Presented at the Int. Conf. on Solid Films and Surfaces. Aug. 23–27. Hamamatsu.

McMullan, W. G., Charbonneau, S., and Thewalt, M. L. W. (1987). Simultaneous subnanosecond timing information and 2D spatial information from imaging photomultiplier tubes. *Rev. Sci. Instrum. 58*, 1626–1628.

Menzel, E. R. (1989). Detection of latent fingerprints by laser excited luminescence. *Anal. Chem. 61*, 557A–561A.

Miller, J. B. (1991). Techniques for NMR imaging of solids. *Trends in Anal. Chem. 10*, 59–64.

Mitchell, K. E., and Menzel, E. R. (1989). Time-resolved luminescence imaging: application to latent fingerprint detection. *In* "Fluorescence Detection III." (E. R. Menzel, ed.), pp. 191–195, Vol. 1054, SPIE Proceedings, Los Angeles, CA.

Moore, E. D. W., Becker, P. L., Fogarty, K. E., Williams, D. A., and Fay, F. S. (1990). $Ca^{2+}$ imaging in single living cells: Theoretical and practical issues. *Cell Calcium 11*, 157–179.

Morgan, C. G., Mitchell, A. C., and Murray, J. G. (1990). Fluorescence decay time imaging using an imaging photon detector with a radiofrequency photon correlation system. *In* "Time-Resolved Laser Spectroscopy in Biochemistry II." (J. Lakowicz, ed.), pp. 798–807, Vol. 1204, SPIE Proceedings, Los Angeles, CA.

Murray, J. G., Cundall, R. B., Morgan, C. G., Evans, G. B., and Lewis, C. (1986). A single-photon-counting Fourier transform microfluorometer. *J. Phys. E: Sci. Instrum. 19,* 349–355.

Niggli, E., and Lederer, W. J. (1990). Real-time confocal microscopy and calcium measurements in heart muscle: Towards the development of a fluorescence microscope with high temporal and spatial resolution. *Cell Calcium 11,* 121–130.

Nordlund T. M., (1988). Streak camera methods in nucleic acid and protein fluorescence spectroscopy. *In* "Time-Resolved Laser Spectroscopy in Biochemistry." (J. Lakowicz, ed., pp. 35–50, Vol. 909, SPIE Proceedings, Los Angeles, CA.

O'Connor, D. V., and Phillips, D. (1984). "Time-Correlated Single Photon Counting." Academic Press, New York.

Pawley, J. B. (1989). "Handbook of Biological Confocal Microscopy." Plenum Press, New York.

Piston, D. W., and Webb, W. W. (1991). Three dimensional imaging of intracellular calcium activity, using two-photon excitation of the fluorescent indicator dye INDO-II in laser scanning microscopy. *Biophys. J. 59,* 156a.

Piston, D. W., Marriott, G., Radivoyevich, T., Clegg, R. M., Jovin, T. M., and Gratton, E. (1989). Wide-band acousto-optic light modulator for frequency-domain fluorometry and phosphorimetry. *Rev. Sci. Instrum. 60,* 2596–2600.

Piston, D. W., Sandison, D. R., and Webb, W. W. (1992). Time-resolved fluorescence imaging and background rejection by two-photon excitation in laser scanning microscopy. *In* "Time-Resolved Laser Spectroscopy in Biochemistry III" (J. Lakowicz, ed.), *SPIE Proceedings, Los Angeles, Ca.,* Vol. 1640, 379–389.

Ramponi, R., and Rodgers, M. A. J. (1987). An instrument for simultaneous acquisition of fluorescence spectra and fluorescence lifetimes from single cells. *Photochem. Photobiol. 45,* 161–165.

Rees, D., McWirther, I., Rounce, P. A., and Barlow, F. E. (1981). Miniature imaging photon detectors II. Devices with transparent photocathodes. *J. Phys. E: Sci. Instrum. 14,* 229–233.

Robert-Nicoud, M., Arndt-Jovin, D. J., Schormann, T., and Jovin, T. M. (1989a). 3-D imaging of cells and tissues using confocal laser scanning microscopy and digital processing. *Eur. J. Cell Biol., 48* (Sup.25), 49–52.

Robert-Nicoud, M., Arndt-Jovin, D. J., Schormann, T., and Jovin, T. M. (1989b). 3-D imaging of DNA structures related to functional properties of cells and whole organisms. *Eur. J. Cell Biol. 48* (Sup.26), 55.

Rodgers, M. A. J., and Firey, P. A. (1985). Instrumentation for fluorescence microscopy with picosecond time resolution. *Photochem. Photobiol. 42,* 613–616.

Rumsey, W. L., VanderKooi, J. M., and Wilson, D. F. (1988). Imaging of phosphorescence: A novel method for measuring oxygen distribution in perfused tissue. *Science Reports 241,* 1649–1651.

Ryan, T. A., Millard, P. J., and Webb, W. W. (1990). Imaging [$Ca^{2+}$] dynamics during signal transduction. *Cell Calcium 11,* 145–155.

Sandison, D. R., and Webb, W. W. (1991). Background rejection in fluorescence confocal microscopy. *Biophys. J. 59,* 156a.

Schormann, T., Robert-Nicoud, M., and Jovin, T. M. (1989). Improved stereo-visualization method for confocal laser scanning microscopy. *Eur. J. Cell Biol.* *48* (Sup.25), 53–54.

Spencer, R. D., and Weber, G. (1969). Measurements of subnanosecond fluorescence lifetime with a cross-correlation phase fluorometer. *Ann. N. Y. Acad. Sci.* *158,* 361–376.

Stelzer, E. H. K., Stricker, R., Pick, R., and Storz, C. (1989). Serial sectioning of cells in three dimensions with confocal scanning laser fluorescence microscopy (Fl-CSLM): Microtomoscopy. *Eur. J. Cell Biol. 48* (Sup.25), 55–56.

Szmacinski, H., Nowaczyk, K., Berndt, K., and Lakowicz, J. R. (1992). Fluorescence lifetime imaging. *Biophys. J. 61,* A35.

Takamatsu, T., and Wier, W. G. (1990). High temporal resolution video imaging of intracellular calcium. *Cell Calcium 11,* 111–120.

Tsien, R. Y. (1989). Fluorescent indicators of ion concentrations. *Meth. Cell. Biol. 30,* 127–156.

Tsien, R. Y., and Harootunian, A. T. (1990). Practical design criteria for a dynamic ratio imaging system. *Cell Calcium 11,* 93–109.

Tsien, R. Y., and Waggoner, A. (1989). Fluorophores for confocal microscopy: photophysics and photochemistry. *In* "Handbook of Biological Confocal Microscopy" (J. Pawley, ed.), pp. 53–67, Plenum Press, New York.

Tuft, R. A., Bowman, D. S., Carrington, W., and Fay, F. S. (1992). High spatial and temporal resolution 3D digital imaging microscope. *Biophys. J. 61,* A34.

Velez, M., and Axelrod, D. (1988). Polarized fluorescence photobleaching recovery for measuring rotational diffusion in solutions and membranes. *Biophys. J. 53,* 575–591.

Vergara, J., DiFranco, M., Compagnon, D., and Suarez-Isla, B. A. (1991). Imaging of calcium transients in skeletal muscle fibers. *Biophys. J. 59,* 12–24.

Vigo, J., Salmon, J. M., and Viallet, P. (1987). Quantitative microfluorometry of isolated living cells with pulsed excitation: Development of an effective and relatively inexpensive instrument. *Rev. Sci. Instrum. 58,* 1433–1438.

Wang, X. F., Uchida, T., and Minami, S. (1989). A fluorescence lifetime distribution measurement system based on phase-resolved detection using an image dissector tube. *Appl. Spectroscopy 43,* 840–845.

Wang, X. F., Kitajima, S., Uchida, T., Coleman, D. M., and Minami, S. (1990). Time-resolved fluorescence microscopy using multichannel photon counting. *Appl. Spectrosc. 44,* 25–30.

Wang, X. F., Uchida, T., Coleman, D. M., and Minami, S. (1991). A two-dimensional fluorescence lifetime imaging system using a gated image intensifier. *Appl. Spectroscopy 45,* 360–366.

Wang, X. F., Uchida, T., and Minami, S. (1992). Time-resolved ratio method for multicomponent fluorescence pattern analysis using a fluorescence lifetime imaging system. *In* "Time-Resolved Laser Spectroscopy in Biochemistry III" (J. Lakowicz, ed.), *SPIE Proceedings, Los Angeles, Ca.,* Vol. 1640, 433–439.

Weber, G. (1986). Solution spectroscopy and image spectroscopy. *In* "Applications of Fluorescence in the Biomedical Sciences." (D. Lansing Taylor, A. S. Waggoner, F. Lanni, R. F. Murphy, R. R. Birge, eds.), pp. 601–615, Alan R. Riss Inc., New York.

Wilson, D. F., Rumsey, W. L., and Vanderkooi, J. M. (1988). Oxygen distribution in isolated perfused liver observed by phosphorescence imaging. *In* "Oxygen Transport in Tissue XI." (K. Rakusan, G. P. Biro, T. K. Goldstick, and Z. Turek, eds.), Vol. 248, pp. 109–115, Plenum Press, New York.

Young, B. F. K., Stewart, R. E., Woodworth, J. G., and Bailey, J. (1986). Experimental demonstration of a 100-ps microchannel plate framing camera. *Rev. Sci. Instrum.* 57, 2729–2931.

# 15

# AUTOMATED FLUORESCENCE IMAGE CYTOMETRY AS APPLIED TO THE DIAGNOSIS AND UNDERSTANDING OF CERVICAL CANCER

**Stephen J. Lockett, Majid Siadat-Pajouh, Ken Jacobson, and Brian Herman**

Department of Cell Biology & Anatomy
Laboratories for Cell Biology
University of North Carolina at Chapel Hill

I. SUMMARY
II. HUMAN PAPILLOMA VIRUSES AND CERVICAL CANCER
III. METHODS OF IDENTIFICATION OF HUMAN PAPILLOMA VIRUSES
IV. AUTOMATED FLUORESCENCE IMAGE CYTOMETRY
   A. Introduction
   B. Fluorescent Stains in Image Cytometry
   C. Image Cytometry Systems in the Authors' Laboratories for the Automated Detection of Fluorescent Diagnostic Markers
V. APPLICATION OF IMAGE CYTOMETRY FOR STUDYING HPV AND CERVICAL CANCER
   A. Detection of HPV 18 E6 Antigen in HeLa Cells

B. Detection of HPV 16 and 18 Using Fluorescence *in Situ* Hybridization
VI. FUTURE PROSPECTS FOR AUTOMATED FLUORESCENCE IMAGE CYTOMETRY
   A. Application of Imaging for Investigating Cervical Cancer
   B. New Fluorescence-Imaging Techniques Useful for Cervical Cancer Studies
   C. An Automatic Fluorescence-Imaging System for Screening Clinical Specimens
ACKNOWLEDGMENTS
REFERENCES

## I. SUMMARY

The combination of recent mechanistic evidence demonstrating an association between human papilloma viruses and cervical cancer, new molecular biology techniques, and advances in image-based cytometry should lead to much more accurate and specific methods for diagnosing and accelerating the understanding of this and other diseases. This chapter summarizes the potential for the combination of each of these areas for diagnosis and mechanistic understanding of cervical cancer at the cellular level. Examples are presented illustrating the feasibility and value of this approach.

## II. HUMAN PAPILLOMA VIRUSES AND CERVICAL CANCER

Substantial evidence accumulated over the past decades associates specific human papilloma viruses (HPV) with human anogenital disorders, most notably cervical cancer (Kessler, 1976; zur Hausen, 1977; Howley, 1991). In 1842, Rigoni-Stern, an Italian doctor, reported that "cancer of cervix" was rare among virgins and nuns and quite common among married women and widows. These observations were repeatedly confirmed in the extensive epidemiological studies of cervical carcinoma lesions conducted during the past 50 years (Koss, 1979; Rotkin, 1981). Many of these studies also indicated that the sexual transmission of agents exhibited prolonged latency of 20 to 25 years. The first concrete evidence suggesting a role for HPV infection in cervical cancer was the recognition that morphological abnormalities constituting cervical dysplasia, also known as cervical intraepithelial neoplasia (CIN), were the cytopathic effects of a papilloma virus infection (Meisels and Fortin, 1976; Purola and Savia, 1977). Additional support for a papilloma virus etiology in cervical cancer came from the demonstration of papilloma virus capsid proteins in the nuclei of some cells in approximately 50% of cases of cervical dysplasia examined (Kurman et al., 1981). The presence of certain HPV DNA types in cervical carcinomas is further support for the role of human papilloma viruses in the etiology of cervical cancer (zur Hausen and Schneider, 1987).

So far, about 65 different types of HPVs have been described (De-Villiers, 1989). Only about 20 of these HPVs are associated with anogenital lesions. Further classification of these 20 HPVs into "high risk" or "low risk" is based on their involvement for malignant progression of genital tract lesions. HPV-6 and HPV-11 are considered low risk because they are mainly associated with venereal warts (condyloma acuminata), which rarely progress to malignancy (Nuovo *et al.*, 1990). HPV types 16, 18, 31, 33, 35, 39, 42, 51, 52, and 56, however, are found in cases of moderate and severe dysplasia and invasive cervical carcinoma and therefore, are classified as high-risk types (Werness *et al.*, 1991). It has been shown that about 84% of cervical carcinomas contain high-risk HPV DNA (Riou *et al.*, 1990).

Integration of HPV DNA has been found to be a prevalent phenomenon in cervical cancer cells and cell lines transformed by these viruses (zur Hausen and Schneider, 1987; zur Hausen, 1991). However, it has been demonstrated that HPV DNA is usually extrachromosomal in the premalignant CIN Lesions (Durst *et al.*, 1985). Integration usually occurs in the E1/E2 region of HPV genome disrupting the E2 viral transcriptional regulatory circuit (Baker *et al.*, 1987; Schwartz *et al.*, 1985). Thus, it has been suggested that integration of the HPV genome in cervical cancers, with disruption of E2 gene, contributes to the proliferation of the cells due to the deregulated expression of E6 and E7 genes, which are involved in immortalization of the cell (Baker *et al.*, 1987; Scheider-Gadicke and Schwartz, 1986; Smotkin and Wettstein, 1986; Munger *et al.*, 1989a). In low-grade cervical lesions infected with HPV-18, mRNAs from the E6/E7, E4/E5, and L1 regions are abundantly found in the cytoplasm. However, in severe neoplasia, mainly E6/E7 is expressed (Broker *et al.*, 1989). Palefsky *et al.* (1991), in support of the result, found E4 protein expression in HPV-16 specimens that otherwise appeared normal but did not find expression in HPV-16 positive cervical cancers. Interestingly, the E4 protein was organized into compact, intranuclear arrays of 25–35 nm in diameter.

It has been shown that the oncogenic effects of tumor viruses, such as HPV, SV40, and adenoviruses, are in part a consequence of the specific interactions between oncoproteins generated by these viruses (E7, T-antigen, and E1a, respectively) and important cellular regulatory proteins in the target cells, such as p53, and product of retinoblastoma tumor suppressor gene, pRB (Decaprio *et al.*, 1988; Whyte *et al.*, 1989; Ewen *et al.*, 1989; Munger *et al.*, 1989b; Werness *et al.*, 1990). The HPV E7 proteins form complexes with the RB antioncoprotein and these interactions might be related to the elevated c-myc and c-ras expression found in some cervical carcinomas (Broker *et al.*, 1989). These interactions may lead to deregulated growth, chromosomal instability, transformation, and carcinogenic progression. Further support for this hypothesis comes from studies on cervical carcinoma derived cell lines, which are negative for HPV DNA sequences. Crook *et al.* (1991) have demonstrated that in HPV–negative lines p53 and pRB are mutated.

More studies are necessary to establish a causative role for HPV in human cervical carcinoma. Recent advances in molecular biology and computerized image analysis should enable researchers to evaluate the relationship between HPV oncoproteins with cellular oncogenes and anti-oncogenes.

## III. METHODS OF IDENTIFICATION OF HUMAN PAPILLOMA VIRUSES

Human papilloma viruses (HPV) are small DNA viruses that specifically infect stratified epithelium and are difficult to cultivate. Only with permissive infection are whole viral particles produced. Since HPV DNA is present in both benign and malignant tissue, and may play a causative role in the transformation process, identification of its genotype is critical. The identification of specific types of HPV requires techniques of DNA or RNA hybridization. Various techniques for the detection of HPV may be used, such as cytology, electron microscopy, and immunohistochemistry. However, nucleic acid hybridization is the most sensitive and reliable method for HPV diagnosis and genotyping. Hybridization methods most commonly used for detection of HPV include Southern blot analysis, Northern blot analysis, dot blot hybridization, filter *in situ* hybridization, polymerase chain reaction (PCR) on intact cellular material, and *in situ* hybridization. We will briefly discuss the principles and importance of each technique as applied to HPV detection.

The technique of *in situ* hybridization allows the sensitive detection of HPV DNA or RNA sequences in individual cells of cytological preparations or sections of tissue. It is the only technique for correlating morphology with HPV type in routinely fixed samples. Detection of HPV by observing koilocytosis in the cell nuclei does not determine the specific type and has a sensitivity of only 15% (Koutsky *et al.*, 1988; Choi, 1991). Briefly, *in situ* hybridization involves placing the tissue sections or cells on slides and fixing with paraformaldehyde. In the next step, cellular DNA and RNA are denatured and then certain amounts of labeled viral DNA or RNA probes are applied. Hybridization is then carried out at a temperature appropriate for the level of stringency required, followed by washing and detection. Radiolabeled probes as well as digoxigenin- and biotin-labeled probes and direct-labeled probes have been used to detect HPV-16, 18, and other HPV types in human cervical carcinoma samples (Moris *et al.*, 1990; Grubendorf-Cohen and Cremer, 1990; Heiles *et al.*, 1988). Although *in situ* hybridization for detection of HPV is applicable to paraffin-embedded and frozen tissues, it is not, as yet, applicable to routine diagnostic work. While *in situ* hybridization is in principle a powerful method for detection of HPV, the following factors can affect sensitivity and specificity of the results: (1) probe type: DNA versus RNA versus oligonucleotide and double stranded versus single stranded; (2) length of the probe and its homology to the target sequence affects sensitivity and specificity of the hybridization reaction; (3) probe concentration, stringency, and duration of hybridization and posthybridization washes; (4) tissue-specific factors, which includes fixation, processing, section thickness, proteinase K digestion.

The polymerase chain reaction (PCR) is used to amplify a segment of DNA that lies between two known short sequences. Oligonucleotide primers are synthesized based on the known sequences of DNA to be amplified in two different directions. DNA Taq polymerase, a heat resistant polymerase, is used to amplify the target DNA in a series of heat denaturation, annealing, and extension steps (Sambrook *et al.,* 1989). Although PCR is very sensitive and an efficient technique, it has some disadvantages, such as a relatively high rate of misincorporation and amplification of contaminating DNA sequences (Bauer *et al.,* 1991). General and specific primers have been designed to detect HPV DNA in human specimens. Yoshikawa *et al.,* (1991) have compared the nucleotide sequence at the L1 region from six HPV types and designed a pair of consensus primers to detect HPV 6, 11, 16, 18, 31, 33, 42, 52, and 58. Most recently the PCR *in situ* technique has been described, which can amplify the target sequences and retain the signal within the individual cells. (Hause *et al.,* 1990). Thus the PCR *in situ* technique should be able to detect shorter and/or fewer copies of specific HPV sequences, compared to normal *in situ* hybridization.

Southern blot analysis was described by Southern (1975) and is the one most frequently used techniques to identify HPV in cell or tissue samples. Briefly, total DNA is extracted from tissues or cells and then digested by restriction enzymes to smaller pieces. The DNA is separated by agarose gel electrophoresis and transferred to a hybridization membrane. HPV DNA is labeled with radioactive or nonradioactive compounds and is hybridized to the cellular DNA on the filter at the proper temperature followed by washing and detecting of the signal generated by labeled probe. This method shows the highest specificity and a sensitivity second only to PCR. Different HPV subtypes can be identified, but the procedure is time consuming. An additional drawback is that this approach cannot be conveniently used to observe the heterogeneity of HPV infection amongst the cells of biopsies or smears. Northern blot and dot blot analysis follow the same principle except that messenger RNA (mRNA) is detected in Northern analysis and total DNA is blotted to the membrane directly in the dot blot analysis. In dot blot analysis, large number of samples can be detected in one experiment, but it is less specific compared to Southern analysis. In filter *in situ* hybridization, cells are filtered onto membrane, lysed, and denatured followed by hybridization with labeled HPV probes. Filters are then washed, and signal is detected by autoradiography. This technique is good for examination of many different gynecological samples in one experiment but is less sensitive than dot blot analysis with the same degree of specificity.

Since HPVs are difficult to culture, the availability of antibodies against their viral-coat proteins are limited. Thus, in situ hybridization is the preferred method to detect specific HPV genotypes. However, antibodies against most of the early and late antigens encoded by the viral DNA sequence are available. Although these antibodies are generally not genotype specific, they may be used in combination with antibodies against cellular oncogenes and tumor-suppressor factors to clarify the role of viral and cellular factors in the etiology of cervical cancer.

## IV. AUTOMATED FLUORESCENCE IMAGE CYTOMETRY

### A. Introduction

Image cytometry (IC) is a method based on computerized microscopy for making quantitative measurements on cellular specimens. The specimens are imaged with a microscope and the images are detected with a TV-like camera. The analog output from the camera is digitized into an array of numbers (pixels) that are stored in computer memory for analysis. The method can make a diverse range of quantitative and nondestructive measurements within the individual cells of specimens (Parry and Hemstreet, 1988, Ploem *et al.*, 1986; Rigaut, 1989) and thus, it is a major approach for studying biochemical processes at the (sub) cellular level (DiGuissepi *et al.*, 1985) and diagnosing clinical specimens from patients (Wied *et al.*, 1989). Examples where quantification is important include determining DNA ploidy, the number of HPV-16 viral copies integrated into the genome (Wolber and Clement 1991), and when subtle visual changes in cell morphology and intracellular staining patterns (e.g., nuclear texture [Palcic *et al.*, 1990]) require detection (Wied *et al.*, 1989). Furthermore, IC is able to diagnose specimens containing too few cells required for other assays (i.e., flow cytometry) and it has been shown that IC can detect virus infected cells (Tanke, 1989) and malignant cells (Lee *et al.*, 1989) occurring at frequencies of 1 in $10^6$. This is because in *in vitro* assays the positive signal from rare abnormal cells is significantly diluted by material from other cells.

A prerequisite for the accurate quantification of any parameter relating to individual cells is the *precise* localization of cells in the specimen via their location in the image. The cells can be manually defined in their images (e.g., by drawing around them with a mouse), but this is a very tedious process when cell number is large and thus the approach is impractical for most clinical applications. Also, the *manual* method has been shown to be not as precise as automatic cell detection methods (Palcic *et al.*, 1990) for the measurement of integrated optical density and texture of feulgen-stained nuclei.

## B. Fluorescent Stains in Image Cytometry

The use of fluorescent stains in IC adds several advantages over using conventional colorimetric, light-absorbing stains followed by imaging with bright-field microscopy. These advantages are (1) Noninterfering images of up to five spectrally distinct fluorescent labels simultaneously staining the specimen can be acquired (DeBiasio *et al.*, 1987), allowing the localization and quantification of multiple specific disease markers in the same cells. This is possible because in epifluorescence microscopy only material that excites and emits at specific wavelengths matching the spectral properties of the fluorescent label appear as signal in the image. (2) When used in conjunction with a low-light-level camera (DiGuiseppi *et al.*, 1985; Tanke, 1989), IC is extremely sensitive and can detect 100-fold-lower concentrations of specific fluorescent-labeled biomolecules inside cells compared to using colorimetric stains. (Mellin, 1990). (3) The numbers recorded in the pixels of fluorescence images are linearly related to the stain concentration in the specimen thus making quantification direct.

## C. Image Cytometry Systems in the Authors' Laboratories for the Automated Detection of Fluorescent Diagnostic Markers

We have used two systems for the detection of disease-specific markers. The first system (DiGuiseppi *et al.*, 1985; Tsay *et al.*, 1990) is an inverted epifluorescence microscope (IM-35, Zeiss, Thornwood, NY) coupled to GEN II SYS camera (Dage-MTI, Michigan City, IN). This camera is a video-rate charge-coupled device (CCD) coupled to a removable image intensifier. The analog camera signal is digitized and stored by a Minvideo One board (Datacube, Peabody, MA) that is installed on the back plane of a host Sun 3/160 workstation (Sun Microsystems, Mountain View, CA). Successive video frames are averaged together to form images with reduced random noise. The degree of focus of the images can be objectively measured by an algorithm (expression 16 in Vollath, 1988). Images are preprocessed by subtracting a background image and correcting for shading over the imaging area (Lockett *et al.*, 1991b).

The procedure for automatically detecting and quantifying fluorescent-stained cells is as follows. The specimens are counterstained with a fluorescent DNA stain (for example Hoechst or propidium iodide) in addition to any other spectrally distinct labels that are required for disease detection. Images of the DNA stain are used for the primary detection of nuclei and cells. This is because all nuclei contain abundant DNA, therefore, the nuclei are represented in the images with high signal-to-noise ratio (SNR). The high SNR, in turn, makes the automatic detection of every nucleus highly reliable. The image analysis is done by an algorithm that automatically calculates threshold intensities to segment the images into regions of nuclei and background. The segmentation detects virtually every nucleus (Lockett, *et al.*, 1991b), but touching or overlapping nuclei can be mistakenly detected as single objects. Such clusters are recognized by their morphology and separated by an additional algorithm that searches for valleys of low pixel intensities across them to serve as dividing paths (Lockett *et al.*, 1992c). It should be pointed out that since the nuclei are smaller than the cells, the cells can overlap without the nuclei doing so. The pixel intensities in the regions defined as nuclei are directly integrated to obtain the DNA content–ploidy of the nuclei. The same regions can be mapped onto other images of the same microscopic scene recorded at different wavelengths to quantify other labels located within the nuclear boundary. The quantification of cytoplasmic labels is done either by directly segmenting the images of the other fluorescent labels to detect the labeled cells (Lockett *et al.*, 1991a; Lockett *et al.*, 1991b), or by using the centers of the detected nuclei as the centers of circles approximately corresponding to the cytoplasms (Lockett *et al.*, 1990a).

The system has been extensively assessed (Lockett *et al.*, 1991b; Lockett *et al.*, 1992a; Tsay *et al.*, 1990). It has a linear response over a 30-fold range of fluorescent intensities (Lockett *et al.*, 1990b) and has a limiting sensitivity of 2172 molecules of equivalent soluble fluorescein measured using surface-labeled, 5.8-$\mu$m-diameter beads. Measurements of the areas and total fluorescent intensities from standard beads and Balb 3T3 cells labeled with 4′,6-diamidine-2-phenylindole (DAPI) showed the system had a quantification error of 2% to 3% when used without the intensifier tube coupled to the CCD (Lockett *et al.*, 1992a). The automatic detection algorithm correctly segmented 96% of the visible nuclei in specimens of cultured cells (Lockett *et al.*, 1991b).

The second system is an inverted epifluorescence microscope (Axiovert 10, Zeiss) coupled to a color, light-integrating CCD camera (American Innovision Inc., San Diego, CA). The three analog camera signals representing the red, blue, and green components of the image are converted to hue, saturation, and intensity (HSI) by an analog board (American Innovision) housed in a 486 personal computer. Images are preprocessed by subtracting a background image and correcting for shading over the imaging area. Detection of the fluorescent objects is performed at video rate by automatic selection of pixels within the specified ranges of HSI previously defined as corresponding to the fluorescent stains.

## V. APPLICATION OF IMAGE CYTOMETRY FOR STUDYING HPV AND CERVICAL CANCER

The Sun workstation–based imaging system has been used for several bio-medical studies. Lockett *et al.*, (1991b) describe the use of the system for determining DNA ploidy distributions of mature rat liver nuclei stained with DAPI and cultures of human endometrial stormal cells transfected with a plasmid that contained an origin-of-replication-defective construct of the temperature-sensitive mutant A209 of the simian virus 40. Cells were also stained with propidium iodide for flow cytometry. The DNA distributions of the endometrial cells compared favorably with distributions obtained using flow cytometry. Lockett *et al.* (1991b) also demonstrated that the system could automatically detect Chlamydiae-infected cells labeled with a rhodamine-conjugated antibody against the bacteria. Lockett *et al.* (1990a and 1991a) showed that the system can automatically and success-fully determine the DNA ploidy histogram of Pap smears and count cells positive for the E6 oncoprotein. E6 can be expressed in a deregulated way in cells with HPV 16/18 integrated into their genome. It is produced in the nucleus and migrates into the cytoplasm. The results (Lockett *et al.*, 1991a) suggested that E6 expression was an early cancer or dysplasia indicator be-cause it was expressed before abnormal DNA ploidy distributions were de-tected. We have also shown that the system can be used to automatically detect cells in breast cancer sections expressing the stripped mucin-3 anti-gen, which is a possible marker for malignancy (Lockett *et al.*, 1990a).

### A. Detection of HPV 18 E6 Antigen in HeLa Cells

HeLa cells are a culturable cervical carcinoma cell line containing 40–200 genomic copies of HPV 18 (Heiles *et al.*, 1988) and actively expressing the E6 oncoprotein. A slide-based specimen of HeLa cells was labeled with mouse anti-HPV 18 E6 monoclonal antibody (MAB 875, Chemicon Intl. Inc., Temecula, CA), followed by rhodamine-labeled goat antimouse sec-ond antibody (Cat. #2211-0231, Organon Technica Corp., West Chester, PA). This specimen serves as a positive control since all cells express E6. As a negative control, another specimen was only labeled with the second an-tibody. Both specimens were counterstained with the fluorescent DNA stain, Hoechst 33342 (Molecular Probes Inc., Eugene, OR). A total of 111 cells were imaged using a $40 \times$, 1.3 NA oil objective, an intensified CCD, and the Sun image acquisition and processing system.

The cell nuclei were automatically detected from images of the Hoechst staining. For each specimen, the DNA ploidy histogram was derived from the set of integrated pixel intensities corresponding to each nucleus. The accuracy with which the DNA content of nuclei was measured was deter-mined from the width of the (2N ploidy) peaks in the histograms. Mathe-

FIGURE 15.1

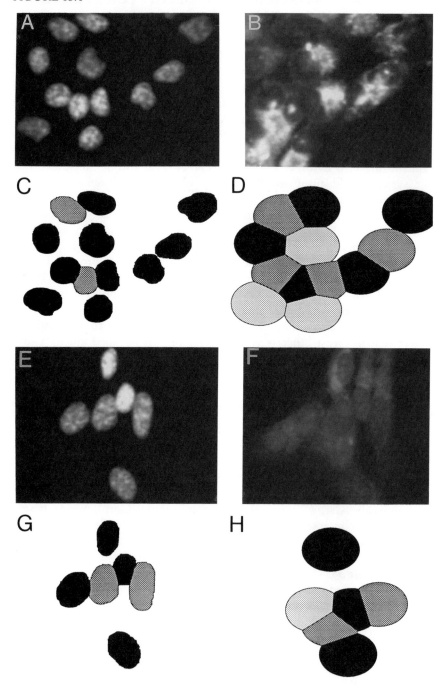

matically, the width was a percentage defined as the full width half maximum of the peak divided by distance of the peak from the origin. The cell cytoplasms were approximated by circles centered on their nuclei and having radii twice the mean nuclear radius. E6 expression for each cell was estimated by integrating the pixel intensities inside each circle from the rhodamine images.

Images of the Hoechst and rhodamine staining in the specimens were also recorded with the color-imaging system and using a $40\times$, 0.75 NA dry objective lens. The CCD integration times for the Hoechst and rhodamine images were 33 ms and 1.33 s, respectively. The fluorescent signals were detected from the images using the automatic color detection facilities of the system.

Figure 15.1 shows typical images of the Hoechst and anti-E6 staining from both specimens. The cells labeled with the primary E6 antibody (Fig. 15.1B) visually showed variable cytoplasmic staining, which in all cases was much more than the cytoplasmic signal from the cells labeled with second antibody alone (Fig. 15.1F). Figures 15.1C and 15.1G show the regions of the images in Figs. 15.1A and 15.1E that were automatically defined as being nuclei. Based on a visual assessment, 94% of nuclei were correctly detected.

Figure 15.2 is a set of histograms showing the distributions of Hoechst and rhodamine (E6) staining from the imaged cells of both specimens. Both DNA ploidy distributions (Figs. 15.2A and 15.2C) show single peaks indicating only cells with 2N ploidy were imaged. The positions of the peaks on the $x$-axis in Figs. 15.2A and 15.2C are different because of slightly different concentrations of Hoechst used for these two slides. The peak widths (see the mathematical definitions of peak width above) were 4% and 7% for Figs. 15.2A and 15.2C, respectively. Although these widths exceed those reported by us using an unintensified CCD camera (Lockett *et al.*, 1992a) and exceed those expected from flow cytometry (Ryan *et al.*, 1988), they are similar to other published results using fluorescence IC (Kamentsky and Kamentsky, 1991; Cowden and Curtis, 1981). The reasons that the widths are greater than those we previously reported (Lockett *et al.*, 1992a) are that an intensifier tube was used and that variable nuclear staining occurred.

Figures 15.2B and 15.2D show the distributions of the rhodamine (E6)

---

Results from screening specimens of HeLa cells using the Sun workstation–based imaging system. (A–D) Images from the positive control specimen. (E–H) Images from the negative control specimen. (A, E) Typical images of the Hoechst-stained nuclei. (B, F) Images recorded with the rhodamine filter set to detect the E6 antibody staining (and any nonspecific staining). (C, G) The regions of images (A) and (E), respectively, that were automatically defined as nuclei. The regions are shaded differently to indicate their association with particular nuclei. (D, H) The circles used for quantifying the rhodamine E6 staining associated with each cell. Circles could not overlap, so instead the straight line equidistant from their centers was used as the division between cells.

FIGURE 15.2

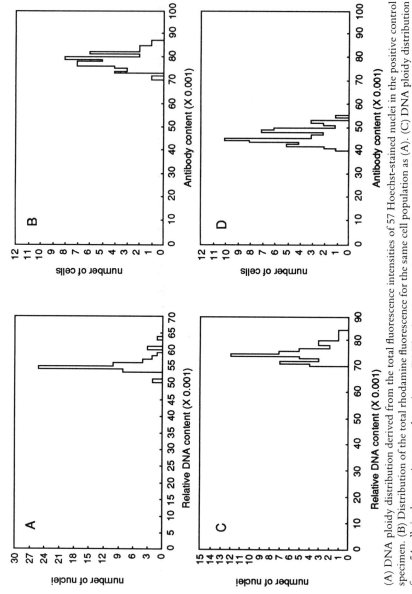

(A) DNA ploidy distribution derived from the total fluorescence intensities of 57 Hoechst-stained nuclei in the positive control specimen. (B) Distribution of the total rhodamine fluorescence for the same cell population as (A). (C) DNA ploidy distribution from 54 cells in the negative control specimen. (D) The distribution of the total rhodamine fluorescence for the same population of cells as (C).

fluorescence from the cells in the positive and negative control specimens, respectively. The peaks widths are much wider than the DNA peaks, which is probably because the amount of E6 and nonspecific binding by the second antibody varied between cells. More importantly, the peak representing the positive control cells did not overlap at all the peak representing the negative control cells. Thus, E6 positive and negative cells can be automatically recognized by this approach. The method of using circles centered on the nuclei for quantifying the cytoplasmic E6 stain is highly approximate. A more accurate method, for example, directly segmenting the cytoplasm (Lockett *et al.*, 1991b), should lead to a much more significant difference between the reported E6 intensities in positive and negative cells.

Figure 15.3 is an example of images acquired using the color-imaging system and the automatically detected fluorescent objects. Figure 15.3 has a similar layout to Fig. 15.1. However, being a color system, the images have true colors corresponding to the emission wavelengths of the fluorescent stains.

## B. Detection of HPV 16 and 18 Using Fluorescence *in Situ* Hybridization

CaSki cells, which are a culturable cervical carcinoma cell line and contain approximately 600 integrated genomic copies of HPV-16 (Smits *et al.*, 1991), and Pap smears from women attending a local clinic were used. The slide-based specimens were stained using DNA probes against HPV-16 and -18 that were directly labeled with Texas Red (TR) dye (Molecular Analysis Inc, Houston, TX). The protocol supplied with the reagents was followed and the specimens were counterstained with Hoechst.

The stained specimens were imaged and analyzed using similar methods to those described previously. However, since the HPV is located in the nucleus, the same regions defined as nuclei from the Hoechst images were used for quantifying the TR fluorescence.

Figures 15.4A and 15.4B are images of CaSki cells from the same microscopic scene acquired using the Sun workstation–based imaging system and Hoechst and TR filters, respectively. Figure 15.4C shows the regions automatically detected as nuclei. Figures 15.5 and 15.6 are in a similar format to Fig. 15.4, but show images of cells from Pap smears. The intense nuclear staining in Fig. 15.5B suggests that this cell is HPV 16/18 positive, and the absence of significant staining in Fig. 15.6B suggests that this cell is HPV 16/18 negative or has too few copies of the viral DNA for detection. Furthermore, the nucleus in Fig. 15.5B shows diffuse HPV staining, whereas the CaSki cell nucleus (Fig. 15.4B) shows punctate staining. Wolber *et al.* (1991) suggested that the punctate staining arises because the HPV was integrated into the host cell genome and that the diffuse staining arises because the virus was episomal.

S. J. Lockett / M. Siadat-Pajouh / K. Jacobson / B. Herman

FIGURE 15.3

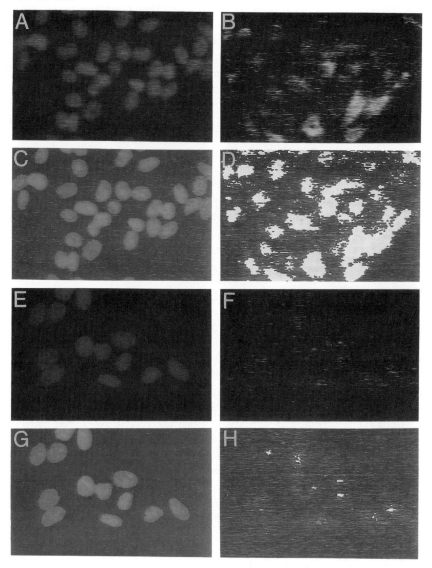

Results from screening specimens of HeLa cells using the color-imaging system. (A–D) Images from the positive control specimen. (E–H) Images from the negative control specimen. (A, E) Typical images of the Hoechst-stained nuclei. (B, F) Images recorded with the rhodamine filter set to detect the antibody staining or nonspecific staining. (C, G) The regions of images (A) and (E), respectively, where Hoechst staining was automatically detected. (D, H) The regions of images (B) and (F) where rhodamine staining was automatically detected. (*See Plate 11 for a color version of this figure.*)

FIGURE 15.4

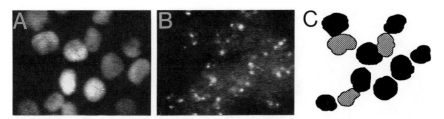

(A) Image of Hoechst-stained nuclei of CaSki cells. (B) Image of the same microscopic scene as (A) recorded with the Texas Red filters to detect the fluorescent probe directed against HPV-16. (C) The regions detected as nuclei that were used for quantification of both the Hoechst and TR staining.

FIGURE 15.5

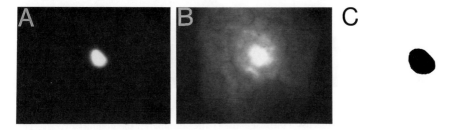

(A) Image of a Hoechst-stained nucleus from a Pap smear showing positivity for HPV-16/18. (B) Image of the same microscopic scene as (A) recorded with the Texas Red filters to detect the HPV-DNA probe. (C) The region in (A) automatically detected as the nucleus.

FIGURE 15.6

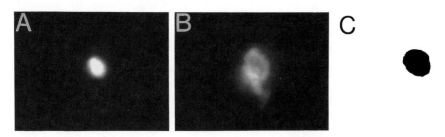

(A) Image of a Hoechst-stained nucleus from the Pap smear showing negativity for HPV-16/18. (B) Image of the same microscopic scene as (A) recorded with the TR filters. (C) The region in (A) automatically detected as the nucleus.

S. J. Lockett / M. Siadat-Pajouh / K. Jacobson / B. Herman

FIGURE 15.7

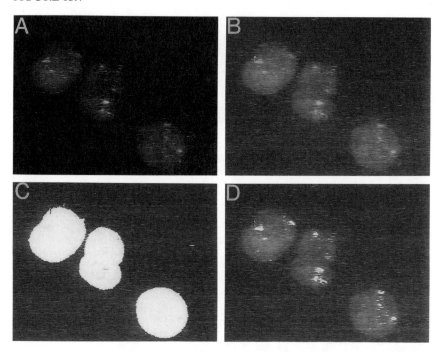

(A) Image of three CaSki cell nuclei costained with the TR-conjugated DNA probes against HPV-16/18 and with Hoechst. (B) Image in (A) after turning on the graphics overlay. This whitens the image but does not change the original RGB image data. (C) Yellow regions illustrating where Hoechst staining was automatically detected. (D) White regions illustrating where TR staining was automatically detected. (*See Plate 12 for a color version of this figure.*)

Figure 15.7A is a composite image acquired using our color imaging system of CaSki cell nuclei showing both the TR and Hoechst staining. It consists of the blue channel from the image acquired using the Hoechst filters and the red channel from the TR image. Figure 15.7B is Fig. 15.7A after turning on the graphics overlay, which whitens the image but otherwise leaves the composite image intact. The overlay is where the regions automatically detected as being positive are stored. Figures 15.7C and 15.7D show, respectively, the regions automatically detected by the imaging system as being Hoechst-stained nuclei (yellow overlay) and as being TR *in situ* hybridization signals (white overlay). (The original image remains partially visible through the yellow and white overlays.) Thus, the *in situ* hybridization signal per cell nucleus can be quantified by integrating the red pixel intensities overlayed by the white color for each nucleus overlayed by the yellow color.

## VI. FUTURE PROSPECTS FOR AUTOMATED FLUORESCENCE IMAGE CYTOMETRY

### A. Application of Imaging for Investigating Cervical Cancer

The previously described imaging techniques offer powerful approaches for disease investigation. There are two main advantages of imaging over molecular biology and biochemical techniques. The first advantage is its ability to directly image the molecular events happening in live cells or have happened in fixed cells. Thus, the results should be closely correlated to the actual disease processes in patients. The second advantage is the capability of imaging to measure multiple parameters nondestructively in cell specimens. In other words, many (dozens) of parameters can be quantitatively and simultaneously measured in the same (intact) cells. Since IC leaves specimens intact, other assays (e.g., Southern blot hybridization or flow cytometry) can be subsequently applied to the same specimens either to substantiate the IC results or to make additional measurements.

The types of parameters that can be measured by IC include (1) the quantity of specific antigens and nucleic acid sequences, (2) the localization of the biomolecules at subcellular levels, (3) the coexistance of at least three different biomolecules (and their spatial relationships) in the same cells, (4) cell and nuclear morphology, and (5) the spatial relationships and interactions between cells. Measurement of these parameters makes it much easier to determine the various transformation mechanisms happening in abnormal (e.g., precancer) cells compared to normal cells and makes it possible to address the following kinds of questions: (1) Which cell types undergo a particular transformation process? (2) Is more than one transforming agent present in the same cell at the same time? (3) Do the actions of one cell cause reactions in neighboring cells?

Quantitative imaging links results from *in vitro* biochemical assays with cytological and histological observation. This is because imaging does both types of measurements on the same cells and thus supplies the information relating specific molecules, cell morphology, and cytoarchitecture. The link offered by IC is important for accurately correlating the evidence that HPV causes cervical cancer to the cytological classifications made in clinical diagnostics (Singer and Jenkins, 1991; Choi, 1991) and to nuclear texture (Wied *et al.*, 1989). In certain situations, quantitative imaging can detect subtler changes in morphology and cytoarchitecture than visual observation. This is because, although visual observation is sensitive and consistent when comparing cells in the same microscopic scene, visual observation cannot make absolute measurements or easily compare cells far apart or in different specimens. On the other hand, IC can make a variety of morphological measurements (e.g., nuclear to cytoplasmic area ratio, nuclear texture, cell circularity) on every manually or automatically defined cell and save the results for subsequent comparisons. Furthermore, multivariate analysis (Weber, 1988) can combine the results into new parameters that optimally discriminate between normal and abnormal cells.

IC can automatically detect cells, thus, permitting the examination of large numbers of cells (from $10^4$ to $10^6$) per specimen. Present ICs detect virtually all nucleated cells (Lockett et al., 1991b), but not every overlapping cell is separately detected. However, such errors can be checked by subsequent visual verification.

## B. New Fluorescence-Imaging Techniques Useful for Cervical Cancer Studies

The following sections describe some new fluorescence-imaging techniques that are applicable to diagnosing and understanding cervical cancer. Most of these techniques can be directly carried out with our existing imaging systems while others would require minor modifications.

### 1. Three-Dimensional Imaging

The images from conventional two-dimensional (2D) microscopy do not fully represent the properties of a specimen because they are projections through an inherently three-dimensional (3D) object. An important example where 2D imaging proves inadequate is in the analysis of clustered cells in tissue specimens (Atkin, 1991) because the cells appear overlapping and out of focus. Although the remarkable image-processing capabilities of the eye–brain combination can usually discern the individual cells in these images, computer analysis, on the other hand (that is highly reliable and quantitatively precise when analyzing isolated cells) (Lockett et al., 1992a), normally performs unacceptably (Banda-Gamboa et al., 1992). A solution to this limitation is to use 3D microscopy for imaging such specimens (Rigaut et al., 1991) because the resulting 3D images show clustered cells as separate entities and in focus. In this situation, computer analysis is able to reliably detect the cells (Lockett et al., 1992b).

There are three ways to obtain the 3D images. The first is to physically thin section the specimens ($\sim$3 $\mu$m thickness), acquire 2D images of each section, and reconstruct the images into a 3D representation. The approach has several drawbacks because the sectioning is labor intensive, each 2D image has to be accurately aligned and the spatial resolution in the direction parallel to the optical axis is limited by the specimen thickness. The second way is to use confocal microscopy, which employs two pinholes in the image planes of the excitation and emission light paths to directly acquire 3D images. The approach does not have the drawbacks previously mentioned, but it is restricted to only acquiring 3D images. This has a practical limitation because when screening an entire clinical specimen, such as a Pap smear, a vast quantity of data is acquired ($\sim 10^{11}$ bytes, most of which contain no useful information). There is a third way to acquire 3D images and it offers a solution to the drawbacks of the previously described methods. The approach involves using conventional 2D imaging to automatically preselect scenes of clustered cells. Then, multiple images at different focal planes for each scene of clusters are acquired. Since each image is acquired using conventional (epi-fluorescence) 2D microscopy, they contain the sum of in-focus signal from the focal plane and out-of-focus signal from planes above and below focus. Therefore, deconvolution algorithms are applied to remove the out-of-focus fluorescence signal but leave the in-focus signal (Carrington *et al.,* 1989; Agard and Sedat, 1983). The resulting set of 2D images constitutes the 3D image. The achievable spatial resolution is lower than confocal microscopy, although this approach is probably more practical when screening tissue specimens. This is because the limited angular aperture of the objective lens causes loss of a biconic region of frequencies in the 3D Fourier spectrum (Macías-Garza *et al.,* 1988).

Three-dimensional imaging has many other useful applications when studying cervical cancer. For example, the sensitivity for detecting specific signals is increased. This is because in 3D imaging, the signal at a particular point can only be obscured by irrelevant signals originating at the same point. However, in 2D imaging, the specific signal can additionally be obscured by signal from other focal planes (White *et al.,* 1987). Multiple DNA probes can be more accurately localized in interphase nuclei using 3D imaging. A particular example is the simultaneous localization of three probes using only two fluorescent labels (Nederlof *et al.,* 1990). This is done by labeling probe one with label one, probe two with label two, and probe three with both labels. Each probe can be uniquely detected using 2D epi-fluorescence imaging provided they are not located along the same line parallel to the optical axis. In this case, probes one and two would mistakenly appear as the same as probe three. However, 3D imaging removes this ambiguity by providing the necessary spatial resolution parallel to the optical axis.

A specific application of 3D imaging to understanding cervical cancer is to detect the integration of HPV into the host genome of intact cells. The cells would be double labeled with a fluorescent DNA stain (e.g., DAPI) and a fluorescent DNA probe against the HPV. The 3D imaging would be used to visualize the organization of nuclear chromatin in 3D. If the HPV is integrated, then the fluorescent dots from the HPV label would be localized along the nuclear chromation. The study could be extended to detect a preferential integration of HPV onto particular chromosomes by replacing the DNA stain with probes against specific chromosomes.

## 2. First Image Plane Microscope

Compound microscopes consist of an objective lens and projection lens. The objective magnifies and focuses the image onto the primary image plane (PIP). The projection lens further magnifies the image and corrects it for chromatic and geometric aberrations but at the expense of degrading spatial resolution, attenuating the light intensity, and reducing the field of view. However, new lens manufacturing technologies now produce objectives that project corrected images at the PIP, thus, the detection camera can be placed here and the projection lens done away with (Jaggi et al., 1991). This instrument is called a first-image plane microscope. The (CCD) camera must have both small pixels (7-$\mu$m dimension) to capture details in the smaller image at the PIP and be composed of a large array (1000 × 1000) to utilize the wider field of view.

The microscope is useful in cervical cancer studies because the wider field of view means an entire specimen can be scanned with fewer image acquisitions, thereby speeding up screening of the specimen. The higher light transmission improves the sensitivity for detecting weak fluorescent markers.

## 3. Imaging with a Color Camera

Color cameras report spectral as well as intensity information and this has two advantages in fluorescence microscopy. First, using the camera in conjunction with triple bandpass optical filters (Omega Optical Inc., Brattleboro, VM) allows specimens labeled with three fluorescent stains to be imaged in a single acquisition. This reduces the time for screening a specimen and avoids image registration shifts caused by changing filter sets (Galbraith et al., 1991). Second, the camera improves discrimination between signals from specific staining, light scatter, and autofluorescence because these signals normally differ spectrally and therefore appear as different colors in the images.

Color cameras have not been widely used in fluorescence imaging because of their low quantum efficiency (QE). However, the recent addition of a light-integration mode (American Innovision Inc., San Diego, CA) has circumvented the low QE by allowing low-light signals to accumulate on the face plate of the camera prior to readout. Sensitivity is still limited by the spectral filters coating the face plate because they only transmit photons in the correct wavelength range to the underlying sensors. An alternative that circumvents this limitation is to use chevron-type beam splitters to spatially separate the colors into red, green, and blue images and then employ three monochrome CCDs to detect each color (Kinosita et al., 1991).

An application of color imaging to cervical cancer understanding would be to verify the hypothesis that the E2 gene is disrupted upon HPV integration leading to E6 and E7 overexpression. E2, E6, and E7 mRNA would each be labeled by *in situ* hybridization using spectrally distinct fluorescence-conjugated cDNA probes. If the postulate is true, then the imaging should show cells with only integrated HPV as being E2 negative and E6 and E7 positive; cells with nonintegrated HPV positive for E6, E7, and E2; and uninfected cells less positive for E6, E7, and E2 mRNAs. Thus, in this application, quantification is important to determine the presence and form of HPVs in the cells.

## 4. Charge Injection Device Cameras

Charge injection device (CID) cameras (Williams and Carta, 1989) have never been extensively applied in fluorescence microscopy despite the fact that they could have several advantages over tube-based and CCD cameras. Their wide spectral response (from 200 to 1000 nm), high quantum efficiency (35%, 200–400 nm) (Kaplan, 1990), and almost 100% fill factor (which means the entire face plate is sensitive to light) offer high sensitivity. In CID cameras, each pixel is essentially under independent control and its signal can be read nondestructively. This means, for example, that a subset of the image can be read out at high speed for automatic-focusing purposes. Since pixels can be read out in any sequence, the Fourier transform of the image can be directly obtained. In low-light-level applications, the image can be monitored during acquisition and the light integration stopped as soon as adequate signal strength has been reached. Furthermore, pixels need not have the same integration time, so dim areas of a scene can be imaged without pixels in bright areas being overexposed. Potentially, these cameras could provide more flexible and sensitive detection of fluorescence-labeled cervical specimens.

## 5. Time-Resolved Imaging

Sensitivity in fluorescence microscopy is mainly limited by relatively high autofluorescence from cellular material, which is $\sim10^4$ fluorescein equivalents per cell (Tanke, 1989). However, time-resolved microscopy used in conjunction with delayed luminescence labels [lifetime $>500$ $\mu$s (Tanke, 1989; Beverloo *et al.*, 1990)] offers a solution to this limitation. This is because the technique allows the autofluorescence to decay before collection of the signal begins. [Most common fluorescent labels cannot be used in this application because their lifetime is similar to the lifetime of autofluorescence ($<100$ ns).] Time resolution is achieved by inserting two phase-locked rotating chopper blades into the excitation and emission light paths of a microscope (Marriott *et al.*, 1991). The blades are timed so that the specimen is exposed to a pulse of excitation light (~20 $\mu$s), followed by delay (~50 $\mu$s), and then the emission light is collected by the camera for ~100 $\mu$s. The quantity of light detected per cycle is usually low, therefore light from many cycles of the blades is integrated at the (CCD) camera face plate. This approach reduces detected autofluorescence by 100-fold, because nearly all the autofluorescence has decayed before collection of the signal from the label commences. As a consequence, the sensitivity for detecting low levels of labeled biomolecules is increased by 100-fold, making the technique as sensitive as radioactive markers. The use of this technology may enable the detection of single copies of HPV per cell as well as detection of much lower concentrations of other biomarkers, compared to the present techniques.

## 6. Resonance Energy-Transfer Imaging

Resonance energy-transfer (RET) imaging (Herman, 1989) is a microscopic technique for quantifying the binding of two types of molecules. One type is labeled with fluorophores (the donors) that have their emission spectra overlapping the excitation spectrum of the fluorophores labeling the other type (the acceptors). If donors and acceptors are within 1 to 10 nm, which occurs when the molecules are bound, then energy from the excited donors is nonradiatively transferred to the acceptors. The phenonema is observed by exposing the specimen to light of the donor excitation wavelengths and collecting light at the acceptor emission wavelengths. The emission light intensity is proportional to the concentration of bound molecules and is also a function of the distances between the acceptors and donors. (The fluorophores are independently quantified using normal epifluorescence microscopy.) Furthermore, since this is an imaging-based technique, binding can be localized within the cells.

An application of RET imaging could be to quantify the binding of E6 and p53 antigens *in vivo*. Such binding studies have been done *in vitro* (Werness *et al.*, 1990), but binding has not been observed directly in cells.

## 7. Nanovid Microscopy

Nanovid microscopy (Gelles *et al.*, 1988) is a technique for localizing point objects with precisions (from 20 to 200 nm) that are much higher than the limiting spatial resolution of epifluorescence microscopy (~0.2 $\mu$m). In this technique, images of the objects (usually fluorescence molecules or a gold particle of 20–40-nm diameter) are acquired. The images show the objects as intensity distributions that correspond to the point spread function (PSF) of the system. The objects are localized by calculating the centroid of the distribution, either directly or by fitting the expected PSF, to localize the objects. The accuracy of the method is much greater than the spatial resolution, but the approach only works for objects separated by at least the spatial resolution. However, this limitation can be circumvented, by labeling the objects with different (spectrally distinct) fluorophores. In this situation, noninterfering images of each object can be acquired by using the appropriate filter sets for each fluorophore and their centroids separately determined.

An application of nanovid microscopy would be in determining the distance between the integration sites of HPVs and host genes (e.g., tumor suppressor genes and oncogenes). If HPV is affecting the expression of host genes (Couturier *et al.*, 1991), then it is may be reasonable to expect that the two genes are close to each other in the genome. In normal epifluorescence microscopy, neighboring genes must be at least 1 megabase apart in metaphase spreads and 50 kilobases apart in interphase nuclei (Lichter and Ward, 1990; Trask *et al.*, 1991) for their *in situ* hybridization spots to be resolved. The nanovid microscopy should allow genes tenfold closer together to be resolved and the number of bases between them measured.

## C. An Automatic Fluorescence-Imaging System for Screening Clinical Specimens

This section outlines a proposed automatic fluorescence IC for prescreening clinical specimens. The purpose of this prescreening is to automatically identify the cells that are positive for specific cancer markers (e.g., HPV genotypes). The IC will report to the operator the proportion of positive cells and their locations in the specimen. The operator will then need only to select the cells considered positive by the IC and not have to inspect every cell in the specimen. The specimens are stained with two specific cancer markers and counterstained with a DNA stain (e.g., Hoechst). The counterstain is for detecting all cells in the specimen and determining the DNA ploidy distribution. The detector is a large area (1024 × 1024 pixels) color camera.

The steps in the screening are as follows: (1) Image the entire specimen at low resolution (1.33 $\mu$m per pixel) using a 10× objective lens and filters to detect the DNA label. Each image is automatically focused prior to acquisition. (2) Automatically analyze the images to locate every cell in the specimen and determine whether cells are isolated or clustered. (3) Rescan the parts of the specimen containing cells using a 40× (high NA) objective lens (pixel resolution = 0.33 $\mu$m). This can be set up so that a fixed number (e.g., 4 × 10⁴) of cells are screened or a fixed ratio of isolated to clustered cells are screened. The imaging is done using a pulsed xenon–mercury burner, which provides higher intensities than continuous illuminators and using a triple band-pass filter set so that the three labels are simultaneously detected. Also, each image is automatically focused before acquisition. (4) Analyze the images using the 2D segmentation algorithms. Clustered nuclei will be divided by the cluster division algorithms. If this approach fails then the 3D imaging approach could be applied. (5) Quantify the intensity of each of the three stains per cell. (6) Display the results either as separate histograms of the intensity distribution for each stain or as a combined 3D scattergram for all three stains. (7) Allow the user to confirm results by visualizing cells within selected ranges of fluorescence intensities (e.g. with high DNA ploidy or positive for HPV) with suspicious morphologies or from specimen areas with abnormal tissue cytoarchitecture.

## ACKNOWLEDGMENTS

These studies were supported by grants from the Gustavas and Louise Pfeiffer Research Foundation and the Whitaker Foundation, a contract with American Inovision Inc. (San Diego, CA.), and a donation of reagents from Molecular Analysis Inc. (Houston, TX.).

## REFERENCES

Agard, D. A., and Sedat, J. W. (1983). Three-dimensional architecture of a polytene nucleus. *Nature 302,* 676–681.

Atkin, N. B. (1991). The clinical usefulness of determining ploidy patterns in human tumors as measured by slide-based feulgen microspectrophotometry. *Analyt. Quant. Cytol. Histol. 13,* 75–79.

Baker, C. C., Phelps, W. C., Lindgren, V., Braun, M. J., Gonda, M. A., and Howley, P. M. (1987). Structural and transcriptional analysis of human papillomavirus type 16 sequences in cervical carcinoma cell lines. *J. Virol. 61,* 962–971.

Banda-Gamboa, H., Ricketts, I., Cairns, A., Hussein, K., Tucker, J. H., and Husain, N. (1992). Automation in cervical cytology: An overview review. *Analyt. Cell. Pathol. 4,* 25–48.

Bauer, H. M., Ting, Y., Greer, C. E., Chambers, J. C., Tashiro, C. J., Chimira, J., Reingold, A., and Maros, M. (1991). Genital human papilloma virus infection in female university students as determined by a PCR-based method. *JAMA 265,* 472–477.

Beverloo, H. B., van Schadewijk, A., van Gelderen-Boele, S., and Tanke, H. J. (1990). Inorganic phosphors as new luminescent labels for immunocytochemistry and time-resolved microscopy. *Cytometry 11,* 784–792.

Broker, T. R., Chow, L. T., Chin, M. T., Rhodes, C. R., Wollinsky, S. M., Whitbeck, A., and Stoler, M. H. (1989). A molecular portrait of human papillomavirus carcinogenesis. *Cancer Cells 7,* 197–208.

Carrington, W. A., Fogarty, K. E., Lifschitz, L., and Fay, F. S. (1989). Three-dimensional imaging on confocal and wide-field microscopes. *In* "The Handbook of Biological Confocal Microscopy." (J. Pawley, ed.), pp. 137–146, IMR Press, Madison, WI.

Choi, Y. J. (1991). Detection of human papillomavirus DNA on routine Papanicolaou's smears by *in situ* hybridization with the use of biotinylated probes. *Am. J. Clin. Pathol. 95*, 475–480.

Couturier, J., Sastre-Garau, X., Scheider-Maunoury, S., Labib, A., and Orth, G. (1991). Integration of Papillomavirus DNA near myc genes in genital carcinomas and its consequences for proto oncogene expression. *J. Virol 65*, 4534–4538.

Cowden, R. R., and Curtis, S. K. (1981). Microfluorometric investigations of chromatin structure I. Evaluation of nine DNA-specific fluorochromes as probes of chromatin organization. *Histochem. 72*, 11–23.

Crook, T., Wrede, D., and Vousden, R. H. (1991). p53 point mutations in HPV negative human cervical carcinoma cell lines. *Oncogene 6*, 873–875.

DeBiasio, R., Bright, G. R., Ernst, L. A., Waggoner, A. S., and Taylor, D. L. (1987). Five-parameter fluorescence imaging: Wound healing of living swiss 3T3 cells. *J. Cell Biol. 105*, 1613–1622.

DeCaprio, J. A., Ludlow, J. W., Figge, J., Shew, J.-Y., Huang, C-M., Lee, W-H., Marsillio, E., Paucha, E., and Livingston, D. M. (1988). SV40 large tumor antigen forms a specific complex with the product of the retinoblastoma susceptibility gene. *Cell 54*, 275–283.

DeVilliers, E.-M. (1989). Heterogeneity of the human papillomavirus group. *J. Virol. 53*, 4898–4903.

DiGuiseppi, J., Inman, R., Ishihara, A., Jacobson, K., and Herman, B. (1985). Applications of digitized fluorescence microscopy to problems in cell biology. *Biotechniques 3*, 394–403.

Durst, M., Kleinheinz, A., Hotz, M., and Gissmann, L. (1985). The physical state of human papillomavirus type 16 DNA in benign and malignant genital tumours. *J. Gen. Virol. 66*, 1515–1522.

Ewen, M. B., Ludlow, J. W., Marsilio, E., DeCaprio, J. A., Milikan, R. C., Cheng, S. H., Paucha, E., and Livingston, D. M. (1989). An N-terminal transformation-governing sequence of SV40 large T antigen contributes to the binding of both p110Rb and a second cellular protein, p120. *Cell 58*, 257–267.

Galbraith, W., Wagner, M. C. E., Chao, J., Abaza, M., Ernst, L. A., Nederlof, M. A., Hartsock, R. J., Taylor, D. L., and Waggoner, A. S. (1991). Imaging cytometry by multiparameter fluorescence. *Cytometry 12*, 579–596.

Gelles, J., Schnapp, B. J., and Sheetz, M. P. (1988). Tracking kinesin-driven movements with nanometre-gold particles. *Nature 331*, 450–453.

Grubendorf-Cohen, E. I., and Cremer, S. (1990). The demonstration of human papillomavirus 16 genomes in the nuclei of genital cancers using two different methods of *in situ* hybridization. *Cancer 65*, 238–241.

Hause, A. T., Retzel, E. F., and Stakus, K. A. (1990). Amplification and detection of centiviral DNA inside cells. *Proc. Natl. Acad. Sci. U.S.A. 87*, 4971–4975.

Heiles, H. B. J., Genersch, E., Kessler, C., Neumann, R., and Eggers, H. J. (1988). *In situ* hybridization with digoxigenin-labeled DNA of human papillomaviruses (HPV 16/18) in HeLa and SiHa cells. *BioTechniques. 6*, 978–981.

Herman, B. (1989). Resonance energy transfer microscopy. *Meth. Cell Biol. 30*, 219–243.

Howley, P. M. (1991). Role of human Papillomaviruses in human cancer. *Cancer Res. 51* (suppl.), 5019s–5022s.

Jaggi, B., Poon, S., Pontifex, B., Fengler, J., Marquis, J., and Palcic, B. (1991). A quantitative microscope for image cytometry. *Proc. SPIE. 1448,* 89–97.

Kamentsky, L. A., and Kamentsky, L. D. (1991). Microscope-based multiparameter laser scanning cytometer yielding data comparable to flow cytometry data. *Cytometry 12,* 381–387.

Kaplan, H. (1990). New jobs for charge-transfer devices. *Photonics Spectra. 24,* 11, 86–7.

Kessler, I. I. (1976). Human cervical cancer as a venereal disease. *Cancer Res. 36,* 783–791.

Kinosita, K., Itoh, H., Ishiwata, S., Hirano, K., Nishizaka, T., and Hayakawa, T. (1991). Dual-view microscopy with a single camera: real-time imaging of molecular orientations and calcium. *J. Cell Biol. 115,* 67–73.

Koss, L. G. (1979). "Diagnostic Cytology and Its Histological Bases, 3rd Ed." Lippincott, Philadelphia, PA.

Koutsky, L. A., Galloway, D. A., and Holmes, K. K. (1988). Epidemiology of genital human papillomavirus infection. *Epidem. Rev. 10,* 122–163.

Kurman, R. J., Shah, K. H., Lancaster, W. D., and Jenson, A. B. (1981). Immunoperoxidase localization of papillomavirus antigens in cervical dysplasia and vulva condylomas. *Am. J. Obstet. Gynecol. 140,* 931–935.

Lee, B. R., Haseman, D. B., and Reynolds, C. P. (1989). A digital image microscopy system for rare-event detection using fluorescent probes. *Cytometry 10,* 256–262.

Lichter, P., and Ward, D. C. (1990). Is non-isotopic *in situ* hybridization finally coming of age? *Nature 345,* 93–94.

Lockett, S. J., Jacobson, K., O'Rand, M., and Herman, B. (1990a). Automated clinical image cytometry of fluorescence stained Pap smears and breast tissue sections. *Trans. Royal. Micros. Soc. 1,* 539–542.

Lockett, S. J., O'Rand, M., Rinehart, C., Kaufman, D., Jacobson, K., and Herman, B. (1990b). Automatic measurement of DNA ploidy using fluorescent microscopy. *In* "Optical Microscopy for Biology." B. Hermans and K. Jacobson, (eds.), pp. 603–613, Wiley-Liss Inc., New York.

Lockett, S. J., Jacobson, K., O'Rand, M., Kaufman, D. G., Corcoran, M., Simonsen, M. G., Taylor, H., and Herman, B. (1991a). Automated image-based cytometry with fluorescence-stained specimens. *Biotechniques 10,* 514–519.

Lockett, S. J., O'Rand, M., Rinehart, C. A., Kaufman, D. G., Herman, B., and Jacobson, K. (1991b). Automated fluorescence image cytometry: DNA quantification and detection of chlamydia infections. *Analyt. Quant. Cytol. Histol. 13,* 27–44.

Lockett, S. J., Jacobson, K., and Herman, B. (1992a). Quantitative precision of an automated, fluorescence-based image cytometer. *Analyt. Quant. Cytol. Histol. 14,* 187–202.

Lockett, S. J., Jacobson, K., and Herman, B. (1992b). Application of 3D digital deconvolution to optically sectioned images for improving the automatic analysis of fluorescent-labeled tumor specimens. *Proc. SPIE. 1660,* 130–139.

Lockett, S. J., Jacobson, K., and Herman, B. (1992c). An algorithm for the automatic separation of clustered, fluorescent-stained nuclei in digitized images. (In preparation.)

Macías-Garza, F., Bovik, A. C., Diller, K. R., Aggarwal, S. J., and Aggarwal, J. K. (1988). Digital reconstruction of three-dimensional serially sectioned optical images. *IEEE Trans. Acoust. Speech Sig. Proc. 36*, 1067–1075.

Marriott, G., Clegg, R. M., Arnt-Jovin, D. J., and Jovin, T. M. (1991). Time resolved imaging microscopy. *Biophys. J. 60*, 1374–1387.

Meisels, A., and Fortin, R. (1976). Condylomatous lesions of the cervix and vagina. 1. Cytologic patterns. *Acta. Cytol. 20*, 505–509.

Mellin, W. (1990). Cytophotometry in tumor pathology: A critical review of methods and applications, and some results of DNA analysis. *Path. Res. Pract. 186*, 37–62.

Moris, R. G., Arends, M. J., Bishop, P. E., Sizer, K., Duval, E., and Bird, C. C. (1990). Sensitivity of Digoxygenin and biotin labeled probe for detection of human papillomavirus by *in situ* hybridization. *J. Clin. Path. 43*(10), 800–805.

Munger, K., Phelps, W. C., Bubb, V., Howley, P. M., and Schlegel, R. (1989a). The E6 and E7 genes of the human papillomavirus type 16 together are necessary and sufficient for transformation of primary human keratinocytes. *J. Virol. 63*, 4417–4421.

Munger, K., Werness, B. A., Dyson, N., Phelps, W. C., and Howley, P. M. (1989b). Complex formation of human papillomavirus E7 proteins with the retinoblastoma tumor suppressor gene product. *EMBO J. 8*, 4099.

Nederlof, P. M., van der Flier, S., Wiegant, J., Raap, A. K., Tanke, H. J., Ploem, J. S., and Van der Ploeg, M. (1990). Multiple fluorescence *in situ* hybridization. *Cytometry 11*, 126–131.

Nuovo, G. J., Friedman, D., and Richart, R. M. (1990). *In situ* hybridization analysis of human papillomavirus DNA segregation patterns in lesions of the female genital tract. *Gynecol. Oncol. 36*, 256–262.

Palcic, B., Jaggi, B., and MacAulay, C. (1990). The importance of image quality for computing texture features in biomedical specimens. *Proc. SPIE 1205*, 155–162.

Palefsky, J. M., Winkler, B., Rabanas, J.-P., Clark, C., Chan, S., Nizet, V., and Schoolnik, G. K. (1991). Characterization of *in vivo* expression of the human papillomavirus type 16 E4 protein in cervical biopsy tissues. *J. Clin. Invest. 87*, 2132–2141.

Parry, W. L., and Hemstreet, G. P. (1988). Cancer detection by quantitative fluorescence image analysis. *J. Urol. 139*, 270–274.

Ploem, J. S., van Driel-Kulker, A. M. J., Goyarts-Veldstra, L., Ploem-Zaaijer, J. J., Verwoerd, N. P., and van der Zwan, M. (1986). Image analysis combined with quantitative cytochemistry. *Histochem. 84*, 549–555.

Purola, E., and Savia, E., (1977). Cytology of gynecologic condyloma acuminatum. *Acta. Cytol. 21*, 26–31.

Rigaut, J. P. (1989). Image analysis in histology hope, disillusion and hope again. *Acta Stereol. 8*, 3–12.

Rigaut, J. P., Vassy, J., Herlin, P., Duigou, F., Masson, E., Briane, D., Foucrier, J.,

Carvajal-Gonzalez, S., Downs, A. M., and Mandard A.-M. (1991). Three-dimensional DNA image cytometry by confocal scanning laser microscopy in thick tissue blocks. *Cytometry 12*, 511–524.

Riou, G., Favre, M., Jeannel, D., Bourhis, J., Le Doussal, V., and Orth, G. (1990). Association between poor prognosis in early-stage invasive cervical carcinomas and non-detection of HPV DNA. *Lancet. 335*, 1171–1174.

Rotkin, I. D. Etiology and Epidemiology of cervical cancer. *In* "Cervical Cancer." G. Dallenbach-Hellweg (ed.), Springer, Berlin Heidelberg, New York.

Ryan, D. H., Fallon, and M. A., Horan, P. K. (1988). Flow cytometry in the clinical laboratory. *Clin Chim. Acta. 171*, 125–174.

Schneider-Gadicke, A., and Schwartz, E. (1986). Different human cervical carcinoma cell lines show similar transcription patterns of human papillomavirus type 18 early genes. *EMBO J. 5*, 2285–2292.

Schwartz, E., Freese, U. K., Gissmann, L., Mayer, W., Roggenbuck, B., Stremlau, A., and zur Hausen, H. (1985). Structure and transcription of human papillomavirus sequences in cervical carcinoma cells. *Nature* (Lond.) *314*, 111–114.

Singer, A., and Jenkins, D. (1991). Viruses and cervical cancer. *Brit. Med. J. 302*, 251–252.

Smits, H. L., Cornelissen, M. T. E., Jebbink, M. F., van den Tweel, J. G., Struyk, A. P. H. B., Briët, M., and Schegget, J. T. (1991). Human papillomavirus type 16 transcripts expressed from viral-cellular junctions and full-length viral copies in CaSki cells and in a cervical carcinoma. *Virology 182*, 870–873.

Smotkin, D., and Wettstein, F. O. (1986). Transcription of human papillomavirus-type 16 early genes in a cervical cancer and a cancer-derived cell line and identification of the E7 protein. *Proc. Natl. Acad. Sci. U.S.A. 83*, 4680–4684.

Southern, E. M. (1985). Detection of specific sequences among DNA fragments separated by gel electrophoresis. *J. Mol. Biol. 98*, 503.

Tanke, H. J. (1989). Does light microscopy have a future? *J. Micros. 155*, 405–418.

Trask, B. J., Massa, H. F., Christensen, M., Fertitta, A., Carrano, A. V., Patel, P. I., Lupski, J. R., Kenwrick, S., and Gitschier, J. (1991). Genome mapping by multicolor fluorescence *in situ* hybridization. *Cytometry* Suppl. *5*, 281A.

Tsay, T.-T., Inman, R., Wray, B., Herman, B., and Jacobson, K. (1990). Characterization of low-light-level cameras for digitized video microscopy. *J. Micros. 160*, 141–159.

Vollath, D. (1988). The influence of scene parameters and of noise on the behavior of automatic focusing algorithims. *J. Micros. 151*, 133–146.

Weber, J. E. (1988). Applications of multivariate analysis in diagnostic cytology. *Analyt. Quant. Cytol. Histol. 10*, 54–72.

Werness, B. A., Levine, A. J., and Howley, P. M. (1990). Association of human papillomavirus types 16 and 18 E6 proteins with p53. *Science* (Washington, DC) *248*, 76–79.

Werness, B. A., Munger, K., and Howley, P. M. (1991). The role of the human papillomavirus oncoproteins in transformation and carcinogenic progression. *In* "Important Advances in Oncology." (V. T. DeVita, Jr., S. Hellman, and S. A. Rosenberg eds.), pp. 2–18, J. B. Lippincott Company, Philadelphia, PA.

White, J. G., Amos, W. B., and Fordham, M. (1987). An evaluation of confocal versus conventional imaging of biological structures by fluorescence light microscopy. *J. Cell Biol. 105,* 41–48.

Whyte, P., Williamson, N. M., and Harlow, E. (1989). Cellular targets for transformation by the adenovirus E1A proteins. *Cell 56,* 67–75.

Wied, G. L., Bartels, P. H., Bibbo, M., and Dytch, H. E. (1989). Image analysis in quantitative cytopathology and histopathology. *Human Pathol. 20,* 549–571.

Williams, B., and Carta, D. (1989). CID cameras: More than an alternative to CCDs. *Advanced Imaging. 4*(1), 48–50.

Wolber, R. A., and Clement, P. B. (1991). *In situ* DNA hybridization of cervical small cell carcinoma and adenocarcinoma using biotin-labeled human *papillomavirus* probes. *Mod. Pathol. 4,* 96–100.

Yoshikawa, H., Kawana, T., Mizuno, M., Yoshikura, H., and Iwamoto, A. (1991). Detection and typing of multiple genital human papillomaviruses by DNA amplification with consensus primers. *JPN. J. Cancer Res. 82*(5), 524–31.

zur Hausen, H. (1977). Human papillomaviruses and their possible role in squamous cell carcinomas. *Cur. Top. Microbiol. Immunol. 78,* 1–30.

zur Hausen, H. (1991). Minireview Human papillomaviruses in the pathogenesis of anogenital cancer. *Virology. 184,* 9–13.

zur Hausen, H., and Schneider, A. (1987). The role of papillomaviruses in human anogenital cancer. *In* "The Pappillomaviruses." (P. M. Howley and N. P. Salzman eds.), Vol. 2. pp. 245–263, Plenum Publishing Corp., New York.

# INDEX

Acid loading, 207
Adenoviruses, 405
Amplitude modulation, 378
Analog PMTs, 310
Anaphase, 67
Anion transporter, 19
Antibody binding, 320
Audio track, 227
Autocorrelation function, 316
Autofluorescence, 375
Autoluminescence, 424
Automated fluorescence image cytometry, 408

Background, 195, 197, 226, 227
Barrier-free pathlength (BFP), 292, 293
Basic proteins, 265
Binding constants, 318

Caged compounds
    amino acid neurotransmitters, 58
    ATP, 28
    benzylic carbon, 33
    calcium, 47
    cAMP, 28
    carbamoycholine, 57
    cyclic nucleotides, 46
    diacyl glycerol, 56
    diazo, 30, 53, 54
    diazo-2, 54, 55
    diazo-3, 55, 57
    diazo-4, 55
    DM-nitrophen, 51, 53, 62, 66, 229, 230, 231
    enzymes, 60

flash photolysis, 214, 230, 231
fluorophores, 9, 59
4,5-dimethoxy-2-nitrobenzyl (DMNB), 35, 46, 56, 58
4-formyl-6-methoxy-3-nitrophenoxyacetate, 56
inositol phosphate, 55
IP$_3$, 63
neurotransmitters, 57
nitr, 48, 53
nitr-5, 16, 39, 49, 50, 51, 62, 63, 65, 66, 70, 229
nitr-7, 48, 69
nitr-8, 49, 50
nitr-9, 50
nitrobenzyl, 46
nitronate, 33
nitrosoketone, 33
nucleotides, 44
1-(4',5'-dimethoxy-2'-nitrophenyl)ethyl (DMNPE), 35, 36, 47, 58
1-(2'-nitrophenyl)ethyl, 34
phenylephrine, 58
photoactivation, 9
photolysis, 28, 30, 39, 232
Calcium
    BAPTA, 14, 54, 67, 145
    binding kinetics, 146
    calcium-pH calculations, 199, 201
    calcium-pH kinetics, 180, 182–183, 189, 202–207
    cardiac current modulation, 60–62
    cell shrinkage, 252–253
    Dansyl-PC distribution, 271
    fluorescence lifetime imaging microscopy (FLIM), 384
    indicators, 14, 223, 224–225, 350

433

Calcium (*continued*)
  induced calcium release (CICR), 62, 136,
      228, 230
  *in situ* calibration, 145
  intracellular free transient, 116
  ionomycin, 205
  laser-scanning confocal microscope, 216
  membrane domains, 265, 280
  multiparameter imaging, 181–182
  NBD-PS enrichment, 270, 271
  noise sources, 104–105
  oscillations, 67, 88, 135–137, 139
  oscillatory behavior, 69, 108, 134–135
  prothrombin, 324
  quantal (all-or-none) Ca$^{2+}$ release, 74
  standards, 138–139, 145
  waves, 147, 148, 179
Calibration, procedures, 93, 141, 220, 332
Cameras
  charge-coupled device (CCD), 3, 103,
      115, 117–118, 120, 150, 161, 196,
      216, 256, 310, 380, 386, 393, 409, 412
  color, light integrating, 102, 160, 243
  cooled CCD, 103
    frame-transfer (FT), 410
    intensified, 116, 127–128, 160
    masked cameras, 385
    thermoelectrically cooled, 155
  charge injection device (CID), 216, 423
  charge transfer, 119, 122
  charge-transfer period, 129
  color, light integrating CCD, 410
  color camera, 422
  dark current, 390
  detectors, 216
  diode arrays, 380, 393
  dual detectors, 347
  dynamic range, 217, 376
  electrical interference, 109
  framing camera, 392, 393
  full-frame, 160
  gain, 194
  gamma, 194
  gamma circuits, 348
  gateable microchannel plate (MCP), 380
  gating, 377, 384
  Genllsys intensifier, 156
  geometric distortion, 156, 159, 160, 198,
      199, 256
  image dissection tube, 388
  image intensifier, 380, 409

image photon detectors, 388, 390
image-tube-type cameras, 155
integration, 405, 421
integration time, 201
intensified charge-coupled devices, 155
interlacing, 123
interlacing fields, 197
interlacing format, 102
interline transfer, 116, 125
intrascene dynamic range, 157, 160
*intravital* microscopic methods, 355
ISIT cameras, 156, 380
lag, 155, 156, 158, 159
light interaction mode, 422
linearity, 159, 227, 376
microchannel plate, 159, 217
microchannel plate detectors, 381
microchannel plate electron multipliers,
    393
multianode MCPs, 388
newvicons, 243
noise, 104
nonlinear camera responses, 156
on-chip-lens (OCL) technology, 123
one dimensional position sensitive device,
    389
photocathode, 102
photodiode, 310
photoelectronic streak camera, 390, 391
position sensitive detectors (PSD), 389
pseudointerlace, 124, 128
random pixel access, 386
raster scan and scanning, 217, 250
readout speed, 376
shading, 156, 159, 160
shading error, 195, 197
shot noise, 105
silicon intensified target (SIT) camera,
    3, 156
spatially resolved detection, 375, 384, 386
synchroscan streak camera, 390, 391
thermal noise, 105, 217
thermionic emission, 105
two-dimensional resistive anode detector,
    389
vertical blanking interval, 117
Cancer. *See* Cervical cancer
Cell-membrane contacts, 326–327
Cell-substrate contacts, 311–313
Cell viability, 181
Cell volume, 181, 251, 252, 253, 345

Centrifugal transport, 291
Centripetal transport, 291
Centroid, 290, 425
Cervical cancer, 404, 411, 419–421
  carinoma, 405
  dysplasia, 404
  intraepithelial neoplasia (CIN), 404
Chemical noise, 105
Chevron-type beam splitters, 422
Chronic implantation, 363
Compartmentalization, 18, 143
Constant power, 300
Cross-correlation principle, 381
Cytoskeleton
  actin filaments, 73, 74, 312
  cell adhesion, 326
  cell-substrate contacts, 311, 332
  cytoarchitecture, 419
  cytoskeletal proteins, 8
  focal contacts, 311

Deconvolution, 421
Demodulation ratio, 382
Domain radiance value distribution, 268,
  270, 274–275, 278
Dot blot, 406, 407
Dual-emission fluorescence imaging, 179,
  258

EGTA, 145
Electrical potential, 349, 350
Endocytosis, 12
Environmental factors, 4, 94
Evanescent wave, 297–299
Excitation-contraction coupling, 63

$F_c$ receptors, 321–322
Film, 216
Filter *in situ* hybridization, 406, 407, 408
Flow cytometry, 408, 414
Fluorescence, 2, 237
  analog chemistry, 5, 9
  anisotropy, 377
  cellular autofluorescence, 140
  collisional-quenching, 254
  correlation spectroscopy (FSC), 314
  delayed fluorescence, 384, 423
  dipole, 300

emission, 95
energy transfer (FET), 91
excitation, 95
extinction-coefficients, 44
ground-state depletion, 215
*in situ* hybridization, 415
lifetime, 2, 330, 377
lifetime imaging, 214, 374
lifetime imaging microscopy (FLIM),
  384, 392
native fluorescence, 226
orientation factor, 331
pathlength, 15, 137
phosphorescence, 384, 386
quantum efficiency (QE), 30, 422, 423
quantum yield, 330, 376
resonance energy-transfer, 318
solvent polarity, 4
Stokes shift, 2
triplet, 30, 32, 149
Fluorescent probes, 89
  acetoxymethyl (AM) esters, 19, 38, 50,
    92, 219
  arsenazo III, 39
  BCECF, 6, 19, 92, 93, 182, 184, 207, 351
  bis-NB-caged carboxyfluorescein (CF), 72
  BODIPY-ceramide, 12, 347
  calcein, 345, 346
  calcium-crimson, -green, -orange, -red,
    16
  calcium-sensitive fluorescent probes, 161
  carbocyanine dyes, 347
  dextran, 17, 19, 91
  direct loading, 222
  dye
    embuffering, 140
    compartmentalization, 142
    extrusion, 142, 144
    leakage, 183
  ester
    deesterification, 92
    incomplete cleavage, 142, 144
    loading, 219, 220
  excitation ratio-type dyes, 184
  fluorescein, 11, 277
  fluorescence photoactivation and dissipa-
    tion, 59
  fluorescent analogs, 238
  fluorescent indicator, 225
  fluorescent peptides, 9
  fluorescent stains, 409

Fluorescent probes (*continued*)
fluorgenic enzyme, 4
fluo-3, 16, 39, 50, 67, 70, 75, 148, 219,
220, 221, 350
fura-2
calcium oscillation, 137–149
calcium-pH measurements, 180–181,
182, 190
calcium triggers in mitosis, 66–67
deesterification of dyes, 92
development and uses, 14–16
DIC-fluorescence microscope, 247, 252
excitation filters, 154
*in vitro* spectra, 93
ion imaging, 350
noise sources, 104, 105
oscillatory behavior, 134
patch clamp method, 223, 224, 225
photoreleased calcium, 39, 50
Hoechst, 409, 411, 412–13, 418
indo, 180, 188, 201
indo-1
calcium oscillation, 137–149
calcium-pH measurements, 180–181,
182, 190
development and uses, 14–16
excitation-contraction coupling experiments, 230
excitation filters, 154
ion imaging, 350, 351
oscillatory behavior, 134
patch clamp method, 223, 224, 225
injection loading, 223
ion indicators, 13
JC-1, 13
mag-fura-2, 94
MeroCaM, 11
NBD-PC, 267, 272, 275
NBD-PS, 270
1-acyl-2-[*N*-(4-nitrobenzo-2-oxa-1,3-
diazole)-aminocaproyl] phosphatidic
acid, 265
PBFI, 19
phalloidin, 5
propidium iodide, 409, 411
quin-1, 14
quin-2, 350
resorufin, 60, 73
rhod-2, 75
rhodamine, 123, 222, 343, 347, 350
SBFI, 18, 19, 254

SNAFL, 17
SNARF, 17, 202
SNARF-1, 183, 190, 351, 352
SPQ, 17, 254
styryl pyridinium dyes, 12
tetramethylrhodamine, 11, 349, 352
Fourier transform, 423
FPR, 287, 288
Free diffusion, 291
Free-radical scavenger, 196
Frequency domain, 395
Fresnel calculation, 302

Genotype, 406
Göos-Hanchen shift, 297

Hamster cheek-pouch chamber, 364, 365
Hardware
A/D (analog-to-digital) converter, 102,
120, 124, 158, 162
array detector, 383, 393
CD ROM, 165
chambers, 363
chopper, 185
electronic shutters, 154
erasable optical drives, 164
fiber optic, 186, 196, 191, 357, 365
filter wheel, 97, 108, 153, 154, 185
grating monochromators, 154
modulators, 378
monochromatic, 360
optical memory disk recorder (OMDR),
164, 194
RAM, 164, 166
shutter, 231
U-matic VCR, 194
write once, read many (WORM), 164
Heart, 230
HeLa cells, 411
Heterodyne detection, 382, 383, 388
Hindered diffusion, 286
Human papilloma virus (HPV)
E1/E2 region, 404, 405, 411, 419
E2, 423
E6, 405, 411, 412, 415, 423, 424
E7, 405
episomal, 415
genotypes, 408
HPV-6, 404

HPV-11, 404
HPV-16, 408, 415
HPV-18, 405, 411
koiliocytosis, 406
ploidy, 408
venereal warts, 404
Hybridization, 406

Image analysis
 arithmetic logical units (ALUs), 163
 automatic detection, 410
 binning, 161
 cluster division algorithms, 425
 contrast enhancement, 383
 digital-image processing, 214
 digital-image throughput, 245
 emission ratioing, 190
 histogram, 266
 image distortion, 192
 image header information, 167
 image processors, 162
 image registration shifts, 100, 107, 198,
  256, 422
 look-up tables (LUTs), 163
 optical slices, 352
 point spread function (PSF), 424
 pseudocolor, 167
 regions of interest (ROI), 121, 167, 195, 204
 segmentation, 410
 smoothing, 195
 spatial filter, 250
 three-dimensional imaging, 395, 420
 2D segmentation algorithms, 425
Image acquisition
 automatically focused, 425
 digitizers, 163
 frame averaging, 156
 frame integration period, 129
 register buffers, 126
 shift registers, 126
Image cytometry, 408, 409, 411
Imaging, 88, 309
Incident intensities, 298, 299
Incident plane waves, 299
*In situ* hybridization, 418, 423
Intact organs *in situ*, 356
Interference fringes, 304, 307
Intracellular-free magnesium, 230
Ionic strength, 142
Iontophoresis, 39

Keratocytes, 73
Kinetochore, 72
Kupffer cells, 362

Lamellipodium, 74
Laminin, 322–324
Lasers, 36, 37, 376
 argon ion lasers, 304
 cavity-dumped laser, 378
 diode lasers, 378
 femtosecond laser pulses, 393
 krypton ion lasers, 304
 liquid crystal tunable laser, 155
 mode-locked, 378
 modulating frequency, 384
 $N_2$-laser pumped dye lasers, 378
 $N_2$ lasers, 378
 phase-locked loop, 383
 power, 344
 total internal reflectance fluorescence
  microscopy, 303–306
Lifetime, 374, 382
Lifetime images, 377
Light sources, 375
 flash lamps, 36, 231
 mercury-arc lamps, 36
 monochromatic light, 359
 three-wave illuminator, 191
 transillumination, 357, 359, 360, 365
 transmitted light, 240
 ultraviolet light, 30, 32, 139
 wavelength tunability, 376
 wavelength tuning, 95
 xenon arc lamp, 231
 xenon flashlamps, 36
 xenon light source, 151
Lipids
 dansyl-phosphatidyl choline, 267, 270,
  275, 278
 dansyl-phosphatidylethanolonine, 272
 dioleylphosphatidylcholine (DOPC), 272
 gel-phase, 277
 phase transition, 277
 phospholipid monolayer, 328
 phospholipids, 268, 271, 274
 planar bilayers, 328
 sulfatide lipids, 322
 supported phospholipid monolayers, 316
Liver, 362
Living organs, 356

Magnesium, 18, 52, 238
MDCK cells, 207
Melanotropes, 204, 206
Membrane
    $Ca^{2+}$-induced domains, 265
    contacts, 326
    domains, 264, 265, 267, 274, 277, 279
    fluidity, 181
    fluid mosaic model, 263
    glycoproteins, 286, 292
    integral membrane proteins, 333
    lipids, 12, 280
    phase separations, 265
    plasma membrane potential, 12
    potential, 13, 181
    proteins, 277, 280
    receptors, 318
    solid-liquid interfaces, 313
    supported membranes, 332
    supported phospholipid bilayers, 317, 324
    supported planar bilayers, 321, 333
Metabolic inhibitor, 346
Metaphase, 71
Microcirculation, 355
Microinjection, 39, 56, 92
Microscopy
    analyzer, 242, 246
    angle of incidence, 297
    back focal plane, 245
    chromatic aberrations, 198
    chronic *in vivo* microscopy, 365
    compound trinocular microscopy, 357
    confocal, 74, 103, 148, 214, 215, 220, 340,
        341, 342, 344, 352, 363, 395, 420
    darkfield, 238
    differential interference microscopy
        (DIC), 101, 152, 180, 239, 243,
        246–248, 251–252, 254–256, 258,
        288
    double-view video microscopy, 140
    epiillumination, 361
    first image plane, 422
    fluorescence, 3, 149
    focal plane, 306, 352, 421
    focusable condenser, 357
    four channel video, 188
    gold particles, 290, 292
    high-numerical aperture objectives, 196,
        300, 359, 361
    image plane, 340
    *in vivo* microscopy, 356

laser tweezers, 292
luminescence digital-imaging microscopy
    (LDIM), 386
microscope objectives, 150
multiparameter approach, 179, 181, 419
nanovid microscopy, 238, 286, 288, 424
Nipkow disk, 340
numerical aperture, 239
parfocal, 242
phase contrast, 180
phase rings, 240
picosecond fluorescence microscope
    (PFM), 392
pinhole, 393
pinhole aperture, 340, 342, 344
primary image plane (PIP), 422
resolution, 123, 124
resonance energy transfer, 296, 424
single particle tracking, 285, 287
slit-scanning confocal microscopy, 340, 341
spatial resolution, 215, 216, 243, 422, 425
tandem-scanning, 340
trinocular, 242
two-photon excitation microscopy, 214
two-photon fluorescence detection, 376
two-photon time-resolved imaging, 393
Microtubule, 71
Microvasculature, 356, 359, 361
Mitosis, 66
Modulation fluorometry (MPF), 381
Monoclonal antibody-lipid hapten binding,
    318–321
Monomer transfer, 272
Monovalent ligands, 292
Morphological measurements, 419
Motion artifacts, 107
Multichannel photon counting, 388
Multifrequency, cross-correlated phase, 381
Multifrequency phase-sensitive detection,
    380
Multi-image experiments, 186
Multiple pass DR, 100
Muscle, 60
Myocytes, 61, 62, 147, 345–346
Myotubes, 312


NADH, 384
Nernst equation, 347, 349
Neuron, 65
Neurotransmitter release, 64

Normalization, 266
Northern blot analysis, 406, 407
NPE, 46, 47
Nuclear morphology, 419
Nuclear ploidy, 410, 411, 413
Nuclear texture, 408
Nucleation release, 74
Nucleic acid hybridization, 406

1/4-λ plate, 241
Oocytes, 75, 147
Optical devices
  acoustooptic (AO) device, 98
    modulator, 378, 387
    scanning, 218
    tunable filter (AOTF), 98, 154
  bandpass, 97, 422
  beam splitters, 98, 191, 422
  dichroic filter (DM2), 95, 248
    mirror, 186, 190, 198, 242, 256
    reflector, 99
  dove prism, 250
  dual monochromators, 97
  electrooptic, 378
  image splitters, 192
  infrared filters, 196
  parabolic mirror, 308
  Pockels cell, 390
  prisms, 241, 245, 246, 306, 307
  quartz collector, 152
  quartz-rod method, 357
  Wollaston prism, 241, 245, 246
Order parameters, 329
Organelles, 347
Out-of-focus
  fluorescence, 266
  images, 339, 340
  light, 352
  signal, 421
Oxidation, 225
Oxygen-free radical, 239
Ozone-free lamps, 152

Pap smear, 411, 415, 421
Particle diffusion, 289
Particle trajectory, 287
Patch-clamp pipettes, 224
Patch pipette, 230
PCR *in situ* technique, 407

Permeabilization, 38
p53, 405, 424
pH
  BCECF, 351
  caged compounds, 30
  calcium-pH calculations, 199
  calcium-pH measurements, 182, 188, 190, 203
  calcium standards, 145
  calibration procedures, 142
  commercially prepared media, 153
  depolarization of pituitary intermediate lobe melanotropes, 203–207
  DM-nitrophen, 52, 53
  fluorescein fluorescence, 202
  indicators, 16–17
  multiparameter approach, 179, 181
  SNARF fluorescence, 203
Phase-sensitive detection, 378
Phase shift, 382
Phosphorescence lifetime, 376
Photobleaching, 9–10, 232
  biotinylated tubuline injection experiments, 72
  bleaching, 17, 375, 376, 393
  confocal microscopy, 344
  fluorescence recovery after photobleaching (FRAP), 59, 264, 296, 314, 316
  fluorochromes, 215
  indo-1, 140
  ISIT camera, 159
  lateral diffusion, 264, 287, 292, 327
  lipid distribution measurement, 266
  low-excitation light levels, 178, 195, 202
  polarization, 330
  probe characteristics, 91
  ratio-type ion indicator dyes, 183
  SNARF-1, 351
  spot photobleaching, 327
  UV-light-induced damage, 155
Photodamage, 92, 148, 344
Photodynamic, 202
Photometry, 309, 310
Photomultiplier, 103
Photomultiplier gain, 344
Photomultiplier tube, 217, 250
Photon counting, 310
Photon noise, 289
Photorelease, 77
Phototoxicity, 92

Pixel, 102, 104, 119, 423
  arrays, 161
  clock, 116
  flicker, 193
  size, 166
Plants, 70
Polarization, 376
  anisotropic fluorescence emission, 299
  depolarization, 203
  fluorescence photobleaching recovery, 329
  fluorescence polarization, 89
  image, 91
  microscopy, 296
  photobleaching, 330
  polarizers, 240, 241
  total internal reflectance (TIR), 328
Polymerase chain reaction (PCR), 407
Polyvalency, 287
Power supplies, 152
$P$ plane, 256
Protein aggregation, 277
Protein kinase C, 5, 279
Protein oligomerization, 280
Prothrombin, 324

Quantification error, 410
Quin-1, 14
Quin-2, 14

Radiance values, 272, 274
Ratio emission, 101
  imaging, 15, 99, 182, 194
  indicator dyes, 183
  spectroscopy, 89
Real time, 106
Red blood cells, 312
Refractive index, 296, 301, 302
Resonance energy transfer, 10, 279
Retinoblastoma tumor suppressor gene, 405
Rotational diffusion, 329, 377
Ryanodine, 62, 63

Sample chamber, 308
Sampling artifacts, 106
  behavior, 117
  phase difference, 129
  rates, 159
  theorem, 147
  time, 186

Saponin, 38
Sarcoplasmic reticulum, 62
Scattered light, 375
Second messengers
  ADP, 44
  ATP, 44, 45
  ATP$\beta$, 45
  ATP$\gamma$S, 45
  caffeine, 62, 63
  cAMP, 10, 11, 46, 47, 61, 91
  carbomoylcholine, 58
  cGMP, 47
  diacylglycerol (DAG), 5, 135
  GABA ($\gamma$-aminobutyric acid), 58
  glycine, 58
  GTP$\gamma$S, 29
  H$^+$, 64, 70, 75, 238, 254
  InsP$_3$, 136
  InsP$_4$, 136
  phorbol esters, 5
  phospholipase C, 135
  $sn$-1,2-dioctanoylglycerol, 69
Sensing columns, 125
Sensitivity, 375
Signal-to-noise ratio (SNR), 139, 217, 310, 344, 376, 410
Simultaneous excitation, 180, 190, 253
Snell's Law, 297
Sodium, 18, 181, 238, 254
Sodium-calcium exchanger, 231
Software, 166
Southern blot analysis, 406, 407
Spatial information, 228
Spatial-sampling characteristics, 123
Spectral response, 423
Spectrofluorimeter, 178
Spectroscopy, 31
Spillover, 197, 198
$S$ plane, 256
SPT, 287, 288, 289, 291, 293
Stochastic process, 290
Stomatal pores, 70, 71
Stringency, 407
Substrates, 4
Sulfatides, 323
Surface diffusion, 302, 327–328, 332
SV40, 405

Temperature, 142, 153, 276
Temporal band width, 116
  disparity, 249

resolution, 290
sampling characteristics, 125
sampling error, 187
sampling process, 126, 127
Texture, 409
Thapsigargin, 136
Thermotropic phase transition, 280
Time-correlated single photon counting
  (TCSPC), 381
Time domain, 381
Time-lapse experiments, 167
Time-resolution, 215, 216
Time-resolved emission, 375
Time-resolved fluorescence imaging, 378
Time-resolved imaging, 377, 395, 423
Time-resolved methods, 375
Time-resolved photobleaching anisotropy,
  330
Time-resolved spectroscopy, 376
Time-resolved two-photon fluorescence de-
  cay, 376
Tissue specimens, 420
Total internal reflectance fluorescence microscopy (TIRFM), 297
  antibody-$F_c$-receptor binding, 321–322
  cell-substrate contacts, 311–313
  cell-supported membrane contacts,
    326–327
  development and applications, 296–297,
    331–333
  fluorescence emission at interface,
    299–300
  fluorophore orientation and rotational dif-
    fusion, 328–330
  imaging, 309–310
  intermediate layers, 300–303
  laminin-sulfatide binding, 322–324
  lasers, 303
  monoclonal antibody-lipid hapten bind-
    ing, 318–321
  optical geometries for fluorescence excita-
    tion, 306–308
  photometry, 310–311
  polarized evanescent wave intensities,
    297–299

protein adsorption to solid-liquid interfaces, 313–316
prothrombin-binding and negatively
  charged lipid bilayers, 324
resonance energy transfer, 330–331
sample compartments, 308–309
supported membranes, 316–318, 324–325
surface diffusion, 327–328
Transport, 291
Treadmilling, 74
Tryptophan, 4
Two-dimensional rotational diffusion, 330
Two-dimensional time-resolved image, 387


UNIX, 162, 164, 166


Van't Hoff equation, 251
Vertical-storage buffers, 125
Vibration, 153, 186
Video
  amplifiers, 120
  camera, 155
  detectors, 240
  field rates, 178
  image processor, 245
  imaging, 214
  RS-170, 116, 117, 120, 128, 160, 192
  splitter-inserter, 248
  switching device (VSDI), 245
  tape, 227
  wipers, 193
Viscosity, 4, 36, 105, 142, 144
Voltage-clamp, 225, 231


Whole-cell patch-clamp methods, 214


Yeast, 12


Zwitterionic, 19